MYCORRHIZAE: SUSTAINABLE AGRICULTURE AND FORESTRY

MYCORRHIZAE: SUSTAINABLE AGRICULTURE AND FORESTRY

Edited by

Zaki Anwar Siddiqui
Aligarh Muslim University, Aligarh, India

Mohd. Sayeed Akhtar
Aligarh Muslim University, Aligarh, India

and

Kazuyoshi Futai
Kyoto University, Kyoto, Japan

Editors
Zaki Anwar Siddiqui
Aligarh Muslim University
Aligarh, India

Mohd. Sayeed Akhtar
Aligarh Muslim University
Aligarh, India

Kazuyoshi Futai
Kyoto University
Kyoto, Japan

ISBN: 978-1-4020-8769-1   e-ISBN: 978-1-4020-8770-7

Library of Congress Control Number: 2008931059

© 2008 Springer Science + Business Media B.V.
No part of this work may be reproduced, stored in a retrieval system, or transmitted
in any form or by any means, electronic, mechanical, photocopying, microfilming, recording
or otherwise, without written permission from the Publisher, with the exception
of any material supplied specifically for the purpose of being entered
and executed on a computer system, for exclusive use by the purchaser of the work.

Printed on acid-free paper.

9 8 7 6 5 4 3 2 1

springer.com

CONTENTS

Preface vii

Contributors ix

1. **Mycorrhizae: An Overview** 1
   Zaki A. Siddiqui and John Pichtel

2. **The Molecular Components of Nutrient Exchange in Arbuscular Mycorrhizal Interactions** 37
   Ruairidh J.H. Sawers, Shu-Yi Yang, Caroline Gutjahr, and Uta Paszkowski

3. **Arbuscular Mycorrhizal Fungi as Potential Bioprotectants against Plant Pathogens** 61
   Mohd. Sayeed Akhtar and Zaki A. Siddiqui

4. **Arbuscular Mycorrhizae and Alleviation of Soil Stresses on Plant Growth** 99
   Philippe Giasson, Antoine Karam, and Alfred Jaouich

5. **Arbuscular Mycorrhizal Fungi Communities in Major Intensive North American Grain Productions** 135
   M.S. Beauregard, C. Hamel, and M. St.-Arnaud

6. **Arbuscular Mycorrhizae: A Dynamic Microsymbiont for Sustainable Agriculture** 159
   Jitendra Panwar, R.S. Yadav, B.K. Yadav, and J.C. Tarafdar

7. **Indirect Contributions of AM Fungi and Soil Aggregation to Plant Growth and Protection** 177
   Kristine A. Nichols

8. **Arbuscular Mycorrhizae and their Role in Plant Restoration in Native Ecosystems** 195
   Krish Jayachandran and Jack Fisher

9. **Effects of Interactions of Arbuscular Mycorrhizal Fungi and Beneficial Saprophytic Mycoflora on Plant Growth and Disease Protection** — 211
   M.G.B. Saldajeno, W.A. Chandanie, M. Kubota, and M. Hyakumachi

10. **The Mycorrhizosphere Effect: A Multitrophic Interaction Complex Improves Mycorrhizal Symbiosis and Plant Growth** — 227
    R. Duponnois, A. Galiana, and Y. Prin

11. **Ectomycorrhizae and their Importance in Forest Ecosystems** — 241
    Kazuyoshi Futai, Takeshi Taniguchi, and Ryota Kataoka

12. **Ectomycorrhizal Associations Function to Maintain Tropical Monodominance** — 287
    Krista L. McGuire

13. **The Use of Mycorrhizal Biotechnology in Restoration of Disturbed Ecosystem** — 303
    Ali M. Quoreshi

14. ***In vitro* Mycorrhization of Micropropagated Plants: Studies on *Castanea sativa* Mill** — 321
    Anabela Martins

15. **Effective and Flexible Methods for Visualizing and Quantifying Endorhizal Fungi** — 337
    Susan G.W. Kaminskyj

Subject index — 351

PREFACE

Mycorrhizae are indigenous to soil and plant rhizosphere and potential tools for sustainable agriculture. They enhance the growth of a root system and even of an entire plant and often control certain plant pathogens. It is a fascinating subject, multidisciplinary in nature, and concerns scientists involved in plant heath and plant protection. There have been marked advances in this field during the last few decades. This book stresses the need to document the information, developing a unifying theme which treated mycorrhizae in a holistic manner. Mycorrhizal fungi improve plant vigor and soil quality by using the greater surface area. The hyphae of these fungi extend out into the soil, secrete extracellular enzymes and efficiently absorb the maximum amount of available nutrients and deliver these nutrients back to the plant inside the root cell wall. They play a crucial role in plant nutrient uptake, water relations, ecosystem establishment, plant diversity, and the productivity of plants. Scientific research in this field involves multidisciplinary approaches to understand adaptation of mycorrhizae to the rhizosphere, mechanism of root colonization, effect on plant physiology and plant growth, biofertilization, plant resistance, biocontrol of plant pathogens. The main purpose of the book is to provide a comprehensive source of current literature and future prospects for research.

The book has 15 chapters and attempts to present balanced accounts on various aspects of mycorrhizae. Chapter 1 gives an overview of mycorrhizal fungi and their role in reduction in chemicals and sustainable agriculture, while Chapter 2 describes the molecular basis of nutrient exchange between arbuscular mycorrhizae (AM) and plants. Other chapters deal with AM fungi as potential bioprotectants against plant pathogens, role of AM fungi in alleviation of soil stresses on plant growth, AM fungi communities in major intensive North American grain productions and indirect contributions of AM fungi and soil aggregation to plant growth and protection. Chapters on AM fungi and their role in plant restoration in native ecosystems and interactions of AM fungi and beneficial saprophytic mycoflora and mycorrhizosphere effect: a multitrophic interaction complex is also discussed. Ectomycorrhizae (ECM) and their importance in forest ecosystem and ECM associations in tropical rain forests, function to maintain tropical monodominance and the use of mycorrhizal biotechnology in restoration of disturbed ecosystem are presented in other chapters. A separate chapter has been devoted to *in vitro* mycorrhization of micropropagated plants while effective and flexible methods for visualizing and quantifying endorhizal fungi are dealt in detail in the last chapter.

The book is not an encyclopedic review. However, an international emphasis has been placed on trends and probable future developments. The chapters incorporate both theoretical and practical aspects, and may serve as base line information for future research through which significant developments can be expected. This book will be useful to students, teachers and researchers, both in universities and research institutes, especially working in areas of agricultural microbiology, plant pathology, and agronomy.

With great pleasure, we extend our sincere thanks to all the contributors for their timely response, excellent and up to date contributions and consistent support and cooperation. We are thankful to Professors Ainul Haq Khan, Aqil Ahmad, R.P. Singh, Department of Botany, A.M.U. Aligarh, Professor Wasim Ahmad, Department of Zoology, A.M.U. Aligarh, and Professor C.M.M. Bandara Visiting Professor, Graduate School of Agriculture, Kyoto University, Japan for their moral support during this project. We acknowledge with thanks the valuable assistance from my friends, well wishers and students. Special thanks are extended also to Drs. S. Hayat, Department of Botany, A.M.U. Aligarh, Yuko Takeuchi, Graduate School of Agriculture, Kyoto University and Abdul Rajjak, Department of Chemistry, Kyoto University, Kyoto, Japan for their encouragement, courtesy and help as this book progressed.

We are extremely thankful to Springer, Dordrecht, the Netherlands for completing the review process expeditiously to grant acceptance for publication. Cooperation and understanding of its staff especially of Maryse Walsh, Publishing Editor and Melanie van Overbeek, Senior Publishing Assistant is also thankfully acknowledged.

We express sincere thanks to our family members, for all the support they provided, and regret the neglect and loss they suffered during the preparation of this book.

Finally, we must be gracious to the Providence who helped us to develop and complete a book on **Mycorrhyzae: Sustainable Agriculture and Forestry**.

Zaki A. Siddiqui
M. Sayeed Akhtar
Kazuyoshi Futai

CONTRIBUTORS

Akhtar, M. Sayeed
Department of Botany
Aligarh Muslim University
Aligarh-202 002,
INDIA

Beauregard, M.S.
Institut de recherche en biologie végétale,
Jardin botanique de Montréal, 4101,
Sherbrooke St Est, Montréal, QC,
CANADA, H1X 2B2

Chandanie, W. Arachchige
United Graduate School of Agricultural Science,
Gifu University,
Gifu, JAPAN

Duponnois, Robin
IRD. UMR 113
CIRAD/INRA/IRD/SUPAGRO/UM2
Laboratoire des Symbioses Tropicales et Méditerranéennes (LSTM)
Montpellier
FRANCE

Fisher, Jack
Fairchild Tropical Botanical Garden,
Miami, FL, 33199
USA

Futai, Kazuyoshi
Graduate School of Agriculture
Kyoto University
Sakyo-ku 606-8502
Kyoto, JAPAN

Galiana, Antoine
CIRAD. UMR 113
CIRAD/INRA/IRD/SUPAGRO/UM2
Laboratoire des Symbioses Tropicales et Méditerranéennes (LSTM)
Montpellier, FRANCE

Giasson, Philippe
Earth Sciences Department
University of Quebec at Montreal
P.O. 8888, Succursale Centre-Ville
Montreal, Quebec, H3C 3P8
CANADA

Gutjahr, Caroline
Department of Plant Molecular Biology
University of Lausanne
Biophore Building
CH-1015 Lausanne
SWITZERLAND

Hamel, C.
Semiarid Prairie Agricultural Research Centre
Agriculture and Agri-Food Canada
1 Airport Road, Box 1030
Swift Current, SK
CANADA, S9H 2X3

Hyakumachi, Mitsuro
Laboratory of Plant Pathology
Faculty of Applied Biosciences
Gifu University
Gifu
JAPAN

Jaouich, Alfred
Earth Sciences Department
University of Quebec at Montreal
P.O. 8888, Succursale Centre-Ville
Montreal, Quebec, H3C 3P8
CANADA

Jayachandran, Kris
Environmental Studies Department
Florida International University
University Park, Miami
FL 33199
USA

Kaminskyj, Susan G.W.
Department of Biology
University of Saskatchewan
112 Science Place
Saskatoon, SK S7N 5E2
CANADA

Karam, Antoine
Department of Soil Science & Agri-Food
Engr. College of Agricultural & Food Sciences
Laval University Quebec, PQ G1K 7P4
CANADA

Kataoka, Ryota
Graduate School of Agriculture
Kyoto University
Sakyo-ku 606-8502
Kyoto, JAPAN

Kubota, Mayumi
Laboratory of Plant Pathology
Faculty of Applied Biosciences
Gifu University
Gifu, JAPAN

Martins, Anabela
Department of Biology and Biotechnology
Escola Superior Agrária
Instituto Politécnico de Bragança
5301-855 Bragança
PORTUGAL

McGuire, Krista L.
Department of Ecology and Evolutionary Biology
University of California
Irvine CA92697
USA

Nichols, Kristine
USDA, Agricultural Research Service
Northern Great Plains Research Laboratory
Mandan, ND 58554
USA

Panwar, Jitendra
Biological Sciences Group
Birla Institute of Technology and Science
Pilani-333031,
INDIA

Paszkowski, Uta
Department of Plant Molecular Biology,
University of Lausanne,
Biophore Building,
CH-1015 Lausanne,
SWITZERLAND

Pichtel, John
Ball State University
Natural Resources and Environmental Management,
Muncie, IN 47306
USA

Prin, Yves
CIRAD. UMR 113
CIRAD/INRA/IRD/SUPAGRO/UM2
Laboratoire des Symbioses Tropicales et Méditerranéennes (LSTM)
Montpellier
FRANCE

Quoreshi, Ali M.
Symbiotech Research Inc. Alberta, Canada T9E 7N5 and
Department sciences of wood and the forest
Pavillon Marchand, Université Laval, Québec
Québec G1K 7P4
CANADA

Saldajeno, M.G. Barcenal
United Graduate School of Agricultural Science
Gifu University
Gifu, JAPAN

Sawers, Ruairidh J.H.
Department of Plant Molecular Biology
University of Lausanne
Biophore Building
CH-1015 Lausanne
SWITZERLAND

Siddiqui, Zaki Anwar
Department of Botany
Aligarh Muslim University
Aligarh-202 002
INDIA

St-Arnaud, M.
Institut de recherche en biologie végétale
Jardin botanique de Montréal, 4101
Sherbrooke St Est, Montréal, QC, H1X 2B2
CANADA

Taniguchi, Takeshi
Graduate School of Agriculture
Kyoto University
Sakyo-ku 606-8502
Kyoto
JAPAN

Tarafdar, Jagdish Chandra
Central Arid Zone Research Institute
Jodhpur- 342003
INDIA

Yadav, Brijesh Kumar
Central Arid Zone Research Institute,
Jodhpur- 342005
INDIA

Yadav, Ranjeet Singh
Main Wheat Research Station
S. D. Agricultural University
Vijapur-382870
INDIA

Yang, Shu-Yi
Department of Plant Molecular Biology
University of Lausanne
Biophore Building
CH-1015 Lausanne
SWITZERLAND

Chapter 1

MYCORRHIZAE: AN OVERVIEW

ZAKI A. SIDDIQUI[1] AND JOHN PICHTEL[2]
[1]*Department of Botany, Aligarh Muslim University, Aligarh 202002, INDIA;* [2]*Ball State University, Natural Resources and Environmental Management, Muncie, IN 47306 USA*

Abstract: Mycorrhizae establish symbiotic relationships with plants and play an essential role in plant growth, disease protection, and overall soil quality. Of the seven types of mycorrhizae described in current scientific literature (arbuscular, ecto, ectendo, arbutoid, monotropoid, ericoid and orchidaceous mycorrhizae), the arbuscular and ectomycorrhizae are the most abundant and widespread. This chapter presents an overview of current knowledge of mycorrhizal interactions, processes, and potential benefits to society. The molecular basis of nutrient exchange between arbuscular mycorrhizal (AM) fungi and host plants is presented; the role of AM fungi in disease protection, alleviation of heavy metal stress and increasing grain production is also reviewed. Use of mycorrhizae, primarily AM and ectomycorrhizae (ECM), on plant growth promotion and disease suppression are discussed and their implications on sustainable agriculture are considered. The effect of co-inoculation of AM fungi and beneficial saprophytic mycoflora, in terms of plant growth promotion and root colonization, are discussed. The role of AM fungi in the restoration of native ecosystems and the mycorrhizosphere effect of multitrophic interactions are briefly outlined. The mechanisms by which mycorrhizae transform a disturbed ecosystem into productive land are briefly discussed. The importance of reintroduction of mycorrhizal systems in the rhizosphere is emphasiszed and their impact in landscape regeneration and in bioremediation of contaminated soils are discussed. The importance of ECM in forest ecosystems, and associations of ECM in tropical rainforests and their function in maintaining tropical monodominance are discussed. *In Vitro* mycorrhization of micropropagated plants and visualizing and quantifying endorhizal fungi are briefly explained.

Keywords: Agriculture; forestry; mycorrhizae; reclamation; sustainable.

1 INTRODUCTION

In 1885 Albert Bernard Frank (Frank, 1885), in his study of soil microbial-plant relationships, introduced the Greek term 'mycorrhiza', which literally means 'fungus roots'. Mycorrhizal fungi form symbiotic relationships with plant roots in a fashion similar to that of root nodule bacteria within legumes. Of the seven types of mycorrhizae described (arbuscular, ecto, ectendo-, arbutoid, monotropoid, ericoid and orchidaceous mycorrhizae), arbuscular mycorrhizae and ectomycorrhizae are the most abundant and widespread (Smith and Read, 1997; Allen et al., 2003). Arbuscular mycorrhizal (AM) fungi comprise the most common mycorrhizal association and form mutualistic relationships with over 80% of all vascular plants (Brundrett, 2002). AM fungi are obligate mutualists belonging to the phylum Glomeromycota and have a ubiquitous distribution in global ecosystems (Redecker et al., 2000). Ectomycorrhizal (ECM) fungi are also widespread in their distribution but associate with only 3% of vascular plant families (Smith and Read, 1997). These fungi are members of the phyla Ascomycota and Basidiomycota, and the ECM mutualism is thought to have been derived multiple times independently from saprophytic lineages (Hibbett et al., 2000).

Ectendomycorrhizae possess characteristics of both ECM and AM fungi (Table 1). As with ECM, both a hartig net and mantle structures are produced in ectendomycorrhizae, although the mantle may be reduced compared with ECM. The hartig net is defined as an inward growth of hyphae which penetrates the root structure. Intracellular penetration of healthy plant cells by these fungi also occurs, a characteristic unlike that of ECM but consistent with AM. Ectendomycorrhizae can be formed with roots of many angiosperm and gymnosperm species; fungal symbionts include members of the Basidiomycota, Ascomycota, or Zygomycota. In fact, the same fungal species can form either ECM or ectendomycorrhizae depending upon the plant species with which it is associated. Similarly, arbutoid mycorrhizae possess characteristics of both ECM and AM fungi, i.e., there is a well developed mantle, a hartig net, and prolific extrametrical mycelium. Additionally, intracellular penetration occurs and hyphal coils are produced in autotrophic cells. These mycorrhizae are associated with members of the Ericales; namely, Arbutus and Arctostaphylos species. The fungal symbionts are exclusively Basidiomycete species, which may form ECM with other autotrophic hosts.

Monotropoid and orchid mycorrhizae are formed between Basidiomycete fungi and achlorophyllous plant species. Monotropoid mycorrhizae are formed between plants of the Monotropaceae family and a specific subset of fungi in the Russulaceae or the Boletaceae family. Orchid mycorrhizae have only been found in association with Basidiomycete species. In the other mycorrhizal symbioses plants are usually generalists and associate with a wide array of fungal species. In contrast, plants that participate in monotropoid and orchid mycorrhizal associations are highly specific, associating with only a narrow range of fungal species. It had been thought that these mycorrhizal associations are formed exclusively with Basidiomycete fungal species; however, it has recently been discovered that several species of tropical achlorophyllous epiphytes form mycorrhizal associations with AM fungal species in the Glomeromycota (Bidartondo et al., 2002). Ericoid mycorrhizae are known to form between autotrophs in the Ericaceae and fungi in the Ascomycota. Intracellular penetration of root cells occurs and there is no mantle or hartig net development.

Table 1. Major categories of mycorrhizae and their attributes. (Adapted from Smith and Read, 1997).

Mycorrhizal type	Arbuscular	Ecto	Ectendo	Arbutoid	Monotropoid	Ericoid	Orchidaceous
Fungal taxa	Glomero	Basidio Asco Zygo	Basidio Asco	Basidio	Basidio	Asco	Basidio
Plant taxa	Bryo Pterido Gymno Angio	Gymno Angio	Gymno Angio	Ericales	Monotropoideae	Ericales Bryo	Orchidaceae
Intracellular colonization	+	–	+	+	+	+	+
Fungal sheath	–	+	+/–	+/–	+	–	–
Hartig net	–	+	+	+	+	–	–
Vesicles	+/–	–	–	–	–	–	–
Achlorophylly	–	–	–	–	+	–	+

+ = present; – = absent

2 MOLECULAR BASIS OF NUTRIENT EXCHANGE IN AM FUNGI

The driving force behind AM interactions is an exchange of nutrients between fungus and plant. Glomeromycotan fungi are obligate symbionts and rely on the carbon provided by their plant hosts to complete their life cycle. In return, the fungus provides nutritional benefits to the plant by delivering minerals, including the biologically essential nutrients phosphorus (P) and nitrogen (N). The majority of this nutrient exchange is believed to occur within root cortical cells containing highly-branched hyphal structures termed *arbuscules*. As arbuscules develop they become enveloped by newly synthesised host membrane tissue; the arbuscules never enter the host cytoplasm.

The plant and fungal arbuscular membranes define a space, the interfacial apoplast, into which nutrients can be delivered and from which they can be extracted (Harrison, 1999; Balestrini and Bonfante, 2005). The molecular components of the interface and the enclosing membranes facilitate and regulate the processes of nutrient exchange. Plant and fungal encoded membrane localised ATPase proteins maintain the interfacial apoplast at low pH, providing an electrochemical gradient that powers the action of various membrane-localised proton:nutrient symporters. A combination of physiological, biochemical and genetic analyses has begun to define the components of the arbuscular membranes, the best characterized of which are the PHT1-type phosphate transporter proteins (Rausch et al., 2001; Javot et al., 2007b). Gene expression studies have demonstrated that specific members of these protein families are expressed in the roots of colonized plants. For example, the medic gene *Medtu:Pht1;4* is expressed exclusively in colonized roots and the protein product is specifically localized to the peri-arbuscular membrane within arbusculated cells (Harrison et al., 2002; Javot et al., 2007a).

Phosphate acquisition via the mycorrhizal pathway begins with the uptake of free phosphate from soil by fungal extra-radical hyphae (Bucher, 2007). These fungal hyphae extend beyond the host root system, allowing a greater soil volume to be exploited for phosphate uptake. Uptake at the soil-hypha interface is mediated by fungal high-affinity phosphate transporters of the Pht1 family (Harrison and Buuren, 1995). Following fungal uptake, phosphate is transferred to the fungal vacuole where it is polymerized to form polyphosphate chains and translocated through the vacuolar compartment to the intraradical hyphae. The polyphosphate is then hydrolysed and phosphate released to the interfacial apoplast. From the interfacial apoplast, plant mycorrhizal Pht1 transporters guide the phosphate across the peri-arbuscular membrane. Once in the plant cytosol, phosphate is translocated into the vasculature for delivery to all parts of the plant.

The extraradical mycelium of mycorrhizal fungi also absorb ammonium, nitrate and amino acids (Hodge et al., 2001), and the role of mycorrhizal nitrogen delivery is becoming better understood (Chalot et al., 2006). The majority of nitrogen is thought to be taken up in the form of ammonia via the action of fungal-encoded AMT1 family transporters such as the protein GintAMT1 characterized from *Glomus intraradicies* (Lopez-Pedrosa et al., 2006). There is no evidence for fungal translocation of either ammonium or nitrate and it is thought that nitrogen transport occurs in the form of the amino acid arginine (Govindarajulu et al., 2005). Having been translocated to the intra-radical hyphae, amino acids may be delivered directly to the interfacial apoplast for plant absorption. However, there is also evidence for an alternative route whereby arginine is broken down by ornithine aminotransferase and urease to release free ammonium. It has been proposed that ammonium is exported by protein-mediated mechanisms and a candidate fungal AMT transporter has been identified that is highly expressed in the internal hyphae. Additionally, studies of ectomycorrhizal fungi have identified proteins homologous to the yeast Ato proton:ammonium antiporter. Once in the apoplast, ammonium is taken up by the plant. Gene expression analyses of medic and rice have identified mycorrhiza-induced transcripts that putatively encode ammonium transporters that are candidates for this function. There is also the possibility for passive ammonia uptake across the peri-arbuscular membrane, perhaps facilitated by the presence of aquaporin proteins (Uehlein *et al.*, 2007).

The use of radiolabeled substrates has demonstrated that AM fungi take up plant carbohydrates in the form of hexose. The route of this transport, and whether or not it is specific to the arbuscule, remains to be determined. There is little molecular evidence for the presence of hexose export proteins in the peri-arbuscular membrane itself. Although a number of mycorrhiza-responsive sugar transporter genes have been identified in medic, they are thought to act as proton:sugar symporters in sugar import rather than export, possibly in support of high metabolic activity in arbusculated cells (Harrison, 1996).

One simple mechanistic alternative to explain the transport of sugars to the apoplast is passive movement. Once in the apoplast, hexose is thought to be absorbed by the fungus via specific transport proteins. Although such fungal transporters remain to be identified, the recent characterization of the GpMST1 hexose transporter from *Geosiphon pyriformis*, a Glomeromycotan fungus that associates with single-celled algae, provides a promising direction for further investigation (Schussler *et al.*, 2006). In the intraradical hyphae, much of the carbon is converted to storage lipids, predominantly triacylglycerides. Lipids not only act to store carbon but are also the main form of carbon translocated from intra- to extra-radical hyphae where they provide the major respiratory substrate.

AM fungi provide other benefits to their plant hosts. In addition to enhancing mineral nutrition, they increase tolerance to water stress, induce greater resistance to pathogens and reduce sensitivity to toxic substances present in the soil. However, the costs of colonization can be high, with up to 20% of the host's fixed carbon being delivered to the microbial symbiont. Nonetheless, under experimental conditions when nutrients are limiting, mycorrhizal crop plants typically exhibit a performance advantage over non-colonized siblings. However, under the nutrient-saturated conditions occurring in high-input agricultural systems, the relative advantages are reduced while the carbon costs remain, and the performance of colonized plants can fall below that of non-colonized plants (Janos, 2007). Although mycorrhizal fungi possess a limited potential to improve on present levels of crop performance, profitable agricultural application of AM symbioses demands that the inevitable loss of carbon to the fungus is compensated by a reduction in the overall cost of production on a per unit yield basis.

Studies of plant performance have identified variation in the capacity of plants to benefit from mycorrhizal colonization, both within and between species. Molecular characterization of mycorrhizal nutrient uptake allows a more detailed interpretation of such performance variation (Sawers *et al.*, 2008). Expression studies have revealed that the induction of the mycorrhizal state in the plant is not a simple superimposition of novel gene activity on the non-colonized state but a distinct switch from an asymbiotic to a symbiotic growth regime. One important consequence of this distinction is the partial genetic independence of plant performance between non-colonized and colonized plants and the subsequent implications this has for the interpretation of phenotypic variation and the selection of future varieties. Clearly, an understanding of the molecular basis of nutrient exchange has great potential to benefit diverse aspects of mycorrhizal research and to contribute to the future application of AM fungi in agriculture, forest management, site remediation, and other settings.

3 AM FUNGI AND PLANT DISEASE CONTROL

Plant diseases can be controlled by manipulation of indigenous microbes or by introducing antagonists to reduce the disease-producing propagules (Linderman, 1992). AM fungi and their associated interactions with plants reduce the damage caused by plant pathogens (Siddiqui and Mahmood, 1995; Siddiqui *et al.*, 1999; Harrier and Watson, 2004). With the increasing cost of pesticides and the environmental and public health hazards associated with pesticides and pathogens resistant to chemical pesticides, AM fungi may provide a more suitable and environmentally acceptable alternative for sustainable agriculture and forestry. The interactions between different AM fungi and plant pathogens vary with the host plant and the cultural system. Moreover, the protective effect of AM inoculation may be both systemic and localized.

Plant parasitic nematodes occur in agricultural soils worldwide, and most crops are susceptible to damage by these parasites. Nematode parasitism on host plants may cause up to 50% yield losses, and these losses may be aggravated when the plant is predisposed to other pathogens. Diseases caused by fungal pathogens persist in the soil matrix and in residues on the soil surface. Damage to root and crown tissue is often concealed in the soil; thus, diseases may not be noticed until the above-ground parts of the plant are severely affected.

Colonization of the root by AM fungi generally reduces the severity of diseases caused by plant pathogens. Reduced damage in mycorrhizal plants may be due to changes in root growth and morphology; histopathological changes in the host root; physiological and biochemical changes within the plant; changes in host nutrition; mycorrhizosphere effects which modify microbial populations; competition for colonization sites and photosynthates; activation of defense mechanisms; and nematode parasitism by AM fungi (Siddiqui and Mahmood, 1995). Of the various mechanisms proposed for biocontrol of plant diseases, effective bioprotection is a cumulative result of all mechanisms working either separately or together. The challenges to achieving biocontrol via use of AM fungi include the obligate nature of AM fungi, limited understanding of the mechanisms involved, and the role of environmental factors in these interactions.

AM fungi are rarely found in commercial nurseries due to the use of composted soil-free media, high rates of fertilizer application and regular application of fungicide drenches. The potential advantages of AM fungi in horticulture, agriculture, and forestry are not perceived by these industries as significant. This perception may be due in part to inadequate methods for large-scale inoculum production.

Cropping sequences, fertilization, and plant pathogen management practices affect both AM fungal propagules in soil and their effects on plants (Bethlenfalvay and Linderman, 1992). In order to apply AM fungi in sustainable agriculture, knowledge of factors such as fertilizer inputs, pesticide use, and soil management practices which influence AM fungi is essential (Allen, 1992; Bethlenfalvay and Linderman, 1992). In addition, efficient inoculants should be identified and employed as biofertilizers, bioprotectants, and biostimulants for sustainable agriculture and forestry.

4 AM FUNGI AND ALLEVIATION OF SOIL HEAVY METAL STRESS

Some heavy metal elements such as Cu, Fe, Mn, Ni and Zn are essential for normal growth and development of plants. These metals are required in numerous enzyme-catalyzed or redox reactions, in electron transfer, and have structural function in nucleic acid metabolism (Gohre and Paszkowski, 2006). In contrast, metals like Cd, Pb, Hg, and As are not essential (Mertz, 1981) and may be toxic to plants at very low concentrations in soils. Heavy metals occur in terrestrial and aquatic ecosystems from both natural and anthropogenic sources, and are also emitted into the atmosphere.

The roots of terrestrial plants are in immediate contact with soil metal ions. Essential heavy metals are transferred into the root by specific uptake systems, but at high concentrations they also enter the cell via nonspecific transporters. At high concentrations heavy metals interfere with essential enzymatic activities by modifying protein structure or by replacing an essential element, resulting in deficiency symptoms. As a consequence toxicity symptoms such as chlorosis, growth retardation, browning of roots, effects on both photosystems, cell cycle arrest, and others are observed.

Anthropogenic soil contamination resulting from mining activities, industrial processes, agriculture, and military activities have resulted in high localized concentrations of heavy metals. Conventional soil remediation practices in most countries rely primarily on the excavation of the contaminated soil. However, physical displacement, transport and storage, or alternatively soil washing are expensive procedures which leave a site devoid of soil microflora.

AM fungi are significant in the remediation of contaminated soil as accumulation (Jamal et al., 2002). The external mycelium of AM fungi allows for wider exploration of soil volumes by spreading beyond the root exploration zone (Khan et al., 2000), thus providing access to greater quantities of heavy metals present in the rhizosphere. Higher concentrations of metals are also stored in mycorrhizal structures in the root and in fungal spores. AM fungi can also increase plant establishment and growth despite high levels of soil heavy metals due to improved nutrition (Taylor and Harrier, 2001), water availability (Auge, 2001), and soil aggregation properties (Kabir and Koide, 2000) associated with this symbiosis.

AM fungi occur in the soil of most ecosystems, including polluted soils. By acquiring phosphate, micronutrients and water and delivering a proportion to their hosts they enhance the host nutritional status. Similarly, heavy metals are taken up via the fungal hyphae and can be transported to the plant. Thus, in some cases mycorrhizal plants experience enhanced heavy metal uptake and root-to-shoot transport while in other cases AM fungi contribute to heavy metal immobilization within the soil. The result of mycorrhizal colonization on remediation of contaminated soils depends on the plant–fungus–heavy metal combination and is influenced by soil chemical and physical conditions.

The significance of AM fungi in soil remediation has been recognized (Gaur and Adholeya, 2004; Khan, 2005). A vast amount of literature is available on the effects of mycorrhizal colonization on plants under heavy metal stress but contradictory observations and wide variations in results are reported (Khan, 2005). Enhanced understanding of heavy metal tolerance of plants and AM fungi has defined valuable parameters for improving phytoremediation, i.e., the engineered use of green plants to remediate an affected site. The utility of AM fungi in soil remediation is also important for sustainable agriculture. Application of these fungi is generally useful to overcome heavy metal problems and to alleviate soil stress, and ultimately increases agricultural production.

In many cases AM fungi serve as a filtration barrier against transfer of heavy metal ions from roots to shoots. The protection and enhanced capability of mineral uptake result in greater biomass production, a prerequisite for successful remediation.

AM isolates existing naturally in heavy metal-polluted soils are more metal-tolerant than isolates from non-polluted soils, and are reported to efficiently colonize plant roots in heavy metal-stressed environments. Thus, it is important to screen indigenous and heavy metal-tolerant isolates in order to guarantee the effectiveness of AM symbiosis in restoration of contaminated soils. The potential of phytoremediation of contaminated soil can be enhanced by inoculating metal hyperaccumulating plants with mycorrhizal fungi at the contaminated site. However, there is a need to optimize the conditions to grow AM fungi in large quantities, and to characterize and screen a large number of AM fungal species for tolerance to metals.

5 AM FUNGAL COMMUNITIES AND GRAIN PRODUCTION

With population increase, urban sprawl and the growing interest in the use of biofuels, significant pressures are occurring on some of the highest quality agricultural soils in many nations. Growth of grain and oilseed crops such as barley, corn, soybean and wheat have been an important part of the agricultural economy for years and the continuous increases in demand and prices have led farmers to apply highly intensive agricultural management practices, with the aim of increasing crop productivity. Tillage, crop rotation, fallows, changes in plant cultivars and pesticide application are often used with broadacre field crops, and all these practices influence the surrounding environment (Mozafar *et al.*, 2000; Carter and Campbell, 2006).

Fertilizer use represents a common agricultural management practice, but a growing body of evidence has demonstrated an array of negative impacts on ecosystems from their use. No matter which form of fertilizer is applied (organic or mineral), conventional farming generates large N and P surpluses, which can lead to N leaching through the soil profile and P losses in runoff (Brady and Weil, 2002). Not only is there a high financial cost to farmers associated with this loss, but the phenomenon also resulted in soil contamination. In addition, excess fertilizer inputs can be a major threat to aquatic ecosystems through surface and groundwater degradation (Kirchmann and Thorvaldsson, 2000). Recently, fertilizer runoff from agricultural fields was emphasized among the causes of excessive cyanobacterial growth and increasing of potentially harmful blooms leading to restricted access to lakes.

Low-input agricultural systems have gained attention in many Industrialized countries due to increased interest in the conservation of natural resources, reduction of environmental degradation, and the escalating costs of fertilizers. Conventional farming systems using lower application rates of fertilizers and pesticides have been developed, but are used only minimally in North American grain production, perhaps due to insufficient understanding of agricultural soils dynamics (Ryan and Graham, 2002).

Numerous biological, chemical and physical factors influence soil quality. Among them, rhizosphere microbial communities have been shown to directly affect soil fertility by carrying out essential processes that contribute to nutrient cycling, and enhancing soil structure and plant growth and health (Mader et al., 2002; Wu et al., 2005; Miransari et al., 2007; St-Arnaud and Vujanovic, 2007). The extent to which these communities interact is thus of great importance and involves phenomena such as hormone production, enhancement of nutrient availability, and decrease of root diseases. Arbuscular mycorrhizal symbioses have been shown to benefit growth of many field crops in large part due to the extensive hyphal network development in soil, more efficient exploitation of nutrients, and enhanced plant uptake (Smith and Read, 1997). AM symbiosis also increases resistance to biotic and abiotic stresses and reduces disease incidence, representing a key component of sustainable agriculture (St-Arnaud and Vujanovic, 2007; Subramanian and Charest, 1999; Aliasgarzad et al., 2006). Appropriate management of mycorrhizae in agriculture should ultimately result in a substantial reduction in chemical use and production costs.

Soils generally contain indigenous AM fungi that colonize plant roots (Covacevich et al., 1995). The growth enhancement and P uptake of plants colonized by AM fungi is a well-known process (Pfleger and Linderman, 1996; Schweiger and Jakobsen, 1999; Jeffries et al., 2003). Not all plants are dependent on mycorrhizal associations (Azcón and Ocampo, 1981; Trouvelot et al., 1982; Hetrick et al., 1993); however, most increase in yield following inoculation with AM fungi (Jakobsen and Nielsen, 1983; Baon et al., 1992; Talukdar and Germida, 1994; Xavier and Germida, 1997; Al-Karaki et al., 1998) particularly in low-P soils (Thompson, 1990; Rubio et al., 2003). With the current tendency for reduced use of agrochemicals, research is being directed at crop yield improvement and yield sustainability. The efficient use of AM fungi may allow for the attainment of acceptable yield levels with minimum fertilizer dose, while also reducing costs and environmental pollution risk (Covacevich et al., 2007). This is a promising approach for obtaining high yields with low fertilizer inputs in order to support sustainable agricultural systems.

6 SUSTAINABLE AGRICULTURE

Sustainable agricultural systems employ natural processes to achieve acceptable levels of productivity and food quality while minimizing adverse environmental impacts (Harrier and Watson, 2004). Sustainable agriculture must, by definition, be ecologically sound, economically viable, and socially responsible. Similarly, sustainable forestry refers to an overall commitment to environmental conservation that integrates the production of trees for useful products with reforestation and conservation of soil, air, water quality, wildlife and aesthetics. Sustainable agriculture relies on long-term solutions using proactive rather than reactive measures at system levels.

Several soil fertility factors contribute to sustainable agriculture through control of soil-borne diseases, including increased soil microbial activity leading to increased competition and parasitism within the rhizosphere (Jawson et al., 1993; Knudsen et al., 1995). Research and development strategies are presently focused on the search for suitable alternatives to the use of commercial synthetic pesticides. Progress has also been made, however, in exploring the use of microorganisms to improve soil fertility. Greater emphasis is being placed on enhanced exploitation of indigenous soil microbes which contribute to soil fertility, increased plant growth and plant protection.

Mycorhizal fungi, particularly AM, are ubiquitous in soil and create symbiotic associations with most terrestrial plants including agricultural crops, cereals, vegetables, and horticultural plants. In agriculture, several factors such as host crop dependency to mycorrhizal colonization, tillage system, fertilizer application, and the potential of mycorrhizal fungi inocula, affect plant response and plant benefits from mycorrhizae. Interest in AM fungi propagation for sustainable agriculture is increasing due to its role in the promotion of plant health, and improvements in soil fertility and soil aggregate stability. These fungi can be utilized effectively for increasing yields while minimizing use of pesticides and inorganic fertilizers.

To improve crop production in infertile soils, chemical fertilizers have been intensively used, organic matter is incorporated and soil management technologies such as fallow or legume cultivation have also been used. Reliance should be on biological processes by adapting germplasm to advance soil conditions, enhance soil biological activity and optimise nutrient cycling to minimise external inputs and maximise the efficiency of their use (Sanchez, 1994). This approach has been developed for soil biota management using earthworms and microsymbionts (Woomer and Swift, 1994; Swift, 1998).

These soil organisms may represent more than 90% of soil biological activity and thus contribute to nutrient cycling, soil fertility and symbiotic processes in the rhizosphere. Soil fungal diversity and activity have not been adequately studied and understood (Hawksworth, 1991). Mycorrhizae represent an important group because they have a wide distribution, and may contribute significantly to microbial biomass and to soil nutrient cycling processes in plants (Harley and Smith, 1983). Mycorrhizal associations are beneficial to plants and thus crop productivity for sustainable agriculture (Gianinazzi-Pearson and Diem, 1982; Bethlenfalvay, 1992). They improve nutrient uptake, especially P, and also uptake of micronutrients such as zinc or copper; they stimulate the production of growth substances and may reduce stresses, diseases or pest attack (Sylvia and William, 1992; Davet, 1996; Smith and Read, 1997). For appropriate use of this technology, it is necessary to select the best inocula adapted to the specific limiting environmental factors for crop productivity.

7 INDIRECT CONTRIBUTION OF AM FUNGI IN SOIL AGGREGATION AND PLANT GROWTH

Mycorrhizal symbiosis has evolved to assist plants in colonizing the land. In the environment of early Earth, ecological pressures resulted in a highly efficient symbiotic relationship where plants traded photosynthetic carbon for fungally-acquired nutrients and water.

The formation of a biomolecule such as glomalin would have served as an evolutionary advantage to the fungus. Glomalin is a glycoprotein produced on hyphae of AM in the soil. Originally, glomalin production might have arisen to protect fungal hyphae from losses of water or nutrients when being carried from microsites in the soil back to the plant, from fluctuations in turgor pressure due to wet/dry cycles, and from decomposition by microbial attack. The indirect or 'secondary' impacts of glomalin on the formation and stabilization of soil aggregates further improved the efficiency of the symbiotic relationship and the growth environment (Andrade *et al.*, 1998b; Rillig and Mummey, 2006).

Modern agricultural practices have placed new pressures on plant-mycorrhizal symbiosis. Tillage practices physically disrupt soil aggregates and AM hyphal networks. This action deteriorates soil structure, lessens fertility and nutrient cycling ability, and results in more C allocation within the fungal hyphae to reestablishing these networks and less C to glomalin formation (Nichols and Wright, 2004). No-tillage (NT) practices along with continuous cropping systems (by eliminating fallow periods and/or growing cover crops), using mycorrhizal host crops, and reducing synthetic inputs (especially P), enhance the plant-mycorrhizal symbiotic relationship (Nichols and Wright, 2004; Preger *et al.*, 2007; Roldan *et al.*, 2007; Rillig, 2004; Rillig *et al.*, 2007). These practices also increase the percentages of water-stable aggregates within the soil by increasing hyphal length, root and microbial exudates in the mycorrhizosphere, and allocating more C to glomalin production. In addition, higher levels of C sequestration are possible in these systems, since not only is C being allocated below-ground to hyphal networks and formation of the highly stable glomalin molecule, but organic matter occluded within aggregates appears to have a turnover time double that of free organic matter (Six et al., 2001; Nichols and Wright, 2004; Preger et al., 2007; Roldan et al., 2007; Rillig, 2004; Rillig, et al., 2007). Therefore, effective management of soil organisms and, as a consequence, agricultural systems, will maintain a consistent supply of plant-available nutrients to meet the demands of food, feed, fiber and biofuels production for a growing world population while maintaining optimal ecosystem function.

8 AM FUNGI AND THEIR ROLE IN RESTORATION OF NATIVE ECOSYSTEMS

Desertification of terrestrial ecosystems claims several million hectares annually (Warren et al., 1996). Disturbance of natural plant communities is the first visible symptom, but is often accompanied or preceded by loss of key physicochemical and biological soil properties (Skujins and Allen, 1986). These properties largely determine soil quality and fertility, and thus plant establishment and productivity. Hence their degradation results in a loss of sustainability. Soil degradation limits the potential for reestablishment of native plants (Agnew and Warren, 1996; Warren et al., 1996), and erosion and desertification are accelerated. Desertification reduces the inoculum potential of mutualistic microbial symbionts that are key ecological factors in governing the cycles of major plant nutrients and hence in sustaining vegetative covers in natural habitats.

AM fungi enhance the ability of plants to establish and cope in stressful situations including nutrient deficiency, drought, and soil disturbance (Barea et al., 1997; Schreiner et al., 1997). Those regions characterized by long, dry, hot summers, with scarce, erratic, but torrential rainfall, together with anthropogenic activities like overgrazing, nonregulated cultivation techniques, and deforestation, are major threats to ecosystem sustainability (López-Bermúdez and Albaladejo, 1990; Vallejo et al., 1999). Susceptibility to desertification is generally increasing worldwide (Warren et al., 1996).

Desertified and desertification-threatened areas are common and there are many representative areas where reclamation or rehabilitation programs are being attempted to restore sustainable ecosystems (Francis and Thornes, 1990; Herrera et al., 1993). Shrub communities, associated with other small woody plants, are characteristic of many semiarid ecosystems. Thus, reestablishing a shrubland is a key step in revegetation strategies. All the woody legumes involved also form a symbiotic association with AM fungi (Herrera et al., 1993). The fungal mycelium which extends from the mycorrhizal roots forms a three-dimensional network which links the roots and the soil environment. It constitutes an efficient system for nutrient uptake (particularly P) and scavenging in nutrient-poor conditions. The mycelium also contributes to the formation of water-stable aggregates necessary for good soil tilth (Jeffries and Barea, 2000). Loss of microsymbiont propagules from degraded ecosystems can preclude either natural or artificial processes of revegetation; therefore, augmentation of the inoculum may be needed (Requena et al., 1996).

In revegetation schemes, inoculation of plants with microsymbionts should not only help plant establishment (Herrera et al., 1993) but also improve soil physical, chemical, and biological properties contributing to soil quality (Carrillo-García et al., 1999). The introduction of target indigenous species of plants associated with a managed community of microbial symbionts is a successful biotechnological tool to aid the recovery of desertified ecosystems.

9 INTERACTION OF AM FUNGI AND SAPROPHYTIC MYCOFLORA

Interactions of AM fungi with other soil microorganisms are diverse. AM fungi interact with almost all organisms in the mycorrhizosphere but few studies have been conducted on their interactions with beneficial saprophytic fungi (Calvet et al., 1992, 1993; Green et al., 1999). Saprophytic fungi living on the rhizoplane (McAllister et al., 1996; Garcia-Romera et al., 1998) and mycorrhizosphere of plants generally procure their nutritional requirements from organic matter and other elements in the soil (Garcia-Romera et al., 1998). AM fungi and saprophytic organisms like the plant growth promoting fungi (PGPF) have equally good potential in both plant growth promotion and plant disease control (Hyakumachi and Kubota, 2004b). Since both are beneficial microorganisms, their synergistic or additive effects could be more valuable than their individual effects.

In vitro studies on AM fungi and saprophytic fungi demonstrated stimulatory, inhibitory or no effect on spore or conidial germination and hyphal growth of either fungi. Pot and field studies indicate that saprophytic fungi can affect AM development inside the root. The results of the interaction of saprophytic fungi on AM colonization differ widely. The effect of AM fungi on root colonization of saprophytic fungi is generally known and is determined via measurement of the population of saprophytic fungi in rhizosphere soil. Moreover, interaction of saprophytic fungi and AM fungi differ with respect to the species of AM fungi and saprophytic fungi involved.

Mycorrhizal plants produce compounds which interfere with rhizosphere microorganisms and modify the microbial community around the mycorrhizal roots (Linderman, 1988; Linderman and Paulitz, 1990). In addition, the extraradical mycelium of AM may also impact microbial populations around mycorrhizal roots (Filion et al., 1999). The interaction effect of AM fungi and other beneficial soil microorganisms on plant growth has been demonstrated, and trends are similar. PGPF may increase mineralization and suppress deleterious microorganisms (Hyakumachi and Kubota, 2004b). PGPF-mediated induced systemic resistance (ISR) has been demonstrated by increased lignin deposition at the point of attempted penetration by the pathogen (Hyakumachi and Kubota, 2004a) and also by conspicuous superoxide generation by culture filtrates of respective PGPF isolates (Koike et al., 2001). Biochemical analysis has revealed systemic accumulation of salicylic acid and increased activities of chitinase, ß-1,3-glucanases and peroxidase in plants inoculated by PGPF (Hyakumachi and Kubota, 2004a; Yedidia et al., 1999). Hossain et al. (2007) hypothesized that multiple defense mechanisms are involved in disease suppression.

The interaction of saprophytic fungi and AM fungi may vary depending on the inherent characteristics of the fungi being tested. The effect could be contradictory within species of the same genus and even within strains of the same species. Generally, interactions of these groups of microorganisms are synergistic or additive in plant growth promotion and disease suppression. Suitable combinations of these organisms may increase plant growth and enhance resistance to plant pathogens. Combinations generally increase genetic diversity in the rhizosphere microrganisms that resultantly persist, and which utilize a wider array of mechanisms to increase plant growth. Combinations of these organisms may be effective over a wide range of environmental conditions. In particular, combinations of AM fungi and PGPF may provide protection at different times, under different conditions, and occupy different or complementary niches.

10 MYCORRHIZOSPHERE EFFECT: A MULTI-TROPHIC INTERACTION

The mycorrhizosphere includes the region around the mycorrhizal roots. Due to the ubiquity of mycorrhizal symbioses in terrestrial ecosystems, most of the actively absorbing rootlets are connected with the surrounding soil through an interface called the mycorrhizosphere (Johansson et al., 2004). In addition, because mycorrhizal symbiosis modifies root morphology (Linderman, 1988), the volume of the mycorrhizosphere soil is larger than the rhizosphere soil. From a biochemical point of view, root exudation in the mycorrhizosphere is quantitatively and qualitatively different from that in the rhizosphere because mycorrhizal fungi use some of the root exudates and modify root metabolic functions (Rambelli, 1973; Leyval and Berthelin, 1993; Rygiewicz and Andersen, 1994). Furthermore, mycorrhizal fungi associated with plant roots can produce antibiotics (Olsson et al., 1996). These differences between rhizosphere and mycorrhizosphere explain the so-called 'mycorrhizosphere effect' defined by Linderman (1988), which qualifies modifications of the microbial equilibrium induced by the mycorrhizal symbiosis.

Mycorrhizal fungi alter root exudation both quantitatively and qualitatively (Leyval and Berthelin, 1993), as they catabolise some root exudates and modify root metabolic functions. The microbial communities of the soil surrounding mycorrhizal roots and extrametrical mycelium differ from those of the rhizosphere of non-mycorrhizal plants and bulk soil (Katznelson et al., 1962; Garbaye, 1991). Specific relationships occur between mycorrhizal fungi and mycorrhizosphere microbiota, and there is abundant literature attesting that mycorrhizal symbiosis is largely influenced by soil microorganisms (De Oliveira and Garbaye, 1989). These interactions mainly have focused on the effects of microbial communities on mycorrhizal formation, and on mycorrhizal efficiency on host plant growth.

In addition to their known direct effect on plant growth, mycorrhizal symbionts could positively act on host plant development through a selective effect on bacterial communities involved in soil functioning and soil fertility. The loss or reduction of activity of mycorrhizal fungi has often been detected (Bethlenfalvay and Schüepp, 1994). Hence, management of soil mycorrhizal potential is of great importance since mycorrhizal symbiosis determines plant biodiversity, ecosystem variability and productivity directly from its influence on plant mineral nutrition, but also indirectly from its impact on soil microbial functioning.

Mycorrhizal and rhizobial symbioses often act synergistically in terms of infection rate, mineral nutrition and plant growth (Amora-Lazcano et al., 1998). The fungal effect on plant P uptake is beneficial for the functioning of the nitrogenase enzyme of the rhizobial symbiont leading to higher N fixation and, consequently, to better root growth and mycorrhizal development (Johansson et al., 2004). The fungal effect on rhizobial development is dependent on mycorrhizal colonization of root systems but also on the fungal symbiont. However, the influence of mycorrhizal symbiosis on nodule development is not limited to a quantitative effect; i.e., fungi can modify the structure of rhizobial bacteria along the root system. Below-ground diversity of AM fungi is a major factor contributing to the maintenance of plant diversity and to ecosystem functioning (van der Heijden et al., 1998). This fungal diversity also has a beneficial effect on the nodulation process.

The mycorrhizosphere effect exerted a significant stimulating effect on populations of fluorescent pseudomonads in soil (Andrade et al., 1998a).

The hyphae of ectomycorrhizal fungi could be the source of C to soil microbial communities, as fungal exudates (Sun *et al.*, 1999) and/or from senescence of hyphae (Bending and Read, 1995) that are used by fluorescent pseudomonas. In addition to this quantitative effect, ectomycorrhizal symbiosis has also modified the distribution of phosphate-solubilizing and lipase producing fluorescent pseudomonads, especially in the hyphosphere soil compartment. Extramatrical mycelium can absorb and translocate to the host plant, soluble P from mineral and organic matter, through the excretion of organic acids and phosphatase, respectively (Landeweert *et al.*, 2001). These results corroborate those of Frey-Klett *et al.* (2005) who demonstrated that phosphate-solubilizing fluorescent pseudomonads were more abundant in the hyphosphere than in bulk soil.

Culture-independent methods for the analysis of soil microbial community structure such as fatty acid extraction (Cavigelli *et al.*, 1995; Ibekwe and Kennedy, 1998) and PCR-DGGE (polymerase chain reaction-denaturing gradient gel electrophoresis) (Ferris *et al.*, 1996; Muyzer and Smalla, 1998; Assigbetse *et al.*, 2005) are increasingly being used. In contrast, little is known of the importance of the functional diversity of soil microbial communities (Pankhurst *et al.*, 1996) resulting from limited access to suitable techniques. The functional diversity of microbial communities includes the range and relative expression of activities involved in decomposition, nutrient transformation, and plant growth promotion (Giller *et al.*, 1997).

Decomposition functions performed by heterotrophic microbes is one component of microbial functional diversity. An assay has been developed to provide a measure of the catabolic functional diversity in soil. This assay provides catabolic response profiles (patterns of *in situ* catabolic potential, ISCP) by measuring the short-term utilization of a range of readily available substrates added to soils (Degens and Vojvodic-Vukovic, 1999). Patterns of ISCP provide a real-time measure of microbial functional diversity, since they allow direct measurement of substrate catabolism by microbial communities in soils without prior culturing of microorganisms. This methodological approach has been widely used to describe the mycorrhizal effect on the functional diversity of soil microbial communities.

11 ECTOMYCORRHIZAE AND FOREST ECO-SYSTEMS

Forest trees are in general completely dependent upon a symbiotic association of their roots with ectomycorrhizal fungi. These fungi mobilize minerals from soil and transfer them to the plant. In exchange the trees deliver assimilated C to the fungi. Ectomycorrhizal fungi have a limited capability to enzymatically degrade the complex carbohydrates of most organic detritus and, instead, rely upon their tree hosts for their energy needs. In return, they take up P, N, sulfur, and zinc from soil and translocate them to their host and greatly extend the functional root system of the host (Allen, 1991). An ectomycorrhizal fungus can connect to and integrate roots of several trees, such that fungi and roots grow as one intact unit. Most ectomycorrhizal fungi are basidiomycetes, with *Amanita, Cortinarius, Lactarius, Russula,* and *Suillus* among the best known ectomycorrhizal genera (Hacskaylo, 1972). Ectomycorrhizal associations are widespread, particularly in temperate regions, and involve many of the ecologically important tree species such as *Pseudotsuga, Picea, Pinus, Abies, Salix, Quercus, Betula* and *Fagus*.

Fundamental knowledge about the biodiversity of soil microbial communities and their functional impact, especially for the ectomycorrhizal fungi, is essentially non-existent. However, maintaining this below-ground biodiversity is essential for the maintenance of a healthy forest and for successful reforestation programs. Ectomycorrhizal fungi are, economically, one of the most important groups of soil fungi. These organisms form a symbiotic relationship with a plant, forming a sheath around the root tip. The fungus then forms a hartig net. The fungus then gains C and other essential organic substances from the tree and in return support the tree in taking up water, mineral salts and metabolites. The fungus also repels parasites, nematodes, and soil pathogens. Indeed, most forest trees are highly dependent on their fungal partners and in areas of poor soil quality could possibly not exist in their absence. Thus, in optimal forest husbandry, a lack of management of mycorrhizal fungi could result in damage to trees and forest crops.

The importance of ectomycorrhiza in forest plantations has received much attention when it was observed that trees often fail to establish at new sites if the ectomycorrhizal symbiont is absent. This effect has been observed in exotic pine transplantation in different parts of the world. In Western Australia, *Pinus radiata* and *P. pinaster* failed to establish in nursery beds in the absence of mycorrhizal fungi (Lakhanpal, 2000). Even the addition of fertilizer had no effect on the establishment of seedlings on such sites. Addition of forest soil produced normal and healthy seedlings, however, because the forest soil contained propagules of mycorrhizal fungi.

High ectomycorrhizal diversity is important in the healthy functioning of woodlands. Different fungi appear to occupy different niches. Some may be more proficient at supporting the tree in taking up particular nutrients, others may be specialized at protecting against pathogens, and others may assist in enzyme production. Intensive study is needed to determine the ectomycorrhizal diversity which will optimize forest husbandry.

Pine wilt disease (PWD) is a globally serious forest disease and demonstrates the importance of tree-ectomycorrhizal relationships. Pines planted on a mountain slope of Japan (Yamaguchi Prefecture) were killed by

PWD, but some trees managed to survive on top of the slope, where mycorrhizal relationships had developed better than on lower slopes. The abundant mycorrhizae found in the upper slope enhanced water uptake by the pines, mitigated drought stress, and thereby decreased the mortality from pine wilt disease (Akema and Futai, 2005). Moreover, under laboratory conditions, inoculation of pine seedlings with ECM fungi confirmed their enhanced resistance to pine wood nematode infection. Pine seedlings are also known to tolerate environmental stresses such as acid mist, when infected with ECM fungi (Asai and Futai, 2001).

ECM fungi also make a significant contribution to forest ecosystems by increasing their network among trees through which nutrients may be transported. In addition ECM fungi improve the growth of host plants at the seedling stage. Many pioneer plants in barren tips and other waste lands are facilitated in their establishment by ECM. This association has been successfully applied to reforestration in tropical forests by inoculating mycorrhizae on nursery seedlings (Lakhanpal, 2000). In forest nursery management it is well known that pine seedlings could not be replanted from the nursery to any other location once they start to expand new lateral roots in spring, though it is easy to replant them in winter. This effect can be attributed to damage of the mycorrhizal association to newly developed lateral roots in spring. Thus, mycorrhizal association is essential for pine seedlings. Generally, we ignore the importance of the mycorrhizal relationship, because mycorrhizae occur underground and are invisible. When trees are exposed to biotic or abiotic stresses, the importance of the mycorrhizal association is noticed, as in the case of pine wilt disease. More than 90% of land plants are associated with mycorrhizal fungi, and two thirds of them are arbuscular mycorrhizae. But tree species predominant in temperate forests are ectomycorrhizal. Why have ECM fungi established the special mycorrhizal relationship with such trees? Does the ECM relationship bring about the prosperity of the trees, or does the prosperity of the trees ensure the establishment of ECM associations? There are many questions to be resolved on the ECM relationship, but the beneficial effects of ECM on these trees is an established fact.

12 ECTOMYCORRHIZAL ASSOCIATIONS TO MAINTAIN TROPICAL MONODOMINANCE

Tropical rainforests harbor the highest known tree diversity on the planet, and a wealth of ecological studies has attempted to explain the intimate association of so many co-occurring species (Leigh *et al.*, 2004; Valencia *et al.*, 1994). Tropical tree diversity, however, is not uniformly diverse, and the existence of tropical 'monodominance', where a single tree species dominates the canopy (Connell and Lowman, 1989; Torti *et al.*, 2001), is just one example of the extreme variation found in rainforests. While other traits likely play a role in the maintenance of tropical monodominance, the particular mechanisms by which ECM associations operate in the maintenance of tropical monodominance have not been fully explored (Torti and Coley, 1999).

Using a combination of field and laboratory experiments, transplanted *Dicymbe corymbosa* seedlings were found to survive better in a monodominant forest compared to a mixed forest. Poor ECM colonization of transplanted *D. corymbosa* seedlings in the mixed forest suggested the ECM inoculum limitation may prevent rapid expansion of the monodominant forest into the mixed forest (McGuire, 2007c). Incorporation of seedling roots to the common ECM network served as an effective mechanism for higher *D. corymbosa* seedling survival in the monodominant forest, potentially providing seedlings with photosynthate from overstory individuals of the same species (McGuire, 2007a). The leaf litter layer on the forest floor was found to be significantly deeper in the monodominant forest than in the mixed forest, although averaged over two years, above-ground litter production was higher in the mixed forest relative to monodominant forest (McGuire, 2007b). This observation suggests that slower decomposition in the monodominant forest was responsible for a greater floor mass. A reciprocal litter bag transplant experiment demonstrated that leaf litter decomposition was slower in the monodominant forest compared to the mixed forest, and that leaf type (*D. corymbosa* or mixed species leaf litter) had no effect on decomposition rate. Moreover, microbial biomass and biomarkers for broad saprotrophic bacteria were higher in the mixed forest compared to the monodominant forest. DGGE analysis revealed that fungal community composition was different between forest types and that forest type explained about 80% of the variation (McGuire, 2007b). These observations suggest that ECM fungi were suppressing saprotrophic decomposition in the monodominant forest to gain preferential access to nutrients contained in forest floor. Together, these studies provide evidence that positive feedbacks between ECM fungi and ECM monodominant trees function to maintain tropical monodominance.

13 MYCORRHIZAL BIOTECHNOLOGY IN RESTORATION OF DISTURBED ECOSYSTEMS

Surface mining activities generate huge areas of disturbed land in many parts of the world and there is an urgent need for soil reconstruction and restoration of productive and functional soil-plant systems on these degraded lands. Ecological restoration is the process of supporting the reclamation of an ecosystem that has been disturbed, degraded, and destroyed (SER, 2004). Soil disturbance can have a tremendous impact and alters the composition and activity of mycorrhizal fungi as well as the host plants (Allen et al., 2005). Numerous studies have documented the fundamental importance of mycorrhizal symbiosis and other microbial systems in reclamation and restoration of contaminated and disturbed ecosystems (Miller, 1987; Danielson and Visser, 1989; Meharg and Cairney, 2000; Khan, 2006; Robertson et al., 2007). Microsymbionts provide benefits not only to individual plants, but most essential ecosystem processes depend upon different mycorrhizal fungi and associated microorganisms. Mycorrhizal fungi can play a critical role in ecosystem dynamics and productivity.

Of the various types of mycorrhizae, four main types, i.e., arbuscular mycorrhizae (AM), ectomycorrhizae (ECM), ectendo-mycorrhizae, and ericoid mycorrhizae (ERM) are important for returning disturbed sites to productive agricultural and forested lands. The ECM, which are generally associated with forest plants, and formed by basidiomycetes, ascomycetes, and several species of zygomycetes comprising some 5,000–6,000 species (Brundrett et al., 1996). The ECM fungi have the ability to provide buffering capacity to plant species against various environmental stresses (Malajczuk et al., 1994).

The majority of vascular plants form arbuscular mycorrhizae. The AM are formed by a small group of fungi in the new phylum Glomeromycota and class Glomeromycetes (Schüβler et al., 2001) containing eight known genera. These symbiotic fungi promote plant growth (Klironomos, 2003) and enhance a number of essential ecosystem processes such as plant productivity, plant diversity and soil structure, and act as bio-ameliorators of stressed soil conditions (van der Heijden et al., 1998, 2006; Vogelsang et al., 2006; Al Karaki et al., 2001).

The ectendomycorrhizae exhibit some structural similarities to both ecto and endomycorrhizae. They are frequently found on the roots of plants growing on disturbed lands. The ericoid mycorrhizae (ERM) associate with plants belonging to the order Ericales. The ERM are associated with Ericaceous plants that have very fine root systems and typically grow in acid, peaty, and infertile soils. The fungus enables access to recalcitrant sources of minerals and provides protection from adverse soil conditions. Ericoid plants withstand extremely difficult environments and can become established on various eroded lands. Ericoid mycorrhizae have demonstrated a special role in the mineralization of soil organic N (Read et al., 1989).

In, northeastern Alberta, the oil sands industry produces vast areas of degraded land that requires reclamation and revegetation (Fung and Macyk, 2000). However, creating functional and self-sustaining boreal-forested lands and wildlife habitat on oil sands mines is a challenging task. Besides the physical and chemical properties of disturbed sites, the soil disturbance event often destroys mycorrhizal fungal networks and other microbial activities in the soil system. Consequently, vegetation establishment is limited by the harsh conditions in areas affected by mining activities.

Successful establishment of forest tree seedlings at reforestation sites is often dependent on mycorrhizal association and on the ability of seedlings to acquire site resources early in the plantation establishment period (Amaranthus and Perry, 1987; van den Driessche, 1991). Mycorrhizal inoculation has been proven beneficial in a wide range of situations. Of the various approaches to enhance reforestation success, inoculation of nursery seedlings with site-adapted mycorrhizal fungi is a successful and environmentally-friendly approach. Inoculation of nursery seedlings has the potential for selective inoculation with stress-tolerant microsymbiont strains, adapted to both target host species and site. Consequently, more effective and greater success is expected in the reclamation effort.

The goal of any reclamation program is to convert disturbed land to its comparable pre-disturbance ecosystem. Therefore, re-introduction of soil microbial processes is recommended for successful reclamation. Careful consideration is required in isolation, identification, selection of site-adapted microsymbionts and understanding of the ecophysiology best suited for reclamation of disturbed lands. Ecologically-based technologies, such as use of mycorrhizal and actinorhizal plants, are expected to provide sustainable use of natural resources by converting severely disturbed lands into productive forested lands. Application of mycorrhizal biotechnology has great potential and can play an essential role in restoration of degraded lands in many surface-mined areas.

14   *IN VITRO* MICORRHIZATION OF MICROPROPAGATED PLANTS

Ectomycorrhizal (ECM) fungi bring several advantages to plants, including increased root absorbing area (Bowen, 1973; Harley and Smith, 1983), enhanced uptake of nutrients (Harley and Smith, 1983), increased resistance to plant pathogens (Marx, 1969), and drought tolerance (Duddridge *et al.*, 1980; Boyd *et al.*, 1986; Meyer, 1987; Feil *et al.*, 1988; Marx and Cordell, 1989). ECM can also increase growth and nutrient content of plants growing in low-nutrient soils (Jones *et al.*, 1991). Water stress appears to be one of the major causes for failure of micropropagated plants during acclimation. The use of compatible mycorrhizal fungi in the soil substrate during the weaning process could not only improve the nutritional state of the plants, but also increase resistance to the water stress of *ex vitro* conditions, increasing weaning rates.

The methods of axenic synthesis have been subject to criticism, because working under conditions where: (1) interacting factors are eliminated; (2) C sources are provided to allow fungal growth before infection sets in; and (3) substrates are sterilized; may significantly change the determinants of efficiency and type of infection (Piché and Peterson, 1988). Non-axenic systems allow detailed studies of root colonization by the fungus (Fortin et al., 1983).

Fungal attachment to the root epidermis takes place via root polysaccharide secretion (Nylund, 1980). The translocation of photosynthetic products to the root increases the concentration of carbon compounds in root exudates. These exudates are composed primarily of amino acids, proteins, organic acids and plant growth regulators. Mineral balance and plant growth regulator concentrations directly control cell permeability and fungal adhesion to the root when mycorrhization occurs (Barea, 1986). Axenic and non-axenic mycorrhizal syntheses mainly differ in terms of time and degree of infection (Duddridge and Read, 1984). These findings validate the use of *in vitro* mycorrhization techniques. Furthermore, Brunner (1991) showed that mycorrhiza obtained by different methods of *in vitro* synthesis had mantles and hartig nets with similar structures. Micropropagated plants develop under high moisture and low lighting conditions. They often experience low lignification levels and decreased functionality of root systems that cause low survival rates during weaning. Mycorrhization of micropropagated plants before acclimation increases survival, enhancing the functionality of the root system and the mineral nutrition of the plant (Rancillac, 1982; Grellier et al., 1984; Heslin and Douglas, 1986; Poissonier, 1986; Tonkin et al., 1989; Martins, 1992, 2004; Martins et al., 1996; Herrmann et al., 1998; Díez et al., 2000). Similarly, *in vitro* mycorrhization of micropropagated plants can be used to increase survival and growth during *ex vitro* weaning (Nowak, 1998). From the point of view of mycorrhization studies, on the other hand, the use of seedlings with *in vitro* mycorrhizal systems must take into account the genetic differences among plants that may condition plant response to mycorrhization. Plant micropropagation techniques allow controlling both biotic and abiotic mycorrhizal conditions, while genetic uniformity is guaranteed.

In vitro mycorrhization (using endo and ectomycorrhizae) of micropropagated plants can be used to increase survival and growth during *ex vitro* weaning (Martins et al., 1996; Nowak, 1998). In the case of fruit trees, inoculation with AM fungi facilitated *in vitro* plant adaptation to *ex vitro* conditions (Sbrana et al., 1994). *In vitro* ectomycorrhization can improve microcutting rooting (Normand et al., 1996) and enables *vitro* plants to acclimate more readily (Martins et al., 1996; Díez et al., 2000). The *in vitro* mycorrhization of micropropagated plants like *Helianthemum* spp. (Morte et al., 1994) and *Cistus* spp. (Díez and Manjón, 1996) has been obtained only in very few Mediterranean species. Even acclimation of somatic embryos can be improved through mycorrhization, such as that reported for cork oak (Díez *et al.*, 2000).

Mycorrhizal plants normally have a different morphology of the root system that corresponds to a different soil exploitation strategy. Fresh and dry weights of mycorrhizal roots are higher than in non-mycorrhizal roots while the length of the major root is lower. The number of roots per unit length and per unit weight is consequently higher in mycorrhizal plants as reported by Biggs and Alexander (1981). Under *in vitro* conditions, mycorrhization increases the growth parameters of plants. Mycorrhization also leads to the morphological development of mycorrhizal structures. Micropropagated plant performance improves and survival capacities increase accordingly. Micropropagation and mycorrhization can be combined as a tool to impart viability to the production of difficult to propagate species, thus increasing their survival and growth. Mycorrhization can provide a sustainable method for plant production, either by micropropagation or through traditional propagation methods.

15 VISUALIZATION OF ENDORHIZAL FUNGI

Few of the interactions between plant roots and endorhizal and rhizosphere fungi produce macroscopic phenotypes, particularly in the early stages. AM fungi interactions predominate, which are found in about 80% of terrestrial plant families and do not produce a visible change in root tissues. In addition, roots harbor a diversity of endophytic fungi that now are shown to be widely distributed and have roles including stress tolerance. In order to fully understand the nature of plant-fungal interactions and their relative importance to the plant, it is essential that endophytic fungi be clearly visualized and accurately quantified.

Fine roots measure several cell layers thick and may have pigmented surface layers, whereas most fungi are disseminated mycelia of hyaline hyphae. In order to conduct transmitted light microscopy of endorhizal fungi, roots are typically cleared of cytoplasm and then stained with chlorazole black E, trypan blue, or lactofuchsin. However, these stains may provide insufficient contrast for high resolution imaging, or for examining fine endophyte hyphae, which typically measure less than 1.5 µm in diameter.

In order to study plants collected from sites where endorhizal colonization has not been well explored, and for species where endorhizal fungi had been thought rare or absent, a sensitive method for fungal visualization has been developed, along with a secure method for specimen preservation. This method is considered reliable for quantifying potentially multiple endorhizal morphotypes. Endorhizal visualization methods using fluorescence microscopy can be applied to studying roots with multiple fungal morphotypes and from herbarium specimens (Ormsby *et al.*, 2007).

For samples harvested from field sites where there is often little control of the fungal root symbiont(s), lactofuchsin staining viewed with epifluorescence microscopy provides a convenient combination of relative simplicity of preparation and imaging quality. Confocal epifluorescence optics are often superior for documentation, whereas widefield optics are more efficient for quantitation.

Assessing endorhizal colonization using the Multiple Quantitation Method (MQM) is straightforward and reproducible. MQM is well-suited for plant roots harvested from natural locations where the endorhizal fungi are not necessarily well described. Furthermore, MQM can describe the relative contributions of different endorhizal interactions, which must intrinsically be related to their importance to the plant's physiology.

16 CONCLUSIONS

Mycorrhizal fungi are now known to provide a wide range of significant benefits to their plant hosts. In addition to enhancing mineral nutrition, they induce greater resistance to soil pathogens, enhance tolerance to drought stress, and reduce sensitivity to toxic substances occurring in the soil.

Introduction of mycorrhizal fungi do not appear to offer much advantage to enhanced nutrition or disease resistance in native species. Optimization of the ability of native fungi to colonize hosts in their natural habitat or to minimize loss of these fungi with disturbance is required. Highly dependent crop hosts should be selected over mycorrhizal- independent hosts in crop rotations or in multiple cropping systems. Traditional methods of breeding and producing crop plants in soils with high nutrient contents may select against the most efficient fungal communities or even against the mycorrhizal association.

Many efforts have been made in recent years to accrue benefits from mycorrhizae for agriculture, horticulture, forestry, and site remediation. The results have been consistently positive, with some difficulties due to complications from diverse variables under field conditions. Mycorrhizal interactions between plants, fungi, and the environment are complex and often inseparable. Mycorrrhizae are an essential below-ground component in the establishment and sustainability of plant communities, but thorough knowledge is required to achieve maximum benefits from these microorganisms and their associations.

REFERENCES

Agnew, C., and Warren, A., 1996, A framework for tackling drought and land degradation. *J. Arid. Environ.* **33**: 309–320.

Akema, T., and Futai, K., 2005, Ectomycorrhizal development in a *Pinus thunbergii* stand in relation to location on a slope and effect on tree mortality from pine wilt disease. *J. For. Res.* **10**: 93–99.

Aliasgarzad, N., Neyshabouri, M.R., and Salimi, G., 2006, Effects of arbuscular mycorrhizal fungi and *Bradyrhizobium japonicum* on drought stress of soybean. *Biologia* **61**: 324–328.

Al-Karaki, G.N., Al-Raddad, A., and Clark, R.B., 1998, Water stress and mycorrhizal isolate effects on growth and nutrient acquisition of wheat. *J. Plant Nutr.* **21**: 891–902.

Al-Karaki, G.N., Hammad, R., and Rusan, M., 2001, Response of two tomato cultivars differing in salt tolerance to inoculation with mycorrhizal fungi under salt stress. *Mycorrhiza* **11**: 43–47.

Allen, M.F., 1991, *The Ecology of Mycorrhiza*. Cambridge: Cambridge University Press, p. 184.

Allen, M.F., 1992, *Mycorrhizal Functioning: An Integrative Plant-Fungal Process*. Routledge, NY: Chapman & Hall, p. 534.

Allen, M.F., Swenson, W., Querejeta, J.I., Egerton-Warburton, L.M., and Treseder, K.K., 2003, Ecology of mycorrhizae: a conceptual framework for complex interactions among plants and fungi. *Ann. Rev. Phytopathol.* **41**: 271–303.

Allen, M.F., Allen, E.B., and Gómez-Pompa, A., 2005, Effects of mycorrhizae and non-target organisms on restoration of a seasonal tropical forest in Quintano Roo, Mexioco: factors limiting tree establishment. *Restor. Ecol.* **13**: 325–533.

Amaranthus, M.P., and Perry, D.A., 1987, Effect of soil transfer on ectomycorrhizal formation and the survival and growth of conifer seedlings on old, reforested clear-cuts. *Can. J. For. Res.* **17**: 944–950.

Amora-Lazcano, E., Vazquez, M.M., and Azcon, R., 1998, Response of nitrogen-transforming microorganisms to arbuscular mycorrhizal fungi. *Biol. Fert. Soils* **27**: 65–70.

Andrade, G., Linderman, R.G., and Bethlenfalvay, G.J., 1998a, Bacterial associations with the mycorrhizosphere and hyphosphere of the arbuscular mycorrhizal fungus *Glomus mosseae*. *Plant Soil* **202**: 79–87.

Andrade, G., Mihara, K.L., Linderman, R.G., and Bethlenfalvay, G.J., 1998b, Soil aggregation status and rhizobacteria in the mycorrhizosphere. *Plant Soil* **202**: 89–96.

Asai, E., and Futai, K., 2001, Retardation of pine wilt disease symptom development in Japanese black pine seedlings exposed to simulated acid rain and inoculated with *Bursaphelenchus xylophilus*. *J. For. Res.* **6**: 297–302.

Assigbetse, K., Gueye, M., Thioulouse, J., and Duponnois, R., 2005, Soil bacterial diversity responses to root colonization by an ectomycorrhizal fungus are not root-growth dependent. *Microb. Ecol.* **50**: 350–359.

Auge, R.M., 2001, Water relations, drought and vesicular-arbuscular mycorrhizal symbiosis. *Mycorrhiza* **11**: 3–42.

Azcón, R., and Ocampo, J.A., 1981, Factors affecting the vesicular-arbuscular infection and mycorrhizal dependency of thirteen wheat cultivars. *New Phytol.* **87**: 677–685.

Balestrini, R., and Bonfante, P., 2005, The interface compartment in arbuscular mycorrhizae: a special type of plant cell wall? *Plant Biosyst.* **139**: 8–15.

Baon, J.B., Smith, S.E., Alston, A.M., and Wheeler, R.D., 1992, Phosphorus efficiency of three cereals as related to indigenous mycorrhizal infection. *Aust. J. Agric. Res.* **43**: 479–491.

Barea, J.M., 1986, Importance of hormones and root exudates in mycorrhizal phenomena. In: *Mycorrhizae: Physiology and Genetics*. 1st Eur. Symp. Mycor. Paris: ESM, Dijon, INRA, pp. 177–187.

Barea, J.M., Azcón-Aguilar, C., and Azcón, R., 1997, Interactions between mycorrhizal fungi and rhizosphere microorganisms within the context of sustainable soli-plant systems.

In: Gange, A.C., and Brown, V.K. (eds.), *Multitrophic Interactions in Terrestrial Systems*. Cambridge: Backwell Science, pp. 65–77.

Bending, G.D., and Read, D.J., 1995, The structure and function of the vegetative mycelium of ectomycorrhizal plants. V. Foraging behavior and translocation of nutrients from exploited litter. *New Phytol.* **130**: 401–409.

Bethlenfalvay, G.J., 1992, Mycorrhizae and crop productivity. In: Bethlenfalvay, G.J., and Linderman, R.G. (eds.), *Mycorrhizae in Sustainable Agriculture*. Madison, WI: ASA Special Publication No. 54, pp. 1–27.

Bethlenfalvay, G.J., and Linderman, R.G., 1992, *Mycorrhizae in Sustainable Agriculture*. St Paul, MN: The American Phytopathological Society Special Publication No. 54, p. 124.

Bethlenfalvay, G.J., and Schüepp, H., 1994, Arbuscular mycorrhizas and agrosystem stability. In: Gianinazzi, S. and Schüepp, H. (eds.), *Impact of Arbuscular Mycorrhizas on Sustainable Agriculture and Natural Ecosystems*. Basel, Switzerland: Birkhäuser Verlag, pp. 117–131.

Bidartondo, M.I., Redecker, D., Hijri, I., Wiemken, A., Bruns, T.D., Domínguez, L., Sérsic, A., Leake, J.R., and Read, D.J., 2002, Epiparasitic plants specialized on arbuscular mycorrhizal fungi. *Nature* **419**: 389–392.

Biggs, W.L., and Alexander, I.J., 1981, A culture unit for the study of nutrient uptake by intact mycorrhizal plants under aseptic conditions. *Soil Biol. Biochem.* **13**: 77–78.

Bowen, G.D., 1973, Mineral nutrition of mycorrhizas. In: Marks, G.C., and Kozlowski, T.C. (eds.), *Ectomycorrhizas*. . New York/London: Academic, pp. 151–201.

Boyd, R., Furbank, R.T., and Read D.J., 1986, Ectomycorrhiza and the water relations of trees. In: Gianinazzi-Pearson, V., and Gianinazzi, S. (eds.), *Mycorrhizae, Physiology and Genetics*. Paris: INRA, pp. 689–693.

Brady, N.C., and Weil, R.R., 2002, *The Nature and Properties of Soils*. New Jersey: Prentice Hall, p. 960.

Brundrett, M.C., 2002, Coevolution of roots and mycorrhizas of land plants. *New Phytol.* **154**: 275–304.

Brundrett, M., Bougher, N., Dell, B., Grove, T., and Malajczuk. N., 1996, *Working with Mycorrhizas in Forestry and Agriculture*. Canberra, Australia: ACIAR.

Brunner, I., 1991, Comparative studies on ectomycorrhizae synthesized with various *in vitro* techniques using *Picea abies* and two *Hebeloma* species. *Trees* **5**: 90–94.

Bucher, M., 2007, Functional biology of plant phosphate uptake at root and mycorrhiza interfaces. *New Phytol.* **173**: 11–26.

Calvet, C., Barea, J.M., and Pera, J., 1992, *In vitro* interactions between the vesicular-arbuscular mycorrhizal fungus *Glomus mosseae* and some saprophytic fungi isolated from organic substrates. *Soil Biol. Biochem.* **24**: 775–780.

Calvet, C., Pera, J., and Barea, J.M., 1993, Growth response of marigold (*Tagetes erecta* L.) to inoculation with *Glomus mosseae*, *Trichoderma aureoviride* and *Pythium ultimum* in a peat-perlite mixture. *Plant Soil* **148**: 1–6

Carrillo-García, A., Leon de la Luz, J.L., Bashan, Y., and Bethlenfalvay, G.J., 1999, Nurse plants, mycorrhizae, and plant establishment in a disturbed area of the Sonoran Desert. *Restor. Ecol.* **7**: 321–335.

Carter, M.R., and Campbell, A.J., 2006, Influence of tillage and liquid swine manure on productivity of a soybean-barley rotation and some properties of a fine sandy loam in Prince Edward Island. *Can. J. Soil Sci.* **86**: 741–748.

Cavigelli, M.A., Robetson, G.P., and Klug, M.J., 1995, Fatty acid methyl ester (FAME) profiles as measures of soil microbial community structure. *Plant Soil* **170**: 99–113.

Chalot, M., Blaudez, D., and Brun, A., 2006, Ammonia: a candidate for nitrogen transfer at the mycorrhizal interface. *Tren. Plant Sci.* **11**: 263–266.

Connell, J.H., and Lowman, M.D., 1989, Low-diversity tropical rain forests: some possible mechanisms for their existence. *Amer. Natural.* **134**: 88–119.

Covacevich, F., Echeverría, H.E., and Andreoli, Y.E., 1995, Micorrización vesículo arbuscular espontánea en trigo en función de la disponibilidad de fósforo. *Ciencia del Suelo* **13**: 47–51.

Covacevich, F., Echeverría, H.E., and Aguirrezabal, L.A.N., 2007, Soil available phosphorus status determines indigenous mycorrhizal colonization of field and glasshouse-grown spring wheat from Argentina. *Appl. Soil Ecol.* **35**: 1–9.

Danielson, R.M., and Visser, S., 1989, Host response to inoculation and behaviour of induced and indigenous ectomycorrhizal fungi of jack pine grown on oil-sands tailings. *Can. J. For. Res.* **19**: 1412–1421.

Davet, P., 1996, *Vie microbienne du sol et production végétale*. Paris: INRA.

Degens, B.P., and Vojvodic-Vukovic, M., 1999, A sampling strategy to assess the effects of land use on microbial functional diversity in soils. *Austr. J. Soil Res.* **37**: 593–601.

Del Val, C., Barea, J.M., and Azcon-Aguilar, C., 1999, Diversity of arbuscular mycorrhizal fungus populations in heavy-metal-contaminated soils. *Appl. Environ. Microbiol.* **65**: 718–723.

De Oliveira, V.L., and Garbaye, J., 1989, Les microorganismes auxiliaires de l'établissement des symbioses ectomycorhiziennes. *Eur. J. Pathol.* **19**: 54–64.

Díez, J., and Manjón, J.L., 1996, Mycorrhizal formation by vitroplants of *Cistus albidus* L. and *C. salvifolius* L. and its interest for truffle cultivation in poor soils. In: Azcón-Aguilar, B. and Barea, J.M. (eds.), *Mycorrhizas in Integrated Systems from Genes to Plant Development*. Brussels: European Commission, pp. 528–530.

Díez, J., Manjón, J.L., Kovács, G.M., Celestino, C., and Toribio, M., 2000, Mycorrhization of vitroplants raised from somatic embryos of cork oak (*Quercus suber* L.) *Appl. Soil Ecol.* **15**: 119–123.

Duddridge, J.A., and Read, D.J., 1984, Modification of the host-fungus interface in mycorrhizas synthesized between *Suillus bovinus* (Fr.) O. Kuntz and *Pinus sylvestris* L. *New Phytol.* **96**: 583–588.

Duddridge, J.A., Malibari, A., and Read, D.J., 1980, Structure and function of mycorrhizal rhizomorphs with special reference to their role in water transport. *Nature* **287**: 834–836.

Feil, W., Kottke, I., and Oberwinkler, F., 1988, The effect of drought on mycorrhizal production and very fine root system development of norway spruce under natural and experimental conditions. *Plant Soil* **108**: 221–231.

Ferris, M.J., Muyzer, G., and Ward, D.M., 1996, Denaturing gradient gel electrophoresis profiles of 16S rRNA-defined populations inhabiting a hot spring microbial mat community. *Appl. Envir. Microbiol.* **62**: 340–346.

Filion, M., St-Arnaud, M., and Fortin, J.A., 1999, Direct interaction between the arbuscular mycorrhizal fungus *Glomus intraradices* and different rhizosphere microorganisms. *New Phytol.* **141**: 525–533.

Fortin, J.A., Piché, Y., and Godbout, C., 1983, Methods for synthesizing ectomycorrhizas and their effect on mycorrhizal development. *Plant Soil* **71**: 275–284.

Francis, C.F., and Thornes, J.B., 1990, Matorral: erosin and reclamation. In: Albadalejo, J., Stocking, M.A., and Díaz, E. (eds.), *Soil Degradation and Rehabilitation in Mediterranean Environmental Conditions*. Murcia, Spain: CSIC, pp. 87–115.

Frank, A.B., 1885, Uber di auf werzelsymbiose beruhende Ernahrung gewisser Baume durch unterirdischeplize. *Ber. Dtsch. Bot. Ges.* **3**: 128–145.

Frey-Klett, P., Chavatte, M., Clausse, M.L., Courrier, S., Le Roux, C., Raaijmakers, J., Martinotti, M.G., Pierrat, J.C., and Garbaye, J., 2005, Ectomycorrhizal symbiosis affects functional diversity of rhizosphere fluorescent pseudomonads. *New Phytol.* **165**: 317–328.

Fung, M.Y.P., and Macyk, T.M., 2000, Reclamation of oil sand mining areas. In: R.I., Barnhisel, R.I., Darmody, R.G., and Daniels, W.L. (eds.), *Reclamation of Drastically Disturbed Lands*. American Society of Agronomy monographs, 2nd edn. **41**: 755–744.

Garbaye, J., 1991, Biological interactions in the mycorrhizosphere. *Experientia* **47**: 370–375.

Garcia-Romera, I., Garcia-Garrido, J.M., Martin, J., Fracchia, S., Mujica, M.T., Godeas, A., and Ocampo, J.A., 1998, Interaction between saprophytic *Fusarium* strains and arbuscular mycorrhizas of soybean plants. *Symbiosis* **24**: 235–246.

Gaur, A., and Adholeya, A., 2004, Prospects of arbuscular mycorrhizal fungi in phytoremediation of heavy metal contaminated soils. *Curr. Sci.* **86**: 528–534.

Gianinazzi-Pearson, V., and Diem, H.G., 1982, Endomycorrhizae in the tropics. In: Dommergues, Y.R., and Diem, H.G. (eds.), *Microbiology of Tropical Soil and Plant Productivity*. The Hague: Martinus Nijhoff/Dr Junk W. Publishers.

Giller, K.E., Beare, M.H., Lavelle, P., Izac, A.-M.N., and Swift, M.J., 1997, Agricultural intensification, soil biodiversity and agroeco-system function. *Appl. Soil Ecol.* **6**: 3–16.

Gohre, V., and Paszkowski, U., 2006, Contribution of the arbuscular mycorrhizal symbiosis to heavy metal phytoremediation. *Planta* **223**: 1115–1122.

Govindarajulu, M., Pfeffer, P., Jin, H., Abubaker, J., Douds, D.D., Allen, J.W., Bücking, H., Lammers, P.J., and Shachar-Hill, Y., 2005, Nitrogen transfer in the arbuscular mycorrhizal symbiosis. *Nature* **435**: 819–823.

Green, H., Larsen, J., Olsson, P.A., Jensen, D.F., and Jakobsen, I., 1999, Suppression of the biocontrol agent *Trichoderma harzianum* by mycelium of the arbuscular mycorrhizal fungus *Glomus intraradices* in root-free soil. *Appl. Environ. Microbiol.* **65**: 1428–1434.

Grellier, B., Letouzé, R., and Strullu, D.G., 1984, Micropropagation of birch and mycorrhizal formation *in vitro*. *New Phytol.* **97**: 591–599.

Hacskaylo, E., 1972, Mycorrhiza: the ultimate in reciprocal parasitism? *BioScience* **22**: 577–582.

Harley, J.L., and Smith, S.E. 1983, *Mycorrhizal symbiosis*. London: Academic.

Harrier, L.A., and Watson, C.A., 2004, The potential role of arbuscular mycorrhizal (AM) fungi in the bioprotection of plants against soil-borne pathogens in organic and/or other sustainable farming systems. *Pest Manag. Sci.* **60**: 149–157.

Harrison, M., 1999, Biotrophic interfaces and nutrient transport in plant/fungal interfaces. *J. Exp. Bot.* **50**: 1013–1022.

Harrison, M.J. 1996, A sugar transporter from Medicago truncatula: altered expression pattern in roots during vesicular-arbuscular (VA) mycorrhizal associations. *Plant J.* **9**: 491–503.

Harrison, M.J., and Buuren, M.Lv., 1995, A phosphate transporter from the mycorrhizal fungus *Glomus versiforme*. *Nature* **378**: 626–629.

Harrison, M.J., Dewbre, G.R., and Liu, J., 2002, A phosphate transporter from Medicago truncatula involved in the acquisition of phosphate released by arbuscular mycorrhizal fungi. *Plant Cell* **14**: 2413–2429.

Hawksworth, D.L., 1991, The fungal dimension of biodiversity: magnitude significance and conservation. *Mycol. Res.* 95: 641–655.

Herrera, M.A., Salamanca, C.P., and Barea, J.M., 1993, Inoculation of woody legumes with selected arbuscular mycorrhizal fungi and rhizobia to recover desertified Mediterranean ecosystems. *Appl. Environ. Microbiol.* **59**: 129–133.

Herrmann, S., Munch, J.-C., and Buscot, F., 1998, A gnotobiotic culture system with oak microcuttings to study specific effects of mycobionts on plant morphology before, and in the early phase of, ectomycorrhiza formation by *Paxillus involutus* and *Piloderma croceum*. *New Phytol.* **138**: 203–212.

Heslin, M.C., and Douglas, G.C., 1986, Effects of ectomycorrhizal fungi on growth and development of poplar plants derived from tissue culture. *Sci. Hort.* **30**: 143–149.

Hetrick, B.A.D., Wilson, G.W.T., and Cox, T.S., 1993, Mycorrhizal dependence of modern wheat cultivars and ancestors: a synthesis, *Can J. Bot.* **71**: 512–517.

Hibbett, D.S., Gilbert, L.B., and Donoghue, M.J., 2000, Evolutionary instability of ectomycorrhizal symbioses in basidiomycetes. *Nature* **407**: 506–508.

Hodge, A., Campbell, C.D., and Fitter, A.H., 2001, An arbuscular mycorrhizal fungus accelerates decomposition and acquires nitrogen directly from organic material. *Nature* **413**: 297–299.
Hossain, M.M., Sultana, F., Kubota, M., Koyama, H., and Hyakumachi, M., 2007, The plant growth-promoting fungus *Penicillium simplicissimum* GP17-2 induces resistance in *Arabidopsis thaliana* by activation of multiple defense signals. *Plant Cell Physiol.* **48**: 1724–1736.
Hyakumachi, M., and Kubota, M., 2004a, Biological control of plant diseases by plant growth promoting fungi. *Proc. Int. Sem. Biological Cont. Soilborne Plant Dis.*, Japan-Argentina Joint Study, pp. 87–123.
Hyakumachi, M., and Kubota, M., 2004b, Fungi as plant growth promoter and disease suppressor. In: Arora, D.K. (ed.), *Fungal Biotechnology in Agricultural, Food, and Environmental Applications*. New York: Marcel Dekker, pp. 101–110.
Ibekwe, A.M., and Kennedy, A.C., 1998, Fatty acid methyl ester (FAME) profiles as tool to investigate community structure of two agricultural soils. *Plant Soil* **206**: 151–161.
Jakobsen, I., and Nielsen, N.E., 1983, Vesicular-arbuscular mycorrhiza in field-grown crops. I. Mycorrhizal infection in cereals and peas as various times and soil depths. *New Phytol.* **93**: 401–413.
Jamal, A., Ayub, N., Usman, M., and Khan, A.G., 2002, Arbuscular mycorrhizal fungi enhance zinc and nickel uptake from contaminated soil by soyabean and lentil. *Int. J. Phytoremed.* **4**: 205–221.
Janos, D.P., 2007, Plant responsiveness to mycorrhizas differs from dependence upon mycorrhizas. *Mycorrhiza* **17**: 75–91.
Javot, H., Penmetsa, R.V., Terzaghi, N., Cook, D.R., and Harrison, M.J., 2007a, A Medicago truncatula phosphate transporter indispensable for the arbuscular mycorrhizal symbiosis. *Proc. Natl. Acad. Sci. USA* **104**: 1720–1725.
Javot, H., Pumplin, N., and Harrison, M.J., 2007b, Phosphate in the arbuscular mycorrhizal symbiosis: transport properties and regulatory roles. *Plant Cell Environ.* **30**: 310–322.
Jawson, M.D., Franzlubbers, A.J., Galusha, D.K., and Aiken, R.M., 1993, Soil fumigation within monoculture and rotations—response of corn and mycorrhizae. *Agron. J.* **85**: 1174–1180.
Jeffries, P., and Barea, J.M., 2000, Arbuscular mycorrhiza—a key component of sustainable plant-soil ecosystems. In: Hock, B. (ed.), *The Mycota. IX. Fungal Associations*. Berlin: Springer KG, pp. 95–113.
Jeffries, P., Gianinazzi, S., Perotto, S., Turnau, K., and Barea, J.M., 2003, The contribution of arbuscular mycorrhizal fungi in sustainable maintenance of plant health and soil fertility, *Biol. Fert. Soils* **37**: 1–16.
Johansson, J.F., Paul, L.R., and Finlay, R.D., 2004, Microbial interactions in the mycorrhizosphere and their significance for sustainable agriculture. *FEMS Microbiol. Ecol.* **48**: 1–13.
Jones, M.D., Durall, D.M., and Tinker, P.B., 1991, Fluxes of carbon and phosphorus between symbionts in willow ectomycorrhizas and their changes with time. *New Phytol.* **119**: 99–106.
Kabir, Z., and Koide, R.T., 2000, The effect of dandelion or a cover crop on mycorrhiza inoculum potential, soil aggregation and yield of maize. *Agric. Eco. Environ.* **78**: 167–174.
Katznelson, H., Rouatt, J.W., and Peterson, E.A., 1962, The rhizosphere effect of mycorrhizal and nonmycorrhizal roots of yellow birch seedlings. *Can. J. Bot.* **40**: 377–382.
Khan, A.G., 2005, Role of soil microbes in the rhizospheres of plants growing on trace metal contaminated soils in phytoremediation. *J. Trace Elem. Med. Biol.* **18**: 355–364.
Khan, A.G., 2006, Mycorrhizoremediation- an enhanced form of phytoremidiation. *J Zhejiang Univ. Science B* **7**: 503–514.

Khan, A.G., Kuek, C., Chaudhry, T.M., Khoo, C.S., and Hayes, W.J., 2000, Role of plants, mycorrhizae and phytochelators in heavy metal contaminated land remediation. *Chemosphere* **41**: 197–207.

Kirchmann, H., and Thorvaldsson, G., 2000, Challenging targets for future agriculture. *Eur. J. Agron.* **12**: 145–161.

Klironomos, J.N., 2003, Variation in plant response to native and exotic arbuscular mycorrhizal fungi. *Ecology* **84**: 2292–2301.

Knudsen, I.M.B., Debosz, K., Hockenhull, J., Jensen, D.F., and Elmholt, S., 1995, Suppressiveness of organically and conventionally managed soils towards brown foot rot of barley. *Appl. Soil Ecol.* **12**: 61–72.

Koike, N., Hyakumachi, M., Kageyama, K., Tsuyumu, S., and Doke, N., 2001, Induction of systemic resistance in cucumber against several diseases by plant growth promoting fungi: lignification and superoxide generation. *Eur. J. Plant Pathol.* **107**: 523–533.

Lakhanpal, T.N., 2000, Ectomycorrhiza-an overview. In: Mukerji, K.G., Chamola, B.P., and Singh, J. (eds.), *Mycorrhizal Biology*. New York: Kluwer Academic/Plenum, 336 pp., pp. 101–118.

Landeweert, R., Hoffland, E., Finlay, R.D., Kuyper, T.W., and van Breemen, N., 2001, Linking plants to rocks: ectomycorrhizal fungi mobilize nutrients from minerals. *Trends Ecol. Evol.* **16**: 248–254.

Leigh, E.G., Davidar, P., Dick, C.W., Puyravaud, J., Terborgh, J., ter Steege, H., and Wright, S.J., 2004, Why do some tropical forests have so many species of trees? *Biotropica* **36**: 445–473.

Leyval, C., and Berthelin, J., 1993, Rhizodeposition and net release of soluble organic compounds of pine and beech seedlings inoculated with rhizobacteria and ectomycorrhizal fungi. *Biol. Fert. Soils* **15**: 259–267.

Linderman, R.G., 1988, Mycorrhizal interactions with the rhizosphere microflora: the mycorrhizosphere effect. *Phytopathology* **78**: 366–371.

Linderman, R.G., 1992, VA mycorrhizae and soil microbial interactions. In: Bethelenfalvay, G.J., and Linderman, R.G. (eds.), *Mycorrhizae in Sustainable Agriculture*. Madison, WI: ASA Special Publication No. 54, pp. 45–70.

Linderman, R.G., and Paulitz, T.C., 1990, Mycorrhizal-rhizobacterial interactions. In: Hornby, D., Cook, R.J., Henis, Y., Ko, W.H., Rovira, A.D., Schippers, B., and Scott, P.R. (eds.), *Biological Control of Soil-Borne Plant Pathogens*. Wallingford, UK: CAB International, pp. 261–283.

López-Bermúdez, F., and Albaladejo, J., 1990, Factores ambientales de la degradación del suelo en el area mediterránea. In: Albadalejo, J., Stocking, M.A., and Díaz, E. (eds.), *Soil Degradation and Rehabilitation in Mediterranean Environmental Conditions*. Murcia, Spain: CSIC, pp. 15–45.

Lopez-Pedrosa, A., Gonzalez-Guerrero, M., Valderas, A., Azcon-Aguilar, C., and Ferrol, N., 2006, GintAMT1 encodes a functional high-affinity ammonium transporter that is expressed in the extraradical mycelium of Glomus intraradices. *Fungal Genet. Biol.* **43**: 102–110.

Mader, P., Fliessbach, A., Dubois, D., Gunst, L., Fried, P., and Niggli, U., 2002, Soil fertility and biodiversity in organic farming. *Science* **296**: 1694–1697.

Malajczuk, N., Redell, P., and Brundrett. M., 1994, The role of ectomycorrhizal fungi in minesite reclamation. In: Pfleger, F.L., and Linderman, R.G. (eds.), *Mycorrhizae and Plant Health*. St Paul, MN: The Amer. Phytopathol. Soc.

Martins, A., 1992, Micorrização *in vitro* de plantas micropropagadas de *Castanea sativa* Mill. Dissertação para obtenção do grau de mestre. Faculdade de Ciências de Lisboa, 124 pp.

Martins, A., 2004, Micorrização controlada de Castanea sativa Mill.: Aspectos fisiológicos da micorrização in vitro e ex vitro. Dissertação de doutoramento em Biologia/Biotecnologia Vegetal. Faculdade de Ciências de Lisboa. Universidade Clássica de Lisboa, 506 pp.

Martins, A., Barroso, J., and Pais, M.S., 1996, Effect of ectomycorrhizal fungi on survival and growth of micropropagated plants and seedlings of *Castanea sativa* Mill. *Mycorrhiza* **6**: 265–270.
Marx, D.H., 1969, The influence of ectotrophic mycorrhizal fungi on the resistance of pine roots to pathogenic infections. I. Antagonism of mycorrhizal fungi to root pathogenic fungi and soil bacteria. *Phytopathology* **59**: 153–163.
Marx, D.H., and Cordell, C.E., 1989, Use of ectomycorrhizas to improve forestation practices. In: Whipps, J.M., and Lumsden, R.D. (eds.), *Biotechnology of Fungi for Improving Plant Growth*. Cambridge: Cambridge University Press, pp. 1–25.
McAllister, C.B., Garcia-Garrido, J.M., García-Romera, I., Godeas, A., and Ocampo, J.A., 1996, *In vitro* interactions between *Alternaria alternata*, *Fusarium equiseti* and *Glomus mosseae*. *Symbiosis* **20**: 163–174.
McGuire, K.L., 2007a, Common ectomycorrhizal networks may maintain monodominance in a tropical rain forest. *Ecology* **88**: 567–574.
McGuire, K.L., 2007b, *Ectomycorrhizal Associations Function to Maintain Tropical Monodominance: Studies from Guyana*. Ann Arbor, MI: Ph.D. dissertation, University of Michigan.
McGuire, K.L., 2007c, Recruitment dynamics and ectomycorrhizal colonization of *Dicymbe corymbosa*, a monodominant tree in the Guiana Shield. *J. Trop. Ecol.* **23**: 297–307.
Meharg, A.A., and Cairney, J.W.G., 2000, Ectomycorrhizas- extending the capabilities of rhizosphere remediation? *Soil Biol. Biochem.* **32**: 1475–1484.
Mertz, W., 1981, The essential trace elements. *Science* **213**: 1332–1338.
Meyer, F.H., 1987, Extreme site conditions and ectomycorrhizae (especially *Cenococcum geophyllum*). *Mycorrhiza and Plant Stress. Mykorrhiza und bei Pflanzen. Angew. Bot.*, pp. 39–46.
Miller, R.M., 1987, Mycorrhizae and succession. In: Jordan III, W.R., Gilpin, M.E., and Aber, J.D. (eds.), *Restoration Ecology: A Synthetic Approach to Ecological Research*. New York: Cambridge University Press pp. 205–219.
Miransari, M., Bahrami, H.A., Rejali, F., Malakouti, M.J., and Torabi, H., 2007, Using arbuscular mycorrhiza to reduce the stressful effects of soil compaction on corn (*Zea mays* L.) growth. *Soil Biol. Biochem.* **39**: 2014–2026.
Morte, M.A., Cano, A., Honrubia, M., and Torres, P., 1994, In vitro mycorrhization of micropropagated *Helianthemum almeriense* plantlets with *Terfezia claveryi* (desert truffle). *Agric. Sci. Finland* **3**: 309–314.
Mozafar, A., Anken, T., Ruh, R., and Frossard, E., 2000, Tillage intensity, mycorrhizal and nonmycorrhizal fungi, and nutrient concentrations in maize, wheat, and canola. *Agron. J.* **92**: 1117–1124.
Muyzer, G., and Smalla, K., 1988, Application of denaturing gradient gel electrophoresis (DGGE) and temperature gradient gel electrophoresis (TGGE) in microbial ecology. *Antonie van Leeuwenhoek* **73**: 127–141.
Nichols, K.A., and Wright, S.F., 2004, Contributions of soil fungi to organic matter in agricultural soils. In: Magdoff, F., and Weil, R. (eds.), *Functions and Management of Soil Organic Matter in Agroecosystems*. Washington, DC: CRC, pp. 179–198
Normand, L., Bartschi, H., Debaud, J.-C., and Gay, G., 1996, Rooting and acclimatization of micropropagated cuttings of *Pinussylvestris* are enhanced by the ectomycorrhizal fungus *Hebeloma cylindrosporum*. *Physiol. Plant.* **98**: 759–766.
Nowak, J., 1998, Benefits of *in vitro* "biotization" of plant tissue cultures with microbial inoculants. *In vitro Cell. Dev. Biol. Plant* **34**: 122–130.
Nylund, J.E., 1980, Symplastic continuity during Hartig net formation in Norway Spruce ectomycorrhizae. *New Phytol.* **86**: 373–378.
Olsson, P.A., Chalot, M., Bååth, E., Finlay, R.D., and Söderström, B., 1996, Ectomycorrhizal mycelia reduce bacterial activity in sandy soil. *FEMS Microbiol. Ecol.* **21**: 77–86.

Ormsby, A., Hodson, E., Li, Y., Basinger, J., and Kaminskyj, S., 2007, Arbuscular mycorrhizae associated with Asteraceae in the Canadian High Arctic: the value of herbarium archives. *Can. J. Bot.* **85**: 599–606.

Pankhurst, C.E., Ophel-Keller, K., Doube, B.M., and Gupta V.V.S.R., 1996, Biodiversity of soil microbial communities in agricultural systems. *Biodiv. Conser.* **5**: 197–209.

Pfleger, F.L., and Linderman, R.G., 1996, *Mycorrhiza and Plant Health*, 2nd edn.. USA: APS.

Piché, Y., and Peterson, R.L., 1988, Mycorrhiza initiation: an example of plant microbial interactions In: Fredrick, A., and Valentine, E.D. (eds.), *Forest and Crop Biotechnology. Progress and Prospects*. New York: Springer.

Poissonier, M., 1986, Mycorhization *in vitro* de clones d'eucalyptus. Annales AFOCEL. Note de laboratoire: 81–93.

Preger, A.C., Rillig, M.C., Johns, A.R., Du Preez, C.C., Lobe, I., and Amelung, W., 2007, Losses of glomalin-related soil protein under prolonged arable cropping: a chronosequence study in sandy soils of the South African Highveld. *Soil Biol. Biochem.* **39**: 445–453.

Rambelli, A., 1973, The rhizosphere of mycorrhizae. In: Marks, G.C., and Kozlowski, T.T. (eds.), *Ectomycorrhizae: Their Ecology and Physiology*. New York: Academic, pp. 299–343.

Rancillac, M., 1982, Multiplication végétative *in vitro* et synthèse mycorhizienne: pin maritime, Hebelome, Pisolithe. Les colloques de l'*INRA*. **13**: 351–355.

Rausch, C., Daram, P., Brunner, S., Jansa, J., Laloi, M., Leggewie, G., Amrhein, N., and Bucher, M., 2001, A phosphate transporter expressed in arbuscule-containing cells in potato. *Nature* **414**: 462–470.

Read, D.J., Leake, J.R., and Langdale, A.R., 1989, The nitrogen nutrition of mycorrhizal fungi and their host plants. In: Boddy, L., Marchant, R., and Read, D.J. (eds.), *Nitrogen, Phosphorus, and Sulphur Utilization by Fungi*. Cambridge: Cambridge University Press, pp. 181–204.

Redecker, D., Morton, J.B., and Bruns, T.D., 2000, Ancestral lineages of arbuscular mycorrhizal fungi (*Glomales*). *Mol. Phylogen. Evol.* **14**: 276–284.

Requena, N., Jeffries, P., and Barea, J.M., 1996, Assessment of natural mycorrhizal potential in a desertified semiarid ecosystem. *Appl. Environ. Microbiol.* **62**: 842–847.

Rillig, M.C., 2004. Arbuscular mycorrhizae, glomalin, and soil aggregation. *Can. J. Soil Sci.* **84**: 355–363.

Rillig, M.C., and Mummey, D.L., 2006, Tansley review – mycorrhizas and soil structure. *New Phytol.* **171**: 41–53.

Rillig, M.C., Caldwell, B.A., Wosten, H.A.B., and Sollins, P., 2007, Role of protein in soil carbon and nitrogen storage: controls on persistence. *Biogeochem.* **85**: 25–44.

Robertson, S.J., McGill, W.B., Massicotte, H.B., and Rutherford, P.M., 2007. Petrolium hydrocarbon contamination in boreal forest soils: a mycorrhizal ecosystems perspective. *Biol. Res.* **82**: 213–240.

Roldan, A., Salinas-Gracia, J.R., Alguacil, M.M., and Caravaca, F., 2007, Soil sustainability indicators following conservation tillage practices under subtropical maize and bean crops. *Soil Till. Res.* **93**: 273–282.

Rubio, R., Borie, F., Schalchli, C., Castillo, C., and Azcón, R., 2003, Occurrence and effect of arbuscular mycorrhizal propagules in wheat as affected by the source and amount of phosphorus fertilizer and fungal inoculation. *Appl. Soil Ecol.* **23**: 245–255.

Ryan, M.H., and Graham, J.H., 2002, Is there a role for arbuscular mycorrhizal fungi in production agriculture? *Plant Soil* **244**: 263–271.

Rygiewicz, P.T., and Andersen, C.P., 1994, Mycorrhizae alter quality and quantity of carbon allocated below ground. *Nature* **369**: 58–60.

Sanchez, P.A., 1994, Properties management of soils in the tropics. New York: Wiley-Interscience.
Sawers, R., Gutjahr, C., and Paszkowski, U., 2008, Cereal mycorrhiza: an ancient symbiosis in modern agriculture. *Tren. Plant Sci.* **13**: 93–97.
Sbrana, C., Giovannetti, M., and Vitagliano, C., 1994, The effect of mycorrhizal infection on survival and growth renewal of micropropagated fruit rootstocks. *Mycorrhiza* **5**: 153–156.
Schreiner, R.P., Milhara, K.L., McDaniel, H., and Bethlenfalvay, G.J., 1997, Mycorrhizal fungi influence plant and soil functions and interactions. *Plant Soil* **188**: 199–209.
Schüβler, A., Schwarzott, D., and Walker, C., 2001, A new fungal phylum, the glomeromycota: phylogeny and evolution. *Mycol. Res.* **105**: 1413–1421.
Schussler, A., Martin, H., Cohen, D., Fitz, M., and Wipf, D., 2006, Characterization of a carbohydrate transporter from symbiotic glomeromycotan fungi. *Nature* **444**: 933–936.
Schweiger, P.F., and Jakobsen, I., 1999, Direct measurement of arbuscular mycorrhizal phosphorus uptake into field-grown winter wheat. *Agron. J.* **91**: 998–1002.
SER (Society for Ecological Restoration International Science and Policy Working Group), 2004, *The SER International Primer on Ecological Restoration* (available from http://www.ser.org) accessed in July 2005. Tucson, AZ: Society for Ecological Restoration International.
Siddiqui, Z.A., and Mahmood, I., 1995, Role of plant symbionts in nematode management: a Review. *Bioresource Technol.* **54**: 217–226.
Siddiqui, Z.A., Mahmood, I., and Khan, M.W., 1999, VAM fungi as prospective biocontrol agents for plant parasitic nematodes. In: Bagyaraj, D.J., Verma, A., Khanna, K.K., and Kehri, H.K. (eds.), *Modern Approaches and Innovations in Soil Management*. Meerut, India: Rastogi, pp. 47–58.
Six, J., Carpenter, A., van Kessel, C., Merck, R., Harris, D., Horwath, W.R., and Lüscher, A., 2001. Impact of elevated CO_2 on soil organic matter dynamics as related to changes in aggregate turnover and residue quality. *Plant Soil* **234**: 27–36.
Skujins, J., and Allen, M.F., 1986, Use of mycorrhizae for land rehabilitation. *MIRCEN J.* **2**: 161–176.
Smith, S.E., and Read, D. J., 1997, *Mycorrhizal Symbiosis*, 2nd edn. London: Academic.
St-Arnaud, M., and Vujanovic, V., 2007, Effects of the arbuscular mycorrhizal symbiosis on plant diseases and pests. In: Hamel, C., and Plenchette, C. (eds.), *Mycorrhizae in Crop Production*. New York: Haworth, pp. 67–122.
Subramanian, K.S., and Charest, C., 1999, Acquisition of N by external hyphae of an arbuscular mycorrhizal fungus and its impact on physiological responses in maize under drought-stressed and well-watered conditions. *Mycorrhiza* **9**: 69–75.
Sun, Y.P., Unestam, T., Lucas, S.D., Johanson, K.J., Kenne, L., and Finlay, R.D., 1999, Exudation reabsorption in mycorrhizal fungi, the dynamic interface for interaction with soil and other microorganisms. *Mycorrhiza* **9**: 137–144.
Swift, M.J., 1998, Toward the second paradigm: integrated biological management of soil. Paper presented for the FERTBIO Conference, Brazil.
Sylvia, D.M., and Williams, S.E., 1992, Mycorrhizae and environmental stresses. In: Bethlenfalvay, G.J., and Linderman, R.G. (eds.), *Mycorrhizae in Sustainable Agriculture*. Madison, WI: ASA Special Publication No. 54, pp. 101–124.
Talukdar, N.C., and Germida, J.J., 1994, Growth and yield of lentil and wheat inoculated with three *Glomus* isolates form Saskatchewan soils. *Mycorrhiza* **5**: 145–152.
Taylor, J., and Harrier, L.A., 2001, A comparison of development and mineral nutrition of micropropagated *Fragaria* × *ananassa* cv. Elvira (strawberry) when colonized by nine species of arbuscular mycorrhizal fungi. *Appl. Soil Ecol.* **18**: 205–215.
Thompson, P., 1990, Soil sterilization methods to show VA mycorrhizae aid P and Zn nutrition of wheat in vertisols. *Soil Biol. Biochem.* **22**: 229–240.

Tonkin, C.M., Malajczuk, N., and McComb, J.A., 1989, Ectomycorrhizal formation by micropropagated clones of *Eucalyptus marginata* inoculated with isolates of *Pisolithus tinctorius*. *New Phytol.* **111**: 209–214.
Torti, S.D., and Coley, P.D., 1999, Tropical monodominance: a preliminary test of the ectomycorrhizal hypothesis. *Biotropica* **31**: 220–228.
Torti, S.D., Coley, P.D., and Kursar, T.A., 2001, Causes and consequences of monodominance in tropical lowland forests. *Amer. Natur.* **157**: 141–153.
Trouvelot, A., Gianinazzi-Pearson, V., and Gianinazzi, S., 1982, Les endomycorrhizes en agriculture: recherches sur le blé. In: INRA, Les Colloques de l'INRA, Editor, *Les mycorrhizes: biologie et utilisation* vol. 13. Paris: INRA.
Uehlein, N., Fileschi, K., Eckert, M., Bienert, G.P., Bertl, A., and Kaldenhoff, R., 2007, Arbuscular mycorrhizal symbiosis and plant aquaporin expression. *Phytochemistry* **68**: 122–129.
Valencia, R.H., Balslev, H., Paz, H., and Mino, C.G., 1994, High tree alpha-diversity in Amazonian Ecuador. *Biodiv. Conserv.* **3**: 21–28.
Vallejo, V.R., Bautista, S., and Cortina, J.R., 1999, Restoration for soil protection after disturbances. In: Trabaud, L. (ed.), *Life and Environment in the Mediterranean*. Wessex, UK: Wit, pp. 301–343.
Van den Driessche, R., 1991, Effects of nutrients on stock performance in the forest. In: *Mineral Nutrition of Conifer Seedlings*. Boca Raton, FL/Ann Arbor, MI/Boston, MA: CRC, pp. 229–260.
Van der Heijden, M.G.A., Klironomos, J.N., Ursic, M., Moutoglis, P., Streitwolf-Engel, R., Boller, T.A., Wiemken, A., and Sanders, I.R., 1998, Mycorrhizal fungal diversity determines plant biodiversity, ecosystem variability and productivity. *Nature* **396**: 69–72.
Van der Heijden, M.G.A., Streitwolf-Engel, R., Riedl, R., Siegrist, S., Neudecker, A., Ineichen, K., Boller, T., Wiemken, A., and Sanders, I.R., 2006, The mycorrhizal contribution to plant productivity, plant nutrition, and soil structure in experimental grassland. *New Phytol.* **172**: 739–752.
Vogelsang, K.M., Reynolds, H.L., and Bever, J.D., 2006, Mycorrhizal fungal identity and richness determine the diversity and productivity of tallgrass prairie system. *New Phytol.* **172**: 554–562.
Warren, A., Sud, Y.C., and Rozanov, B., 1996, The future of deserts. *J. Arid. Environ.* **32**: 75–89.
Woomer, P.L., and Swift, M.J., 1994, *The Biological Management of Tropical Soil Fertility*. Chichester, UK: Wiley/UK: TSBF and Sayce.
Wu, S.C., Cao, Z.H., Li, Z.G., Cheung, K.C., and Wong, M.H., 2005, Effects of biofertilizer containing N-fixer, P and K solubilizers and AM fungi on maize growth: a greenhouse trial. *Geoderma* **125**: 155–166.
Xavier, L.J.C., and Germida, J.J., 1997, Growth response of lentil and wheat at *Glomus clarum* NT4 over a range of P levels in a Saskatchewan soil containing indigenous AM fungi. *Mycorrhiza* **7**: 3–8.
Yedidia, I., Benhamou, N., and Chet, I., 1999, Induction of defense responses in cucumber plants (*Cucumis sativus* L.) by the biocontrol agent *Trichoderma harzianum*. *Appl. Environ. Microbiol.* **65**: 1061–1070.

Chapter 2

THE MOLECULAR COMPONENTS OF NUTRIENT EXCHANGE IN ARBUSCULAR MYCORRHIZAL INTERACTIONS

RUAIRIDH J.H. SAWERS, SHU-YI YANG, CAROLINE GUTJAHR, AND UTA PASZKOWSKI
Department of Plant Molecular Biology, University of Lausanne, Biophore Building, CH-1015 Lausanne, Switzerland

Abstract: The driving force behind arbuscular mycorrhizal (AM) interactions is an exchange of nutrients between fungus and plant. Glomeromycotan fungi are obligate symbionts and rely on the carbon provided by their plant hosts to complete their life cycle. In return, the fungus provides nutritional benefits to the plant, notably by delivering minerals. The majority of this nutrient exchange is thought to occur in root cortical cells containing the highly-branched fungal arbuscules. In this chapter, we describe the molecular components of the arbusculated cell and the proteins involved in the transfer of nutrients between fungus and plant. We consider, in detail, the passage of phosphorous and nitrogen from the soil to the arbusculated cell and the concomitant delivery of carbon to the fungal symbiont. In natural conditions, the exchange of nutrients does not need to be completely equitable and selective pressure may act on both partners to push the balance in their favour. In cultivated plants, the artificial environment may further distort the balance. We discuss how a better understanding of the molecular regulation of nutrient transfer benefits attempts to optimise AM associations for agricultural use.

Keywords: Arbuscular mycorrhiza; nutrient exchange; phosphate; nitrogen; carbohydrate.

1 INTRODUCTION

Arbuscular mycorrhizal (AM) symbioses are formed between plant roots and members of the Glomeromycota (Schüßler *et al.*, 2001). These fungi obtain carbon from their hosts and, in return, provide nutritional benefits to the plant (Smith and Read, 1997). Mycorrhizal development begins with the germination of fungal spores present in the soil. Initially, simple, un-branched hyphae are produced and growth is sustained by reserves of triglycerides and glycogen present in the spore (Bago et al., 2000). Hyphae that enter the vicinity of a suitable host respond to spatial cues provided by root exudates, of which the best characterized are the strigolactones, and begin to branch more vigorously (Akiyama et al., 2005). On contact with the host, a hyphopodium is formed and the root epidermis penetrated in a process tightly coordinated between fungus and plant (Genre et al., 2005).

2 DEVELOPMENT AND ANATOMY OF ARBUSCULAR MYCORRHIZAL ASSOCIATIONS

Once inside the root cortex, the fungus begins symbiotic growth and produces longitudinal hyphae from which extend the highly branched, tree-like arbuscules (Smith and Read, 1997). Arbuscules are formed by dichotomous branching of intra-radical fungal hyphae. As the hyphae become increasingly ramified, the width of the fungal wall decreases (from ~500 nm thick in intra-radical longitudinal hyphae to ~30 nm thick in the terminal arbuscular branches) and the wall structure becomes more open (Bonfante-Fasolo et al., 1990). During arbuscule development, the plant vacuole fragments, organelles multiply and move to surround the arbuscule (Lohse et al., 2005) and the nucleus expands and moves into the centre of the cell (Balestrini et al., 1992). Although the arbuscule eventually expands to largely fill the cortical cell, the fungus never penetrates the host cell plasma membrane. Increased biosynthetic activity within the host cell allows the production of additional membrane components to keep pace with arbuscule growth and, as the arbuscule expands and branches, it is enveloped by newly synthesised host membrane. In the mature arbusculated cell, the host membrane will have increased in surface area several times to completely surround the fungal structure (Alexander et al., 1988). Subsequently, the fungus develops an extensive network of extraradical hyphae that extends beyond the plant root system and provides an increased soil volume for nutrient acquisition (Smith and Read, 1997). In certain mycorrhizal interactions the pattern of fungal growth is somewhat different with the production of coiled structures taking the place of the arbuscules (Dickson et al., 2007). However, we will not specifically consider these structures here.

Arbuscules are short-lived structures and begin to senesce after 4–10 days of activity (Strack et al., 2003). As the arbuscule collapses, the hyphal remnants are encapsulated by the cell wall components and degraded. Subsequently, the plant cell returns to the pre-arbuscular state and can be re-colonized at a later time. Mycorrhizal colonization is an asynchronous process and a mature colonized root system will harbour cells at all stages of development. The regulation of the cycle of colonization is clearly important to the balance of nutrient exchange. The plant partner has the ability to partially regulate the overall level of colonization; for example, maize root colonization decreases as phosphate availability increases (Braunberger et al., 1991; Kaeppler et al., 2000), and barley has been shown to resist further AM colonization once a certain critical level of fungal growth is reached (Vierheilig, 2004). Similarly, if genetic disruption is used to prevent plant uptake of fungal phosphate, the plant will not support the symbiosis and the fungus dies (Maeda et al., 2006; Javot et al., 2007a).

The plant peri-arbuscular membrane and the arbuscular surface enclose a narrow space (80–100 nm) between microbial symbiont and host that is known as the interface compartment or interfacial apoplast (Balestrini and Bonfante, 2005). The interfacial apoplast connects plant host and symbiotic fungus and represents a space into which nutrients can be delivered and from which they may be acquired. Although, the peri-arbuscular membrane is continuous with the cell plasma membrane and, consequently, the interfacial apoplast continuous with the cellular apoplast, the composition of the symbiotic interface is unique (Strack et al., 2003; Balestrini and Bonfante, 2005). During arbuscule development, the peri-arbuscular membrane continues to produce cell wall materials, including pectins, xyloglucans, polygalacturonans and hydroxyproline-rich glycoproteins (HRGPs). These components are secreted into the interfacial apoplast but do not further assemble into a wall structure, possibly due to the action of fungal lytic enzymes (Balestrini and Bonfante, 2005). Gene expression studies have identified transcripts encoding proline rich proteins in arbusculated cells of medic (van Buuren et al., 1999) and maize (Balestrini et al., 1997). The medic transcript (*MtAM1*) is predicted to encode a protein containing both an N-terminal signal sequence and a C-terminal glycosyl phosphatidylinositol (GPI) anchor addition site, consistent with localisation attached to the external phase of the peri-arbuscular membrane (van Buuren et al., 1999). The mature MtAM1 protein exhibits features typical of the heavily glycosylated arabinogalactan HRGPs (AGPs) and contains two potential Src-homology (SH-3) ligand sites. The SH-3 ligand motif was identified by its ability to interact with the SH-3 domain and is implicated in mediating protein-protein interactions during signalling events (Ren et al., 1993). However, beyond these sequence predictions little is known about the role of HRGPs in AM symbioses.

The permeability and transfer properties of the arbuscular membranes are thought to be enhanced by the presence of aquaporin proteins (Uehlein et al., 2007). Aquaporins function in animals, plants and fungi to greatly increase the permeability of lipid membranes (Kruse et al., 2006) and are known to play a significant role in the regulation of plant water relations

(Javot and Maurel, 2002). Plant aquaporins are assigned to four sub-families, tonoplast intrinsic proteins (TIPs), plasma membrane intrinsic proteins (PIPs), Nodulin-26 like intrinsic proteins (NIPs) and small basic intrinsic proteins (SIPs). Aquaporin genes expressed to higher levels following AM colonization have been characterized from e.g. parsley (Roussel et al., 1997), bean (Aroca et al., 2007) and medic (Krajinski et al., 2000; Journet et al., 2002; Manthey et al., 2004; Brechenmacher et al., 2005; Uehlein et al., 2007). In addition, gene expression profiling has identified a NIP aquaporin from rice that is up-regulated following colonization (Güimil et al., 2005). In medic, accumulation of PIP proteins in the root cortex has been demonstrated by immunological staining, although the localization of aquaporins to the peri-arbuscular membrane never been unequivocally demonstrated (Uehlein et al., 2007). Characterization of the expression of additional aquaporin genes has added confusion to the interpretation of their role in AM symbiosis. Under drought stress, mycorrhizal colonization has been observed to reduce expression of PIP aquaporin genes in the roots of soybean and lettuce (Porcel et al., 2006). Similarly, colonization of salt stressed tomato by AM fungi reduced the accumulation of transcripts encoding a PIP protein (Ouziad et al., 2005). Furthermore, the use of a medic split-root inoculation system has demonstrated that a single aquaporin gene can be induced in the colonized parts of the root system but down-regulated in adjacent non-colonized portions (Liu et al., 2007). Intriguingly, it has been shown that aquaporins may also facilitate the movement of small solutes suggesting that they could play a wider role in mycorrhizal nutrient exchange than originally proposed (Jahn et al., 2004).

Acidotropic staining techniques have shown that the interfacial apoplast is strongly acidified with respect to plant and fungal cytoplasm (Guttenberger, 2000). Plant and fungal P-type H^+-ATPases are thought to maintain a high proton concentration within the apoplast (Gianinazzi-Pearson et al., 1991). Plant-encoded P-type H^+-ATPase genes that are induced by mycorrhizal colonization have been identified from e.g., barley (Murphy et al., 1997), tobacco (Gianinazzi-Pearson et al., 2000), tomato (Ferrol et al., 2002) and medic (Krajinski et al., 2002). Immunocytological techniques have confirmed the localisation of tobacco H^+-ATPase proteins to the peri-arbuscular membrane (Gianinazzi-Pearson et al., 2000). On the fungal side, expression of the H^+-ATPase gene *GmHA5* from *Glomus mossae* is induced at the onset of symbiotic growth and remains highly expressed during intra-radical development (Requena et al., 2003). The electrochemical gradient established by ATPase proteins is thought to be essential to the functioning of diverse proton:nutrient transporters (Gianinazzi-Pearson et al., 1991) as will be discussed further below.

It has been hypothesised that plants employ components of their pathogen defence mechanism to regulate symbiotic fungal growth (discussed in Paszkowski, 2006). Among the classes of defence-related genes that have been found to respond to mycorrhizal expression are chitinases (Salzer *et al.*, 2000), lectins (Frenzel *et al.*, 2005, 2006) and germin-like proteins (Doll *et al.*, 2003). These protein classes are functionally diverse and, although many hypotheses have been advanced, their role during mycorrhizal colonization remains poorly defined. The most direct evidence for a regulatory role is available for the chitinases. In addition to gene expression data, enzymatic assays have demonstrated chitinase activity in mycorrhizal roots of e.g., leek (Spanu *et al.*, 1989), tobacco (Dumas-Gaudot *et al.*, 1992), bean (Lambais and Mehdy, 1993), pea (Dumas-Gaudot *et al.*, 1994), alfalfa (Volpin *et al.*, 1994) and soybean (Xie *et al.*, 1999). Chitinases catalyse the hydrolysis of the glycosidic bonds that link glucosamine residues in the chitin polymer, the primary constituent of fungal hyphal walls. Chitinases do not degrade mature fungal walls and act primarily on exposed chitin chains, such as those found at the growing hyphal tips of newly forming arbuscular branches (Collinge *et al.*, 1993; Bestel-Corre *et al.*, 2002). Reducing the levels of a barley chitinase by antisense transformation has been shown to result in an increase in mycorrhizal colonization and it has been proposed that chitinase activity promotes arbuscule turnover (Troedsson *et al.*, 2005). However, in further studies, over-expression of chitinase genes in tobacco had no influence on mycorrhizal colonization (Vierheilig *et al.*, 1995) and over-expression of a medicago chitinase in root culture had no effect on the level of colonization, although there was a promotion of fungal spore germination (Elfstrand *et al.*, 2005).

3 ARBUSCULAR MYCORRHIZAL PHOSPHATE ACQUISITION

Phosphate is an essential nutrient and is limiting for plant growth in many environments (Bucher, 2007). Phosphate is present in the soil in the form of inorganic orthophosphate (P_i) and is readily sequestered by cations, especially in acidic conditions, of which the most abundant are iron, aluminium and calcium. The mobility of sequestered phosphate is reduced and, as a consequence, plant uptake rapidly exhausts the phosphate available in the vicinity of the root system and creates a localised depletion zone (Bucher, 2007). In modern agriculture, the problem of phosphate limitation has been addressed by the extensive use of phosphate-additions, more than 4,000,000 tons annually in the USA alone (www.fao.org). However, as supplies are reduced, phosphate becomes increasingly difficult and costly to extract. Furthermore, the efficiency of phosphate uptake may be as low as 20% (Zhu *et al.*, 2003) and much of the added phosphates will pass to adjacent water courses with detrimental environmental consequences. It has been demonstrated that, in wild ecosystems, plants derive much of their phosphate via mycorrhizal fungi (Smith *et al.*, 2004). Investigating the current importance, and potential future benefits, of mycorrhizal colonization to crop phosphate uptake remains one of the major concerns of current mycorrhiza research.

Two distinct plant phosphate uptake pathways have been described; a direct pathway functions in the absence of colonization and a second, mycorrhiza-associated pathway operates in colonized plants (Smith *et al.*, 2003; Fig. 1). These two pathways are not complementary but function, to an extent, as physiological alternatives (Smith *et al.*, 2004). The onset of mycorrhizal colonization results in the expression of specific phosphate transporter genes encoding members of the Phosphate transporter 1 (Pht1) family (Bucher, 2007; Javot *et al.*, 2007b). A second set of *Pht1* family transcripts accumulates in the absence of mycorrhizal fungi and, in certain cases, is reduced in accumulation following colonization (Bucher, 2007; Javot *et al.*, 2007b). Pht1 proteins are proton:phosphate symporters belonging to the major facilitator superfamily (Bucher, 2007). They are homologous to the yeast PHO84 phosphate transporter and share a topology of 12 membrane spanning helices (Bun-Ya *et al.*, 1991). The family is defined by a highly conserved signature motif (GGDYPLSATIxSE) present in the fourth transmembrane domain (Karandashov and Bucher, 2005). Functional testing of Pht1 proteins, predominantly by heterologous expression in yeast, has revealed a broad range of affinity for phosphate, although the functional importance of this variation is unresolved (Javot *et al.*, 2007b).

It is now widely accepted that plant Pht1 family proteins play a key role in the transport of P_i across the peri-arbuscular membrane (Karandashov and Bucher, 2005; Javot *et al.*, 2007b; Table 1). Expression analysis has identified genes encoding candidate arbuscular Pht1 transporters from a number of plant species; *Pht1* genes that are specifically expressed in AM colonized roots have been identified from medic (*MEDtr:Pht1;4*. Harrison *et al.*, 2002), rice (*ORYsa:Pht1;11*. Paszkowski *et al.*, 2002), potato (*SOLtu:Pht1;4* and *SOLtu:Pht1;5*. Nagy *et al.*, 2005), wheat (*TRIac:Pht1:myc*. Glassop *et al.*, 2005) and tomato (*LYCes:Pht1;4*. Nagy *et al.*, 2005); *Pht1* genes encoding transcripts that accumulate constitutively and then accumulate to higher levels following AM colonization have been identified from potato (*SOLtu:Pht1;3*. Rausch *et al.*, 2001), tomato (*LYCes:Pht1;3* and *LYCes:Pht1;5*. Nagy *et al.*, 2005), lotus (*LOTja:Pht1;3*. Maeda *et al.*, 2006), barley (*HORvu:Pht1;8*. Rae *et al.*, 2003; Glassop *et al.*, 2005) and maize (*ZEAma:Pht1;6*. Nagy *et al.*, 2006). Phylogenetic analysis of the *Pht1* gene family (Karandashov *et al.*, 2004) has grouped *MEDtr:Pht1;4*, *ORYsa:Pht1;11*, *SOLtu:Pht1;4*, *SOLtu: Pht1;5*, *LYCes:Pht1;4*, *HORvu:Pht1;8*, *ZEAma:Pht1;6* into a single sub-family. The genes *SOLtu:Pht1;3*, *LYCes:Pht1;3* and *LOTja:Pht1;3* form a distinct subfamily that, to date, consists solely of sequences from dicotyledonous plants and which has no AM-associated representatives in the fully sequenced rice genome (Paszkowski *et al.*, 2002). The use of *in situ* hybridization has localised mRNAs encoding LOTja:Pht1;3 (Maeda *et al.*, 2006), TRIac:Pht1:myc, HORvu:Pht1;8 and ZEAma:Pht1;6 (Glassop *et al.*, 2005) exclusively to arbusculated cells. Promoter-reporter fusions have given similar results for *MEDtr:Pht1;4* (Harrison *et al.*, 2002), *LYCes:Pht1;4* and *SOLtu:Pht1;4* (Nagy *et al.*, 2005). Laser micro-dissection has confirmed the localization of *LYCes:Pht1;4* transcripts and shown that *LYCes:Pht1;3* transcripts also accumulate uniquely in arbusculated cells (Balestrini *et al.*, 2007).

Mycorrhizal *Pht1* gene induction is thought to require the action of the lipid messenger-molecule lyso-phosphatidyl-choline (LPC), as demonstrated by the ability of mycorrhizal lipid fractions and purified LPC to induce activity of the *SOLtu:Pht1;3* and *SOLtu:Pht1;4* promoters in root culture (Drissner *et al.*, 2007). At this time it is not known whether LPC in the root lipid fraction is of plant or fungal origin. LPC acts through, or in conjunction, with *cis*-acting promoter elements to drive arbuscule specific expression. The promoter regions of the three *Pht1* family genes *SOLtu:Pht1;3*, *MEDtr:Pht1;4* and *ORYsa:Pht1;11* have been characterized in detail by fusion to the *β-glucuronidase* (*GUS*) reporter gene and subsequent transformation into root culture for testing in the presence of AM fungi (Karandashov *et al.*, 2004). Native *MEDtr:Pht1;4* and *ORYsa:Pht1;11* transcripts accumulate exclusively in colonized plants. The native *SOLtu:Pht1;3* gene is constitutively active but transcript accumulation is higher following colonization. Following transfer into cultured roots derived from a number of dicotyledenous species, the promoter regions of *SOLtu:Pht1;3*, *MEDtr:Pht1;4* drove GUS reporter gene expression in response to mycorrhizal colonization (Karandashov *et al.*, 2004). In contrast, the *ORYsa:Pht1;11* promoter was found to be inactive in these cultures. Promoter-deletion analysis defined a 129 bp region of the *SOLtu:Pht1;3* promoter as sufficient to condition mycorrhiza-specific expression. Further computational phylogenetic footprinting analysis identified a cluster of six putative *cis* elements within this 129 bp region (Karandashov *et al.*, 2004). Of these elements, the motif CTTC, present in the promoter of *SOLtu:Pht1;3*, could also be found in the promoters of *MEDtr:Pht1;4* and *LYCes:Pht1;4*. The CTTC motif was also detected in the promoter of a previously characterised medic mycorrhiza-induced glutathione transferase (GST) (Karandashov *et al.*, 2004). A second motif, TAAT, was common to the promoters of *SOLtu:Pht1;3*, *MEDtr:Pht1;4* and *LYCes:Pht1;4* but absent from the medic GST promoter. Notably, neither the CTTC nor the TAAT motif were present in the promoter of the rice gene *ORYsa:Pht1;11*. This is consistent with the absence of activity of the *ORYsa:Pht1;11* promoter in eudicot root culture and suggests divergence of regulatory elements between monocotyledonous and dicotyledenous plants.

Immunocytochemical analysis using a specific antibody has localized the MEDtr:Pht1;4 protein to the peri-arbuscular membrane (Harrison *et al.*, 2002). Detection was strongest around the fine branches of the arbuscule while the thicker branches showed little staining. No staining was observed in the cell plasma membrane, indicating a specific localisation to the peri-arbuscular membrane. Cell fractionation confirmed the presence of MEDtr:Pht1;4 in the membrane fraction. Interestingly, the distinction in staining between trunk and fine branches suggests a degree of spatially heterogeneity within the peri-arbuscular membrane. Mature arbuscules were more strongly stained than young arbuscules and little staining was observed in senescing or collapsed arbuscules, suggesting that MEDtr:Pht1;4 accumulation is co-ordinated with the life of the arbuscule. With the exception of MEDtr:Pht1;4, the subcellular localisation of plant mycorrhizal Pht1 transporters has not been unequivocally determined but homology to *MEDtr:Pht1;4*, suggests a similar peri-arbuscular localization.

Phosphate acquisition by the AM pathway begins with the uptake of P_i free in the soil by fungal extra-radical hyphae. These fungal hyphae extend beyond the host root system allowing a greater soil volume to be exploited for phosphate uptake. In addition, AM colonization promotes physiological responses in the host, such as root branching and phosphatase secretion that indirectly promote phosphate uptake (Ezawa *et al.*, 2005). There is also at least one report that AM fungi can hydrolyse an organic phosphate source, thereby liberating phosphate that would not otherwise be available to the host (Koide and Kabir, 2000). Uptake of inorganic phosphate at the soil-hypha interface is mediated by fungal high affinity phosphate transporters of the Pht1 family (Harrison and van Buuren, 1995; Maldonado-Mendoza *et al.*, 2001; Benedetto *et al.*, 2005). Following fungal uptake, phosphate is transferred to the fungal vacuole where it is polymerized to form poly-phosphate chains (Ezawa *et al.*, 2001). Poly-phosphate is translocated through the vacuolar compartment to the intraradical hyphae (Ohtomo and Saito, 2005) where it is presumed to be hydrolysed prior to release of free phosphate into the interfacial apoplast of arbusculated cells. The mechanism of poly-phosphate breakdown has not been characterized but is hypothesized to require the action of fungal phosphatase enzymes present in the arbuscule (Javot *et al.*, 2007b). It is not known if release of P_i requires the action of fungal phosphate transporters, although transcripts encoding the phosphate transporter GmosPT of *Glomus mosseae* have recently been found to accumulate in arbusculated cells (Benedetto *et al.*, 2005; Balestrini *et al.*, 2007). From the interfacial apoplast, the plant mycorrhizal Pht1 transporters are thought to transport P_i across the peri-arbuscular membrane. Reverse genetics techniques have been used to confirm the functional importance of LOTja:Pht1;3 (Maeda *et al.*, 2006) and MEDtr:Pht1;4 (Javot *et al.*, 2007a) in mycorrhizal phosphate uptake. Lotus plants with an RNA interference

(RNAi)-mediated reduction in LOTja:Pht1;3 protein accumulated less phosphate when colonized by AM fungi (Maeda et al., 2006). Similar results were observed in medic plants in which *MEDtr:Pht1;4* was either silenced using an RNAi construct or the endogenous gene mutated (Javot et al., 2007a), providing strong evidence that Pht1 proteins are required for mycorrhiza-associated P_i uptake. Once in the plant cytosol, phosphate is translocated into the vasculature for delivery to all parts of the plant (Bucher, 2007).

4 ARBUSCULAR MYCORRHIZAL NITROGEN UP-TAKE

Similarly to phosphate, nitrogen is a major limiting nutrient of plant growth, especially during the production of cereal crops. Consequently nitrogen additions also feature heavily in modern high-input agricultural systems. Nitrogen is available in the soil in the form of ammonium (NH_4^+) and nitrate (NO_3^-). Although the concentration of ammonium in the soil is 10–1,000 times lower than that of nitrate, ammonium is the preferential form of nitrogen absorbed when plants are subjected to nitrogen deficiency (Lee and Rudge, 1986; Gazzarrini et al., 1999) or grown in water-logged or acid soils (Marschner, 1995). Ammonium has low mobility in the soil and a depletion zone is formed in the vicinity of the roots in a fashion similar to that observed with phosphate. The extraradical mycelium of mycorrhizal fungi can absorb ammonium (Johansen et al., 1992, 1996; Johnson et al., 1997) nitrate (Johansen et al., 1996) and amino acids (Hodge et al., 2001) and the role of mycorrhizal nitrogen delivery is becoming increasingly recognized (Govindarajulu et al., 2005; Cruz et al., 2007). The majority of nitrogen is thought to be taken up in the form of ammonium (Villegas et al., 1996; Hawkins et al., 2000; Toussaint et al., 2004) via the action of fungal-encoded AMT1 family transporters such as the protein GintAMT1 characterized from *Glomus intraradices* (Lopez-Pedrosa et al., 2006; Fig. 1).

There is no evidence for fungal translocation of either ammonium or nitrate and it is more likely that nitrogen is translocated in the form of amino acids (Bago et al., 2001). Arginine is by far the most abundant amino acid in the extraradical mycelium and is thought to be the major form of translocated nitrogen (Johansen et al., 1996; Govindarajulu et al., 2005; Jin et al., 2005). Within the extraradical mycelium, ammonium is thought to be first combined with glutamate to form glutamine by the enzymes of the glutamine synthetase/glutamate synthase (GS/GOGAT) cycle (Johansen et al., 1996; Breuninger et al., 2004). Arginine can then be readily synthesized from glutamine by the enzyme argininosuccinate synthase (ASS) (Cruz et al., 2007). Having been translocated to the intraradical hyphae, arginine is broken down by ornithine aminotransferase and urease to release free ammonium. Both of these enzyme activities have been shown to be higher in the intraradical hyphae than in the extraradical mycelium (Govindarajulu et al., 2005; Cruz et al., 2007). De-protonated ammonia (NH_3) has the potential to passively diffuse into the apoplast provided a concentration gradient is maintained. Alternatively, either NH_3 or NH_4^+ could be exported by protein-mediated mechanisms. A candidate fungal AMT transporter has been identified that is highly expressed in the internal hyphae (Govindarajulu et al., 2005). However, the role for AMT proteins in the export of ammonia against a proton gradient is unclear and it has been hypothesised that AMT proteins actively translocate ammonia into intracellular vesicles for sub-sequent release into the interfacial apoplast by an exocytotic mechanism. In addition, studies of ectomycorrhizal fungi have identified proteins homologous to the yeast Ato proton:ammonium antiporter (Chalot et al., 2006) and such proteins may yet be found to function in AM symbiosis. Once in the apoplast, ammonium is potentially taken up directly by the plant. Gene expression analyses of medic and rice have identified mycorrhizainduced transcripts that putatively encode ammonium transporters that are candidates for this function (Frenzel et al., 2005; Güimil et al., 2005; Hohnjec et al., 2005). Additionally, putative mycorrhiza induced nitrate transporters have been identified in tomato, medic and rice that could play a role (Hildebrandt et al., 2002; Güimil et al., 2005; Hohnjec et al., 2005). There is also the possibility for passive ammonia uptake across the peri-arbuscular membrane, perhaps facilitated by the presence of aquaporin proteins (Uehlein et al., 2007).

5 CARBON DELIVERY TO THE FUNGAL PARTNER IN ARBUSCULAR MYCORRHIZAL SYMBIOSIS

Arbuscular mycorhizal fungi are obligate symbionts entirely dependent on their plant host for the supply of carbon (Bago et al., 2000). The use of radio-labeled substrates has demonstrated that AM fungi take up plant carbohydrate in the form of hexose (Shachar-Hill et al., 1995; Solaiman and Saito, 1997; Fig. 1). The route of this transport, and whether or not it is specific to the arbuscule, remains to be determined (Harrison, 1999). Within the plant, carbon is predominantly translocated in the form of sucrose and the action of invertase and sucrose synthase enzymes is required to release hexoses (Sturm and Tang, 1999). Such activities have not been identified from AM fungal species and it is assumed that sucrose breakdown is catalysed by plant-encoded enzymes. Mycorrhiza-responsive sucrose synthase and invertase genes have been identified in maize (Ravnskov et al., 2003), bean (Blee and

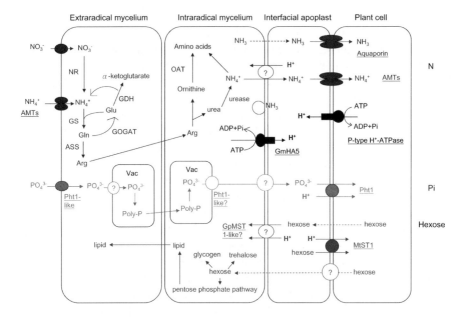

Fig. 1. Transport of nitrogen, phosphate and carbohydrate between arbuscular mycorrhizal fungi and plant. Nitrate or ammonium is taken up by the extraradical mycelium and becomes assimilated into amino acids, especially Arginine, via the GS/GOGAT pathway. Arginine is transferred from the extraradical into the intraradical mycelium and broken down into ammonium. Ammonium may be transported into the interfacial apoplast by Ato-like proton: ammonium antiporters. The acidic apoplast provides a proton-driven force for taking up ammonium by plant ammonium transporters. Alternatively, ammonia passively diffuses into the interfacial apoplast and may be taken up by plant aquaporins. Phosphate is acquired by the extraradical mycelium and becomes polymerized into poly-phosphate chains inside the vacuole. The poly-phosphate is transferred through the vacuolar compartment to the intraradical mycelium, where it is hydrolyzed into phosphate and transported into the interfacial apoplast. It is not clear how the efflux of phosphate is mediated. In the peri-arbuscular membrane, plant Pht1 transporters, such as MtPT4, take up phosphate from the interfacial apoplast. In the carbohydrate transport pathway, GpMST1-like transporters may be possible candidates for hexose uptake from the interfacial apoplast. Passive diffusion could also play a role. The AM symbiosis-induced plant hexose transporter Mtst1 is predicted to act in sugar import. Fungal and plant P-type H^+-ATPase pump protons out of the cell resulting in the generation of a proton gradient. The electrochemical gradient established by ATPase proteins is essential to the functioning of diverse proton:nutrient transporters. Abbreviations: NR, nitrate reductase; Arg, arginine; GS, glutamine synthetase; Gln, glutamine; Glu, Glutamate; ASS, argininosuccinate synthetase; GOGAT, Glutamate synthase; GDH, Glutamate dehydrogenase; OAT, ornithine aminotransferase.

Anderson, 2002), medic (Hohnjec *et al.*, 2003) and tomato (Schaarschmidt *et al.*, 2006). Although transcripts and promoter activities of several of these genes have been localized to arbusculated cells, accumulation and activity have additionally been observed in adjacent cells and generally associated with intraradical hyphae (Blee and Anderson, 2002; Hohnjec *et al.*, 2003;

Schaarschmidt *et al.*, 2006). It remains unclear whether these genes encode activities directly involved in carbon delivery to the fungus or are required to support a general increase in metabolic activity associated with colonization (Harrison, 1999). The mechanism of hexose delivery to the apoplast is similarly unresolved. There is little molecular evidence for the presence of hexose export proteins in the peri-arbuscular membrane itself. Although a number of mycorrhiza responsive sugar transporter genes have been identified in medic, of which the best characterized is *Mtst1* (Harrison, 1996), they are thought to act as proton:sugar symporters in sugar import not in export, perhaps to support high metabolic activity in arbusculated cells (Harrison, 1996). One simple mechanistic alternative to explain the transport of sugars to the apoplast is passive movement (Bago *et al.*, 2000). Once in the apoplast, hexose is thought to be absorbed by the fungus via specific transport proteins. Although such fungal transporters remain to be identified, the recent characterization of the GpMST1 hexose transporter from *Geosiphon pyriformis*, a Glomeromycotan fungus that associates with single celled algae, provides a promising direction for further investigation (Schussler *et al.*, 2006). It is also possible that passive transport plays a role in fungal sugar uptake, with rapid conversion of hexose to trehalose or glycogen, or metabolism via the pentose phosphate pathway, maintaining a favorable concentration gradient (Shachar-Hill *et al.*, 1995). Once in the intraradical hyphae, much of the carbon is converted to storage lipids, predominantly triacylglycerides. Lipids not only act to store carbon but are also the main form of carbon translocated from intra- to extra- radical hyphae where they provide the major respiratory substrate (Pfeffer *et al.*, 1999).

6   OPTIMIZATION OF ARBUSCULAR MYCORRHIZAL NUTRIENT EXCHANGE FOR AGRICULTURAL APPLICATION

AM fungi provide many benefits to their plant hosts. In addition to enhancing mineral nutrition, they can increase tolerance to water stress, induce greater resistance to pathogens and reduce sensitivity to toxic substances present in the soil (Smith and Read, 1997). However, the costs of colonization can be high, with up to 20% of the host's fixed carbon being delivered to the microbial symbiont (Bago *et al.*, 2000). Nonetheless, under experimental conditions when nutrients are limiting, mycorrhizal crop plants typically exhibit a performance advantage over non-colonized siblings (Smith and Read, 1997). However, under the nutrient saturating conditions found in high-input agricultural systems, the relative advantages are reduced while the carbon costs remain and the performance of colonized plants can fall below that of those that are non-colonized (Janos, 2007). Although mycorrhizal fungi possess a limited potential to improve on present levels of crop performance, profitable agricultural application of AM symbioses demands only that the inevitable loss of carbon to the fungus is compensated by a reduction in the overall cost of production on a per unit yield basis. With the rising costs of nutrient additions, and an increased awareness of the broader environmental costs of their use, the potential real-world value of mycorrhiza continues to increase. Perhaps the most attractive scenario for application is the potential use of AM fungi to increase the uptake-efficiency of nutrient additions, thus allowing a reduction in their use without compromising total yields. An increase in available data now makes it possible to consider such uses at the molecular level; analysis of existing variation can be facilitated and candidate genes identified for future manipulation.

Table 1. Genes discussed in the text. AM ind, AM induced – constitutive expression enhanced in colonized plants. AM spe, AM specific – transcript only accumulates in colonized plants. AM/D rep, AM and drought repressed – constitutive expression reduced in drought stress, reduction promoted by AM colonization. Arb, specific expression in arbusculated cells. Ext, transcripts abundant in external mycelium. Accession numbers reference Genbank with the exception of those marked * that reference tentative EST contigs or microarray identifiers. There is potential redundancy between these less well-characterized sequences. (1, Rae *et al.*, 2003; 2, Glassop *et al.*, 2005; 3, Maeda *et al.*, 2006; 4, Nagy *et al.*, 2005; 5, Harrison *et al.*, 2002; 6, Paszkowski *et al.*, 2002; 7, Güimil *et al.*, 2005; 8, Rausch *et al.*, 2001; 9, Nagy *et al.*, 2006; 10, Harrison and van Buueren, 1995; 11, Benedetto *et al.*, 2005; 12, Balestrini *et al.*, 2007; 13, Maldonado-Mendoza *et al.*, 2001; 14, Hohnjec *et al.*, 2005; 15, Lopez-Pedrosa *et al.*, 2006; 16, Hildebrandt *et al.*, 2002; 17, Gianinazzi-Pearson *et al.*, 2000; 18, Porcel *et al.*, 2006; 19, Ouziad *et al.*, 2005; 20, Uehlein *et al.*, 2007; 21, Krajinski *et al.*, 2000; 22, Brechenmacher *et al.*, 2005; 23, Journet *et al.*, 2002; 24, Manthey *et al.*, 2004; 25, Roussel *et al.*, 1997; 26, Aroca *et al.*, 2007; 27, Murphy *et al.*, 1997; 28, Ferrol *et al.*, 2002; 29, Krajinski *et al.*, 2002; 30, Requena *et al.*, 2003; 31, Van Buueren *et al.*, 1999; 32, Balestrini *et al.*, 1997; 33, Salzer *et al.*, 2000; 34, Bestel-Corre *et al.*, 2002; 35, Harrison *et al.*, 1996; 36, Schussler *et al.*, 2006).

Phosphate transporters

Species	Gene name	Expression	Accession	Ref.
Hordeum vulgare	*HORvu;Pht1;8* or *HvPT8*	AM ind	AAO72440	1, 2
Lotus japonicus	*LOTju;Pht1;3* or *LjPT3*	AM ind	BAE93353	3
Lycopersicon esculentum	*LYCes;Pht1;4* or *LePT4*	AM spe	AAX85192	4
Lycopersicon esculentum	*LYCes;Pht1;5* or *LePT5*	AM ind	AAX85194	4
Medicago truncatula	*MEDtr;Pht1;4* or *MtPT4*	AM spe	AAM76744	5

(continued)

Table 1. *(Cont.)*

Oryza sativa	ORYsa;Pht1;11 or OsPT11	AM spe	AAN39052	6
Oryza sativa	ORYsa;Pht1;13 or OsPT13	AM spe	AAN39054	6, 7
Solanum tuberosum	SOLtu;Pht1;3 or StPT3	AM ind	CAC87043	8
Solanum tuberosum	SOLtu;Pht1;4 or StPT4	AM spe	AAW51149	4
Solanum tuberosum	SOLtu;Pht1;5 or StPT5	AM spe	AAX85195	4
Triticum aestivum	TRIae;Pht1;myc	AM spe	CAH25730	2
Zea mays	ZEAma;Pht1;6 or ZmPT6	AM ind	CAH25731	9, 2
Glomus versiforme	GvPT	Hyphae	AAC49132	10
Glomus mosseae	GmosPT	Hyphae/arb	DQ074452	11, 12
Glomus intraradices	GiPT	Hyphae	AAL37552	13

Ammonium transporters

Species	Gene name	Expression	Accession	Ref.
Medicago truncatula	MtAMT1;1	AM ind	TC77463[*]	14
Oryza sativa	OsAMT3;1	AM ind	NP_001044932	7
Glomus intraradices	GintAMT1	Ext	CAI54276	15

Nitrate transporters

Species	Gene name	Expression	Accession	Ref.
Lycopersicon esculentum	LeNRT2;3	AM ind	AY038800	16
Medicago truncatula	Uncharacterized Nitrate transporter	AM ind	TC78158[*]	14
Medicago truncatula	Uncharacterized Nitrate transporter	AM ind	TC78157[*]	14
Medicago truncatula	Uncharacterized Nitrate transporter	AM ind	TC80954[*]	14
Medicago truncatula	Uncharacterized Nitrate transporter	AM ind	TC84545[*]	14
Medicago truncatula	Uncharacterized Nitrate transporter	AM ind	AL383332	17
Oryza sativa	OsPTR2	AM ind	NP_001067564	7

(continued)

Aquaporins

Species	Gene name	Expression	Accession	Ref.
Glycine max	GmPIP1	AM/D rep	CAI79102	18
Glycine max	GmPIP2	AM/D rep	CAI79103	18
Lactuca sativa	LsPIP1	AM/D rep	CAI79104	18
Lactuca sativa	LsPIP2	AM/D rep	CAI79105	18
Lycopersicon esculentum	Nramp2	AM ind	AAS67887	19
Medicago truncatula	MtNIP1	AM ind	AY059381	20
Medicago truncatula	MtPIP1;1	AM ind	AF386739	20
Medicago truncatula	MtPIP2;1	AM ind	AY059380	20
Medicago truncatula	MtAQP1	AM ind	AJ251652	21
Medicago truncatula	Uncharacterized NIP	AM ind	AJ311232	22
Medicago truncatula	Uncharacterized PIP	AM ind	MtC00007*	23
Medicago truncatula	Uncharacterized NIP	AM ind	MtC10430*	23
Medicago truncatula	Uncharacterized NIP	AM ind	BG582879	24
Medicago truncatula	Uncharacterized NIP	AM ind	BG583224	24
Oryza sativa	Uncharacterized NIP	AM spe	BAD53665	7
Oryza sativa	Uncharacterized NIP	AM ind	AAV44140	7
Petroselinum crispum	PcRb7	AM ind	CAA88267	25
Phaseolus vulgaris	PvPIP1;2	AM ind	Z48232	26

H^+-ATPases

Species	Gene name	Expression	Accession	Ref.
Hordeum vulgare	BMR78	AM ind		27
Lycopersicon esculentum	LHA2	AM ind	AAF98344	28
Medicago truncatula	MtHA1	AM spe	CAB85494	29
Nicotiana tabacum	PMA2	AM ind	M80492	17
Nicotiana tabacum	PMA4	AM ind	X66737	17
Oryza sativa	OsAM43	AM ind	NP_001048647	7
Glomus mossae	GmHA5		AAO91802	30

Hydroxyproline-rich glycoproteins

Species	Gene name	Expression	Accession	Ref.
Medicago truncatula	MtAM1	AM ind	AAD39890	31
Zea mays	ZmHRGP	AM ind	AAB23539	32

Chitinases

Species	Gene name	Expression	Accession	Ref.
Medicago truncatula	MtChitinase III-2	AM ind	AAF67827	33
Medicago truncatula	MtChitinase III-3	AM ind	AAQ21404	33
Medicago truncatula	MtChitinase I	AM ind	AAP68451	34

(continued)

Table 1. *(Cont.)*

Hexose transporters				
Species	Gene name	Expression	Accession	Ref.
Medicago truncatula	*Mtst1*	AM ind	AAB06594	35
Geosiphon pyriformis	*GpMST1*		CAJ77495	36

Studies of plant performance have identified variation in the capacity of plants to benefit from mycorrhizal colonization both within and between species (e.g. Hetrick *et al.*, 1992; Kaeppler *et al.*, 2000; Smith *et al.*, 2004). In addition, there is evidence that genetic variability in the fungal partner also influences the level of responsiveness and that different plant-fungal combinations alter the degree of benefit derived by the host (Smith *et al.*, 2004). Typically, the impact of colonization has been measured as the performance difference between colonized and non-colonized plants under limiting nutrient conditions. Molecular characterization of mycorrhizal nutrient uptake has begun to allow a more detailed interpretation of mycorrhizal performance variation. For example, on the basis of characterization of phosphate uptake, it has been recognized that a lack of responsiveness does not have to be linked to a lack of importance of mycorrhizal associations; it can be that irrespective mycorrhizal phosphate uptake dominates in a field setting (Smith *et al.*, 2003). Furthermore, it can be recognized that variation in responsiveness results from a combination of variation in both non-mycorrhizal and mycorrhizal growth (discussed in Sawers *et al.*, 2008). The characterization of transporter families suggests that these two traits are, to some extent, genetically independent. For example, allelic variation in a myorrhiza-specific phosphate transporter gene can exist in a population independently of variation in a gene encoding a transporter required for direct phosphate uptake; there does not have to be a phenotypic correlation between the efficiency of mycorrhizal and direct phosphate uptake. Consequently, a given species or variety could be identified as more responsive than another on the basis of exceptionally poor performance in the absence of AM symbiosis, rather than a greater capacity to benefit from the association. It is clearly important to distinguish these two alternatives when attempting to identify beneficial genetic variation for the optimization of crop mycorrhizal associations. Optimization of mycorrhizal symbiosis for agricultural application can be considered as the attempt to extract the maximum plant benefit from colonization for the minimum loss of resources. Plants possess the capacity to influence the balance of this exchange. As the mechanisms of this regulation are better understood and the responsible genes identified there is the potential to directly push this balance further towards the plant partner.

7 PERSPECTIVES

Increasing characterization of the molecular components of mycorrhizal nutrient exchange is providing an important complement to physiological studies. However, the current picture is based predominantly on correlative gene expression information. Two main approaches will no doubt shape future progress; first, an increased biochemical characterization, notably of the plant-fungal interface and constituent proteins; second, the use of reverse genetics and genetic manipulation to directly assess the functional role of proteins implicated in nutrient exchange. However, the available gene expression data provide important insights in their own right. Expression studies have revealed that the induction of the mycorrhizal state in the plant is not a simple superimposition of novel gene activity on the non-colonized state but a switch from an asymbiotic to a symbiotic growth program. One important consequence of this distinction is the partial genetic independence of performance between non-colonized and colonized plants and the subsequent implications this has for interpretation of phenotypic variation and the selection of future varieties. Clearly, an understanding of the molecular basis of nutrient exchange has great potential to benefit diverse aspects of mycorrhizal research and to contribute to the future application of AM fungi in an agricultural setting.

REFERENCES

Akiyama, K., Matsuzaki, K., and Hayashi, H., 2005, Plant sesquiterpenes induce hyphal branching in arbuscular mycorrhizal fungi. *Nature* **435**: 824–827.

Alexander, T., Meier, R., Toth, R., and Weber, H., 1988, Dynamics of arbuscule development and degeneration in mycorrhizas of *Triticum aestivum* L. and *Avena sativa* L. with reference to *Zea mays* L. *New Phytol.* **110**: 363–370.

Aroca, R., Porcel, R., and Ruiz-Lozano, J.M., 2007, How does arbuscular mycorrhizal symbiosis regulate root hydraulic properties and plasma membrane aquaporins in *Phaseolus vulgaris* under drought, cold or salinity stresses? *New Phytol.* **173**: 808–816.

Bago, B., Pfeffer, P.E., and Shachar-Hill, Y., 2000, Carbon metabolism and transport in arbuscular mycorrhizas. *Plant Physiol.* **124**: 949–958.

Bago, B., Shachar-Hill, Y., and Pfeffer, P.E., 2001, Could the urea cycle be translocating nitrogen in the arbuscular mycorrhizal symbiosis? *New Phytol.* **149**: 4–8.

Balestrini, R., and Bonfante, P., 2005, The interface compartment in arbuscular mycorrhizae: a special type of plant cell wall? *Plant Biosyst.* **139**: 8–15.

Balestrini, R., Berta, G., and Bonfante, P., 1992, The plant nucleus in mycorrhizal roots: positional and structural modifications. *Biol. Cell* **75**: 235–243.

Balestrini, R., Jose-Estanyol, M., Puigdomenech, P., and Bonfante, P., 1997, Hydroxyproline-rich glycoprotein mRNA accumulation in maize root cells colonized by an arbuscular mycorrhizal fungus revealed by *in situ* hybridization. *Protoplasma* **198**: 36–42.

Balestrini, R., Gomez-Ariza, J., Lanfranco, L., and Bonfante, P., 2007, Laser microdissection reveals that transcripts for five plant and one fungal phosphate transporter genes are contemporaneously present in arbusculated cells. *Mol. Plant Microbe Inter.* **20**: 1055–1062.

Benedetto, A., Magurno, F., Bonfante, P., and Lanfranco, L., 2005, Expression profiles of a phosphate transporter gene (*GmosPT*) from the endomycorrhizal fungus *Glomus mosseae*. *Mycorrhiza* **15**: 620–627.

Bestel-Corre, G., Dumas-Gaudot, E., Gianinazzi-Pearson, V., and Gianinazzi, S., 2002, Mycorrhiza-related chitinase and chitosanase activity isoforms in *Medicago truncatula*. *Symbiosis* **32**: 173–194.

Blee, K., and Anderson, A., 2002, Transcripts for genes encoding soluble acid invertase and sucrose synthase accumulate in root tip and cortical cells containing mycorrhizal arbuscules. *Plant Mol. Biol.* **50**: 197–211.

Bonfante-Fasolo, P., Faccio, A., Perroto, S., and Schubert, D., 1990, Correlation between chitin distribution and cell wall morphology in the mycorrhizal fungus *Glomus versiforme*. *Mycol. Res.* **94**: 157–165.

Braunberger, P.G., Miller, M.H., and Peterson, R.L., 1991, Effect of phosphorous nutrition on morphological characteristics of vesicular-arbuscular mycorrhizal colonization of maize. *New Phytol.* **119**: 107–113.

Brechenmacher, L., Weidmann, S., van Tunien, D., Chatagnier, O., Gianinazzi, S., Franken, P., and Gianinazzi-Pearson, V., 2004, Expression profiling of up-regulated plant and fungal genes in early and late stages of *Medicago truncatula-Glomus mosseae* interactions. *Mycorrhiza* **14**: 253–262.

Breuninger, M., Trujillo, C., Serrano, E., Fischer, R., and Requena, N., 2004, Different nitrogen sources modulate activity but not expression of glutamine synthetase in arbuscular mycorrhizal fungi. *Fungal Gen. Biol.* **41**: 542–552.

Bucher, M., 2007, Functional biology of plant phosphate uptake at root and mycorrhiza interfaces. *New Phytol.* **173**: 11–26.

Bun-Ya, M., Nishimura, M., Harashima, S., and Oshima, Y., 1991, The PHO84 gene of *Saccharomyces cerevisiae* encodes an inorganic phosphate transporter. *Mol. Cell Biol.* **11**: 3229–3238.

Chalot, M., Blaudez, D., and Brun, A., 2006, Ammonia: a candidate for nitrogen transfer at the mycorrhizal interface. *Trends Plant Sci.* **11**: 263–266.

Collinge, D.B., Kragh, K.M., Mikkelsen, J.D., Nielsen, K.K., Rasmussen, U., and Vad, K., 1993, Plant chitinases. *Plant J.* **3**: 31–40.

Cruz, C., Egsgaard, H., Trujillo, C., Ambus, P., Requena, N., Martins-Loucao, M.A., and Jakobsen, I., 2007, Enzymatic evidence for the key role of arginine in nitrogen translocation by arbuscular mycorrhizal fungi. *Plant Physiol.* **144**: 782–792.

Dickson, S., Smith, F.A., and Smith, S.E., 2007, Structural differences in arbuscular mycorrhizal symbioses: more than 100 years after Gallaud, where next? *Mycorrhiza* **17**: 375–393.

Doll, J., Hause, B., Demchenko, K., Pawlowski, K., and Krajinski, F., 2003, A member of the germin-like protein family is a highly conserved mycorrhiza-specific induced gene. *Plant Cell Physiol.* **44**: 1208–1214.

Drissner, D., Kunze, G., Callewaert, N., Gehrig, P., Tamasloukht, M.B., Boller, T., Felix, G., Amrhein, N., and Bucher, M., 2007, Lyso-phosphatidylcholine is a signal in the arbuscular mycorrhizal symbiosis. *Science* **318**: 265–268.

Dumas-Gaudot, E., Furlan, V., Grenier, J., and Asselin, A., 1992, New acidic chitinase isoforms induced in tobacco roots by vesicular arbuscular mycorrhizal fungi. *Mycorrhiza* **1**: 113–136.

Dumas-Gaudot, E., Asselin, A., Gianinazzi-Pearson, V., Gollotte, A., and Gianinazzi, S., 1994, Chitinase isoforms in roots of various pea genotypes infected with arbuscular mycorrhizal fungi. *Plant Sci.* **99**: 27–37.

Elfstrand, M., Feddermann, N., Ineichen, K., Nagaraj, V.J., Wiemken, A., Boller, T., and Salzer, P., 2005, Ectopic expression of the mycorrhiza-specific chitinase gene *Mtchit 3-3* in *Medicago truncatula* root-organ cultures stimulates spore germination of glomalean fungi. *New Phytol.* **167**: 557–570.

Ezawa, T., Smith, S.E., and Smith, F.A., 2001, Differentiation of polyphosphate metabolism between the extra- and intraradical hyphae of arbuscular mycorrhizal fungi. *New Phytol.* **149**: 555–563.

Ezawa, T., Hayatsu, M., and Saito, M., 2005, A new hypothesis on the strategy for acquisition of phosphorus in arbuscular mycorrhiza: up-regulation of secreted acid phosphatase gene in the host plant. *Mol. Plant Microbe Inter.* **18**: 1046–1053.

Ferrol, N., Pozo, M.J., Antelo, M., and Azcón-Aguilar., 2002, Arbuscular mycorrhizal symbiosis regulates plasma membrane H^+-ATPase gene expression in tomato plants. *J. Exp. Bot.* **53**: 1683–1687.

Frenzel, A., Manthey, K., Perlick, A.M., Meyer, F., Puhler, A., Kuster, H., and Krajinski, F., 2005, Combined transcriptome profiling reveals a novel family of arbuscular mycorrhizal-specific *Medicago truncatula* lectin genes. *Mol. Plant Microbe Inter.* **18**: 771–782.

Frenzel, A., Tiller, N., Hause, B., and Krajinski, F., 2006, The conserved arbuscular mycorrhiza-specific transcription of the secretary lectin *MtLec5* is mediated by a short upstream sequence containing specific protein binding sites. *Planta* **224**: 792–800.

Gazzarrini, S., Lejay, L., Gojon, A., Ninnemann, O., Frommer, W.B., and von Wiren, N., 1999, Three functional transporters for constitutive, diurnally regulated, and starvation-induced uptake of ammonium into *Arabidopsis* roots. *Plant Cell* **11**: 937–948.

Genre, A., Chabaud, M., Timmers, T., Bonfante, P., and Barker, D.G., 2005, Arbuscular mycorrhizal fungi elicit a novel intracellular apparatus in *Medicago truncatula* root epidermal cells before infection. *Plant Cell* **17**: 3489–3499.

Gianinazzi-Pearson, V., Smith, S.E., Gianinazzi, S., and Smith, F.A., 1991, Enzymatic studies on the metabolism of vesicular-arbuscular mycorrhizas, V. Is H^+-ATPase a component of ATP-hydrolyzing enzyme activities in plant-fungus interfaces? *New Phytol.* **117**: 61–74.

Gianinazzi-Pearson, V., Arnould, C., Oufattole, M., Arango, M., and Gianinazzi, S., 2000, Differential activation of H^+-ATPase genes by an arbuscular mycorrhizal fungus in root cells of transgenic tobacco. *Planta* **211**: 609–613.

Glassop, D., Smith, S.E., and Smith, F.W., 2005, Cereal phosphate transporters associated with the mycorrhizal pathway of phosphate uptake into roots. *Planta* **222**: 688–698.

Govindarajulu, M., Pfeffer, P., Jin, H., Abubaker. J., Douds, D.D., Allen, J.W., Bücking, H., Lammers, P.J., and Shachar-Hill, Y., 2005, Nitrogen transfer in the arbuscular mycorrhizal symbiosis. *Nature* **435**: 819–823.

Güimil, S., Chang, H., Zhu, T., Sesma, A., Osbourn, A., Roux, C., Ioannidis, V., Oakely, E.J., Docquier, M., Descombes, P., Briggs, S.P., and Paszkowski, U., 2005, Comparative transcriptomics of rice reveals an ancient pattern of response to microbial colonization. *Proc. Nat. Acad. Sci. USA* **102**: 8066–8070.

Guttenberger, M., 2000, Arbuscules of vesicular-arbuscular mycorrhizal fungi inhabit an acidic compartment within plant roots. *Planta* **211**: 299–304.

Harrison, M., 1999, Biotrophic interfaces and nutrient transport in plant/fungal interfaces. *J. Exp. Bot.* **50**: 1013–1022.

Harrison, M., Dewbre, G., and Liu, J., 2002, A phosphate transporter of *Medicago truncatula* involved in the acquisition of phosphate released by arbuscular mycorrhizal fungi. *Plant Cell* **14**: 2413–2429.

Harrison, M.J., 1996, A sugar transporter from *Medicago truncatula*: altered expression pattern in roots during vesicular-arbuscular (VA) mycorrhizal associations. *Plant J.* **9**: 491–503.

Harrison, M.J., and van Buuren, M.L., 1995, A phosphate transporter from the mycorrhizal fungus *Glomus versiforme*. *Nature* **378**: 626–629.

Hawkins, H.J., Johansen, A., and George, E., 2000, Uptake and transport of organic and inorganic nitrogen by arbuscular mycorrhizal fungi. *Plant Soil* **226**: 275–285.

Hetrick, B.A.D., Wilson, G.W.T., and Cox, T.S., 1992, Mycorrhizal dependence of modern wheat varieties, landraces, and ancestors. *Can. J. Bot.* **70**: 2032–2040.

Hildebrandt, U., Schmelzer, E., and Bothe, H., 2002, Expression of nitrate transporter genes in tomato colonized by an arbuscular mycorrhizal fungus. *Physiol. Plant.* **115**: 125–136.

Hodge, A., Campbell, C.D., and Fitter, A.H., 2001, An arbuscular mycorrhizal fungus accelerates decomposition and acquires nitrogen directly from organic material. *Nature* **413**: 297–299.

Hohnjec, N., Perlick, A., Pühler, A., and Küster, H., 2003, The *Medicago truncatula* sucrose synthase gene *MtSuc1* is activated both in the infected region of root nodules and in the cortex of roots colonized by arbuscular mycorrhizal fungi. *Mol. Plant Microbe Inter.* **16**: 903–915.

Hohnjec, N., Vieweg, M., Pühler, A., Becker, A., and Küster, H., 2005, Overlaps in the transcriptional profiles of *Medicago truncatula* roots inoculated with two different *Glomus* fungi provide insights in to the genetic program activated during arbuscular mycorrhiza. *Plant Physiol.* **137**: 1283–1301.

Jahn, T.P., Moller, A.L., Zeuthen, T., Holm, L.M., Klaerke, D.A., Mohsin, B., Kuhlbrandt, W., and Schjoerring, J.K., 2004, Aquaporin homologues in plants and mammals transport ammonia. *FEBS Lett.* **574**: 31–36.

Janos, D.P., 2007, Plant responsiveness to mycorrhizas differs from dependence upon mycorrhizas. *Mycorrhiza* **17**: 75–91.

Javot, H., and Maurel, C., 2002, The role of aquaporins in root water uptake. *Ann. Bot.* (Lond) **90**: 301–313.

Javot, H., Penmetsa, R.V., Terzaghi, N., Cook, D.R., and Harrison, M.J., 2007a, A *Medicago truncatula* phosphate transporter indispensable for the arbuscular mycorrhizal symbiosis. *Proc. Nat. Acad. Sci. U S A* **104**: 1720–1725.

Javot, H., Pumplin, N., and Harrison, M.J., 2007b, Phosphate in the arbuscular mycorrhizal symbiosis: transport properties and regulatory roles. *Plant Cell Environ.* **30**: 310–322.

Jin, H., Pfeffer, P.E., Douds, D.D., Piotrowski, E., Lammers, P.J., and Shachar-Hill, Y., 2005, The uptake, metabolism, transport and transfer of nitrogen in an arbuscular mycorrhizal symbiosis. *New Phytol.* **168**: 687–696.

Johansen, A., Jakobsen, I., and Jensen, E.S., 1992, Hyphal transport of ^{15}N labelled nitrogen by a vesicular-arbuscular mycorrhizal fungus and its effect on depletion of inorganic soil N. *New Phytol.* **122**: 281–288.

Johansen, A., Finlay, R.D., and Olsson, P.A., 1996, Nitrogen metabolism of the external hyphae of the arbuscular mycorrhizal fungus *Glomus intraradices*. *New Phytol.* **133**: 705–712.

Johnson, N., Graham, J., and Smith, F., 1997, Functioning of mycorrhizal associations along the mutualism-parasitism continuum. *New Phytol.* **135**: 575–585.

Journet, E.-P., van Tuinen, D., Gouzy, G., Crespeau, H., Carreau, V., Farmer, M.-J., Niebel, A., Schiex, T., Jaillon, O., Chatagnier, O., Godiard, L., Micheli, F., Kahn, D., Gianinazzi-Pearson, V., and Gamas, P., 2002, Exploring root symbiotic programs in the model legume *Medicago truncatula* using EST analysis. *Nucleic Acids Res.* **30**: 5579–5592.

Kaeppler, S.M., Parke, J.L., Mueller, S.M., Senior, L., Stuber, C., and Tracy, W.F., 2000, Variation among maize inbred lines and detection of quantitative trait loci for growth at low phosphorous and responsiveness to arbuscular mycorrhizal fungi. *Crop Sci.* **40**: 358–364.

Karandashov, V., and Bucher, M., 2005, Symbiotic phosphate transport in arbuscular mycorrhizas. *Trends Plant Sci.* **10**: 22–29.

Karandashov, V., Nagy, R., Wegmuller, S., Amrhein, N., and Bucher, M., 2004, Evolutionary conservation of a phosphate transporter in the arbuscular mycorrhizal symbiosis. *Proc. Nat. Acad. Sci. USA* **101**: 6285–6290.

Koide, R.T., and Kabir, Z., 2000, Extraradical hyphae of the mycorrhizal fungus *Glomus intraradices* can hydrolyse organic phosphate. *New Phytol.* **148**: 511–517.

Krajinski, F., Biela, A., Schubert, D., Gianinazzi-Pearson, V., Kaldenhoff, R., and Franken, P., 2000, Arbuscular mycorrhiza development regulates the mRNA abundance of *Mtaqp1* encoding a mercury-insensitive aquaporin of *Medicago truncatula*. *Planta* **211**: 85–90.

Krajinski, F., Hause, B., Gianinazzi-Pearson, V., and Franken, P., 2002, *Mtha1*, a plasma membrane H^+-ATPase gene from *Medicago truncatula* shows arbuscule-specific induced expression in mycorrhizal tissue. *Plant Mol. Biol.* **4**: 754–761.

Kruse, E., Uehlein, N., and Kaldenhoff, R., 2006, The aquaporins. *Genome Biol.* **7**: 206.

Lambais, M.R., and Mehdy, M.C., 1993, Suppression of *endochitinase*, *β-1-3-endoglucanase*, and *chalcone isomerase* expression in bean vesicular-arbuscular mycorrhizal roots under different soil phosphate conditions. *Mol. Plant Microbe Inter.* **6**: 75–83.

Lee, R.B., and Rudge, K.A., 1986, Effects of nitrogen deficiency on the absorption of nitrate and ammonium by barley plants. *Ann. Bot.* **57**: 471–486.

Liu, J., Maldonado-Mendoza, I., Lopez-Meyer, M., Cheung, F., Town, C.D., and Harrison, M.J., 2007, Arbuscular mycorrhizal symbiosis is accompanied by local and systemic alterations in gene expression and an increase in disease resistance in the shoots. *Plant J.* **50**: 529–544.

Lohse, S., Schliemann, W., Ammer, C., Kopka, J., Strack, D., and Fester, T., 2005, Organization and metabolism of plastids and mitochondria in arbuscular mycorrhizal roots of *Medicago truncatula*. *Plant Physiol.* **139**: 329–340.

Lopez-Pedrosa, A., Gonzalez-Guerrero, M., Valderas, A., Azcon-Aguilar, C., and Ferrol, N., 2006, *GintAMT1* encodes a functional high-affinity ammonium transporter that is expressed in the extraradical mycelium of *Glomus intraradices*. *Fungal Genet Biol.* **43**: 102–110.

Maeda, D., Ashida, K., Iguchi, K., Chechetka, S.A., Hijikata, A., Okusako, Y., Deguchi, Y., Izui, K., and Hata, S., 2006, Knockdown of an arbuscular mycorrhiza-inducible phosphate transporter gene of *Lotus japonicus* suppresses mutualistic symbiosis. *Plant Cell Physiol.* **47**: 807–817.

Maldonado-Mendoza, I.E., Dewbre, G.R., and Harrison, M.J., 2001, A phosphate transporter gene from the extraradical mycelium of an arbuscular mycorrhizal fungus *Glomus intraradices* is regulated in response to phosphate in the environment. *Mol. Plant Microbe Inter.* **14**: 1140–1148.

Manthey, K., Krajinski, F., Hohnjec, N., Firnhaber, C., Pühler, A., Perlick, A.M. and Küster, H., 2004, Transcriptome profiling in root nodules and arbuscular mycorrhiza identifies a collection of novel genes induced during *Medicago truncatula* root endosymbioses. *Mol. Plant Microbe Inter.* **17**: 1063–1077.

Marschner, H., 1995, *Mineral Nutrition in Higher Plants*. Academic, London.

Murphy, P.J., Langridge, P., and Smith, S.E., 1997, Cloning plant genes differentially expressed during colonization of *Hordeum vulgare* by the vesicular arbuscular mycorrhizal fungus *Glomus intraradices*. *New Phytol.* **135**: 291–301.

Nagy, R., Karandashov, V., Chague, V., Kalinkevich, K., Tamasloukht, M., Xu, G., Jakobsen, I., Levy, A.A., Amrhein, N., and Bucher, M., 2005, The characterization of novel mycorrhiza-specific phosphate transporters from *Lycopersicon esculentum* and *Solanum tuberosum* uncovers functional redundancy in symbiotic phosphate transport in solanaceous species. *Plant J.* **42**: 236–250.

Nagy, R., Vasconcelos, M.J.V., Zhao, S., McElver, J., Bruce, W., Amrhein, N., Raghothama, K.G., and Bucher, M., 2006, Differential regulation of five *Pht1* phosphate transporters from maize (*Zea mays* L.). *Plant Biol.* **8**: 186–197.

Ohtomo, R., and Saito, M., 2005, Polyphosphate dynamics in mycorrhizal roots during colonization of an arbuscular mycorrhizal fungus. *New Phytol.* **167**: 571–578.

Ouziad, F., Hildebrandt, U., Schmelzer, E., and Bothe, H., 2005, Differential gene expressions in arbuscular mycorrhizal-colonized tomato grown under heavy metal stress. *J. Plant Physiol.* **162**: 634–649.

Paszkowski, U., 2006, Mutualism and parasitism: the yin and yang of plant symbioses. *Curr. Opin. Plant Biol.* **9**: 364–370.

Paszkowski, U., Kroken, S., Roux, C., and Briggs, S.P., 2002, Rice phosphate transporters include an evolutionarily divergent gene specifically activated in arbuscular mycorrhizal symbiosis. *Proc. Nat. Acad. Sci. U S A* **99**: 13324–13329.

Pfeffer, P., Douds, D., Bécard, G., and Shachar-Hill, Y., 1999, Carbon uptake and the metabolism and transport of lipids in an arbuscular mycorrhiza. *Plant Physiol.* **120**: 587–598.

Porcel, R., Aroca, R., Azcon, R., and Ruiz-Lozano, J.M., 2006, *PIP* aquaporin gene expression in arbuscular mycorrhizal *Glycine max* and *Lactuca sativa* plants in relation to drought stress tolerance. *Plant Mol. Biol.* **60**: 389–404.

Rae, A.L., Cybinski, D.H., Jarney, J.M., and Smith, F.W., 2003, Characterization of two phosphate transporters from barley; evidence foe diverse function and kinetic properties among members of the Pht1 family. *Plant Mol. Biol.* **53**: 27–36.

Rausch, C., Daram, P., Brunner, S., Jansa, J., Laloi, M., Leggewies, G., Amrhein, N., and Bucher, M., 2001, A phosphate transporter expressed in arbuscule-containing cells in potato. *Nature* **414**: 462–470.

Ravnskov, S., Wu, Y., and Graham, J., 2003, Arbuscular mycorrhizal fungi differentially affect expression of genes coding for sucrose synthases in maize roots. *New Phytol.* **157**: 539–545.

Ren, R., Mayer, B.J., Cicchetti, P., and Baltimore, D., 1993, Identification of a ten-amino acid proline-rich SH3 binding site. *Science* **259**: 1157–1161.

Requena, N., Breuninger, M., Franken, P., and Ocon, A., 2003, Symbiotic status, phosphate, and sucrose regulate the expression of two plasma membrane H^+-*ATPase* genes from the mycorrhizal fungus *Glomus mosseae*. *Plant Physiol.* **132**: 1540–1549.

Roussel, H., Bruns, S., Gianinazzi-Pearson, V., Hahlbrock, K., and Franken, P., 1997, Induction of a membrane intrinsic protein-encoding mRNA in arbuscular mycorrhiza and elicitor-stimulated cell suspension cultures of parsley. *Plant Sci.* **126**: 203–210.

Salzer, P., Bonanomi, A., Beyer, K., Vogeli-Lange, R., Aescherbacher, R.A., Lange, J., Wiemken, A., Kim, D., Cook, D.R., and Boller, T., 2000, Differential expression of eight chitinase genes in *Medicago truncatula* roots during mycorrhiza formation, nodulation and pathogen infection. *Mol. Plant Microbe Inter.* **13**: 763–777.

Sawers, R., Gutjahr, C., and Paszkowski, U., 2008, Cereal mycorrhiza: an ancient symbiosis in modern agriculture. *Trends Plant Sci.* **13**: 93–97.

Schaarschmidt, S., Roitsch, T., and Hause, B., 2006, Arbuscular mycorrhiza induces gene expression of the apoplastic invertase *LIN6* in tomato (*Lycopersicon esculentum*) roots. *J. Exp. Bot.* **57**: 4015–4023.

Schüßler, A., Schwarzott, D., and Walker, C., 2001, A new fungal phylum Glomeromycota: phylogeny and evolution. *Mycol. Res.* **105**: 1413–1421.

Schussler, A., Martin, H., Cohen, D., Fitz, M., and Wipf, D., 2006, Characterization of a carbohydrate transporter from symbiotic glomeromycotan fungi. *Nature* **444**: 933–936.

Shachar-Hill, Y., Pfeffer, P., Douds, D., Osman, S., Doner, L., and Ratcliffe, R., 1995, Partitioning of intermediary carbon metabolism in vesicular-arbuscular mycorrhizal leek. *Plant Physiol.* **108**: 7–15.

Smith, S., Smith, F., and Jacobsen, I., 2004, Functional diversity in arbuscular mycorrhizal (AM) symbioses: the contribution of the mycorrhizal P uptake pathway is not correlated with mycorrhizal responses in growth or total P uptake. *New Phytol.* **162**: 511–524.

Smith, S.E., and Read, D.J., 1997, *Mycorrhizal Symbiosis*, 2nd edn. Academic, London.

Smith, S.E., Smith, F.A., and Jakobsen, I., 2003, Mycorrhizal fungi can dominate phosphate supply to plants irrespective of growth responses. *Plant Physiol.* **133**: 16–20.

Solaiman, M., and Saito, M., 1997, Use of sugars by intraradical hyphae of arbuscular mycorrhizal fungi revealed by radiorespirometry. *New Phytol.* **136**: 533–538.

Spanu, P., Boller, T., Ludwig, A., Wiemken, A., and Facccio, A., 1989, Chitinase in roots of *Allium porrum*: Regulation and localization. *Planta* **177**: 447–455.

Strack, D., Fester, T., Hause, B., Schliemann, W., and Walter, M.H., 2003, Arbuscular mycorrhiza: biological, chemical, and molecular aspects. *J. Chem. Ecol.* **29**: 1955–1979.

Sturm, A., and Tang, G., 1999, The sucrose-cleaving enzymes of plants are crucial for development, growth and carbon-partitioning. *Trends Plant Sci.* **4**: 401–407.

Toussaint, J.P., St-Arnaud, M., and Charest, C., 2004, Nitrogen transfer and assimilation between the arbuscular mycorrhizal fungus *Glomus intraradices* Schenck & Smith and Ri T-DNA roots of Daucus carota L. in an in vitro compartmented system. *Can. J. Microbiol.* **50**: 251–260.

Troedsson, U., Olsson, P.A., and Jarl-Sunesson, C.I., 2005, Application of antisense transformation of a barley chitinase in studies of arbuscule formation by a mycorrhizal fungus. *Hereditas* **142**: 65–72.

Uehlein, N., Fileschi, K., Eckert, M., Bienert, G.P., Bertl, A., and Kaldenhoff, R., 2007, Arbuscular mycorrhizal symbiosis and plant aquaporin expression. *Phytochemistry* **68**: 122–129.

van Buuren, M.L., Maldonado-Mendoza, I.E., Trieu, A.T., Blaylock, L.A., and Harrison, M.J., 1999, Novel genes induced during an arbuscular mycorrhizal (AM) symbiosis formed between *Medicago truncatula* and *Glomus versiforme*. *Mol. Plant Microbe Inter.* **12**: 171–181.

Vierheilig, H., 2004, Further root colonization by arbuscular mycorrhizal fungi in already mycorrhizal plants is suppressed after a critical level of root colonization. *J. Plant Physiol.* **161**: 339–341.

Vierheilig, H., Alt, M., Lange, J., Gutrella, M., Wiemken, A., Boller, T., 1995, Colonization of transgenic tobacco constitutively expressing pathogenesis-related proteins by the vesicular-arbuscular mycorrhizal fungus *Glomus mossae*. *Appl. Env. Microbiol.* **61**: 3031–3034.

Villegas, J., Williams, R.D., Nantais, L., Archambault, J., and Fortin, J.A., 1996, Effects of N source on pH and nutrient exchange of extramatrical mycelium in a mycorrhizal Ri T-DNA transformed root system. *Mycorrhiza* **6**: 247–251.

Volpin, H., Elkind, Y., Okon, Y., and Kapulnik, Y., 1994, A vesicular mycorrhizal fungus (*Glomus intraradices*) induces a defense response in alfalfa roots. *Plant Physiol.* **104**: 683–689.

Xie, Z.P., Staehelin, C., Wiemken, A., Broughton, W.J., Muller, J., and Boller, T., 1999, Symbiosis-stimulated chitinase isoenyzmes of soybean (*Glycine max*). *J. Exp. Bot.* **50**: 327–333.

Zhu, Y.G., Smith, F.A., and Smith, S.E., 2003, Phosphorous efficiency and responses of barley (*Hordeum vulgare* L.) to arbuscular mycorrhizal fungi grown in highly calcereous soil. *Mycorrhiza* **13**: 93–100.

Chapter 3

ARBUSCULAR MYCORRHIZAL FUNGI AS POTENTIAL BIOPROTECTANTS AGAINST PLANT PATHOGENS

MOHD SAYEED AKHTAR AND ZAKI A. SIDDIQUI
Department of Botany, Aligarh Muslim University, Aligarh 202002, INDIA

Abstract: Arbuscular Mycorhizal (AM) fungi are ubiquitous and form symbiotic relationships with roots of most terrestrial plants. Their associations benefit plant nutrition, growth and survival due to their enhanced exploitation of soil nutrients. These fungi play a key role in nutrient cycling and also protect plants against environmental and cultural stresses. The establishment of AM fungi in the plant root has been shown to reduce the damage caused by soil-borne plant pathogens with the enhancement of resistance in mycorrhizal plants. The effectiveness of AM fungi in biocontrol is dependent on the AM fungus involved, as well as the substrate and host plant. However, protection offered by AM fungi is not effective against all the plant pathogens and is modulated by soil and other environmental conditions. AM fungi generally reduce the severity of plant diseases to various crops suggesting that they may be used as potential tool in disease management. AM fungi modify the quality and abundance of rhizosphere microflora and alter overall rhizosphere microbial activity. These fungi induce changes in the host root exudation pattern following host colonization which alters the microbial equilibrium in the mycorrhizosphere. Given the high cost of inorganic fertilizers and health hazards associated with chemical pesticides, AM fungi may be most suitable for sustainable agriculture and also for increasing the yield of several crops through biocontrol of plant pathogens. This chapter provides an overview of mechanisms of interaction which take place between soil-borne plant pathogens and AM fungi on different plants. The availability of new tools and techniques for the study of microbial interactions in the rhizosphere may provide a greater understanding of biocontrol processes in the near-future.

Keywords: Arbuscular mycorrhiza; biocontrol; plant diseases; plant pathogens; rhizosphere.

1 INTRODUCTION

Arbuscular mycorrhizal (AM) fungi occur over a wide range of agro climatic conditions and are geographically ubiquitous. They form symbiotic relationships with roots of about 90% land plants in natural and agricultural ecosystems (Brundrett, 2002). The AM association has been observed in 200 families of plants representing 1,000 genera and about 300,000 plant species (Bagyaraj, 1991). It is as normal for the roots of plants to be mycorrhizal as it is for the leaves to photosynthesize (Mosse, 1986). The AM fungi are included in the phylum Zygomycota, order Glomales (Redecker et al., 2000) but recently they have been placed into the phylum 'Glomeromycota' (Schussler et al., 2001). The Glomeromycota is divided into 4 orders, 8 families, 10 genera and 150 species. The common genera are *Acaulospora*, *Gigaspora*, *Glomus* and *Scutellospora* (Schussler, 2005). They are characterized by the presence of extra radical mycelium branched haustoria-like structure within the cortical cells, termed arbuscules, and are the main site of nutrient transfer between the two symbiotic partners (Hock and Verma, 1995; Smith and Read, 1997). AM fungi colonize plant roots and penetrate into surrounding soil, extending the root depletion zone and the root system. They supply water and mineral nutrients from the soil to the plant while AM benefits from carbon compounds provided by the host plant (Smith and Read, 1997; Siddiqui et al., 1999). AM fungi are associated with improved growth of host plant species due to increased nutrient uptake, production of growth promoting substances, tolerance to drought, salinity and synergistic interactions with other beneficial microorganisms (Sreenivasa and Bagyaraj, 1989). The soil conditions prevalent in sustainable agriculture are likely to be more favorable to AM fungi than are those under conventional agriculture (Bethlenfalvay and Schuepp, 1994; Smith and Read, 1997).

Any agricultural operation that disturbs the natural ecosystem will have repercussions on the mycorrhizal system (Mosse, 1986). The preceding crops affect growth and yield of subsequent crops (Karlen et al., 1994). The inclusions of non-mycorrhizal crops within rotations decrease both AM fungal colonization and yield of subsequent crops (Douds et al., 1997; Arihawa and Karasawa, 2000). In addition to crop sequence, varietals selection, cultivation and fallowing have been shown to affect mycorrhizal activity (Ocampo et al., 1980; Hetrick et al., 1996; McGonigle and Miller, 2000). However, impact of soluble fertilizers on colonization and function of AM fungi is contradictory. The application of soluble phosphorus decreased root colonization (Abbott and Robson, 1984) with occasional reports of increases (Gryndler et al., 1990). Similarly, contradictory results have also been reported with nitrogen fertilizer (Baltruschat and Dehne, 1988; Gryndler et al., 1990; Liu et al., 2000). Therefore, uses of AM fungi in the biocontrol for sustainable agriculture require knowledge of culture systems which may affect their establishment and multiplication in the field.

2 BIOPROTECTANT BEHAVIOR OF AM FUNGI

Plant diseases can be controlled by manipulation of indigenous microbes or by introducing antagonists to reduce the disease-producing propagules (Linderman, 1992). AM fungi and their associated interactions with plants reduce the damage caused by plant pathogens (Harrier and Watson, 2004). These interactions have been documented for many plant species (Tables 1 and 2). With the increasing cost of inorganic fertilizers and the environmental and public health hazards associated with pesticides and pathogens resistant to chemical pesticides, AM fungi may provide a more suitable and environmentally acceptable alternative for sustainable agriculture.

AM fungi are a major component of the rhizosphere of plants and may affect the incidence and severity of root diseases (Linderman, 1992). Comprehensive reviews exploring the possibilities of AM fungi in the biocontrol of plant diseases include Schonbeck (1979), Dehne (1982), Bagyaraj (1984), Smith (1988), Caron (1989), Paulitz and Linderman (1991), Linderman (1992, 1994), Siddiqui and Mahmood (1995a), Azcon-Aguilar and Barea (1996), Smith and Read (1997), Mukerji (1999), Siddiqui et al. (1999), Barea et al. (2005). The primary results which can be drawn from the various observations are (1) AM associations reduce the damage caused by plant pathogens, especially those caused by fungi and nematodes; (2) AM symbiosis enhances resistance or tolerance in roots but is not equal in different crops; (3) protection is not effective against all pathogens, and (4) disease protection is modulated by soil and other environmental conditions. Therefore, the interactions between different AM fungi and plant pathogens vary with the host plant and the cultural system. Moreover, the protective effect of AM inoculation may be both systemic and localized. Actions of AM fungi against phytopathogens have been categorized into the following three sub-categories:

2.1 Interaction of AM fungi with plant parasitic nematodes

2.2 Interaction of AM fungi with fungal plant pathogens

2.3 Interaction of AM fungi with other plant pathogens.

2.1 Interaction of AM fungi with plant parasitic nematodes

Plant parasitic nematodes are found in all agricultural regions of the world and any crop is susceptible to suffer damage by these parasites. Plant parasitic nematodes can be separated into different groups according to feeding habits. Ectoparasites remain outside the roots and are generally epidermal or subepidermal feeders. In the case of semiendoparasites, a portion of the organism enters the root to feed, but a portion remains outside the root tissue; migratory endoparasites enter and migrate within roots, feeding on various tissues and induce lethal responses to root tissues. Sedentary endoparasites become immobile as adults and depend on specialized transfer cells for nutrition. Parasitism of nematodes on the host plant may cause up to 50% yield losses but these losses may further be aggravated when the plant is predisposed to other plant pathogens.

The occurrence of AM fungi and plant parasitic nematodes in the roots of different crops and their dependence for nutrition on the host may result in a common association of AM fungi, plant parasitic nematodes and host plant. Association of these two groups of organisms generally exert opposite effects on the host plant. Thus, it is of utmost importance to determine the effect of interaction of these organisms on plant growth and yield including their mutual effects. The interaction of AM fungi and plant parasitic nematodes has received much attention and a large number of research papers have been published (Table 1).

Table 1. Effect of AM fungi on the plant parasitic nematodes and plant growth.

AM fungi	Nematode	Effect	Reference
Gigaspora margarita	*Pratylenchus brachyurus*	Reproduction was similar on mycorrhizal and non-mycorrhizal cotton plants	Hussey and Roncadori, 1978
Glomus mosseae	*M. incognita*	Nematode development was reduced by mycorrhizal roots on tomato	Sikora, 1978
G. fasciculatum	*M. incognita* *M. javanica*	Reduced number and size of galls on mycorrhizal tomato	Bagyaraj et al., 1979
G. etunicatus	*Radopholus similis*	No effect on nematode population on citrus	O'Bannon and Nemec, 1979
G. mosseae	*Tylenchulus semipenetrans*	Disease symptoms were less severe on mycorhizal citrus	O'Bannon et al., 1979
G. macrocarpum	*M. incognita*	Reduced number of galls per gram root was observed on soybean	Kellam and Schenck, 1980
G. fasciculatum	*M. arenaria*	Nematode population was highest on mycorrhizal grapes	Atilano et al., 1981
G. fasciculatum	*Rotylenchulus reniformis*	Adversely affected nematode reproduction on tomato	Sitaramaiah and Sikora, 1982

(continued)

AM fungi	Nematode	Observation	Reference
G. etunicatum G. mosseae	M. incognita	Nematode eggs per gram root were lower on mycorrhizal cotton	Hussey and Roncadori, 1982
G. etunicatus	M. incognita	No effect on nematode reproduction on peach	Strobel et al., 1982
Gigaspora margarita G. mosseae	M. incognita	AM fungus had no effect on M. incognita eggs per plant or eggs per egg mass on tomato	Thompson Cason et al., 1983
G. fasciculatum	M. incognita	Sixty or more spores of AM fungi had adverse effect on nematode reproduction on cotton	Saleh and Sikora, 1984
G. fasciculatum	M. hapla	Nematode reproduction was enhanced on mycorrhizal onion	Kotcon et al., 1985
G. fasciculatum	M. hapla	No effect on nematode reproduction on onion	MacGuidwin et al., 1985
G. fasciculatum	M. incognita	No difference in juvenile penetration on mycorrhizal and non-mycorrhizal tomato plants	Suresh et al., 1985
G. intraradices	M. incognita	Adverse effect on nematode reproduction at both intervals (planting and after 28 days) on cotton	Smith et al., 1986a
G. intraradices	M. incognita	Penetration was similar on mycorrhizal and non-mycorrhizal cotton roots after 7 days of inoculation.	Smith et al., 1986b
G. etunicatum	M. javanica	Severity of nematode disease reduced on mycorrhizal bean.	Zombolin and Oliveira, 1986
G. manihotis G. margarita G. gigantea	M. javanica	Reproduction was differentially suppressed, most pronounced with G. manihotis, less with G. margarita and slightly with G. gigantea on chickpea.	Diederichs, 1987
G. epigaeus	Heterodera cajani	G. epigaeus stimulates nematode reproduction on cowpea	Jain and Sethi, 1987
G. fasciculatum	M. javanica	Pre-inoculation of AM reduced nematode reproduction on tomato	Morandi, 1987
G. fasciculatum G. epigaeus	H. cajani	Prior inoculation of AM fungi adversely affected nematode penetration more than simultaneous inoculation on cowpea	Jain and Sethi, 1988
G. intraradices	Radopholus similis	Plants inoculated with AM fungus 7 days prior to R. similis reduce the ill effect of nematodes on banana	Umesh et al., 1988
G. intraradices	Radopholus citrophilus	Reduced nematode reproduction on citrus	Smith and Kaplan, 1988
G. margarita	M. incognita	Suppressed nematode reproduction on soybean	Carling et al., 1989
G. intraradices	M. incognita	No effect on degree of root galling and number of nematode eggs per egg mass on muskmelon	Heald et al., 1989

(continued)

Table 1. *(Cont.)*

AM Fungi	Nematode	Effect	Reference
G. mosseae *Acaulospora laevis* *G. fasciculatum* *Gigaspora margarita*	*M. incognita*	Individually all AM fungi reduced nematode reproduction but the greatest reduction was caused by *A. laevis* on black pepper	Anandaraj *et al.*, 1990
G. fasciculatum *G. etunicatum*	*M. incognita*	Significantly reduced galling and nematode population when pre-inoculated with AM fungi on pepper	Sivaprasad *et al.*, 1990
Mixed inoculum	*Heterodera glycines*	Increased tolerance to nematodes in greenhouse experiments and outdoor microplots on soybean	Tylka *et al.*, 1991
G. fasciculatum	*M. incognita*	Prior inoculation of AM fungus reduced nematode population on cowpea	Devi and Goswami, 1992
G. fasciculatum	*M. javanica*	Nematode population inhibited in mycorrhizal tomato	Sundarababu *et al.*, 1993
Glomus sp.	*M. javanica*	Nematode population significantly lower on mycorrhizal *Sahelian acacia* than on non-mycorrhizal plants	Duponnois and Cadet, 1994
G. fasciculatum	*M. incognita*	Prior inoculation of AM fungus reduced nematode population on black gram	Sankaranarayanan and Sundarababu, 1994
G. fasciculatum *G. mosseae*	*M. incognita*	Prior inoculation of AM fungi adversely affects nematode population than does simultaneous inoculation, but greatest reduction was caused by *G. fasciculatum* on brinjal	Sharma and Trivedi, 1994
G. mosseae	*M. javanica*	Inoculation of AM fungus suppressed gall index and number of galls per root system	Al-Raddad, 1995
G. fasciculatum	*M. incognita*	Reduced nematode population on chickpea	Rao and Krishnappa, 1995
G. fasciculatum	*H. cajani*	Reduced nematode multiplication on pigeon pea	Siddiqui and Mahmood, 1995b
G. margarita	*H. cajani*	Adversely affect nematode reproduction on pigeon pea	Siddiqui and Mahmood, 1995c
G. macrocarpum	*M. incognita*	Prior inoculation of AM fungus caused significant reduction in galling and nematode population on Subabul	Sundararaju *et al.*, 1995
G. margarita *G. etunicatum*	*M. arenaria*	Inoculation of AM fungi caused increase in galling and egg production on peanut	Carling *et al.*, 1996
G. fasciculatum	*M. incognita*	Prior inoculation of AM fungus reduced galling and improved NPK uptake on tomato	Mishra and Shukla, 1996
G. fasciculatum	*M. incognita*	Significantly reduced galls and no. of eggs per egg mass on tomato	Rao *et al.*, 1996

(continued)

G. mosseae	M. incognita	Reduced galling and nematode population on Musa	Jaizme-Vega et al., 1997
G. fasciculatum	M. incognita	AM fungus had adverse effect on nematode populations on tomato	Mishra and Shukla, 1997
G. mosseae	M. incognita	Significantly reduced nematode populations on Crossandra undulaefolia	Nagesh and Reddy, 1997
G. intraradices	M. javanica	AM fungus had no effect on nematode population on Musa	Pinochet et al., 1997
G. deserticola	M. incognita	Reduced nematode multiplication on tomato	Rao et al., 1997
G. fasciculatum	M. incognita	Reduced nematode population on black gram	Sankaranarayanan and Sundarababu, 1997a
G. mosseae	M. incognita	AM fungus had adverse effect on nematode population on black gram	Sankaranarayanan and Sundarababu, 1997b
G. mosseae G. fasciculatum	M. incognita	Reduced nematode population but the greatest reduction was caused by G. mosseae on okra	Sharma and Trivedi, 1997
G. mosseae	M. javanica	Reduced galling and nematode multiplication on chickpea	Siddiqui and Mahmood, 1997
G. fasciculatum	M. incognita	Prior inoculation of AM fungus had adverse effect on galling and nematode multiplication on barseem	Jain et al., 1998
G. mosseae	M. incognita	Reduced nematode population on brinjal but use with P. lilacinus gave better results	Rao et al., 1998a
G. fasciculatum	M. incognita	Significantly reduced nematode populations on brinjal but results were more pronounced when used with castor cake	Rao et al., 1998b
G. deserticola	M. incognita	Nematode population inhibited on tomato	Rao and Gowen, 1998
G. fasciculatum	M. incognita	Adverse effect on nematode multiplication on green gram	Ray and Dalei, 1998
G. mosseae	M. incognita	Significantly reduced galling and nematode multiplication on tomato	Reddy et al., 1998
G. mosseae G. fasciculatum	M. incognita	Inoculation of AM fungi reduced gall index and nematode population on black gram	Sankaranarayanan and Sundarababu, 1998
G. fasciculatum	M. incognita	Nematode population inhibited on cowpea	Santhi and Sundarababu, 1998
G. mosseae	M. javanica	Reduced nematode multiplication on tomato	Siddiqui and Mahmood, 1998
G. fasciculatum	M. incognita	Reduced nematode multiplication on tomato	Nagesh et al., 1999a

(continued)

Table 1. *(Cont.)*

AM Fungi	Nematode	Effect	Reference
G. mosseae G. fasciculatum G. intraradices A. laevis	M. incognita	Individually all the AM fungi had adverse effects on nematode population on Crossandra undulaefolia	Nagesh et al., 1999b
G. aggregatum G. fasciculatum G. mosseae	M. incognita	Inoculation of all AM fungi reduced nematode population but the greatest reduction caused by G. aggregatum on Hyoscyamus niger	Pandey et al., 1999
G. mosseae	M. incognita	Reduced galling and nematode multiplication on tomato	Rao et al., 1999
G. mosseae	M. incognita	Adverse effect on nematode multiplication on black gram	Sankaranarayanan and Sundarababu, 1999
G. etunicatum	M. incognita	Reduced nematode multiplication on tomato	Bhagawati et al., 2000
G. mosseae	M. incognita	Adverse effect on nematode multiplication on tomato	Bhat and Mahmood, 2000
G. mosseae	M. incognita	Reduced nematode population on okra and Pennisetum glaucum	Jothi et al., 2000
G. fasciculatum G. mosseae G. intraradices G. fulvum	M. incognita	Had adverse effect on nematode multiplication but the greatest reduction was caused by G. mosseae on brinjal	Jothi and Sundarababu, 2000
G. mosseae	M. javanica	Significantly reduced galling and nematode multiplication but the effect was more pronounced when used with ammonium sulphate on tomato	Siddiqui and Mahmood, 2000
G. mosseae	H. cajani	Application of G. mosseae with B. japonicum caused a greater reduction in cyst formation than use of G. mosseae alone	Siddiqui et al., 2000
G. intraradices G. mosseae G. etunicatum	M. javanica	Inoculation of all the AM fungi had adverse effect on nematode population on peach almond hybrid GF-677	Calvet et al., 2001
G. mosseae	M. incognita	Had adverse effect on nematode population on pearl millet and green gram	Jothi and Sundarababu, 2001
G. mosseae	M. javanica	Reduced nematode multiplication on chickpea	Siddiqui and Mahmood, 2001
G. mosseae	M. incognita	Reduced the nematode multiplication on chilli	Sundarababu et al., 2001
G. mosseae	M. incognita	Had adverse effect on nematode population on tomato	Talavera et al., 2001
G. etunicatum Isolate (KS18) G. mosseae Isolate (KS14)	M. hapla	Significantly reduced nematode multiplication but the greatest reduction was caused by G. mosseae on Pyrethrum	Waceke et al., 2001
G. mosseae G. macrocarpum G. caledonicum	M. javanica	Had adverse effect on nematode multiplication on Musa	Elsen et al., 2002

(continued)

Glomus sp.	*M. incognita*	Reduced nematode population on brinjal	Jothi and Sundarababu, 2002
G. fasciculatum *G. macrocarpum* *G. margarita* *A. laevis* *S. dussii*	*M. incognita*	Inoculation of all AM fungi reduced nematode population but the greatest reduction was caused by *G. fasciculatum*	Labeena *et al.*, 2002
Glomus mosseae	*M. incognita*	Reduced galling and nematode multiplication but use of AM fungus with DAP gave better results	Shafi *et al.*, 2002
Glomus sp. (K14)	*M. hapla*	Significantly suppressed nematode multiplication on *Pyrethrum*	Waceke *et al.*, 2002
G. fasciculatum	*M. incognita*	Reduced galling and nematode population on brinjal	Borah and Phukan, 2003
G. coronatum	*M. incognita*	Prior inoculation of AM fungus reduced nematode infestation on tomato	Diedhiou *et al.*, 2003
G. mosseae *G. intraradices* *G. fasciculatum* *Gigaspora gilmori*	*M. incognita*	Inoculation of all AM fungi reduced galling and nematode population but the greatest reduction was caused by *G. mosseae* on chickpea	Jain and Trivedi, 2003
G. fasciculatum	*M. incognita*	Reduced nematode population on tomato	Pradhan *et al.*, 2003
G. mosseae	*M. incognita*	Reduced galling on okra	Sharma and Mishra, 2003
G. fasculatum *G. mosseae*	*M. incognita*	Reduced nematode population but the greatest reduction was caused by *G. fasciculatum* on ginger	Nehra, 2004
G. intraradices	*M. hapla*	Reduced the no. of galls and egg-sacs on tomato cv. 'Hildares' but biocontrol of nematode was not achieved in cv. 'Tiptop'	Masadeh *et al.*, 2004
G. aggregatum	*M. incognita*	Significantly reduced nematode population on *Mentha arvensis*	Pandey, 2005
G. fasciculatum	*M. incognita*	Reduced nematode population on tomato but SBI-G.f. isolate was most effective compared with CTI-G.f. isolate	Kantharaju *et al.*, 2005
G. intraradices	*M. incognita*	Combined inoculation of AM fungus with *A. niger* and *Bacillus* (B22) caused a geater increase in chickpea growth	Akhtar and Siddiqui, 2006
G. fasciculatum *G. constrictum* *G. mosseae* *G. intraradices* *Acaulospora* sp. *Sclerocystis* sp.	*M. incognita*	Individually all AM fungi reduced nematode reproduction but the greatest reduction was caused by *G. fasciculatum* on chickpea	Siddiqui and Akhtar, 2006

(continued)

Table 1. *(Cont.)*

G. mosseae *G. manihotis*	*M. incognita*	Significantly reduced galling in AM fungus-inoculated papaya	Jaizme-Vega *et al.*, 2006
G. fasciulatum	*M. incognita*	Significantly reduced nematode population on chickpea	Akhtar and Siddiqui, 2007a
G. intraradices	*M. incognita*	Reduced nematode multiplication on chickpea	Akhtar and Siddiqui, 2007b
G. fasciculatum	*M. incognita*	Inoculation of AM fungus significantly reduced nematode population and no. of galls on tomato	Shreenivasa *et al.*, 2007
G. mosseae *Gigaspora margarita*	*M. incognita*	*G. mosseae* was superior in reducing galling and nematode multiplication compared with *G. margarita* on tomato	Siddiqui and Akhtar, 2007
G. intraradices	*M. incognita*	Combined use of AM fungus with *Pseudomonas straita* and *Rhizobium* caused greater increase in chickpea growth	Akhtar and Siddiqui, 2008a
G. intraradices	*M. incognita*	Application of *G. intraradices* with *P. alcaligenes* and *B. pumilus* caused greater increase in shoot dry mass of chickpea	Akhtar and Siddiqui, 2008b
G. intraradices	*M. incognita*	Use of AM fungus with *T. harzianum* caused 37% increase in the growth of nematode inoculated tomato plants	Siddiqui and Akhtar, 2008

2.2 Interaction of AM fungi with fungal plant pathogens

More than 10,000 species of fungi are known to cause diseases of plants and are common in soil, air (spores) and on plant surfaces throughout the world in arid, tropical, temperate and alpine regions (Agrios, 2005). The diseases caused by fungal pathogens persist in the soil matrix and in residues on the soil surface, and are defined as soilborne diseases. The soil is a reservoir of inoculum of these pathogens, the majority of which are widely distributed in agricultural soils. Damage to root and crown tissues is often hidden in the soil; thus, these diseases may not be noticed until the aboveground parts of the plant are severely affected, showing symptoms such as stunting, wilting, chlorosis and death.

Fungal diseases are difficult to control because they are caused by pathogens which can survive for long periods in the absence of the normal crop host, and often have a wide host range including weed species. The occurrence of AM fungi and plant pathogenic fungi in roots of different crops and their dependence for nutrition on the host generally result in the interaction of AM fungi, plant pathogenic fungi and host plant. Association of these organisms generally exert opposite effects on the host. Thus it is desirable to test the mutual effects of these organisms on plant growth and yield. The interaction of AM fungi and plant pathogenic fungi has received considerable attention and a large number of research papers have been published (Table 2).

Table 2. Effects of AM fungi on fungal diseases and plant growth.

AM fungi	Pathogenic fungus	Effect	Reference
Glomus intraradices	Fusarium oxysporum f. sp. radicis-lycopersici	AM fungus significantly reduced Fusarium root rot on tomato	Caron et al., 1985
G. fasciculatum	Aphanomyces euteiches	Reduced root rot on pea	Rosendahl, 1985
G. mosseae	F. oxysporum	Significantly reduced Fusarium wilt on tomato and pepper	Al-Momany and Al-Raddad, 1988
G. etunicatum Glomus sp.	Pythium ultimum	Prior or simultaneous inoculation of AM fungus with P. ultimum reduced disease severity on cucumber	Rosendahl and Rosendahl, 1990
G. fasciculatum	Macrophomina phaseolina	Prior inoculation of AM fungus reduced disease on cowpea	Devi and Goswami, 1992
Glomus sp., G. fasciculatum, G. mosseae	Verticillium albo-atrum F. oxysporum f. sp. medicaginis	Seedlings inoculated with AM fungi had lower incidence of wilt in alfalfa than did non-mycorrhizal species	Hwang et al., 1992
G. fasciculatum	F. oxysporum	Prior inoculation of AM fungus reduced colonization by pathogens and severity of disease on cowpea	Sundaresan et al, 1993
G. intraradices	Pythium ultimum	Reduced populations of P. ultimum on Tagetes patula	St Arnaud et al., 1994
G. intraradices	F. oxysporum f. sp. lycopersici	Significantly reduced disease severity but is most effective when applied with T. harizianum	Datnoff et al., 1995
G. mosseae G. vesiformae Scutellospora sinuosa	Verticillium dahliae	Inoculation of AM fungi reduced disease indices in cotton	Liu, 1995
G. fasciculatum	Fusarium oxysporum	Reduced wilt indices in chickpea	Rao and Krishnappa, 1995
G. fasciculatum	Fusarium udum	Significantly reduced disease severity in pigeon pea	Siddiqui and Mahmood, 1995b
G. margarita	Fusarium udum	Reduced wilt indices in pigeon pea	Siddiqui and Mahmood, 1995c
Glomus sp.	Sclerotium cepivorum	AM fungi reduced white rot incidence and delayed disease development on onion	Torres-Barragan et al., 1996

(continued)

Table 2. *(Cont.)*

G. mosseae	Fusarium udum	Reduced disease severity on pigeon pea	Siddiqui and Mahmood, 1996
G. mosseae	Phytophthora nicotianae var. parasitica	Reduced root necrosis, and necrotic root apices ranged between 63–89 %	Trotta et al., 1996
Glomus intraradices	Aphanomyces euteiches	Reduced disease severity in pea	Kjoller and Rosendahl, 1997
G. mosseae	Fusarium solani	Significantly reduced disease severity in chickpea	Siddiqui and Mahmood, 1997
Glomus intraradices	Aphanomyces euteiches	Reduced disease severity in pea	Bodker et al., 1998
G. mosseae	Fusarium udum	Reduced disease severity in pigeon pea	Siddiqui et al., 1998
G. etunicatum	F. oxysporum f. sp. lycopersici	Reduced disease severity in tomato	Ozgonen et al., 1999
G. etunicatum	F. oxysporum f. sp. lycopersici	Reduced disease severity on tomato	Bhagawati et al., 2000
G. mosseae	F. solani R. solani	Significantly reduced severity of diseases on peanut	Elsayed Abdalla and Abdel-Fattah, 2000
G. intraradices	Rhizoctonia solani	Defense response elicited by R. solani significantly suppressed by AM fungus in alfalfa	Guenoune et al., 2001
G. clarum	Rhizoctonia solani	Significantly reduced root necrosis and number of sclerotia on cowpea	Abdel-Fattah and Shabana, 2002
Glomus sp. G. proliferum, G. intraradices G. versiforme	Cylindroclad-ium spathiphylli	AM fungi significantly increased growth and reduced disease severity in banana. Glomus sp. and G. proliferum caused greatest increase in plant growth compared to that caused by G. intraradices and G. versiforme	Declerck et al., 2002
G. mosseae G. intraradices	P. parasitica	G. mosseae was most effective in reducing disease symptoms produced by P. parasitica on tomato	Pozo et al., 2002
G. etunicatum G. intraradices	R. solani	Significantly reduced disease severity in micro- propagated banana	Yao et al., 2002
G. mosseae	Fusarium chlamydosporium	Reduced disease severity but best management was obtained when used with T. viridae	Boby and Bagyaraj, 2003
G. intraradices G. claroideum	Aphanomyces euteiches	Reduced disease severity on pea but effects were more pronounced in plant inoculated with G. intraradices than with G. claroideum	Thygesen et al., 2004
G. fasciculatum	F. oxysporum f. sp. ciceris	Reduced the disease severity in chickpea	Siddiqui and Singh, 2004

(continued)

Glomus intraradices	*F. oxysporum* f. sp. *lycopersici*	Reduced severity of disease in tomato	Akkopru and Demir, 2005
G. intraradices BEG12	*Rhizoctonia solani*	Significantly decreased epiphytic and parasitic growth of pathogen in tomato	Berta *et al.*, 2005
G. mosseae	*Alternaria triticina*	Reduced percent infected leaf area on wheat	Siddiqui and Singh, 2005
G. etunicatum BEG168	*F. oxysporum* f. sp. *cucumerinum*	AM fungus influenced secondary metabolites and increased wilt resistance in cucumber seedlings	Hao *et al.*, 2005
G. mosseae	*C. orbiculare*	AM fungus had no significant effect on disease development	Chandanie *et al.*, 2006
G. intraradices	*M. phaseolina*	Inoculation of AM fungus with *A. niger* and *Bacillus* (B22) caused a geater reduction in root-rot of chickpea	Akhtar and Siddiqui, 2006
G. fasciculatum	*M. phaseolina*	Reduced disease severity in chickpea	Akhtar and Siddiqui, 2007a
G. intraradices	*M. phaseolina*	Significantly reduced disease severity in chickpea	Akhtar and Siddiqui, 2007b
G. mosseae, G. etunicatum, G. fasciculatum Gigaspora margarita	*Phytophthora capsici*	AM fungi significantly increased plant growth and reduced disease severity in pepper but *G. mosseae* reduced disease severity to a greater extent	Ozgonen and Erkilic, 2007
G. intraradices	*M. phaseolina*	Combined inoculation of AM fungus with *Pseudomonas straita* and *Rhizobium* caused a greater reduction in the root-rot of chickpea	Akhtar and Siddiqui, 2008a
G. intraradices	*M. phaseolina*	Combined application of *G. intraradices* with *P. alcaligenes* and *B. pumilus* caused a greater reduction in the root-rot of chickpea	Akhtar and Siddiqui, 2008b

2.3 Interaction of AM fungi with other plant pathogens

Besides nematodes and fungi some other plant pathogens including plant pathogenic bacteria, phytoplasma and plant viruses also interact with AM fungi on various plants. Disease severity caused by these pathogens was generally reduced the mycorrhizal plants (Table 3).

Table 3. Effects of AM fungi on bacteria, phytoplasma and viral diseases and plant growth.

AM fungi	Pathogen	Effect	Reference
G. etunicatum	Citrus tristeza virus and Citurs urgose virus	Growth of Citurs macrophylla inoculated with tristeza virus (T-3 isolate) and Citurs urgose virus (CLRV-2) was not reduced by virus infection in mycorrhizal plants	Nemec and Myhre, 1984
G. mosseae	P. syringae	Neither growth of tomato nor percentage VA infection was negatively affected by pathogenic bacteria	Garcia-Garrido and Ocampo, 1989
Gigaspora gilmorei Acaulospora marrowae G. fasciculatum G. constrictum	Yellow mosaic virus	Mycorrhizal colonization and spore formation was reduced by yellow mosaic virus on mungbean	Jayaram and Kumar, 1995
G. intraradices	Tobacco mosaic virus	Higher incidence and severity of necrotic lesion in mycorrhizal than in non mycorrhizal plants	Shaul et al., 1999
G. mosseae	Yellow disease	Symptoms induced by the phytoplasma were less severe on tomato when the plants were also harboured AM fungi	Lingua et al., 2002
AM fungi	P. solanacearum	Disease decrease in eucalyptus seedlings injected with AM fungi	MingQin et al., 2004
G. intraradices	Pear decline (PD) Phytoplasma	AM fungus significantly increased shoot length both in non-PD and PD infected pear trees	García-Chapa et al., 2004

3 REASONS FOR REDUCED DAMAGE IN MYCORRHIZAL PLANTS

3.1 Change in root growth and morphology

The colonization by AM fungi results in morphological changes to the root, leading to an increased surface area of root. Roots offer structural support to the plants and function in absorption of water and supply mineral nutrients for a wide range of microorganisms (Curl and Truelove, 1986; Rovira, 1985). Changes in root morphology will ultimately affect the plant's responses to other organisms. AM fungal-colonized roots are more highly branched, i.e., the root system contains shorter, more branched, adventitious roots of larger diameters and lower specific root lengths (Schellenbaum *et al.*, 1991; Berta *et al.*, 1993) The AM inoculated plants possess a strong vascular system, which imparts greater mechanical strength to diminish the effects of pathogens (Schonbeck, 1979).

Dehne *et al.* (1978) observed increased lignifications in the endodermal cells of mycorrhizal tomato and cucumber plants and speculated that such responses may account for reduced incidence of Fusarium wilt (*Fusarium oxysporum* f. sp. *lycopersici*). Becker (1976) reported a similar effect on pink root of onion (*Pyrenochaeta terrestris*). Mycorrhizal plants produced wound-barriers at a faster rate than non-mycorrhizal plants and increased wound barrier formation inhibited *Thielaviopsis* black root-rot of mycorrhizal holly (*Ilex crenata*) plants (Wick and Moore, 1984). The AM fungi reduce disease severity caused by *Aphanomyces euteiches* on pea (Bodker *et al.*, 1998) and *Cylindrospermum destructans* in strawberry (Paget, 1975) and these examples emphasize the significance of AM fungi in bio-protection against fungal pathogens.

3.2 Histopathological changes

Histopathological studies on galls caused by the root-knot nematode *M. incognita* showed that galls in mycorrhizal plants had fewer giant cells, which are needed for the development of nematode larvae, than did non-mycorrhizal plants. Nematodes in mycorrhizal plants were smaller and took longer times to mature to the adult form (Suresh, 1980). Smaller syncytia and fewer giant cells were reported to confer resistance against nematodes on the host plant (Trudgill and Parrott, 1969; Fassuliotis, 1970). Histochemical and immunocytochemical studies provided evidence that decreased pathogen development in mycorrhizal root systems both in parts with or with out mycorrhiza were associated with modifications in the host cells, together with the accumulation of defense-related molecules (Sharma and Adholeya, 2000).

3.3 Physiological and biochemical changes

The physiological and biochemical changes caused by mycorrhizal fungi in the host plant generally reduce the severity of nematode diseases (Dehne, 1982). Phenolic compounds have been shown to be formed after mycorrhizal colonization (Sylvia and Sinclair, 1983) and are thought to play a role in disease resistance (Goodman *et al.*, 1967). Production of phyto-alexin was greater on mycorrrhizal roots than on non-mycorrhizal roots (Morandi, 1987) and phytoalexins are believed to play a major role in the host defense system against pathogens (Kaplan *et al.*, 1980). An increase in lignin and phenols in mycorrhizal plants was observed and was associated with reduced nematode reproduction (Sikora, 1978; Umesh *et al.*, 1988;

Singh et al., 1990). Increased phenylalanine and serine concentrations in tomato roots due to inoculation with AM fungi have been observed (Suresh, 1980). These two amino acids are known to be inhibitory to root-knot nematodes (Prasad, 1971; Reddy, 1974). Suresh and Bagyaraj (1984) reported that AM inoculation increased the quantities of sugars and amino acids in plant tissue which may be responsible for the reduction of nematode infestation. However, inferences based on the absence of galling on segments of roots and split root experiments argue for a more localized effect (Tylka et al., 1991; Fitter and Garbaye, 1994).

Various evidence indicates structural and biochemical changes in the cell walls of plants colonized by AM fungi. Dehne and Schonbeck (1979) and Becker (1976) reported enhanced lignification of endosperm cell walls and vascular tissues. Dehne et al. (1978) also demonstrated an increased concentration of antifungal chitinase in AM roots. They suggested that increased arginine accumulations in AM roots suppress sporulation of *Thielaviopsis*, as reported earlier (Baltruschat and Schonbeck, 1975). Morandi et al. (1984) determined increased concentrations of phytoalexins, for example isoflavonoid compounds, in AM roots of soybean compared to those in non-AM soybean and postulated that these compounds are responsible for the increased resistance to fungal pathogens in AM plants. Krishna and Bagyaraj (1986) reported that higher amounts of catechols inhibit *Sclerotium rolfsii* growth *in vitro*. Cordier et al. (1998) demonstrated that control of *Phytophthora parasitica* in the mycorrhizal tomato root system involved induction of localized resistance in arbuscules containing cells and systemic resistance in non-mycorrhizal tissues. The induction of defense-related enzymes in mycorrhizal roots against *P. parasitica* was also reported by Pozo et al. (1998).

3.4 Changes in host nutrition

Mycorrhizal plants contain higher concentrations of phosphorus than do non-mycorrhizal plants (Hayman, 1978; Bowen, 1980). Improvement of phosphorus nutrition following AM colonization of phosphorus-deficient roots results in a decrease in membrane permeability and reduction in root exudation (Graham et al., 1981). Mycorrhizal-induced decreases in root exudation have been correlated with reduction of soil-borne disease (Graham and Menge, 1982), while improved nutritional status of the host brought about by AM fungus-root colonization may affect quantitative changes in root exudates (Linderman, 1985; Reid, 1984).

The severity of nematode damage of cotton was greater on P-fertilized, non-mycorrhizal plants than non-mycorrhizal plants at a P level deemed adequate to high for cotton (Thompson Cason et al., 1983). This effect was attributed to zinc deficiency induced by nematode infection at high soil P levels. High levels of P fertilization inhibit zinc uptake (Mengel and Kirkby, 1979) but apparently AM fungi can alleviate this P-induced zinc deficiency and thus increase host tolerance to nematode parasitism. AM fungi have been shown to induce responses caused by environmental stress in root growth, root exudation, nutrient absorption and host physiology (Smith, 1988). Changes in exudation due to P nutrition alter the chemotaxic attraction of the nematodes to the roots and affects exclusion of nematode species that require a hatching stimulus (Baker and Cook, 1982). In general, AM fungi infection in P-deficient plants affects membrane permeability and exudation patterns in a fashion similar to that caused by P-fertilization in non-mycorrhizal plants.

The obvious contribution to reduction of root diseases is increased nutrient uptake, particularly of P and other minerals, because AM symbiosis results in more vigorous plants, which thus become more resistant or tolerant to pathogen attacks (Linderman, 1994). Davis (1980) found this type of response on *Thielaviopsis* root rot of citrus, where AM plants were larger than nonmycorrhizal plants until the latter were fertilized with additional P. The mycorrhizal-induced compensation processes may explain the increased tolerance of mycorrhizal and P-fertilized plants because plants may compensate for the loss of root mass or function caused by pathogens (Wallace, 1973). Graham and Menge (1982) suggest a similar effect, where AM fungi or added P reduced wheat take-all disease caused by *Gaeumannomyces graminis*, and speculate that enhanced P status of the plant causes a decrease in root exudates used by the pathogen for spore germination and infection. Declerck *et al.* (2002) suggested a similar effect whereby AM fungi or added P reduced root rot of bananas caused by *C. spathiphyylii*. It has been hypothesized that direct competition between root pathogens require host nutrients for reproduction and development and this competition may be the cause of their inhibition (Dehne, 1982; Smith, 1988). Greater tolerance of AM plants is also attributable to increased root growth and phosphate status of the plant (Cameron, 1986). In addition to P, AM fungi can enhance the uptake of Ca, Cu, Mn, S and Zn (Pacovsky *et al.*, 1986; Smith and Gianinazzi-Pearson, 1988). Host susceptibility to pathogens and tolerance to disease can be influenced by the nutritional status of the host and the fertility status of the soil (Cook and Baker, 1982).

Increase in plant growth after root colonization by AM fungi is due to improvement in the mineral nutrient status of host plant. Depending on the host plant and AM fungus isolate, colonization of the root system can increase phosphorus nutrition and other mineral nutrients (Clark and Zeto, 2000). Host susceptibility to infection can also be influenced by nutritional status of the host and fertility status of the soil (Bodker *et al.*, 1998; Jaizme-Vega and Pinochet, 1997). However, in some cases enhanced mineral nutrition of mycorrhizal plants has no affect against pathogens (Graham and Egel,

1988). Therefore, enhanced mineral nutrition of AM plants does not account for all protection conferred by AM fungi to host plant (Caron *et al.*, 1986a).

3.5 Mycorrhizosphere effect

AM fungal colonized plants differ from non-mycorrhizal roots in terms of microbial community composition of the rhizosphere (Marschner *et al.*, 2001). These differences have been attributed to alterations in root respiration rate and quality and quantity of exudates. Plant root systems colonized by AM fungi differ in their effect on the bacterial community composition within the rhizosphere and rhizoplane. The number of facultative anaerobic bacteria, fluorescent pseudomonads, *Streptomyces* species and chitinase producing actinomycetes differ depending on the host plant and the isolate of AM fungus (Harrier and Watson, 2004). In addition, extra radical hyphae of AM fungi provide a physical or nutritional substrate for bacteria. AM symbiosis can also cause qualitative and quantitative changes in rhizospheric microbial populations; the resulting microbial equilibria could influence the growth and health of plants. These changes may result from AM fungus-induced changes in root exudation patterns (Smith, 1987; Azcon-Aguilar and Bago, 1994; Smith *et al.*, 1994; Bansal *et al.*, 2000). Changes in microbial populations induced by AM formation may lead to stimulation of the microbiota which may be antagonistic to root pathogens. AM establishment can change both total microbial populations and specific functional groups of microorganisms in the rhizoplane or the rhizosphere soil (Meyer and Linderman, 1986; Linderman, 1994). Numbers of pathogen-antagonistic actinomycetes were greater in the rhizosphere of AM plants than in nonmycorrhizal controls (Secilia and Bagyaraj, 1987). The authors showed that pot cultures of *G. fasciculatum* harbored actinomycetes antagonistic to *F. solani* than those of non-mycorrhizal plants. Other studies indicate that pathogen suppression by AM fungi involves changes in mycorrhizosphere microbial populations. Caron *et al.* (1985, 1986a, b, c) showed a reduction in *Fusarium* populations in mycorrhizosphere soil of tomatoes and a corresponding reduction in root-rot in AM plants compared with non-AM plants, probably due to the increased antagonism in the AM mycorrhizosphere.

3.6 Competition of colonization sites and photosynthates

AM fungi and soil-borne plant pathogens occupy similar root tissues and there may be direct competition for space if colonization is occurring at the same time (Smith, 1988). If AM fungi and plant pathogens are colonizing the same host tissues there may be competition for space because both usually develop within different cortical cells of roots (Azcon-Aguilar and Barea, 1996). Davies and Menge (1980) observed localized competition between AM fungi and *Phytophthora*. They observed reduced development of *Phytophthora* in AM-colonized and adjacent uncolonized root systems, and pathogens never penetrated arbuscule-containing cells (Cordier *et al.*, 1996). Similarly *Aphanomyces* was suppressed on pea roots by AM fungi only when the two organisms were present on the same root (Rosendahl, 1985). Vigo *et al.* (2000) observed that the number of infection sites was reduced within mycorrhizal root systems and colonization by the AM fungus had no effect on the spread of necrosis.

AM fungi are dependent on the host as a carbon source and 4–20% of the host net photosynthate is transferred to the AM fungus (Smith and Read, 1997). There is much information to support the competition for host photosynthates and this phenomenon may have an important role in interactions with endoparasitic nematodes because of the obligate nature of both organisms for host-derived compounds (Azcon-Aguilar and Barea, 1996).

3.7 Activation of defense mechanism

The activation of specific plant defense mechanisms as a response to AM colonization is an obvious basis for the protective behavior of AM fungi. The elicitation, via an AM symbiosis of specific plant defense reactions, could predispose the plant to an early response to attack by a root pathogen (Gianinazzi-Pearson *et al.*, 1994). In relation to plant defense relevant compounds include phytoalexins, enzymes of the phenylpropanoid pathway, chitinases, β-1,3-glucanases, peroxidases, pathogenesis-related (PR) proteins, callose, and phenolics (Gianinazzi-Pearson *et al.*, 1994).

Phytoalexins are low-molecular-weight, toxic compounds usually accumulating with pathogen attack and are released at the sites of infection (Morandi *et al.*, 1984; Morandi, 1996). Both phenylalanine ammonium-lyase (PAL), the first enzyme of the phenylpropanoid pathway, and chalcone isomerase, the second enzyme specific for flavonoid/isoflavonoid biosynthesis, increased in amount and activity during early colonization of plant roots by AM fungi (Lambais and Mehdy, 1993; Volpin *et al.*, 1994, 1995). These results suggest that AM fungi initiate a host defense response which is subsequently suppressed. Chitinases are little or only transiently induced by AM colonization (Dumas-Gaudot *et al.*, 1992a, b). It has been reported that increased levels of chitinase activity are only detected in AM roots at the beginning of colonization (Spanu *et al.*, 1989; Bonfonte-Fasolo and Spanu, 1992; Lambais and Mehdy, 1993). A decrease in β-1,3 endoglucanase activity has also been reported at specific stages during mycorrhiza development (Lambais and Mehdy, 1993). These observations suggest a systemic suppression of the defense reaction during the establishment of the AM association. PR proteins are synthesized only locally and in very low amounts during

AM colonization, although these molecules were regularly distributed around the arbuscular hyphae (Balestrini *et al.*, 1994).

The increased lignification of root endodermal cells induced by AM colonization has been suggested to play an important in the plant defense mechanism (Dehne, 1982). However, these compounds could sensitize the root to pathogens and enhance mechanisms of defense to subsequent pathogen infection; the results of Benhamou *et al.* (1994) strengthened this hypothesis. It was evident from their results that mycorrhizal carrot roots afford increased protection against *Fusarium oxysporum* f. sp. *chrysanthemi*. In mycorrhizal roots, growth of the pathogen was usually restricted to the epidermis and cortical tissues, whereas in non-mycorrhizal roots the pathogen developed further, infecting even the vascular stele. Fusarium hyphae within mycorrhizal roots exhibited a high level of structural disorganization, characterized by the massive accumulation of phenolic-like compounds and the production of chitinases. This reaction was not induced by non-mycorrhizal roots, suggesting that the activation of plant defense responses by mycorrhiza formation provides a certain protection against the pathogen (Azcon-Aguilar and Barea, 1996). These results need to be confirmed on different plants, and must clearly show that AM infection makes the root more responsive to pathogen attack, i.e., promoting a quicker and stronger reaction against the pathogen.

In contrast to the weak defence response towards AM fungi found in AM hosts, it is noteworthy that in myc⁻ pea mutants, AM fungi trigger a strong resistance reaction. This suggests that the AM fungi are able to elicit a defense response, but that symbiosis-specific genes somehow control the expression of the genes related to plant defense during AM establishment (Gianinazzi- Pearson *et al.*, 1994, 1995, 1996). It is curious, in this context, that the constitutive expression of several PRs in tobacco plants did not affect either the time course or the final level of colonization by *Glomus mosseae*, which was only reduced in plants constitutively expressing an acidic isoform of tobacco PR-2, a glucanase (Vierheilig *et al.*, 1996).

3.8 Nematode parasitism by AM fungi

In some studies, parasitism of nematodes eggs with AM fungus has been demonstrated, but the level of parasitism was not considered sufficient to negatively affect nematode activities (Francl and Dropkin, 1985). AM fungal chlamydospores have been reported to occupy seeds and dead insects in soil (Rabatin and Rhodes, 1982; Taber, 1982) and have limited saprophytic capabilities (Harley and Smith, 1983). It seems likely that the AM fungi colonize only stressed or weakened nematode eggs. The nematode parasitism by AM fungi is opportunistic and depends on carbon nutrition from autotrophic symbionts, rather than being representative of a true host-parasite relationship.

4 EXPLANATION OF DIVERSE RESULTS

The diversity of interactions reported here shows that each pathogen-AM fungus-plant combination is unique and generalizations regarding such interactions are difficult to make. Pathogen-AM fungus interactions are influenced by the host cultivars, pathogen species, AM fungus species and their combination, pathogen and AM fungus initial inoculum densities, soil fertility and sequence of pathogen inoculation (Smith, 1987). Most of the evidence is however, from laboratory, greenhouse and microplot studies. There is a need for extending these studies into field conditions. Moreover, more studies are needed to confirm earlier results. Increased root growth and function, nutrition effects other than those for P, alteration in root exudation patterns, competition for host photosynthates and competition for space and infection sites are some reasons which indirectly affect the host-pathogen relationship in mycorrhizal plants.

Contradictory results are often obtained with the same plant-AM fungus-pathogen system in different soils or substrates, probably because the efficiency of these mechanisms are affected by soil, biological, chemical and physical conditions. The primary soil factors involved are availability of phosphorus and the microflora present. Mycorrhizal infection does not always increase disease resistance; sometimes mycorrhizal infection may increase disease incidence (Caron, 1989). In addition, sometimes differences between genotype of pathogens and AM fungi determine the interaction (Strobel and Sinclair, 1991). Therefore, a mycorrhizal fungus must express its protective effect under a wide range of environmental conditions, and should be aggressive against pathogens and also colonize the roots of host plants aggressively.

5 PRACTICAL CONTROL SYSTEM

Bioprotection from AM fungi-colonized plants is the outcome of complex interactions between plant, pathogen and AM fungi. Various mechanisms are proposed for conferring bioprotection, but generally effective bioprotection is a cumulative result of all mechanisms working either separately or together. The challenges to obtain biocontrol through AM fungi include the obligate nature of AM fungi, poor understanding of the mechanisms involved and the role of environmental factors in these interactions. Moreover, improved understanding of agricultural practices on AM colonization is required using new techniques like confocal laser scanning microscopy. These techniques may reveal the processes involved in root colonization and also in the biocontrol process. Furthermore, these techniques may provide new ways for increasing benefits of AM fungi by their use with other beneficial microorganisms.

The potential of AM fungi to enhance plant growth is well recognized but not exploited to the fullest extent. These organisms are rarely found in nurseries due to the use of composted soil-less media, high levels of fertilizer and regular application of fungicide drenches. The potential advantages of AM fungi in horticulture, agriculture, and forestry are not perceived by these industries as significant. This may be due in part to inadequate methods for large-scale inoculum production. Monoxenic root-organ in vitro culture methods for AMF inocula production have also been attempted by various workers (Mohammad and Khan, 2002; Fortin et al., 2002) but these techniques, although useful for the study of physiological, biochemical, and genetic relationships, have limitations in terms of producing inocula of AM fungi for commercial purposes. Pot cultures in pasteurized soils have been the most widely used method for producing AM fungi inocula but are time-consuming, bulky, and often not pathogen-free. To overcome these difficulties, soil-free methods such as soil-less growth media, aeroponics, hydroponics and axenic cultures of AM fungi have been used successfully to produce AMF-colonized root inocula (Sylvia and Jarstfer, 1994a, b; Mohammad and Khan, 2002). Substrate-free colonized roots produced by these methods can be sheared and used for large-scale inoculation purposes.

Although AMF are ubiquitous, natural associations of AM fungi are not efficient in increasing plant growth (Fitter, 1985). Cropping sequences, fertilization, and plant pathogen management practices affect both AM fungi propagules in soil and their effects on plants (Bethlenfalvay and Linderman, 1992). The propagation system used for horticultural fruit and micropropagated plants can benefit most from AM biotechnology. Micropropagated plants can withstand transplant stress from *in vitro* to *in vivo* systems if they are inoculated with appropriate AM fungi (Lovato et al., 1996; Azcon-Aguilar et al., 2002). In order to use AM fungi in sustainable agriculture, knowledge of factors such as fertilizer inputs, pesticide use, and soil management practices which influence AM fungi is essential (Bethlenfalvay and Linderman, 1992; Allen, 1991, 1992). In addition efficient inoculants should be identified and used as biofertilizers, bioprotectants, and biostimulants for sustainable agriculture.

In general, a single biocontrol agent is used for biocontrol of plant disease against a single pathogen (Wilson and Backman, 1999). This protocol may account for the inconsistent performance by the biocontrol agent, because a single agent is not active in all soil environments or against all pathogens that attack the host plant. On the other hand, mixtures of biocontrol agents with different plant colonization patterns may be useful for the biocontrol of different plant pathogens via different mechanisms of disease suppression. Moreover, mixtures of biocontrol agents with taxonomically different organisms that require different optimum temperature, pH, and moisture conditions may colonize roots more aggressively, and improve plant growth and efficacy of biocontrol. Greater suppression and enhanced consistency against pathogens was observed using mixtures of biocontrol agents (Akhtar and Siddiqui, 2007a). Consortia of biocontrol agents may better adapt to the environmental changes that occur throughout the growing season and protect against a broader range of pathogens. Mixtures of micro-organisms increase the genetic diversity of biocontrol systems that may persist longer in the rhizosphere and utilize a wider array of biocontrol mechanisms (Pierson and Weller, 1994). Multiple organisms may enhance the level and consistency of biocontrol via a more stable rhizosphere community and effectiveness over a wide range of environmental conditions.

AM fungi, root nodule bacteria, plant growth promoting rhizobacteria, antagonistic fungi and their use with composted manure may provide protection at different times, under different conditions, and occupy different or complementary niches (Siddiqui and Akhtar, 2008). These mixtures may coexist without exhibiting adverse effects on each other (Akhtar and Siddiqui, 2007a, b) and suitable combinations of these biocontrol agents may increase plant growth and resistance to pathogens. More detailed investigations of the relationships in various pathosystems and interactions between these microorganisms and the host plant are needed for developing suitable biocontrol of plant diseases.

REFERENCES

Abbott, L. K., and Robson, A. D., 1984, The effect of VA mycorrhizae on plant growth, In: *VA Mycorrhiza*, eds., Powell, C. L., and Bagyaraj, D. J., CRC, Boca Raton, FL, pp. 113–130.

Abdel-Fattah, G. M., and Shabana, Y. M., 2002, Efficacy of the arbuscular mycorrhizal fungus *Glomus clarum* in protection of cowpea plants against root-rot pathogen *Rhizoctonia solani*. *J. Plant Dis. Protec.* **109**: 207–215.

Agrios, G. N., 2005, *Plant Pathology*, 5th edn., Elsevier-Academic, San Diego, CA, pp. 922.

Akhtar, M. S., and Siddiqui, Z. A., 2006, Effects of phosphate solubilizing microorganisms on the growth and root-rot disease complex of chickpea. *Mikol. Fitopatol.* **40**: 246–254.

Akhtar, M. S., and Siddiqui, Z. A., 2007a, Effects of *Glomus fasciculatum* and *Rhizobium* sp. on the growth and root-rot disease complex of chickpea. *Arch. Phytopathol. Plant Protec.* **40**: 37–43.

Akhtar, M. S., and Siddiqui, Z. A., 2007b, Biocontrol of a chickpea root-rot disease complex with *Glomus intraradices*, *Pseudomonas putida* and *Paenibacillus polymyxa*. *Australas. Plant Pathol.* **36**: 175–180.

Akhtar, M. S., and Siddiqui, Z. A., 2008a, Biocontrol of a root-rot disease complex of chickpea by *Glomus intraradices*, *Rhizobium* sp. and *Pseudomonas straita*. *Crop Protec.* **27**: 410–417.

Akhtar, M. S., and Siddiqui, Z. A., 2008b, *Glomus intraradices*, *Pseudomonas alcaligenes*, *Bacillus pumilus* as effective biocontrol agents for the root-rot disease complex of chickpea (*Cicer arietinum* L.). *J. Gen. Plant Pathol.* **74**: 53–60.

Akkopru, A., and Demir, S., 2005, Biocontrol of Fusarium wilt in tomato caused by *Fusarium oxysporum* f. sp. *lycopersici* by AMF *Glomus intraradices* and some rhizobacteria. *J. Pyhtopathol.* **153**: 544–550.

Allen, M. F., 1991, *The Ecology of Mycorrhizae*. Cambridge University Press, Cambridge, pp. 184.

Allen, M. F., 1992, *Mycorrhizal Functioning: An Integrative Plant-Fungal Process*. Chapman Hall/Routledge, New York, pp. 534.

Al-Momany, A., and Al-Raddad, A., 1988, Effect of vesicular-arbuscular mycorrhizae on Fusarium wilt of tomato and pepper. *Alexand. J. Agric. Res.* **33**: 249–261.

Al-Raddad, A. M., 1995, Interaction of *Glomus mosseae* and *Paecilomyces lilacinus* on *Meloidogyne javanica* on tomato. *Mycorrhiza* **5**: 233–236.

Anandaraj, M., Ramana, K. V., and Sharma, Y. R., 1990, Interaction between vesicular-arbuscular mycorrhizal fungi and *Meloidogyne incognita* in black peeper. In: *Mycorrhizal Symbiosis and Plant Growth*, eds., Bagyaraj, D. J., and Manjunath, A., *Proc. Sec. Nat. Conf. on Mycorrhiza*, 21–23 November, Banglore, India, pp. 110–112.

Arihawa, J., and Karasawa, T., 2000, Effect of previous crops on arbuscular mycorrhizal formation and growth of succeeding maize. *Soil Sci. Plant Nutrit.* **46**: 43–51.

Atilano, R. A., Menge, J. A., and Van Gundy, S. D., 1981, Interaction between *Meloidogyne arenaria* and *Glomus fasciculatum* in grape. *J. Nematol.* **13**: 52–57.

Azcon-Aguilar, C., and Bago, B., 1994, Physiological characteristics of host plant promoting an undistributed functioning of the mycorrhizal symbiosis. In: *Impact of Arbuscular Mycorrhizas on Sustainable Agriculture and Natural Ecosystems*, eds., Gianinazzi, S., and Schuepp, H., Birkhauser, Basel, Switzerland, pp. 47–60.

Azcon-Aguilar, C., and Barea, J. M., 1996, Arbuscular mycorrhizas and biological control of soil borne plant pathogens-an overview of the mechanisms involved. *Mycorrhiza* **6**: 457–464.

Azcón-Aguilar, C., Jaizme-Vega, M. C., and Calvet, C., 2002, The contribution of arbuscular mycorrhizal fungi to the control of soilborne plant pathogens. In: *Mycorrhizal Technology in Agriculture: From Genes to Bioproducts*, eds., Gianinazzi, S., Schuepp, H., Haselwandter, K., and Barea, J. M., ALS Birkhauser Verlag, Basel, Switzerland, pp. 187–197.

Bagyaraj, D. J., 1984, Biological interactions with VA mycorrhizal fungi. In: *VA Mycorrhiza*, eds., Powell, C. L., and Bagyaraj, D. J., CRC, Boca Raton, FL, pp. 131–153, 234.

Bagyaraj, D. J., 1991, Ecology of vesicular arbuscular mycorrhizae. In: *Hand Book of Applied Mycology Vol. I*, eds., Arora, D. K., Rai, B., Mukerji, K. G., and Knudsen, G. R., Marcel Dekker, New York, pp. 3–34.

Bagyaraj, D. J., Manjunath, A., and Reddy, D. D. R., 1979, Interaction of vesicular arbuscular mycorrhizas with root knot nematodes in tomato. *Plant Soil* **51**: 397–403.

Baker, K. F., and Cook, R. J., 1982, *Biocontrol of Pathogens*. The American Phytopathological Society, St. Paul, MN.

Balestrini, R., Romera, C., Puigdomenech, P., and Bonfonte, P., 1994, Location of a cell-wall hydroxyproline-rich glycoprotein, cellulose and β-1,3 glucans in apical and differentiated regions of maize mycorrhizal roots. *Planta* **195**: 201–209.

Baltruschat, H., and Dehne, H. W., 1988, The occurrence of vesicular-arbuscular mycorrhiza in agro-ecosystems. I. Influence of nitrogen fertilizer and green manure in continuous monoculture and in crop rotation on the inoculum potential of winter wheat. *Plant Soil* **107**: 279–284.

Baltruschat, H., and Schonbeck, F., 1975, Studies on the influence of endotrophic mycorrhiza on the infection of tobacco by *Thielaviopsis basicola*. *Phytopath. Z.* **84**: 172–188.

Bansal, M., Chamola, B. P., Sarwar, N., and Mukerji, K. G., 2000, Mycorrhizospher: Interaction between rhizosphere microflora and VAM fungi. In: *Mycorrrhizal Biology*, eds. Mukerji, K. G., Chamola, B. P., and Singh, J., Kluwer Academic/Plenum, New York, pp. 143–152.

Barea, J. M., Pozo, M. J., Azcon, R., and Azcon-Aguilar, C., 2005, Microbial co-operation in the rhizosphere. *J. Exp. Bot.* **56**: 1761–1778.
Becker, W. N., 1976, Quantification of onion vesicular-arbuscular mycorrhizae and their resistance to *Pyrenochaeta terrestris*. Ph.D. dissertataion, University of Illinois, Urbana, IL.
Benhamou, N., Fortin, J. A., Hamel, C., St-Arnould, M., and Shatilla, A., 1994, Resistance responses of mycorrhizal Ri T-DNA-transformed carrot roots to infection by *Fusarium oxysporum* f. sp. *chrysanthemi*. *Phytopathology* **84**: 958–968.
Berta, G., Fusconi, A., and Trotta, A., 1993, VA mycorrhizal infection and the morphology and function of root systems. *Environ. Exp. Bot.* **33**: 159–173.
Berta, G., Sampo, S., Gamalero, E., Musasa, N., and Lemanceau, P., 2005, Suppression of Rhizoctonia root-rot of tomato by *Glomus mosseae* BEG 12 and *Pseudomonas fluorescens* A6RI is associated with their effect on the pathogen growth and on the root morphogenesis. *Eur. J. Plant Pathol.* **111**: 279–288.
Bethlenfalvay, G. J., and Linderman, R. G., 1992, Mycorrhizae and crop productivity. In: *Mycorrhizae in Sustainable Agriculture*, eds., Bethlenfalvay, G. J., and Linderman, R. G., Amer. Soc. Agr., Spec. Pub. No. 54, Madison, WI, pp. 1–27.
Bethlenfalvay, G. J., and Schuepp, H., 1994, Arbuscular mycorrhizas and agrosystem stability. In: *Impact of Arbuscular Mycorrhizas on Sustainable Agriculture and Natural Ecosystems*, eds., Gianinazzi, S., and Schüepp, H., Birkhäuser Verlag, Basel, Switzerland, pp. 117–131.
Bhagawati, B., Goswami, B. K., and Singh, S., 2000, Management of disease complex of tomato caused by *Meloidogyne incognita* and *Fusarium oxysporum* f. sp. *lycopersici* through bioagent. *Indian J. Nematol.* **30**: 16–22.
Bhat, M. S., and Mahmood, I., 2000, Role of *Glomus mosseae* and *Paecilomyces lilacinus* in the management of root knot nematode on tomato. *Arch. Phytopathol. Plant Protec.* **33**: 131–140.
Boby, V. U., and Bagyaraj, D. J., 2003, Biological control of root-rot of *Coleus forskohlii* Briq. using microbial inoculatnts. *World J. Microbiol. Biotechnol.* **19**: 175–180.
Bodker, L., Kjoller, R., and Rosendahl, S., 1998, Effect of phosphate and arbuscular mycorrhizal fungus *Glomus intraradices* on disease severity of root rot of peas (*Pisum sativum*) caused by *Aphanomyces euteiches*. *Mycorrhiza* **8**: 169–174.
Bonfonte-Fasolo, P., and Spanu, P., 1992, Pathogenic and endomycorrhizal associations. In: *Methods and Microbiology, Vol., 24: Techniques for the Study of Mycorrhiza*, eds., Norris, J. R., Read, D. J., and Verma, A. K., Academic, London, pp. 142–167.
Borah, A., and Phukan, P. N., 2003, Effect of interaction of *Glomus fasciculatum* and *Meloidogyne incognita* on growth of brinjal. *Ann. Plant Protec. Sci.* **11**: 352–354.
Bowen, G. D., 1980, Misconceptions, concepts and approaches in rhizospheric biology, In: *Contemporary Microbial Ecology*, eds., Ellwood, D. C., Hedger, J. N., Lathem, M. J., Lynch, J. M., and Slater, J. H., Academic, London, pp. 283–304.
Brundrett, M. C., 2002, Coevolution of roots and mycorrhizas of land plants. *New Phytol.* **154**: 275–304.
Calvet, C., Pinochet, J., Hernandez-Dorrego, A., Estaun, V., and Camprubi, A., 2001, Field microplot performance of the peach-almond GF-677 after inoculation with arbuscular mycorrhizal fungi in a replant soil infested with root-knot nematode. *Mycorrhiza* **10**: 295–300.
Cameron, G. C., 1986, Interactions between two vesicular-arbuscular mycorrhizal fungi, the soybean cyst nematode, and phosphorus fertility on two soybean cultivars. M.S. thesis, University of Georgia, Athens.
Carling, D. E., Roncadori, R. W., and Hussey, R. S., 1989, Interactions of vesicular-arbuscular mycorrhizal fungi, root-knot nematode and phosphorus fertilization on soybean. *Plant Dis.* **73**: 730–733.

Carling, D. E., Roncadori, R. W., and Hussey, R. S., 1996, Interaction of arbuscular mycorrhizae, *Meloidogyne arenaria*, and phosphorus fertilization on peanut. *Mycorrhiza* **6**: 9–13.

Caron, M., 1989, Potential use of mycorrhizae in control of soilborne diseases. *Can. J. Plant Pathol.* **11**: 177–179.

Caron, M., Fortin, J. A., and Richard, C., 1985, Influence of substrate on the interaction of *Glomus intraradices* and *Fusarium oxysporum* f. sp. *radicis-lycopersici* on tomatoes. *Plant Soil* **87**: 233–239.

Caron, M., Fortin, J. A., and Richard, C., 1986a, Effect of phosphorus concentration and *Glomus intraradices* on Fusarium crown and root-rot of tomatoes. *Phytopathology* **76**: 942–946.

Caron, M., Fortin, J. A., and Richard, C., 1986b, Effect of *Glomus intraradices* on the infection by *Fusarium oxysporum* f. sp. *radicis-lycopersici* in tomatoes over a 12-week period. *Can. J. Bot.* **64**: 552–556.

Caron, M., Fortin, J. A., and Richard, C., 1986c, Effect of preinfection of the soil by a vesicular-arbuscular mycorrhizal fungus, *Glomus intraradices* on Fusarium crown and root-rot of tomatoes. *Phytoprotec.* **67**: 15–19.

Chandanie, W. A., Kubota, M., and Hyakumachi, M., 2006, Interactions between plant growth promoting fungi and arbuscular mycorrhizal fungus *Glomus mosseae* and induction of systemic resistance to anthracnose disease in cucumber. *Plant Soil* **286**: 209–217.

Clark, R. B., and Zeto, S. K., 2000, Mineral acquisition by arbuscular mycorrhizal plants. *J. Plant Nutr.* **23**: 867–902.

Cook, R. J., and Baker, K. F., 1982, *The Nature and Practice of Biological Control of Plant Pathogens*. APS, St. Paul, MN.

Cordier C., Gianinazzi, S., and Gianinazzi-Pearson, V., 1996, Colonisation patterns of root tissues by *Phytophthora nicotianae* var *parasitica* related to reduced disease in mycorrhizal tomato. *Plant Soil* **185**: 223–232.

Cordier, C., Pozo, M. J., Gianinazzi, S., and Gianinazzi-Pearson, V., 1998, Cell defence responses associated with localised and systemic resistance to *Phytophthora parasitica* induced in tomato by an arbuscular mycorrhizal fungus. *Mol. Plant Microbe Interc.* **11**: 1017–1028.

Curl, E. A., and Truelove, B., 1986, *The Rhizosphere*. Springer, New York, pp. 288.

Datnoff, L. E., Nemec, S., and Pernezny, K., 1995, Biological control of Fusarium crown and root-rot of tomato in Florida using *Trichoderma harzianum* and *Glomus intraradices*. *Biocontr.* **5**: 427–431.

Davies, R. M., and Menge J. A., 1980, Influence of *Glomus fasciculatus* and soil phosphorus on Phytophthora root-rot of citrus. *Phytopathology* **70**: 447–452.

Davis, R. M., 1980, Influence of *Glomus fasciculatus* on *Thielaviopsis basicola* root rot of citrus. *Phytopathology* **70**: 447–452.

Declerck, S., Risede, J. M., Rufyikiri, G., and Delvaux, B., 2002, Effects of arbuscular mycorrhizal fungi on the severity of root rot of bananas caused by *Cylindrocladium spathiphylli*. *Plant Pathol.* **51**: 109–115.

Dehne, H. W., 1982, Interactions between vesicular-arbuscular mycorrhizal fungi and plant pathogens. *Phytopathology* **72**: 1115–1119.

Dehne, H. W., and Schönbeck, F., 1979, Untersuchungen zum einfluss der endotrophen Mycorrhiza auf Pflanzenkrankheiten: II. Phenolstoffwechsel und lignifizierung. *Phytopath. Z.* **95**: 210–216.

Dehne, H. W., Schönbeck, F., and Baltruschat, H., 1978, Untersuchungen zum einfluss der endotrophen Mycorrhiza auf Pflanzenkrankheiten: 3. Chitinase-aktivitat und ornithinzyklus (The influence of endotrophic mycorrhiza on plant diseases: 3 chitinase-activity and ornithinecycle). *J. Plant Dis. Protec.* **85**: 666–678.

Devi, T. P., and Goswami, B. K., 1992, Effect of VA-mycorrhiza on the disease incidence due to *Macrophomina phaseolina* and *Meloidogyne incognita* on cowpea. *Ann. Agric. Res.* **13**: 253–256.
Diederichs, C., 1987, Interaction between five endomycorrhizal fungi and root-knot nematode *Meloidogyne javanica* on chickpea under tropical conditions. *Trop. Agric.* **64**: 353–355.
Diedhiou, P. M., Hallman, J., Oerke, E. C., and Dehne, H. W., 2003, Effects of arbuscular mycorrhizal fungi and non-pathogenic *Fusarium oxysporum* on *Meloidogyne incognita* infestation of tomato. *Mycorrhiza* **13**: 199–204.
Douds, D. D., Galvez, L., Franke-Snyder, M., Reider, C., and Drinkwater, L. E., 1997, Effect of compost addition and crop rotation point upon VAM fungi. *Agric. Ecos. Environ.* **65**: 257–266.
Dumas-Gaudot, E., Furlan, V., Grenier, J., and Asselin, A., 1992a, Chitinase, Chitosanase and β-1,3- glucanase activities in *Allium* and *Pisum* roots colonized by *Glomus* species. *Plant Sci.* **84**: 17–24.
Dumas-Gaudot, E., Furlan, V., Grenier, J., and Asselin, A., 1992b, New acidic chitinase isoforms induced in tobacco roots by vesicular-arbuscular mycorrhizal fungi. *Mycorrhiza* **1**: 133–136.
Duponnois, R., and Cadet, P., 1994, Interactions of *Meloidogyne javaniva* and *Glomus* sp. on growth and N_2 fixation of *Acacia seyal*. *Afro-Asian J. Nematol.* **4**: 228–233.
Elsayed Abdalla, M., and Abdel-Fattah, G. M., 2000, Influence of endomycorrhizal fungus *Glomus mosseae* on the development of peanut pod rot disease in Egypt. *Mycorrhiza* **10**: 29–35.
Elsen, A., Declerck, S., and De Wasele, D., 2002, Effects of three arbuscular mycorrhizal fungi on root knot nematode (*Meloidogyne* spp.) infection of *Musa*. *Infomusa* **11**: 21–23.
Fassuliotis, G., 1970, Resistance of *Cucumis* spp. to root-knot nematode. *Meloidogyne acrita*. *J. Nematol.* **2**: 174.
Fitter, A. H., 1985, Functioning of vesicular- arbuscular mycorrhizas under field conditions. *New Phytol.* **99**: 257–265.
Fitter, A. H., and Garbaye, J., 1994, Interactions between mycorrhizal fungi and other soil organisms. *Plant Soil* **159**: 123–132.
Fortin, J. A., Bécard, G., Declerck, S., Dalpé, Y., St-Arnaud, M., Coughlan, A. P., Piché, Y., 2002, Arbuscular mycorrhiza on root-organ cultures. *Can. J. Bot.* **80**: 1–20.
Francl, L. J., and Dropkin, V. H., 1985, *Glomus fasciculatum*, a week pathogens of *Heterodera glycines*. *J. Nematol.* **17**: 470–475.
Garcia-Chapa, M., Batlle, A., Lavina, A., Camprubi, A., Estaun, V., and Calvet, C., 2004, Tolerance increase to pear decline phytoplasma in mycorrhizal OHF-333 pear root stock. *XIX Int. Symp. on Virus and Virus Like Diseases of Temperate Fruit Crops-Fruit Tree Diseases*, ed., Llacer, G., Valencia, Spain.
Garcia-Garrido, J. M., and Ocampo, J. A., 1989, Effect of VA mycorrhizal infection of tomato on damage caused by *Pseudomonas syringae*. *Soil Biol. Biochem.* **21**: 163–167.
Gianinazzi-Pearson, V., Gollotte, A., Dumas-Gaudot, E., Franken, P., and Gianinazzi, S., 1994, Gene expression and molecular modifications associated with plant responses to infection by arbuscular mycorrhizal fungi. In: *Advances in Molecular Genetics of Plant-Microbes Interactions*, eds., Daniels, M., Downic, J. A., and Osbourn, A. E., Kluwer, Dordrecht, The Netherlands, pp. 179–186.
Gianinazzi-Pearson, V., Gollotte, A., Lherminier, J., Tisserant, B., Franken, P., Dumas-Gaudot, E., Lemoine, M. C., van Tuinen, D., and Gianinazzi, S., 1995, Cellular and molecular approaches in the characterization of symbiotic events in functional arbuscular associations. *Can. J. Bot.* **73**: S526–S532.
Gianinazzi-Pearson, V., Dumas-Gaudot, E., Gollotte, A., Tahiri-Al-aoui, A., and Gianinazzi, S., 1996, Cellular and molecular defense-related root responses to invasion by arbuscular mycorrhizal fungi. *New Phytol.* **133**: 45–57.

Goodman, R. N., Kiraly, Z., and Zaitlin, M., 1967, The biochemistry and physiology of infections. In: *Plant Disease*. Van Nostrand, Princeton, NJ.

Graham, J. H., and Egel, D. S., 1988, Phytophthora root rot development on mycorrhizal and phosphorous fertilized non-mycorrhizal sweet orange seedlings. *Plant Dis.* **72**: 611–614.

Graham, J. H., and Menge, J. A., 1982, Influence of vesicular-arbuscular mycorrhizal fungi and soil phosporus on take-all disease of wheat. *Phytopathology* **72**: 95–96.

Graham, J. H., Leonard, R. T., and Menge, J. A., 1981, Membrane-mediated decreases in root exudation responsible for phosphorus inhibition of vesicular-arbuscular mycorrhiza formation. *Plant Physiol.* **68**: 548–552.

Gryndler, M., Lestina, J., Moravec, V., Prikyl, Z., and Lipavsky, J., 1990, Colonization of maize roots by VAM under conditions of long-term fertilization of varying intensities. *Agric. Ecos. Environ.* **29**: 183–186.

Guenoune, D., Galili, S., Phillips, D. A., Volpin, H., Chet, I., Okon, Y., and Kapulnik, Y., 2001, The defense response elicited by the pathogen *Rhizoctonia solani* is suppressed by colonization of the AM fungus *Glomus intraradices*. *Plant Sci.* **160**: 925–932.

Hao, Z., Christie, P., Qin, L., Wang, C., and Li, X., 2005, Control of Fusarium Wilt of cucumber seedlings by inoculation with an arbuscular mycorrhical fungus. *J. Plant Nutr.* **28**: 1961–1974.

Harley, J. L., and Smith, S. E., 1983, *Mycorrhizal symbiosis*, Academic, New York.

Harrier, L. A., and Watson, C. A., 2004, The potential role of arbuscular mycorrhizal (AM) fungi in the bioprotection of plants against soil-borne pathogens in organic and/or other sustainable farming systems. *Pest Manag. Sci.* **60**: 149–157.

Hayman, D. S., 1978, Endomycorrhizas. In: *Interactions Between Non-pathogenic Soil Microorganisms and Plants*, eds. Dommergues, Y. R., and Kurupa, S. V., Elsevier, Amsterdam.

Heald, C. M., Bruton, B. D., and Davis, R. M., 1989, Influence of *Glomus intraradices* and soil phosphorus on *Meloidogyne incognita* infecting *Cucumis melo*. *J. Nematol.* **21**: 69–73.

Hetrick, B. A. D., Wilson, G. W. T., and Todd, T. C., 1996, Mycorrhizal response to wheat cultivars: relationship to phosphorus. *Can. J. Bot.* **74**: 19–25.

Hock, B., and Verma, A., 1995, *Mycorrhiza Structure, Function, Molecular Biology and Biotechnology*, Springer, Berlin.

Hussey, R. S., and Roncadori, R. W., 1978, Interacton of *Pratylenchus brachyurus* and *Gigaspora margarita* on cotton. *J. Nematol.* **10**: 16–20.

Hussey, R. S., and Roncadori, R. W., 1982, Vesicular-arbuscular mycorrhizae may limit nematode activity and improve plant growth. *Plant Dis.* **66**: 9–14.

Hwang, S. F., Chang, K. F., and Chakravarty, P., 1992, Effect of vesicular-arbuscular fungi on the development of Verticillium and Fusarium wilt of alfalfa. *Plant Dis.* **76**: 239–243.

Jain, C., and Trivedi, P. C., 2003, Effect of vesicular arbuscular mycorrhiza (VAM) on rootknot infested *Cicer arietinum*. *Proc. Nati. Symp. Biodiv. Manag. Nemat. Crop. Syst. Sustain. Agric.*, Jaipur, India.

Jain, R. K., and Sethi, C. L., 1987, Pathogenicity of *Heterodera cajani* on cowpea as influenced by the presence of VAM fungi, *Glomus fasciculatum* or *G. epigaeus*. *Indian J. Nematol.* **17**: 165–170.

Jain, R. K., and Sethi, C. L., 1988, Influence of endomycorrhizal fungi *Glomus fasciculatum* and *G. epigaeus* on penetration and development of *Heterodera cajani* on cowpea. *Indian J. Nematol.* **18**: 89–93.

Jain, R. K., Hasan, N., Singh, R. K., and Pandey, P. N., 1998, Influence of the endomycorrhizal fungus, *Glomus fasciculatum* on *Meloidogyne incognita* and *Tylenchorhynchus vulgaris* infecting barseem. *Indian J. Nematol.* **28**: 48–51.

Jaizme-Vega, M. C., and Pinochet, J., 1997, Growth response of banana to three mycorrhizal fungi in *Pratylenchus goodeyi* infested soil. *Nematropica* **27**: 69–76.

Jaizme-Vega, M. C., Tenoury, P., Pinochet, J., and Jaumot, M., 1997, Interactions between the root-knot nematode *Meloidogyne incognita* and *Glomus mosseae* in banana. *Plant Soil* **196**: 27–35.
Jaizme-Vega, M. C., Rodriguez-Romero, A., and Nunez, A. B. L., 2006, Effect of the combined inoculation of arbuscular mycorrhizal fungi and plant growth promoting rhizobacteria on papaya (*Carica papaya* L.) infected with root-knot nematode *Meloidogyne incognita. Fruits* **61**: 151–162.
Jayaram, J., and Kumar, D., 1995, Influence of mungbean yellow mosaic virus on mycorrhizal fungi associated with *Vigna radiata* var. PS 16. *Indian Phytopathol.* **48**: 108–110.
Jothi, G., and Sundarababu, R., 2000, Interaction of four *Glomus* spp. with *Meloidogyne incognita* on brinjal (*Solanum melongena* L.). *Int. J. Trop. Plant Dis.* **18**: 147–156.
Jothi, G., and Sundarababu, R., 2001, Management of root knot nematode in brinjal by using VAM and crop rotation with green gram and pearl millet. *J. Biol. Contr.* **15**: 77–80.
Jothi, G., and Sundarababu, R., 2002, Nursery management of *Meloidogyne incognita* by *Glomus mossae* in eggplant. *Nematol. Medit.* **30**: 153–154.
Jothi, G., Mani, M. P., and Sundarababu, R., 2000, Management of *Meloidogyne incognita* on okra by integrating non-host and endomycorrhiza. *Curr. Nematol.* **11**: 25–28.
Kantharaju, V., Krishnappa, K., Ravichardra, N. G., and Karuna, K., 2005, Management of root-knot nematode, *Meloidogyne incognita* on tomato by using indigenous isolates of AM fungus, *Glomus fasciculatum. Indian J. Nematol.* **35**: 32–36.
Kaplan, D. T., Keen, N. T., and Thompson, I. J., 1980, Association of glyceollin with incompatible response of soybean roots to *Meloidogyne incognita. Physiol. Plant Pathol.* **16**: 309–318.
Karlen, D. L., Wollenhaupt, N. C., Erbach, D. C., Berry, E. C., Swan, J. B., Eash, N. S., and Jordahl, J. L., 1994, Crop residue effects on soil quality following 10-years of no-till corn. *Soil Till. Res.* **31**: 149–167.
Kellam, M. K., and Schenck, N. C., 1980, Interaction between vesicular-arbuscular mycorrhizal fungus and root-knot nematode on soybean. *Phytopathology* **70**: 293–296.
Kjoller, R., and Rosendahl, S., 1997, The presence of arbuscular mycorrhizal fungus *Glomus intraradices* influences enzymatic activities of the root pathogen *Aphanomyces euteiches* in pea roots. *Mycorrhiza* **6**: 487–491.
Kotcon, J. B., Bird, G. W., Rose, L. M., and Dimoff, K., 1985, Influence of *Glomus fasciculatum* and *Meloidogyne hapla* on *Allium cepa* in organic soils. *J. Nematol.* **17**: 55–60.
Krishna, K. R., and Bagyaraj, D. J., 1986, Phenoilcs of mycorrhizal and uninfected groundnut var. MGS-7. *Curr. Res.* **15**: 51–52.
Labeena, P., Sreenivasa, M. N., and Lingaraju, S., 2002, Interaction effects between arbuscular mycorrhizal fungi and root-knot nematode *Meloidogyne incognita* on tomato. *Indian J. Nematol.* **32**: 118–120.
Lambais, M. R., and Mehdy, M. C., 1993, Suppression of endochitinase β-1,3-endoglucanase, and chalcone isomerase expression in bean VAM roots under different soil phosphate conditions. *Mol. Plant Microbe Inter.* **1**: 75–83.
Linderman, R. G., 1985, Microbial interaction in the mycorrhizosphere. In: *Proc. 6th N. Am. Conf. on Mycorrhizae*, ed. Molina, R., pp. 117–120.
Linderman, R. G., 1992, VA mycorrhizae and soil microbial interactions. In: *Mycorrhizae in Sustainable Agriculture*, eds., Bethelenfalvay, G. J., and Linderman, R. G., ASA Special Publication No. 54, Madison, WI, pp. 45–70.
Linderman, R. G., 1994, Role of VAM fungi in biocontrol. In: *Mycorrhizae and Plant Health*, eds., Pfleger, F. L., and Linderman, R. G., APS, St. Paul, MN, pp. 1–26.
Lingua, G., D'Agostino, G., Massa, N., Antosiano, M., and Berta, G., 2002, Mycorrhiza-induced differential response to a yellow disease in tomato. *Mycorrhiza* **12**: 191–198.

Liu, A., Hamel, C., Hamilton, R. I., and Smith, D. L., 2000, Mycorrhizae formation and nutrient uptake of new corn (*Zea mays* L.) hybrids with extreme canopy and leaf architecture as influenced by soil N and P levels. *Plant Soil* **221**: 157–166.

Liu, R. J., 1995, Effect of vesicular-arbuscular mycorrhizal fungi on Verticillium wilt of cotton. *Mycorrhiza* **5**: 293–297.

Lovato, P. E., Gianinazzi-Pearson, V., Trouvelot, A., Gianinazzi, S., 1996, The state of art mycorrhizas micropropagation. *Adv. Agric. Sci.* **10**: 46–52.

MacGuidwin, A. E., Bird, G. W., and Safir, G. R., 1985, Influence of *Glomus fasciculatum* on *Meloidogyne hapla* infecting *Allium cepa*. *J. Nematol.* **17**: 389–395.

Marschner, P., Crowley, D. E., and Lieberei, R., 2001, Arbuscular mycorrhizal infection changes the bacterial 16s rDNA community composition in the rhizosphere of maize. *Mycorrhiza* **11**: 297–302.

Masadeh, B., von Alten, H., Grunewaldt-Stoecker, G., and Sikora, R. A., 2004, Biocontrol of root knot nematodes using the arbuscular mycorrhizal fungus *Glomus intraradices* and the antagonistic *Trichoderma viridae* in two tomato cultivars differing in their suitability as hosts for the nematodes. *J. Plant Dis. Protec.* **111**: 322–333.

McGonigle, T. P., and Miller, M. H., 2000, The inconsistent effect of soil disturbance on colonisation of roots by arbuscular mycorrhizal fungi: a test of the inoculum density hypothesis. *Appl. Soil Ecol.* **14**: 147–153.

Mengel, K., and Kirkby, E. A., 1979, *Principles of Plant Nutrition*, International Potash Institute, Worbaufen-Bern, Switzerland. p593.

Meyer, J. R., and Linderman, R. G., 1986, Selective influences on populations of rhizosphere or rhizoplane bacteria and actinomycetes by mycorrhizas formed by *Glomus fasciculatum*. *Soil Biol. Biochem.* **18**: 191–196.

MingQin, G., Yu, C., and FengZhen, W., 2004, Resistance of the AM fungus *Eucalyptus* seedlings against *Pseudomonas solanacearum*. *Forest Res., Beijing* **17**: 441–446.

Mishra, A., and Shukla, B. N., 1996, Interaction between *Glomus fasciculatum*, *Meloidogyne incognita* and fungicide in tomato. *Indian J. Mycol. Plant Pathol.* **26**: 38–44.

Mishra, A., and Shukla, B. N., 1997, Interaction between *Glomus fasciculatum*, *Meloidogyne incognita* on tomato. *Indian J. Mycol. Plant Pathol.* **27**: 199–202.

Mohammad, A., and Khan, A. G., 2002, Monoxenic *in vitro* production and colonization potential of AM fungus *Glomus intraradices*. *Indian J. Exp. Bot.* **40**: 1087–1091.

Morandi, D., 1987, VA mycorrhizae, nematodes, phosphorus and phytoalexins on soybean. In: *Mycorrhizae in the Next Decade, Practical Application and Research Priorities*, eds., Sylvia, D. M., Hung, L. L., and Graham, D. H., *Proc. of 7th N. Am. Conf. on Mycorrhizae*, Institute of Food and Agriculture Sciences, University of Florida, Gaineville, GA.

Morandi, D., 1996, Occurrence of phytoalexins and phenolic compounds in endomycorrhizal interactions, and their potential role in biological control. *Plant Soil* **185**: 241–251.

Morandi, D., Baily, J. A., and Gianinazzi-Pearson, V., 1984, Isoflavonoid accumulation in soybean roots infected with vesicular-arbuscular mycorrhizal fungi. *Physiol. Plant Pathol.* **24**: 357–364.

Mosse, B., 1986, Mycorrhiza in a sustainable agriculture. *Biol. Agric. Hort.* **3**: 191–209.

Mukerji, K. G., 1999, Mycorrhiza in control of plant pathogens: molecular approaches. In: *Bio-technological Approaches in Biocontrol of Plant Pathogens*, eds., Mukerji, K. G., Chamola, B. P., and Upadhyay, R. K., Kluwer Academic/Plenum, New York, pp. 135–155.

Nagesh, M., and Reddy, P. P., 1997, Management of *Meloidogyne incognita* on *Crossandra undulaefolia* using vesicular arbuscular mycorrhiza, *Glomus mosseae*, and oil cakes. *Mycorrh. News* **9**: 12–14.

Nagesh, M., Reddy, P. P., Kumar, M. V. V., and Nagaraju, B. M., 1999a, Studies on correlation between *Glomus fasciculatum* spore density, root colonization and *Meloidogyne incognita* infection on *Lycopersicon esculentum*. *J. Plant Dis. Protect.* **106**: 82–87.

Nagesh, M., Reddy, P. P., and Rao, M. S., 1999b, Comparative efficacy of VAM fungi in combination with neem cake against *Meloidogyne incognita* on *Crossandra undulaefolia*. *Mycorrh. News* **11**: 11–13.

Nehra, S., 2004, VAM fungi and organic amendments in the management of *Meloidogyne incognita* infected ginger. *J. Indian Bot. Soc.* **83**: 90–97.

Nemec, S., and Myhre, D., 1984, Virus-*Glomus etunicatum* interactions in Citrus rootstocks. *Plant Dis.* **68**: 311–314.

O'Bannon, J. H., and Nemec, S., 1979, The response of *Citrus* lemon seedlings to a symbionts, *Glomus eutnicatus*, and a pathogen, *Radopholus similis*. *J. Nematol.* **11**: 270–275.

O'Bannon, J. H., Inserra, R. N., Nemec, S., and Vovlas, N., 1979, The influence of *Glomus mossae* on *Tylenchulus semipenetrans* infected and uninfected *Citrus* lemon seedling. *J. Nematol.* **11**: 247–250.

Ocampo, J. A., Martin, J., and Hayman, D. S., 1980, Influence on plant interactions on vesicular arbuscular mycorrhizal infections. I. Host and non-host plants grown together. *New Phytol.* **84**: 27–35.

Ozgonen, H., and Erkilic, A., 2007, Growth enhancement and Phytophthora blight (*Phytophthora capsici* L.) control by arbuscular mycorrhizal fungal inoculation in pepper. *Crop Protec.* **26**: 1682–1688.

Ozgonen, H., Bicici, M., and Erkilic, A., 1999, The effect of salicylic acid and endomycorrhizal fungus *G. intraradices* on plant development of tomato and fusarium wilt caused by *Fusarium oxysporum* f. sp. *lycopersici*. *Turk. J. Agric. For.* **25**: 25–29.

Pacovsky, R. S., Bethelenfalvay, G. J., and Paul, E. A., 1986, Comparisons between P-fertilized and mycorrhizal plants. *Crop. Sci.* **16**: 151–156.

Paget, D. K., 1975, The effect of *Cylindrocarpon* on plant growth responses to VA mycorrhizal, In: *Endomycorrhizae*, eds., Sanders, F. E., Mosse, B., and Tinker, P. B., Academic, London, pp. 593–606.

Pandey, R., 2005, Field application of bio-organics in the management of *Meloidogyne incognita* in *Mentha arvensis*. *Nematol. Medit.* **33**: 51–54.

Pandey, R., Gupta, M. L., Singh, S. B., and Kumar, S., 1999, The influence of vesicular arbuscular mycorrhizal fungi alone or in combination with *Meloidogyne incognita* on *Hyoscyamus niger* L. *Bioresour. Technol.* **69**: 275–278.

Paulitz, T. C., and Linderman, R. G., 1991, Lack of antagonism between the biocontrol agent *Gliocladium virens* and vesicular arbuscular mycorrhizal fungi. *New Phytol.* **117**: 303–308.

Pierson, E. A., and Weller, D. M., 1994, Use of mixtures of fluorescent pseudomonads to suppress take-all and improve the growth of wheat. *Phytopathology* **84**: 940–947.

Pinochet, J., Fernandez, C., Jaimez, M.De, and Tenoury, P., 1997, Micropropagated banana infected with *Meloidogyne javanica* responds to *Glomus intraradices* and phosphorus. *Hort. Sci.* **32**: 35–49.

Pozo, M. J., Dumas-Gaudot, E., Azcon-Aguilar, C., and Barea, J. M., 1998, Chitosanase and chitinase activities in tomato roots during interactions with arbuscular mycorrhizal fungi or *Phytophthora parasitica*. *J. Exp. Bot.* **49**: 1729–1739.

Pozo, M. J., Cordier, C., Dumas-Gaudot, E., Gianinazzi, S., Barea, J. M., and Azcon-Aguilar, C., 2002, Localized verses systemic effect of arbuscular mycorrhizal fungi on defence responses to *Phytophthora* infection in tomato plants. *J. Exp. Bot.* **53**: 525–534.

Pradhan, A., Ganguly, A. K., and Singh, C. S., 2003, Influence of *Glomus fasciculatum* on *Meloidogyne incognita* infected tomato. *Ann. Plant Protec. Sci.* **11**: 346–348.

Prasad, S. S. K., 1971, Effects of amino acids and plant growth substances on tomato and its root knot nematode *Meloidogyne incognita* Chitwood. M.Sc. (Agric.) thesis, University of Agricultural Sciences, Bangalore, India.

Rabatin, S. C., and Rhodes, L. H., 1982, *Acaulospora bireticulata* inside oribatid mites. *Mycologia* **74**: 859–861.

Rao, K. V., and Krishnappa, K., 1995, Integrated management of *Meloidogyne incognita*, *Fusarium oxysporum* f. sp. *ciceri* wilt disease complex in chickpea. *Int. J. Pest Manag.* **41**: 234–237.
Rao, M. S., and Gowen, S. R., 1998, Bio-management of *Meloidogyne incognita* on tomato by integrating *Glomus deserticola* and *Pasteuria penetrans. J. Plant Dis. Protec.* **105**: 49–52.
Rao, M. S., Reddy, P. P., and Mohandas, M. S., 1996, Effect of integration of *Calotropis procera* leaf and *Glomus fasciculatum* on the management of *Meloidogyne incognita* infesting tomato. *Nematol. Medit.* **24**: 59–61.
Rao, M. S., Kerry, B. R., Gowen, S. R., Bourne, J. M., and Reddy, P. P., 1997, Management of *Meloidogyne incognita* in tomato nurseries by integration of *Glomus deserticola* with *Verticillium chlamydosporium. J. Plant Dis. Protec.* **104**: 419–422.
Rao, M. S., Reddy, P. P., and Mohandas, M. S., 1998a, Bio-intensive management of *Meloidogyne incognita* on eggplant by integrating *Paecilomyces lilacinus* and *Glomus mosseae. Nematol. Medit.* **26**: 213–216.
Rao, M. S., Reddy, P. P., Mohandas, M. S., Nagesh, M., and Pankaj, 1998b, Management of root-knot nematode on eggplant by integrating endomycorrhiza (*Glomus fasciculatum*) and Castor (*Ricinus communis*) cake. *Nematol. Medit.* **26**: 217–219.
Rao, M. S., Reddy, P. P., and Nagesh, M., 1999, Bio-management of *Meloidogyne incognita* on tomato by integrating *Glomus mosseae* with *Pasteuria penetrans. Indian J. Nematol.* **29**: 171–173.
Ray, S., and Dalei, B. K., 1998, VAM for root knot nematode management and increased productivity of grain legumes in Orissa. *Indian J. Nematol.* **28**: 41–47.
Reddy, P. P., 1974, Studies on the action of amino acids on the root-knot nematode *Meloidogyne incognita*. Ph.D. thesis, University of Agricultural Sciences Banglore, India.
Reddy, P. P., Nagesh, M., Devappa, V., and Kumar, M. V. V., 1998, Management of *Meloidogyne incognita* on tomato by integrating endomycorrhiza, *Glomus mosseae* with oil cakes under nursery and field condition. *J. Plant Dis. Protec.* **105**: 53–57.
Redecker, D., Morton, J. B., and Bruns, T. D., 2000, Ancestral lineages of arbuscular mycorrhizal fungi (*Glomales*). *Mol. Phylogen. Evol.* **14**: 276–284.
Reid, C. P. P., 1984, Mycorrhizae: a root-soil interface in plant nutrition. In: *Microbial-Plant Interactions*, ASA Special Publication, Vol. 47, ed., R. L. Todd, and J. E. Giddens, pp. 29–50.
Rosendahl, S., 1985, Interactions between the vesicular-arbuscular mycorrhizal fungus *Glomus intraradices* and *Aphanomyces euteiches* root rot of peas. *Phytopathol. Z.* **114**: 31–40.
Rosendahl, C. N., and Rosendahl, S., 1990, The role of vesicular arbuscular mycorrhizal fungi in controlling damping-off and growth reduction in cucumber caused by *Pythium ultimum. Symbiosis* **9**: 363–366.
Rovira, A. D., 1985, Manipulation of the rhizosphere microflora to increase plant production. In: *Reviews of Rural Science, Vol. 6, Bioteechnology and Recombinant DNA Technology in the Annual Production Industries*, eds., Leong, R. A., Barker, J. S. F., Adams, D. B., and Hutchinson, K. J., University of England, Armidale, Australia, pp. 185–197.
Saleh, H., and Sikora, R. A., 1984, Relationship between *Glomus fasciculatum* root colonization on cotton and its effect on *Meloidogyne incognita*. *Nematologica* **30**: 230–237.
Sankaranarayanan, C., and Sundarababu, R., 1994, Interaction of *Glomus fasciculatum* with *Meloidogyne incognita* inoculated at different timings on blackgram (*Vigna mungo*). *Nematol. Medit.* **22**: 35–36.
Sankaranarayanan, C., and Sundarababu, R., 1997a, Effect of oil cakes and nematicides on the growth of blackgram (*Vigna mungo*) inoculate with VAM fungus (*Glomus fasciculatum*) and root-knot nematode (*Meloidogyne incognita*). *Indian J. Nematol.* **27**: 128–130.

Sankaranarayanan, C., and Sundarababu, R., 1997b, Role of phosphobacteria on the interaction of vesicular arbuscular mycorrhiza (*Glomus mosseae*) and root knot nematode (*Meloidogyne incognita*) on blackgram (*Vigna mungo*). *Int. J. Trop. Plant Dis.* **15**: 93–98.

Sankaranarayanan, C., and Sundarababu, R., 1998, Effect of *Rhizobium* on the interaction of vesicular arbuscular mycorrhizae and root knot nematode on blackgram (*Vigna mungo*). *Nematol. Medit.* **26**: 195–198.

Sankaranarayanan, C., and Sundarababu, R., 1999, Role of phosphorus on the interaction of vesicular-arbuscular mycorrhiza (*Glomus mosseae*) and root-knot nematode (*Meloidogyne incognita*) on blackgram. *Indian J. Nematol.* **29**: 105–108.

Santhi, A., and Sudarababu, R., 1998, Effect of chopped leaves and nematicides on the interaction of *Glomus fasciculatum* with *Meloidogyne incognita* on cowpea. *Indian J. Nematol.* **28**: 114–117.

Schellenbaum, L., Berta, G., Ravolanirina, F., Tisserant, B., Gianinazzi, S., and Fitter A. H., 1991, Infuence of endomycorrhizal infection on root morphology in micropropagated woody plant species (*Vitis vinifera* L.). *Ann. Bot.* **68**: 135–141.

Schonbeck, F., 1979, Endomycorrhiza in relation to plant disease. In: *Soil Borne Plant Pathogens*, eds., Schipper, B., and Gams, W., Academic, New York, pp. 271–280.

Schussler, A., 2005, http://www.tudarmstadt.de/fb/bio/bot/schuessler/amphylo/ amphylogeny.html (accessed 19-Oct-2005).

Schussler, A., Schwarzott, D., and Walker, C., 2001, A new fungal phylum, the Glomeromycota, phylogeny and evolution. *Mycol. Res.* **105**: 1413–1421.

Secilia, J., and Bagyaraj, D. J., 1987, Bacteria and actinomycetes associated with pot cultures of vesicular-arbuscular mycorrhizas. *Can. J. Microbiol.* **33**: 1069–1073.

Shafi, A., Mahmood, I., and Siddiqui, Z. A., 2002, Integrated management of root-knot nematode *Meloidogyne incognita* on chickpea. *Thai. J. Agric. Sci.* **35**: 273–280.

Sharma, H. K. P., and Mishra, S. D., 2003, Effect of plant growth promoter microbes on root-knot nematode *Meloidogyne incognita* on okra. *Curr. Nematol.* **14**: 57–60.

Sharma, M. P., and Adholeya, A., 2000, Sustainable management of arbuscular mycorrhizal fungi in the biocontrol of soilborne plant diseases. In: *Biocontrol Potential and Its Exploitation in Sustainable Agriculture. Vol. I: Crop Disease*, eds., Upadhaya, R. K., Mukerji, K. G., and Chamola, B. P., Kluwer Academic/Plenum, New York, pp. 117–138.

Sharma, R., and Trivedi, P. C., 1994, Interaction of root-knot nematode, *Meloidogyne incognita* and VA mycorrhizae, *Glomus fasciculatum* and *Glomus mosseae* on brinjal (*Solanum melongena* L.). *J. Indian Bot. Soc.* **73**: 221–224.

Sharma, W., and Trivedi, P. C., 1997, Concomitant effect of *Peacilomyces lilacinus* and vesicular arbuscular mycorrhizal fungi on root-knot nematode infested okra. *Ann. Plant Protec. Sci.* **5**: 70–74.

Shaul, O., Galili, S., Volpin, H., Ginzberg, I., Elad, Y., Chet, I., and Kapulnik, Y., 1999, Mycorrhiza induced changes in disease severity and PR protein expression tobacco leaves. *Mol. Plant Microbe Inter.* **12**: 1000–1007.

Shreenivasa, K. R., Krishnappa, K., and Ravichandra, N. G., 2007, Interaction effects of arbuscular mycorrhizal fungus *Glomus fasciculatum* and root-knot nematode, *Meloidogyne incognita* on growth and phosphorous uptake of tomato. *Karnat. J. Agric. Sci.*, **20**: 57–61.

Siddiqui, Z. A., and Akhtar, M. S., 2006, Biological control of root-rot disease complex of chickpea by AM fungi. *Arch. Phytopathol. Plant Protec.* **39**: 389–395.

Siddiqui, Z. A., and Akhtar, M. S., 2007, Effects of AM fungi and organic fertilizers on the reproduction of the nematode *Meloidogyne incognita* and on the growth and water loss of tomato. *Biol. Fert. Soils* **43**: 603–609.

Siddiqui, Z. A., and Akhtar, M. S., 2008, Synergistic effects of antagonistic fungi and a plant growth promoting rhizobacterium, an arbuscular mycorrhizal fungus, or composted cow manure on the populations of *Meloidogyne incognita* and growth of tomato. *Bioc. Sci. Technol.* **18**: 279–290.

Siddiqui, Z. A., and Mahmood, I., 1995a, Role of plant symbionts in nematode management: a Review. *Bioresour. Technol.* **54**: 217–226.

Siddiqui, Z. A., and Mahmood, I., 1995b, Biological control of *Heterodera cajani* and *Fusarium udum* by *Bacillus subtilis*, *Bradyrhizobium japonicum* and *Glomus fasciculatum* on pigeonpea. *Fundam. Appl. Nematol.* **18**: 559–566.

Siddiqui, Z. A., and Mahmood, I., 1995c, Some observations on the management of the wilt disease complex of pigeonpea by treatment with vesicular-arbuscular fungus and biocontrol agents for nematodes. *Bioresour. Technol.* **54**: 227–230.

Siddiqui, Z. A., and Mahmood, I., 1996, Biological control of *Heterodera cajani* and *Fusarium udum* on pigeonpea by *Glomus mosseae*, *Trichoderma harzianum* and *Verticillium chlamydosporium*. *Isr. J. Plant Sci.* **44**: 49–56.

Siddiqui, Z. A., and Mahmood, I., 1997, Interaction of *Meloidogyne javanica*, *Fusarium solani* and plant symbionts on chickpea. *Thai. J. Agric. Sci.* **30**: 379–388.

Siddiqui, Z. A., and Mahmood, I., 1998, Effect of a plant growth promoting bacterium, an AM fungus and soil types on the morphometrics and reproduction of *Meloidogyne javanica* on tomato. *Appl. Soil Ecol.* **8**: 77–84.

Siddiqui, Z. A., and Mahmood, I., 2000, Effects of *Bacillus subtilis*, *Glomus mosseae* and ammonium sulphate on the development of *Meloidogyne javanica* and on growth of tomato. *Thai J. Agri. Sci.* **33**: 29–35.

Siddiqui, Z. A., and Mahmood, I., 2001, Effects of rhizobacteria and root symbionts on the reproduction of *Meloidogyne javanica* and growth of chickpea. *Bioresour. Technol.* **79**: 41–45.

Siddiqui, Z. A., and Singh L. P., 2004, Effects of soil inoculants on the growth, transpiration and wilt disease of chickpea. *J. Plant Dis. Protec.* **111**: 151–157.

Siddiqui, Z. A., and Singh, L. P., 2005, Effects of fly ash and soil micro-organisms on plant growth, photosynthetic pigments and leaf blight of wheat. *J. Plant Dis. Protec.* **112**: 146–155.

Siddiqui, Z. A., Mahmood, I., and Hayat, S., 1998, Biocontrol of *Heterodera cajani* and *Fusarium udum* on pigeonpea using *Glomus mosseae*, *Paecilomyces lilacinus* and *Pseudomonas fluorescens*. *Thai J. Agri. Sci.* **31**: 310–321.

Siddiqui, Z. A., Mahmood, I., and Khan, M. W., 1999, VAM fungi as prospective biocontrol agents for plant parasitic nematodes. In: *Modern Approaches and Innovations in Soil Management*, eds., Bagyaraj, D. J., Verma, A., Khanna, K. K., and Kehri, H. K., Rastogi, Meerut, India, pp. 47–58.

Siddiqui, Z. A., Mahmood, I., and Hayat, S., 2000, Influence of plant symbionts and potassium fertilizer on *Heterodera cajani*, crop growth and yield of pigeon pea under field condition. *Indian J. Bot. Soc.* **79**: 109–114.

Sikora, R. A., 1978, Einfluss der endotrophen mykorrhiza (*Glomus mosseae*) auf das wirt-parasit-verhaltnis *Von Meloidogyne incognita* in tomaten. *Zeitschrift fur Pflanzenkrankheiten und Pflanzenschutz* **85**: 197–202.

Singh, Y. P., Singh, R. S., and Sitaramaiah, K., 1990, Mechanisms of resistance of mycorrhizal tomato against root-knot nematodes. In: *Current Trends in Mycorrhizal Research*, eds., Jalali, B. L., and Chand, H., *Proc. Nat. Conf. Mycorrh.*, H.A.U., Hisar, India, pp. 96–97.

Sitaramaiah, K., and Sikora, R. A., 1982, Effect of mycorrhizal fungus *Glomus fasciculatum* on the host parasite relationship of *Rotylenchulus reniformis* in tomato. *Nematologica* **28**: 412–419.

Sivaprasad, P., Jacob, A., Nair, S. K., and George, B., 1990, Influence of VA mycorhhizal colonization on root-knot nematode infestation in *Piper nigrum* L. In: *Trends in Mycorrhizal Research*, eds., Jalali, B. S., and Chand, H., *Proc. Nat. Conf. Mycorrh.*, H. A.U., Hissar, India, pp. 100–101.

Smith, G. E., and Kaplan, D. T., 1988, Influence of mycorrhizal fungus, phosphorus and burrowing nematode interactions on the growth of rough lemon citrus seedlings. *J. Nematol.* **20**: 539–544.
Smith, G. S., 1987, Interactions of nematodes and mycorrhizal fungi. *Vistas on Nematology*, eds., Veech, J. A., and Dickson, D. W., Soc. Nemat., Hyattsville, MD, pp. 292–300.
Smith, G. S., 1988, The role of phosphorus nutrition in interactions of vesicular-arbuscular mycorrhizal fungi with soilborne nematodes and fungi. *Phytopathology* **78**: 371–374.
Smith, G. S., Hussey, R. S., and Roncadori, R. W., 1986a, Penetration and post infection development of *Meloidogyne incognita* on cotton as affected by *Glomus intraradices* and phosporus. *J. Nematol.* **18**: 429–435.
Smith, G. S., Roncadori, R. W., and Hussey, R. S., 1986b, Interaction of endomycorrhizal fungi, superphosphate and *Meloidogyne incognita* on cotton in microplots and field studies. *J. Nematol.* **18**: 208–216.
Smith, S. E., and Giananizzi-Pearson, V., 1988, Physiological interactions between symbionts in vesicular-arbuscular mycorrhizal plants. *Annue. Rev. Plant Physiol. Mol. Biol.* **39**: 221–244.
Smith, S. E., and Read, D. J., 1997, *Mycorrhizal Symbiosis*, 2nd edn, Academic, London.
Smith, S. E., Gianinazzi-Pearson, V., Koide, R., and Kiany, J. W. G., 1994, Nutrient transports in mycorrhizas: structure, physiology and consequences for efficiency of the symbiosis. In: *Management of Mycorrhiza in Agriculture, Horticulture and Forestry*, eds., Robson, A. D., Abbott, L. K., and Malajczuk, N., Kluwer, Dordrecht, The Netherlands, pp. 103–113.
Spanu, P., Boller, T., Ludwig, A., Wiemken, A., Faccio, A., and Bonfonte-Fasolo, P., 1989, Chitinase in roots of mycorrhizal *Allium porrum*: regulation and localisation. *Planta* **117**: 447–455.
Sreenivasa, M. N., and Bagyaraj, D. J., 1989, Use of pesticides for mass production of vesicular-arbuscular mycorrhizal inoculum. *Plant Soil* **119**: 127–132.
St-Arnaud, M., Hamel, C., Caron, M., and Fortin, J. A., 1994, Inhibition of *Pythium ultimum* in roots and growth substrate of mycorrhizal *Tagetes patula* colonized with *Glomus intraradices*. *Can. J. Plant Pathol.* **16**: 187–194.
Strobel, N. E., and Sinclair, W. A., 1991, Influence of temperature and pathogen aggressiveness on biological control of Fusarium root-rot by *Laccaria bicolor* in Douglas fir. *Phytopathology* **81**: 415–420.
Strobel, N. E., Hussey, R. S., and Roncadori, R. W., 1982, Interactions of vesicular-arbuscular mycorrhizal fungi, *Meloidogyne incognita* and soil fertility on Peach. *Phytopathology* **72**: 690–694.
Sundarababu, R., Sankaranarayanan, C., and Santhi, A., 1993, Interaction between vesicular-arbuscular mycorrhiza and *Meloidogyne javanica* on tomato as influenced by time of inoculation. *Indian J. Nematol.* **23**: 125–127.
Sundarababu, R., Mani, M. P., and Arulraj, P., 2001, Management of *Meloidogyne incognita* in Chilli nursery with *Glomus mosseae*. *Ann. Plant Protec. Sci.* **9**: 117–170.
Sundararaju, P., Sudha, S., and Iyer, R., 1995, Reaction of different Subabul cultivars to rootknot nematode and their interaction of nematode and VA mycorrhiza on Subabul. *Indian J. Nematol.* **25**: 70–75.
Sundaresan, P., Ubalthoose Raja, N., and Gunasekaran, P., 1993, Induction and accumulation of phytoalexins in cowpea roots infected with a mycorrhizal fungus *Glomus fasciculatum* and their resistance to Fusarium wilt disease. *J. Biosci.* **18**: 291–301.
Suresh, C. K., 1980, Interaction between vesicular arbuscular mycorrhizae and root-knot nematodes in tomato. M.Sc. (Agric.) thesis, University of Agricultural Sciences, Banglore, India.

Suresh, C. K., and Bagyaraj, D. J., 1984, Interaction between vesicular-arbuscular mycorrhizae and a root-knot nematode and its effect on growth and chemical composition on tomato. *Nematol. Medit.* **12**: 31–39.

Suresh, C. K., Bagyaraj, D. J., and Reddy, D. D. R., 1985, Effect of vesicular-arbuscular mycorrhiza on survival, penetration and development of root-knot nematode in tomato. *Plant Soil* **87**: 305–308.

Sylvia, D. M., and Jarstfer, A. G., 1994a, Sheared root inoculum of vesicular arbuscular mycorrhizal fungi. *App. Environ. Microbiol.* **58**: 229–232.

Sylvia, D. M., and Jarstfer, A. G., 1994b, Production of inoculum and inoculation with Arbuscular Mycorrhizal fungi. In: *Management of Mycorrhizas in Agricture, Horticulture and Forestry*, eds., Robson, A. D., Abbott, L. K., and Malajczuk, N., Kluwer, Dordrecht, The Netherlands, pp. 231–238.

Sylvia, D. M., and Sinclair, W. A., 1983, Phenolic compounds of resistance to fungal pathogens induced in primary roots of Douglas-fir seedlings by the ectomycorrhizal fungus *Laccaris laccata*. *Phytopathology* **73**: 390–397.

Taber, H. A., 1982, Occurrence of *Glomus* spores in weed seeds in soil. *Mycologia* **74**: 515–520.

Talavera, M., Itou, K., and Mizukubo, T., 2001, Reduction of nematode damage by root colonization with arbuscular mycorrhiza (*Glomus* spp.) in tomato *Meloidogyne incognita* (*Tylenchida:Meloidogynidae*) and carrot *Pratylenchus penetrans* (*Tylenchida: Pratylenchidae*) pathosystem. *Appl. Entomol. Zool.* **36**: 387–392.

Thompson Cason, K. M., Hussey, R. S., and Roncadori, R. W., 1983, Interaction of vesicular-arbuscular mycorrhizal fungi and phosphorus with *Meloidogyne incognita* on tomato. *J. Nematol.* **15**: 410–417.

Thygesen, K., Larsen, J., and Bodker, l., 2004, Arbuscular mycorrhizal fungi reduce development of pea root-rot caused by *Aphanomyces euteiches* using oospores as pathogen inoculum. *Eur. J. Plant Pathol.* **110**: 411–419.

Torres-Barragan, A., Zavaleta-Mejia1, E., Gonzalez-Chavez, C., and Ferrera-Cerrato, R., 1996, The use of arbuscular mycorrhizae to control onion white rot (*Sclerotium cepivorum* Berk.) under field conditions. *Mycorrhiza* **6**: 253–257.

Trotta, A., Varese, G. C., Gnavi, E., Fusconi, A., Sampo, S., and Berta, G., 1996, Interactions between the soilborne root pathogen *Phytophthora nicotianae* var. *parasitica* and the arbuscular mycorrhizal fungus *Glomus mosseae* in tomato plants. *Plant Soil* **185**: 199–209.

Trudgill, D. L., and Parrott, D. M., 1969, The behavior of the population of potato cyst nematode *Heterodera rostochiensis* towards three resistant potato hybrids. *Nematologica* **15**: 381.

Tylka, G. L., Hussey, R. S., and Roncadori, R. W., 1991, Interactions of vesicular-arbuscular mycorrhizal fungi, phosphorus and *Heterodera glyciens* on soybean. *J. Nematol.* **23**: 122–123.

Umesh, K. C., Krishnappa, K., and Bagyaraj, D. J., 1988, Interaction of burrowing nematode, *Radopholus similis* (Cobb, 1983) Thorne, 1949, and VA mycorrhiza, *Glomus fasciculatum* (Thaxt.) Gerd. and Trappe, in banana (*Musa acuminata* Colla.). *Indian J. Nematol.* **18**: 6–11.

Vierheilig, H., Alt, M., Gut-Rella, M., Lange, J., Boller, T., and Wiemken, A., 1996, Colonization of tobacco constitutively expressing pathogenesis-related proteins by arbuscular mycorrhizal fungi. In: *Mycorrhizas in Integrated Systems: From Genes to Plant Development*, eds., Azcon-Aguilar, and C., Barea, J. M., European Commission, EUR 16728, Luxembourg, Europe, pp. 270–273.

Vigo, C., Norman J. R., and Hooker, J. E., 2000, Biocontrol of pathogen *Phytophthora parasitica* by arbuscular mycorrhizal fungi is a consequence of effects on infection loci. *Plant Pathol.* **49**: 509–514.

Volpin, H., Elkind, Y., Okon, Y., and Kalpulnik, Y., 1994, A vesicular arbuscular mycorrhizal fungus (*Glomus intraradices*) induces defence response in alfalfa roots. *Plant Physiol.* **104**: 683–689.

Volpin, H., Phillips, D. A., Okon, Y. and Kalpulnik, Y., 1995, Suppression of an isoflavanoid phytoalexin defense response in mycorrhizal alfalfa roots. *Plant Physiol.* **108**: 1449–1454.

Waceke, J. W., Waudo, S. W., and Sikora, R., 2001, Suppression of *Meloidogyne hapla* by arbuscular mycorrhizal fungi (AMF) on *Pyrethrum* in Kenya. *Int. J. Pest Manag.* **47**: 135–140.

Waceke, J. W., Waudo, S. W., and Sikora, R., 2002, Effect of inorganic phosphatic fertilizers on the efficacy of an arbuscular mycorrhizal fungus against a root-knot nematode on *Pytherum*. *Int. J. Pest Manag.* **48**: 307–313.

Wallace, H. R., 1973, *Nematode Ecology and Plant Disease*, Alden London/Oxford.

Wick, R. L., and Moore, L. D., 1984, Histology of mycorrhizal and non-mycorrhizal *Ilex crenata* 'Helleri' challenged by *Thielaviopsis basicola*. *Can. J. Plant Pathol.* **6**: 146–150.

Wilson, M., and Backman, P. A., 1999, Biological control of plant pathogens. In: *Handbook of Pest Management*, ed., Ruberson, J. R., Marcel Dekker, New York, pp. 309–335.

Yao, M., Tweddell, R., and Desilets, H., 2002, Effect of two vesicular-arbuscular mycorrhizal fungi on the growth of micropropagated potato plantlets and on the extent of disease caused by *Rhizoctonia solani*. *Mycorrhiza* **12**: 235–242.

Zombolin, L., and Oliveira, A. A. R., 1986, Interacacoentre *Glomus etunicatum* and *Meloidogyne javanica* emfeijao (*Phaseolus vulgaris* L.) *Fitopatol. Brasil.* **11**: 217.

Chapter 4

ARBUSCULAR MYCORRHIZAE AND ALLEVIATION OF SOIL STRESSES ON PLANT GROWTH

PHILIPPE GIASSON[1], ANTOINE KARAM[2], AND ALFRED JAOUICH[1]
[1]Department of Earth and Atmospheric Sciences, University of Quebec at Montreal, Quebec, Canada; [2]Department of soils and agrifood engineering, Laval University, Quebec, Canada

Abstract: Within the last decade, inventories of the soil's productive capacity indicate severe degradation and loss of arable lands as a result of soil erosion, cultivation, salinization, over-grazing, land clearing, desertification, soil pollution, and atmospheric pollution. Large areas of land have been, and continue to be, contaminated by trace metals, and petroleum hydrocarbons. Many technologies using physical and chemical treatment methods have been developed to remediate contaminated soils. Recently, phytoremediation has been thought to provide an environmentally friendly alternative for the treatment of polluted soils. In phytoremediation of metal-contaminated soils, bioavailability and metal uptake are important factors. Among soil-plant factors controlling metal uptake, the rhizosphere flora is known to play a special role in the phyto-availability of trace elements. In this regard, arbuscular mycorrhizal fungi (AMF), which are among the most common components of soil rhizosphere flora, is of great interest to soil and environmental scientists, from a phyto-remediation and an environmental standpoint. AMF play important roles in the restoration of contaminated ecosystems and are increasingly used in many countries to improve plant nutrition and fertility of degraded land. As AMF are becoming commercially available, their use will also provide further avenues for reducing pollution from agriculture. This chapter reviews the role, the importance, and the application of AMF in ecologically remediating contaminated soils (mycorrhizoremediation). Emphasis is given to the effects of AMF on growth and yield, and on the uptake of trace metals by plants (rhizo-availability) from agricultural and metal-contaminated soils. The chapter also addresses the AMF's potential for improving or sustaining soil fertility.

Keywords: Arbuscular mycorrhizal fungi; nutrient availability; mycorrhizoremediation; rhizoextraction; metal pollution; heavy metals.

1 INTRODUCTION

Arbuscular mycorrhizal fungi (AMF) are important soil microorganisms (Liu and Lianfeng, 2008) that play a key role in facilitating nutrient uptake by crops in a variety of agroecosystems, particularly in low-input farming systems, and in revegetation and rhizomerediation processes (Barea and Jeffries, 1995; Barea et al., 2002; Atkinson et al., 2002; Lombi et al., 2001; Gadd, 2005; Jansa et al., 2008). Many studies in glasshouse and fields have assessed the positive effects of AMF on plant uptake, and plant growth and yield. Enhancing the mycorrhizal system of a low-fertility or degraded soil helps the root system acquire more nutrients (Roesti et al., 2005). It is widely acknowledged that AMF play an important role in improving the uptake of low mobile ions, in phosphate (PO_4^{3-}) and in ammonium (NH_4^+) phases (Smith and Read, 1997; Marschner, 2007; Martin et al., 2007). AMF not only increase the rate of nutrient transfer from the roots to the host plant, but they also increase resistance to biotic and abiotic stresses (Smith and Read, 1997; Khan, 2006; Singh, 2006; Martin et al., 2007). In polluted soils, AMF adapted to the high toxic metal concentrations can restore the biomass values. This chapter aim to provide a synopsis on the role of AMF in rhizoremediation of low-fertility land and polluted soils.

2 WHAT ARE ARBUSCULAR MYCORRHIZAL FUNGI (AMF)?

2.1 Arbuscular mycorrhizal associations

Arbuscular mycorrhizal fungi (AMF) or endomycorrhizae, including fungi belonging to the recently established phylum Glomeromycota (Schüßler et al., 2001), are a normal part of the root system (Gregory, 2006) in most natural and agroecosystems, including polluted soils (Göhre and Paszkowki, 2006). It is postulated that arbuscular mycorrhizae are the ancestral and predominant form of mycorrhizae (Wang and Qiu, 2006). They occur in the soil rhizosphere as spores, hyphae and propagules (Martin et al., 2007). Arbuscular mycorrhizal fungi are considered as obligate symbiotic biotrophs, in that they cannot grow without a host plant supplying them with carbohydrates (glucose and sucrose) (Muchovej, 2001; Harrison, 2005; Martin et al., 2007; Hamel and Plenchette, 2007). In this symbiotic association, the fungus colonizes the plant's root hairs by entering the cortex cells and acts as an extension of the root system (Douds and Millner, 1999; Muchovej, 2001). This type of association is characterized by the formation of arbuscles (finely branched hyphal structures) in the region of the root cortex that may function as nutrient organs (or nutrient exchange sites between the symbionts) and also for fungal multiplication (Muchovej, 2001; Gregory, 2006). According to Douds and Millner (1999), the AMF genera *Gigaspora* and *Scutellospora* produce only arbuscules and extensive intraradical and extraradical hyphal networks (Smith and Read, 1997), whereas *Glomus*, *Entrophospora*, *Acaulospora*, and *Sclerocystis* also produce vesicles (formerly known as vesicular-arbuscular mycorrhizal [VAM] fungi (Martin *et al.*, 2007)). Kistner and Parniske (2002) suggested that the genes involved in arbuscular mycorrhizae and rhizobial symbioses are common in both infection processes. The formation of mycorrhizae induces great changes in the physiology of the roots, in the internal morphology of the plant, and in the mycorrhizosphere, i.e., the soil surrounding the roots (Leyval and Joner, 2001; Gregory, 2006; Martin *et al.*, 2007). The symbiotic association of AMF and plant roots has been considered to be the oldest symbiosis of plants and is suspected to ecologically be the most important symbiotic relationship between microorganisms and higher plants (Paszkowski, 2006).

Arbuscular mycorrhizal associations are reported to occur in about 80% of terrestrial plants including trees, shrubs, forbs and grasses (Gregory, 2006). Many plants are able to establish symbiotic relationships with AMF. The plants are called mycorrhizal crops. However, crop plants from *Brassicaceae*, *Chenopodiaceae*, and *Polygonaceae* do not form mycorrhizal associations. The reader is referred to Varma and Hock (1999), Brundrett

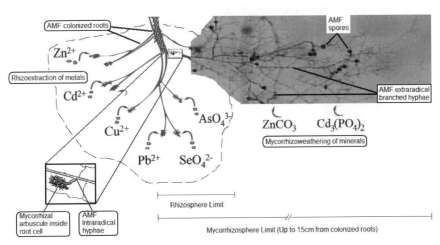

Fig. 1. Rhizosphere and mycorrhizosphere interactions with heavy metals in soils. Mycorrhizal extraradical hyphae release organic acids that weather rocks and minerals in soils. Heavy metals are sequestered and extracted by AMF colonized roots. Nutrients and metals can be exchanged between the fungus and the host plant via mycorrhizal arbuscules inside the root cell.

and Abbott (2002), and Martin *et al.* (2007), for a detailed description and occurrence of AMF.

In the mycorrhizosphere, microscopic fungi naturally occur in soil to form a symbiosis with plant roots and produce a highly elaborated mycelium network (hyphae) (see Fig. 1). These fungal associations could grow into the soil some 5–15 cm from the infected root, reaching farther and into smaller pores than could the plant's own root hairs (Brady and Weil, 2008). AMF have the capability of penetrating extremely small pores in soil and of accessing contaminants contained within (Hutchinson *et al.*, 2003).

2.2 Role of AMF in improving plant metal nutrition

The role of AMF on nutrient uptake (N, P and microelements), on the growth of AM crops, as well as on possible mechanisms of nutrient uptake, have been widely studied, as recently reviewed by Jeffries *et al.* (2003), Al-Karaki (2006), Cardoso and Kuyper (2006), Göhre and Paszkowki (2006), Gregory (2006), Martin *et al.* (2007), and Cavagnaro (2008). It is now generally recognized that AMF enhance the uptake of nitrogen (N) and of relatively immobile soil nutrients such as phosphorus (P), sulfur (S), copper (Cu), zinc (Zn), and boron (B).

AMF increase the plant contact area with soil. They were shown to enhance root absorption area up to 47-fold (Smith and Read, 1997). By colonizing the roots, the fungus enhances plant growth by making soil elements more accessible (George *et al.*, 1992; Nadian *et al.*, 1997; Gregory, 2006; Siddiqui, 2006) and by improving water absorption (Sweat and Davis, 1984; Cui and Nobel, 1992). Accordingly, mycorrhizal colonization improves vegetation establishment and survival particularly in adverse conditions such as in low fertility and arid soils (Jasper *et al.*, 1989; Allen *et al.*, 1996; Smith *et al.*, 1998). Knowing that contaminated sites are generally poor in nutrients and contain a highly altered soil structure, mycorrhizal fungi are suspected to play an important role in vegetation establishment for phytoremediation purposes.

Nutrients are taken up via the fungal hyphae by specific uptake systems and can be mobilized and transported to the plant via continuous fungal extra- and intracellular structures (Göhre and Paszkowki, 2006). It is suggested that constitutive expression or induction of nutrient transporters during symbiosis could improve translocation to the plant (Harrison *et al.*, 2002). However, some studies have reported decreased nutrient uptake or growth of mycorrhizae in certain circumstances (Kucey and Janzen, 1987; Arines *et al.*, 1990). For example, arbuscular mycorrhizal colonization of plants may depend on edaphic properties and environmental factors such as rainfall and sunlight hours. Lingfei *et al.* (2005) found that arbuscular mycorrhizal colonization were negatively correlated with total N, total P, available P and soil organic matter but positively correlated with soil pH. Karanika *et al.* (2008) found, in a field experiment, that AMF colonization was negatively affected by P and positively affected by N addition. However, the response varied among different plant species. In fact, they observed that P addition, in the field experiment, increased the colonization level of the high P demanding annual forb (non-leguminous dicot) such as *Galium lucidum*, decreased hyphal abundance of the forb *Plantago lanceolata* and the grass *Agrostis capillaris*, and appeared to have a negligible effect on the forb *Prunella vulgaris* and on leguminous species.

Other studies have shown a negative impact of AMF on the uptake of some nutrients, probably due to dilution effects (Burleigh *et al.*, 2003) and complex interactions between nutrients (e.g., P and Zn) within AMF at the cellular/sub-cellular levels (Cardoso and Kuyper, 2006; Christie *et al.*, 2004; Cavagnaro, 2008). Antagonistic reactions between nutrients exist under deficiency stress (e.g., P/Zn interaction, Cd/Zn interaction, etc.) (Kabata-Pendias, 2001).

In sum, under low soil nutrient concentrations, improvements in mineral nutrition of mycorrhizal crops can be attributed to the following factors (Burleigh *et al.*, 2003; Christie *et al.*, 2004; Cardoso and Kuyper, 2006; Cavagnaro *et al.*, 2007; Cavagnaro, 2008; Jackson *et al.*, 2008): (1) uptake of available nutrients via the mycorrhizal pathway; (2) differing P uptake kinetics in hyphae from those of roots, possibly through a higher affinity (lower K_m); (3) morphological and physiological changes in roots induced by AMF colonization; (4) differing ways in which roots and hyphae explore microsites, especially small patches of organic matter; (5) changes in edaphic conditions (e.g., pH and others soil variables) favourable to AMF colonization and nutrient solubility and mobility; (6) microbial communities (e.g., activity of mycorrhizal-helper bacteria); (7) nutrient cycling.

3 MYCORRHIZAL RELATIONSHIPS WITH TRACE ELEMENTS

3.1 Heavy metals or trace metal elements

Heavy metals (HM) occur naturally in the environment. Many definitions and interpretations of the term "heavy metal" exist (Duffus, 2002; Karam, 2007). Although imprecise and thoroughly objectionable (Phipps, 1981), the term "heavy metal" has been used increasingly in various publiccations and in legislation related to chemical hazards and the safe use of chemicals (Duffus, 2002) to identify metals with atomic weights greater than 40 (Rand *et al.*, 1995) and densities or specific gravities greater than about

5.0 g/cm^3 (Lozet and Mathieu, 1991; Morris, 1992). This term is often used as a group name for metals and metalloids (semimetals) that have been associated with contamination and potential toxicity (Duffus, 2002). Some authors proposed that this term "heavy metal" be abandoned in favour of "trace element". The later commonly refers to mineral elements that are present in soil in low concentrations, relative to the more abundant element in both the soil solution and the plant (Pandolfini *et al.*, 1997). Here the terms "metal", "heavy metal" and "trace metal" will be used interchangeably to indicate trace metal elements such as arsenic (As), cadmium (Cd), chromium (Cr), Cu, manganese (Mn), mercury (Hg), molybdenum (Mo), nickel (Ni), lead (Pb), selenium (Se), and Zn.

Numerous studies have indicated that agroecosystems receive inputs of heavy metals from the increased use of commercial fertilizers and biocides, from the application of metal-containing wastes such as sewage sludge, pig manure, coal and wood ashes to soils, and from atmospheric deposition (Mhatre and Pankhurst, 1997; Kabata-Pendias, 2001; Kabata-Pendias and Mukherjee, 2007). Although some of these metals are essential plant micronutrients since they are required for plant growth and development (Zn, Cu, Fe, Mn, Ni, Mo, Co), high contents of heavy metals, as well as the long-term presence of potentially toxic metals (Cd, Pb) and metalloids (As) in surface horizon of agricultural soils, are generally considered a matter of concern to society as they may adversely affect the quality of soils and surface water, and compromise sustainable food production (Pandolfini *et al.*, 1997; Kabata-Pendias, 2001; Keller *et al.*, 2002; Voegelin *et al.*, 2003; Kabata-Pendias and Mukherjee, 2007). The soil microbial community is thought to be a sensitive bioindicator of metal pollution effects on bioavailability and biogeochemical processes (Hinojosa *et al.*, 2005).

Metal forms in soils are basically characterized by their differential solubilities in various chemical extractants. The majority of fractionation schemes (Tessier *et al.*, 1979; Ma and Rao, 1997) group soil metal fractions into: "soluble", "exchangeable", "carbonate bound", "sesqui-oxides bound", "organic matter bound/sulfides" and "residual".

All metals present in a soluble form in the soil solution can be taken up by microorganisms and terrestrial plants (Cataldo and Wildung, 1978; Pandolfini *et al.*, 1997; Kabata-Pendias, 2001; Naidu *et al.*, 2003; Boruvka and Drabek, 2004). Many soil and environmental factors influence metal solubility and phytoavailability (Jackson and Alloway, 1992; Pandolfini *et al.*, 1997; Leyval and Joner, 2001; Karam *et al.*, 2003; Kabata-Pendias and Mukherjee, 2007). These factors can be summarized as follows: (1) nature of soil types; (2) nature of the metal species and their interaction with soil colloids and other soil components (sorption-desorption processes; complexation; diffusion; occlusion; precipitation); (3) concentration and chemical form of the metal entering the soil; (4) mineralogical composition (e.g., clay minerals and other aluminosilicates, oxides and hydroxides, carbonates, phosphates, sulphides, sulphates, and chlorides); (5) sorptive properties of soils or binding capacity; (6) physical, chemical, and biological soil properties (e.g., soil texture, soil moisture content and temperature, soil pH, redox potential, cation-exchange capacity, exchangeable cations, salt content, amount and type of clay, organic mater and oxides and hydroxides of Fe and Mn, free carbonates, and microbial activity); (7) biological activity of the rhizosphere; (8) duration of contact with the surface binding these metals; (9) chemical composition of the soil solution; (10) plant type and plant exudate.

Many studies have demonstrated that in neutral or alkaline substrates (soils, mine tailings, etc.) metals are more intensively adsorbed and chelated in unavailable forms relative to acidic substrates. Moreover, in soils rich in calcium carbonate and phosphate, in well-aerated soils with S compounds, and in soils and mine tailings amended with organic materials, metals are less mobile and available, or are associated with substrate constituents in unavailable forms (Kabata-Pendias, 2001; Kabata-Pendias and Mukherjee, 2007; Karam and De Coninck, 2007).

3.2 AMF tolerance and adaptation to heavy metals

The literature presents a range of "classic" ecological principles explaining the processes that increase the tolerance or resistance of a community (Boivin et al., 2002). Resistance refers to the ability of microorganisms to withstand the effects of a pollutant usually effective against them, while tolerance refers to the ability of microorganisms to adapt to the persistent presence of the pollutant. As stated by Leyval and Joner (2001), tolerance and resistance to the toxic effect of heavy metals depends upon the mechanism involved. Briefly, as mentioned in epidemiological studies (Foster and Hall, 1990; Tosun and Gönül, 2005), metal tolerance could be defined as a phenomenon by which microorganisms increase resistance towards stress resulting from exposure to heavy metal toxicity.

Metal tolerance of arbuscular mycorrhizal (AM) and ectomycorrhizal (ECM) fungi have been assessed using several observation methods including: AM spore numbers, root colonization and the abundance of ECM fruiting bodies (Weissenhorn et al., 1993, 1994; Del Val et al., 1999b). Unfortunately, such methods did not give information concerning conditions, limitations and threshold values ensuring the survival and growth of AMF, or about the genetic basis for multi-metal resistance and tolerance. Moreover, AMF coexist with other microbial communities and plant roots that can tolerate and accumulate metals, and this could confound the real interactions between AMF and metals in the medium.

More recently, to evaluate the tolerance of microorganisms in soils polluted with metals, specialists have adopted the concept of pollution-induced community tolerance (PICT) (Niklińska et al., 2006). This perspective stipulates that with time, in an ecosystem, contamination exposure increases tolerance in microbial communities. Davis et al. (2004) used the PICT method to assess the effects of long-term exposure to Zn on the metabolic diversity and tolerance to Zn of soil microbial community. They showed that long-term exposure to Zn imposes stress on soil microbes, resulting in an increased tolerance. They concluded that the long-term accumulation of Zn in soils provides the microbial community with time to adapt to this metal. Indeed, microbial communities are often found to recover after an initial inhibition by high metal inputs (Holtan-Hartwig et al., 2002). This adaptation has been attributed to two factors (Almås et al., 2004). The first one is a gradual decrease in metal availability due to immobilization reactions occurring in the rhizosphere. The other factor is a gradual change in microbial community structure, based on changes in phospholipid fatty acid profiles (Frostegård et al., 1993) which results in more tolerant organisms.

Although metals may induce changes in the microbial community, resulting in microorganisms more resistant to metals (Almås et al., 2004), most essential and non essential metals exhibit toxicity above a certain concentration. This toxicity stress, appreciated by a threshold value (Leyval and Joner, 2001), will vary depending on many factors including the type of microorganism, the physico-chemical properties and concentration of the metal, and the edaphic and environmental conditions (Gadd, 1993).

Even though metals can exhibit a range of toxicities toward soil microorganisms (McGrath, 1994; McGrath et al., 1995; Giller et al.; 1998; Dai et al., 2004; Gadd, 2005; Niklińska et al., 2006), AMF isolates, particularly the ecotypes living in metal-enriched soils, metalliferrous sites and mine spoils heavily polluted with metals, can, depending on intrinsic and extrinsic factors, tolerate and accumulate HM (Gildon and Tinker, 1981, 1983a, b; Weissenhorn et al., 1993, 1994; Joner and Leyval, 1997; Leyval et al., 1997; Smith and Read, 1997; Gadd, 2005). Field investigations have indicated that mycorrhizal fungi can colonize plant in metal contaminated sites (Díaz and Honrubia, 1994; Pawlowska et al., 1996) and in agricultural soils contaminated with metals of different origins, including atmospheric deposition from smelter and sludge amendments (Weissenhorn et al., 1995b, c). Mycorrhizal fungi have also been shown to be associated with metallophyte plants on highly polluted soils. Nevertheless, it should be kept in mind that in some extreme metal conditions, AMF inoculation can be entirely inhibited (Weissenhorn et al., 1994). Del Val et al. (1999b) reported that spore numbers decreased with the increasing amounts of heavy metals, whereas specie richness and diversity increased in soils receiving an intermediate rate of sludge contamination but decreased in soils receiving the highest rate of heavy metal-contaminated sludge.

Several reports and reviews suggested that mycorrhizal fungi (MF) from metal-contaminated sites have developed tolerance against metal toxicity and are well adapted (Weissenhorn et al., 1993, 1994; Del Val et al., 1999a; Leyval and Joner, 2001; Toler et al., 2005; Sudova et al., 2007). The evolution of metal tolerance is showed to be rapid in MF. As stated by Sudova et al. (2007), tolerant strains of some MF may develop within one or two years (Weissenhorn et al., 1994; Tullio et al., 2003). Gonzalez-Chavez et al. (2002a, b) reported that arbuscular mycorrhizal fungi have evolved arsenate resistance and conferred enhanced resistance on *Holocus lanatus*. HM concentration may decrease the numbers and vitality of AMF as a result of HM toxicity (Dixon, 1988; Dixon and Buschena, 1988) or may have no effect on mycorrhizal colonization (Wilkins, 1991; Leyval et al., 1997). Biró et al. (2005) studied the stress buffer effect of the AMF and their colonization behaviour in metal spiked soil on a long-term level in controlled conditions. The soils used were collected after a 12 year metal-adaptation process, where 13 trace element salts, such as Al, As, Ba, Cd, Cr, Cu, Hg, Ni, Pb, Se, Sr and Zn were applied in four gradients (0, 30, 90 and 270 mg/kg dry soil). Barley (*Hordeum vulgare* L.) was used as a test plant. They found a strong dose-dependency at the arbuscular richness in general. The sporulation of the AMF was found as the most sensitive parameter to long-term metal(loid) stress. They reported that Al, As, Ba, Cd, Cr, Cu, Pb, Se, Sr and Zn reduced significantly the spore-numbers of the AMF, while the Ni loadings (at 36 g/soil) increased mycorrhizal sporulation.

At present, potential interaction mechanisms between AMF and metals, and the cellular and molecular mechanisms of HM tolerance in AMF, are poorly understood (Leyval and Joner, 2001; Martin et al., 2007). Metal transporters and plant-encoded transporters are involved in the tolerance and uptake of heavy metals (Göhre and Paszkowski, 2006; Hildebrandt et al., 2007) from extracellular media, or in their mobilization from intracellular stores (Gaither and Eide, 2001). Göhre and Paszkowski (2006) hypothesized that metals could be released at the pre-arbuscular interface and then taken up by plant-encoded transporters.

The ability of an organism to tolerate and to resist metal toxicity may involve more than one of the following mechanisms (Gadd, 1993, 2005; Leyval and Joner, 2001; Lux and Cumming, 2001; Ouziad et al., 2005; Sudová and Vosátka, 2007):

- Fungal gene expression
- Extracellular metal sequestration and precipitation
- Production of metallothioneins (metal binding proteins)

- Avoidance of metals (reduced uptake or increased efflux, formation of complexes outside cells, release of organic acids, etc.)
- Intracellular chelation (synthesis of ligands such as polyphoshates and metallothioneins)
- Compartmentation within leaf vacuoles
- Loss of leafs during dry or cold seasons
- Phosphorus plant status or interaction between P and metals (increased P uptake by host plant)
- Biological sorption via glomalin
- Volatilization.

The expression of several protein encoding genes potentially involved in heavy metal tolerance varied in their response to different heavy metals. Such proteins included a Zn transporter, a metallothionein, a 90 kD heat shock protein and a glutathione S-transferase (all assignments of protein function are putative). Studies on the expression of the selected genes were also performed with roots of *Medicago truncatula* grown in either a natural, Zn-rich heavy metal "Breinigerberg" soil, or in a non-polluted soil supplemented with 100 µM $ZnSO_4$. The transcript levels of the genes analyzed were enhanced up to eightfold in roots grown in the heavy metal-containing soils. The data obtained demonstrate the heavy metal-dependent expression of different AMF genes in the intra- and extraradical mycelium. The distinct induction of gene coding for proteins possibly involved in the alleviation of damage caused by reactive oxygen species (a 90 kD heat shock protein and a glutathione S-transferase) might indicate that heavy metal-derived oxidative stress is the primary concern of the fungal partner in the symbiosis.

In a soil environment, levels and persistence of metal tolerance of the AMF (Leyval and Joner, 2001; Jamal *et al.*, 2002; Turnau and Mesjasz-Przybylowicz, 2003; Toler *et al.*, 2005; Fomina *et al.*, 2005; Biró *et al.*, 2005; Sudová *et al.*, 2007) depends on a number of factors:

- AM community ecotype or diversity of AM fungi
- Specific properties of host plant and conditions of plant growth
- Nature of the metal
- Level of soil metal contamination, particularly available or extractable HM
- Cultivation regime
- Colonization conditions (axenic culture vs symbiotic conditions)
- Activities related to land disturbance
- Seasonal variations.

3.3 Heavy metal uptake by AMF

Many studies have shown that metals are sorbed in the soil system by microbial biomass, such as fungi, yeast, bacteria, algae and cyanobacteria (Lepp, 1992; Mullen et al., 1992; Morley and Gadd, 1995; Kapoor and Viraraghavan, 1998; Zhou, 1999). In general, mobilization of metals by soil microorganisms can be achieved by protonation, chelation, and chemical transformation (Gadd, 2005). The exudates, such as citric acid and other organic compounds, released from both plant roots and soil microorganisms, are very effective in solubilizing and releasing metals from soil components (Murphy and Levy, 1983; Gadd, 1990).

Arbuscular mycorrhizae have often been reported to sequester and to accumulate metals in their biomass as well as in the roots of host plants (Burke et al., 2000; Joner et al., 2000; Leyval and Joner, 2001; Gadd, 2005; Martin et al., 2007). It is reported that intracellular and extraradical mycelium of AM and ectomycorrhizal (ECM) fungi would have potential for metal sorption (Marschner et al., 1998; Joner et al., 2000). Most of the metals were demonstrated to be bound to the cell wall components like chitin, cellulose, cellulose derivatives and melanins of ecto-and endomycorrhizal fungi (Galli et al., 1994). High sorption capacity of fungal mycelium for some metals such as Pb was also confirmed for ECM fungi (Marschner et al., 1998).

Recently, much evidence indicates that AMF exhibit great activity in the mobilization of metals that are bound by soil components (Leyval and Joner, 2001; Gadd, 2005; Göhre and Paszkowski, 2006). AMF can also act as a «barrier» in the uptake or transport of metals. However, little work has been performed to assess the effect of AMF colonization on metal fractionation (metal pools) and labile fractions of metal in soils and mine tailings. The chemical form of metals in the hyphae of AMF has received little investigation. There is no information on the chemical form of many toxic metals in AMF. Besides, all physical parameters inherent to binding sites remain to be elucidated. Much still remains to be learned about factors determining metal uptake by AMF.

Gonzalez-Chavez et al. (2002a, b) designed a set of experiments to investigate the characteristics of sorption and accumulation of Cu by the extraradical mycelium (ERM) of different *Glomus* spp. (*Glomus caledonium* BEG133, *Glomus claroideum* BEG134, *Glomus mosseae* BEG132) isolated from a highly Cu-polluted mine soil and grown on sorghum (*Sorghum vulgare* L.) under controlled conditions. Copper localization and compartmentalization was done using Transmission and Scanning Electron Microscopy equipped with energy dispersive X-ray analysis. They observed that ERM of AMF is able to sorb and accumulate Cu. Their experiments demonstrated and concluded the following:

- ERM of AMF from polluted soils accumulated Cu in the mucilaginous outer hyphal wall zone, cell wall and inside the hyphal cytoplasm.
- The accumulated Cu was mainly associated with Fe in the mucilaginous outer hyphal wall zone and in the cell wall.
- Copper was associated with traces of arsenate inside the cytoplasm of the ERM of *Glomus mosseae* BEG134.
- Arsenate may be accumulated inside the cytoplasm in the same way as polyphosphates.
- Different Cu and arsenate uptake and accumulation strategies (tolerance mechanisms) exist between the three AMF isolated from the same polluted soil.

In another set of experiments with excised mycelium of four *Glomus* spp. with different histories of exposure to heavy metals (Cd and Zn), Joner *et al.* (2000) confirmed the capacity of extraradical hyphae of *Glomus* spp. to fix metal ions. The results showed the following sorption features:

- Sorption was fast and sorbed Cd was achieved within 30 minutes.
- Sorption was concentration dependent and, at the highest solution concentrations, the amounts sorbed seem too high to obey a monolayered Langmuir adsorption model.
- *G. mosseae* P2 (metal-tolerant strain from soil with a 60-year history of industrial metal pollution, and grown on subterranean clover, *Trifolium subterraneum*, c.v. Mount Barker) sorbed significantly more Cd than *G. lamellosum* (from non-contaminated soil, grown on ryegrass, *Lolium perenne*, cv. Barclay) and *G. mosseae* Gm (non-metal tolerant strain, BEG 12, grown on ryegrass).

It would seem likely that AMF behave similarly as ECM and other soil filamentous fungi. AMF have metal binding sites and are able to produce intracellular and extracellular with high affinity for metals. Binding sites vary with AMF species.

Although the mycorrhizal mechanisms for enhancing uptake are not entirely known, some of them could be the following (Gadd, 1990, 1993; Joner *et al.*, 2000; Gonzalez-Chavez *et al.*, 2004):

- Transfer of metals to the hyphae by cation exchange and chelation (non-metabolic binding of metals to cell walls).
- Interacting with hyphal synthetized products or metabolites that act as biosorption agents such as chitin and glomalin, an insoluble glycoprotein. The thin hyaline layer of the spore wall of *Glomus geosporum* AMF is composed mainly of chitin (Sabrana *et al.*, 1995).
- Chelation of metals inside the fungus.
- Intracellular precipitation with phosphate (PO_4).

Uptake of metals is controlled by or depends on different factors (Gadd, 1990, 1993; Laheurte et al., 1990; Joner et al., 2000; Leyval and Joner, 2001), including the following:

- AM species
- Metabolite composition
- Fungal biomass CEC
- Edaphic and environmental conditions
- Metal pools
- Metal electrochemical properties
- Competition between metals for mycorrhizal surface adsorption sites
- Nature of the host plant
- Root exudation patterns.

3.4 Effects of AMF on growth and uptake of trace metals by plants

Recent general reviews concerning the transport of metals to plants by mycorrhizal fungi have been published elsewhere (see Leyval and Joner, 2001; Singh, 2006). The following paragraphs provide a synthesis of the factors that contribute to the divergent influences of AMF on heavy metal status in host plant. As mentioned earlier, an important factor determining the phytoavailability of a trace metal is its binding capacity to soil constituents. Plants readily take up trace metals from soils (or other growth media) through the roots, mainly in a soluble form. The specific properties of the mycorrhizosphere are known to accelerate the immobilization of metals and to accelerate the weathering at the root-soil surface relative to the bulk soil (Mench and Martin, 1991; Courchesne et al., 2001). Mycorrhizal fungi can affect the transformation of trace metals in the soil in several ways (Leyval and Joner, 2001) including: (i) altering the pH of the soil (i.e., acidification), (ii) immobilization (by adsorption, chelation, or absorption of free metallic species in the soil solution) and (iii) modification of root exudation. It is important to note that acidification caused by organic acids secreted by AMF facilitates the mobilization of trace metals.

A number of studies have been carried out on trace metal uptake by mycorrhizal plants and the results vary with each experiment and each host plant. However, it can be generalized that, as demonstrated for ectomycorrhizal and ericoid mycorrhizal fungi, AMF can increase the uptake and accumulation of metals in host plants (Davies et al., 2001, 2002; Hovsepyan and Greipsson, 2004; Rufyikiri et al., 2002, 2003) even when the metals are present at toxic levels. Cheung et al. (2008) found that inoculation of jute (*Corchotus capsulari*, a higher plant) with *G. mosseae* and *G. intraradices* improved plant growth. However, in other situations, where AM fungi exude enzymes that participate in the immobilization process of metals, AMF colonization decreases the uptake and accumulation of metals in host plants (Joner et al., 2000; Leyval et al., 1997; Weissenhorn et al., 1993). Deram et al. (2008) observed that AMF colonization disappeared when Cd concentrations in soil increased. Arbuscular mycorrhizae have also been found to sequester metals in the roots of plants and prevent translocation to the shoot (Burke et al., 2000). In studying the effect of AMF on the accumulation and transport of Pb from an anthropogenically-polluted substrate to root and shoot biomass of maize plants, Sudová and Vosátka (2007) found that Pb concentrations increased in highly colonized root segments, whereas they decreased in the shoots of maize. They hypothesized that Pb was immobilized in the fungal mycelium due to intraradical fungal structures. AM may also protect their host plants from the toxicity of excessive metal or metalloid (Zhu et al., 2001; Bai et al., 2008) through: (i) P nutrition by activating P; (ii) chemical precipitation in the soil; (iii) tissue dilution due to increased plant biomass, (iv) hyphal sequestration of metal; and (v) root immobilization.

The AMF have variable effects on metal uptake (translocation and accumulation in plant tissues) and growth of host plant. Most of these variations could be summarized as follows: (i) metal uptake into the host plant is enhanced or repressed (Kothari et al., 1990; Li et al., 1991; Ietswaart et al., 1992; Bürkert and Robson, 1994; Weissenhorn et al., 1995a; Jamal et al., 2002; Bai et al., 2008); (ii) metal accumulation by plant shoots is reduced under elevated soil metal concentrations while increased under normal metal conditions (Toler et al., 2005); (iii) metal acquisition by plant is reduced and plant growth is enhanced (Weissenhorn et al., 1995b); (iv) metal concentration in shoots is lower at the highest soil metal concentrations (Leyval et al., 1991); (v) metal uptake was either not affected by or not enhanced in mycorrhizal plants, depending on the nature of the metal (Weissenhorn and Leyval, 1995); and (vi) metal accumulation in root and dry matter yield of shoot and root increased (Bai et al., 2008).

Many factors contribute to the divergences of AMF on metal plant uptake, plant growth and plant biomass production (Leyval and Joner, 2001; Citterio et al., 2005; Wang et al., 2005; Audet and Charest, 2006; Deram et al., 2008; Jansa et al., 2008; Piotrowski et al., 2008). These include: (i) fungal genotype; (ii) uptake of metal by plant via AM symbiosis; (iii) root length density, (iv) competition of the AMF communities; (v) seasonal variation in AM; (vi) association with soil microorganisms; (vii) chemical properties of the soil outside the rhizosphere (pH, CEC, etc.); (viii) the metal itself; (ix) concentrations of available metals; (x) soil contamination conditions (contaminated or artificially contaminated vs non-contaminated soil); (xi) interactions between P and metals (addition of P fertilizers); (xii) experimental conditions (light intensity, plant growth stage, available N and P); (xiii) litter inputs; and (xiv) plant species and plant size. Besides, since AMF cannot be grown without a host plant (Leyval and Joner, 2001) and may coexist with other microbial communities (Roesti et al., 2005; Toljander, 2006) that can tolerate and accumulate metals (Lepp, 1992), this would obscure the interaction between AMF and metals in the substrate.

4 AMF FOR MYCORRHIZOREMEDIATION OF CONTAMINATED SOILS AND MINE SITES

4.1 Metal hyperaccumulators

In nature, some plants hyperaccumulate heavy metals. For example, *Viola calaminaria* and *Thlaspi calaminare* grow over calamine deposits in Aachen, Germany and contain over 1% (dry weight) zinc in their tissues. Also, some *Alyssum* species like *A. bertolinii* grow on serpentine soils in Tuscany, Italy and contain over 1% (dry weight) nickel. These species are respectively called calamine and serpentine flora. *Thlaspi caerulescens* from the *Brassicaeae* family can also hyperaccumulate both Zn and Cd (Brooks, 1998). As classified by McIntyre (2003), Zn and Cd hyperaccumulators contain these metals at minimal levels respectively of 10,000 and 100 µg/g.

Heavy metal complexes in hyperaccumulators plants are mainly associated with carboxylic acids like citric, malic and malonic acids. These organic acids are implicated in the storage of heavy metals in leaf vacuoles. Amino acids like cysteine, histidine glutamic acids, and glycine also form heavy metal complexes in hyperaccumulators (Homer et al., 1997). These complexes are more stable than those with carboxylic acids. They are mostly involved in heavy metal transport through xylem. Moreover, hyper-accumulator plants can increase availability of metals like Fe and also Zn, Cu and Mn by releasing chelating phytosiderophores. Hyperaccumulation mechanisms may then be related to rhizosphere processes such as to the release of chelating agents (phytosiderophores and organic acids) and/or to differences in the number or affinity of metal root transporters (Lombi et al., 2001).

Although hyperaccumulator plants are widely used in phytoextraction, they are generally of low biomass, inconvenient for phytoremediation. However, arbuscular mycorrhizae fungi (AMF), especially *Glomus intraradices*, colonized *Festuca* and *Agropyron* species have shown higher heavy metal (Zn, Cd, As and Se) content than non-colonized controls (Giasson et al., 2006). As for hyperaccumulators, fungi can synthesize cysteine-rich metal binding proteins called metallothioneins (Gadd and White, 1989). AMF might therefore be directly implicated in heavy metal hyperaccumulation in plants.

4.2 Mycorrhizosphere and phytoextraction of metals

Phytoremediation has already proven its potential in numerous applications around the world (Baker *et al.*, 1988; Kumar *et al.*, 1995; Giasson and Jaouich, 1998; Salido *et al.*, 2003). There are several processes associated with phytoremediation of heavy metal polluted soils. Phytostabilization is the reduction of the mobility, bioavailability and/or toxicity of the pollutant in the rhizosphere, while the process of phytoaccumulation is the sequestration, by plant roots, of the contaminants, typically heavy metals, and then translocation to their aerial parts. The most common heavy metals found in polluted soils are Pb, As, Cr, Cd, Ni and Zn. In phytoremediation, the contaminant mass is not destroyed but ends up in the plant shoots and leaves, which can then be harvested and disposed of safely.

The relatively low potential cost of phytoremediation allows for the decontamination of many sites that cannot be treated with currently available methods. In addition, it has aesthetic advantages and long term applicability: it preserves the topsoil and reduces the amount of hazardous materials generated during cleanup (Schnoor, 1997; Ensley, 2000). However, research in this field must be pursued to enhance biomass and heavy metals accumulation in plants. In this way, mycorrhizal fungi may be very helpful (see Fig. 1).

Since the early eighties, many researchers have shown that mycorrhizal colonization can have an impact on heavy metal assimilation by plants (Bradley *et al.*, 1981; Gildon and Tinker, 1983a, b). Dehn and Schüepp (1989) have found that mycorrhizal infection enhances heavy metal accumulation in lettuce roots but not in shoots. However, Angle *et al.* (1988), Lambert and Weidensaul (1991), and Jamal *et al.* (2002) have shown that mycorrhizae enhance heavy metal accumulation in legume shoots like soybeans, alfalfa and lentils. Killham and Firestone (1983), Hetrick *et al.* (1994), Mohammad *et al.* (1995), Burke *et al.* (2000), and Bi *et al.* (2003) have found similar results with grasses. In the case of cesium (Cs) and strontium (Sr), Entry *et al.* (1999) have indicated that mycorrhizal plants produce higher biomass and higher Cs and Sr content in plant tissues than non-mycorrhizal plants.

Moreover, Turnau and Mesjasz-Przybylowicz (2003) have found that *Berkheya coddii*, a hyperaccumulator from the Asteraceae family, cultivated with well-developed mycorrhization, which includes arbuscule formation, increased not only the shoot biomass of the plant but also strongly increased the Ni content of shoots. Ni shoot content of *B. coddii* colonized with *Glomus intraradices* was 1.3% of dry weight, while in nonmycorrhizal plants it was below 0.5%.

In a glasshouse experiment, Giasson *et al.* (2006) studied four commonly found AMF species well adapted to North American soils: *Glomus intraradices, Glomus mossae, Glomus etunicatum,* and *Gigaspora gigantea. Glomus* spp. and *Gigaspora* spp. are AMF species identified in metal rich soils (Chaudry et al., 1999). A grass mixture of *Festuca rubra* and *F. eliator* (70%), *Agropyron repens* (25%), and *Trifolium repens* (5%) was used. This vegetation mix is used in land reclamation in Eastern Canada to revegetate mine tailings. *Festuca* species like *F. rubra* are considered characteristic species on metalliferous soils and can accumulate excessive amounts of metals (Smith and Bradshaw, 1979; Pichtel and Salt, 1998). Also, this grass mixture can be harvested several times per year because the articulated stubble can renew itself constantly (Marie-Victorin, 1964). In this study, AMF mycorrhizal root infection varied from 30% to 70% for all heavy metal treatments. Relative arbuscular richness varied from 38% to 84%. Arbuscules are the internal structures in the root cells that facilitate nutrient exchange between the fungus and the host plant. Well developed mycorrhization, which includes arbuscule formation, has shown to increase the metal content in shoots (Turnau and Mesjasz-Przybylowicz, 2003). Absence of arbuscular structures can indicate altered host physiology and carbon allocation, or can be a sign of stress in the mycorrhizal fungus. In their glasshouse study, Giasson et al. (2006) found the following results regarding heavy metal extractions by AMF colonized vegetation:

- There is interspecific variation between AMF regarding translocation of metals to plants.
- Arbuscule relative richness in Zn treatment was the highest (75%) vs other metal treatments.
- Zn, Cd, As, and Se extractions by *Glomus intraradices* colonized plants are generally higher than in non-mycorrhizal plants, depending on the metal concentration in the soil and whether this heavy metal interacts with other metals in that soil.
- Grasses colonized by *Glomus intraradices* had greater Zn, Cd, As, and Se mass extracted than for non inoculated vegetation because of higher plant biomass.
- When in interaction with other metals in the soil, Se is extracted more readily by AMF colonized plants. With time, however, Se in plants is lost in part by volatilization of the dimethyl diselenide form.
- For all four metal treatments (Cd, Zn, As and Se), there is a positive linear correlation between metal in plant tissues and metal content in soils. When soil metal content is increased tenfold metal in plant tissues is also increased by 10, for both colonized and control treatments.
- Metal extraction reaches a plateau after 80 days showing no further phytoaccumulation or sometimes slightly diminishes because of either phytovolatilization (As and Se) or necrosis in plants (Zn) caused by high heavy metal levels. This observation suggests that *G. intraradices* colonized perennial grasses may be harvested after a two-month

period allowing for two to three harvests per year in Canadian latitudes. In this way, phytoremediation can be accelerated two- to threefold.

Lasat (2002) observed that the effect of AMF associations on metal root uptake appears to be metal and plant specific. Greater root length densities and presumably more hyphae enable plants to explore a larger soil volume thus increasing access to cations (metals) not available to non-mycorrhizal plants (Mohammad et al., 1995).

As related by other studies (Shetty et al., 1994), AMF alters the pattern of Zn translocation from root to shoot in Festuca arrundinaceae. Zinc hyphal uptake and translocation are known to be similar to P transport (Cooper and Tinker, 1978; Weissenhorn et al., 1995a). In their in vitro experiment, Giasson et al. (2005b) observed that zinc adsorption at spore propagules was weak – approximately 9.6 µg of Zn per gram of spore in the 500 µg/g Zn treatment because mycorrhizal hyphae vacuoles and arbuscules contain phosphorus in the form of polyphosphate. Additionally, Zn is transferred to the plant host though AMF hyphae and arbuscules. Arbuscules are involved in this transfer by providing a considerable increase in fungus and plant contact surface area (Smith and Read, 1997). Frequent degeneration of fungal arbuscules in the root thus allows Zn content to be transferred directly into the host cell (Gildon and Tinker, 1983a) reducing Zn concentrations in fungi. Turnau and Mesjasz-Przybylowicz (2003) found that well-developed mycorrhization, containing arbuscule formations, increased the metal content in plant shoots. Zn can then be accumulated in leaves as a citrate complex in the vacuole (Salt et al., 1999).

Phosphate is central to mycorrhizal symbiosis. In P deficient soils, plant roots exude chemical signals to attract AMF. In such environments, AMF have developed an active phosphate transporter (Meharg et al., 1994). Arsenate (As(V)) is chemically similar to phosphate and can enter cells via arsenite (As(III)) translocating ATP'ase (Jun et al., 2002). The presence of AMF can therefore enhance both phosphate and arsenate uptake in such conditions (Martin et al., 2007)

Also, at high levels of P, mycorrhizal colonization may be reduced with consequent reductions in uptake and cause deficiencies of essential metals like Cu and Zn. Interactions such as these may be involved in the apparent alleviation of Zn toxicity in polluted sites (Dueck et al., 1986). If the sites are P deficient, then mycorrhizal P uptake can result in increased growth and dilution of Zn in the tissues (Smith and Read, 1997).

In an in vitro study using transformed carrot roots (Daucus carota L.) growing in a phytagel (M media), Giasson et al. (2005b) found that even without pressure, AMF hyphae passed from the proximal to the distal side of the Petri dish into the M media containing low and high concentrations of Zn and Cd. The hyphal network was well developed and sporulation was high in the low heavy metal level side (100 µg/g Zn and 5 µg/g Cd). More than 16,000 spores per half Petri plates were counted for the low Cd and Zn treatments.

In the same experiment, Giasson et al. (2005b) observed that at high heavy metal levels in the media (500 µg/g Zn and 20 µg/g Cd), hyphal network was less developed (taking spiral shapes) and sporulation was weaker. The spore population was approximately 1,500 per half Petri plates for the 20 µg/g Cd treatment and 1,300 for the 500 µg/g Zn treatment. The results are revealing. Essential cation (Zn) and nonessential cation (Cd) translocation from substrate (phytagel) to plant occurred through mycorrhizae hyphae, even at high (toxic) heavy metal concentrations. This is in accordance with Chen et al. (2003), who found that Zn is taken up and transferred to a host plant via extraradical hyphae. Root over growth media accumulation factors reached 5:1 and 18:1 for Zn and Cd, respectively. With over 90 µg/g cadmium and 550 µg/g zinc found in the roots, the presence of *G. intraradices* caused carrots to become cadmium hyperaccumulators and Zn accumulators.

Cadmium, like other nonessential metals, is generally of low abundance in the biosphere and should therefore not compete with specific transport systems for essential metals (Gadd and White, 1989). However, as a result of human activities, nonessential metals are concentrated in certain areas at very high levels. Toxic and nonessential metals, such as Cd, generally bind more strongly to ligands compared with essential metals thereby displacing essential metals from their normal sites, and exerting toxic effects by binding to other sites (Hughes and Poole, 1989).

Furthermore, Cd (0.97 Å) has a similar ionic radius to calcium (Ca) (0.99 Å), and so there is the possibility of metal-for-metal substitution in the predominantly oxygen-containing ligand sites preferred by Ca. Also, because of cadmium's position in the Periodic Table (Group IIB), it bears a chemical resemblance to Zn. Competition among Cd, Ca, and Zn ions for adsorption sites on AM hyphae seem to favour Cd over Ca and Zn (Joner et al., 2000). In microbes, Cd competes with both Mn and Zn transport systems. Cadmium appears to enter via the Mn transport system and is rapidly diffused from resistant cells, via antiporter genes, exchanging cadmium for hydrogen and cation-translocating ATPase (Silver et al., 1989).

Cadmium will also bind at sites normally occupied by Zn containing either a soft ligand, like sulphur (for example, cysteine or metallothionein) or a hard ligand, like nitrogen (for example, histidine) and oxygen (Rayner and Sadler, 1989). A common metal-induced response in fungi is the intracellular synthesis of cysteine rich metal-binding proteins called metallothioneins (MT), which have functions in metal detoxification and also in the storage and regulation of intracellular metal ion concentrations (Gadd and

White, 1989). Fungal cells have certain mechanisms to maintain metal homeostasis and prevent metal toxicity. Glutathione (GSH), metal-binding peptides, metallothionein-like peptides, and sulphide ions play a role in such mechanisms. Cellular metal stress triggers the biosynthesis of some of these molecules, regulated via intracellular metal sensors (Singh, 2006).

There are also small peptides called phytochelatins (PC) in microbes and plants that bind metals such as Cd via cysteinyl residues. These peptides protect plant cells from metal poisoning (Baker et al., 1988). Joner and Leyval (1997) suggested that sequestration of Cd in fungal structures could be responsible for the retention of Cd in the roots. It is likely however that the extent of this retention mechanism is restricted due to the relatively small biomass of the fungi. Giasson et al. (2005b) found Cd adsorption on spore propagules to be at concentrations below the detection limit of a chromatograph detector (HPLC). According to Colpaert (1998), once Cd saturation occurs in the fungi, increased translocation to shoots is thought to occur. Hughes and Poole (1989) found that some heavy metals appear to enter cells directly, possibly through a lesion in the cell membrane, as a result of the strong binding of the cation.

In an *in vitro* study, Giasson et al. (2005b) found that heavy metal accumulation by colonized carrot roots seemed to reach a plateau: 550 µg Zn/g and 90 µg Cd/g, independently of the initial growth media heavy metal concentrations. This could be explained by heavy metal saturation in vegetation after a two-month exposure period (Giasson et al., 2006). Furthermore, Rayner and Sadler (1989) demonstrated that when cadmium levels are increased, adaptation results thereby in increasing the growth rate and reducing the extent of cadmium accumulation from the medium.

These conclusions are worth considering for phytoremediation of heavy metal–contaminated soils enhanced by mycorrhizal inoculation.

4.3 Mycorrhizostabilization of metals

Phytostabilization and mycorrhizostabilization reduce the mobility, bioavailability and/or toxicity of the pollutant in the rhizosphere. Mycorrhizal fungi can enhance soil structure by secreting a glycoprotein slime called glomalin. Fungi glomalin production enhances aggregate formation and may also create larger pores for better growth of hyphae (Thomas et al., 1993; Jastrow et al., 1998). A lack of large pores can restrict fungal growth in soils, however glomalin production was found to be higher in small pores (0.1 mm) than in large ones allowing for more indirect fungal contact with soil (Brady and Weil, 2008). Glomalin can sequester heavy metals such as Cu, Cd, Pb and Mn in polluted soils. Gonzalez-Chavez et al. (2004) found that glomalin from hyphae of an isolate of *Gigaspora rosea* sequestered up to 28 mg Cu/g *in vitro* media.

Mycorrhization can also improve plant resistance towards heavy metal phytotoxicity by biosorption (Dueck *et al.*, 1986; Weissenhorn *et al.*, 1995a). Turnau *et al.* (1993) suggested that sequestration of metals like Cd, titanium (Ti) and barium (Ba) by polyphosphate in fungal structure might be important in minimizing transfer to the plant. Fungal sorption of heavy metals is a passive mechanism of ion immobilization on the surface of microbial cells including processes like adsorption, ion-exchange, complexation, precipitation, and crystallization on and within what may often be a multilaminate, microfibrillar cell wall rich in negatively charged ligands such as phosphoryl, carboxyl, sulfhydryl, hydroxyl, and phenolic groups (Leyval and Joner, 2001).

Lead has low mobility in soil (less than Cd and Zn) (Orlowska *et al.*, 2002) and it seems to form organic complexes with soil organic matter considering it is unavailable for plants. Also, plants have mechanisms to precipitate Pb in highly insoluble forms in the rhizosphere, such as the $PbSO_4$ (Brooks, 1995). Furthermore, sequestration of Pb in roots was found to be correlated with an increase in the number of fungal vesicles in highly colonized species. Fungal vesicles may be involved in storing toxic compounds and, thereby, could provide an additional detoxification mechanism (Göhre and Paszkowski, 2006).

4.4 Mycorrhizae and phytovolatilization of metals

A number of the elements in subgroups II, V and VI of the Periodic Table, like Hg, As and Se, form volatile hydrides or methyl derivatives that can be liberated in the atmosphere, probably as a result of the action of bacteria or soil fungi (Brooks, 1998). Metals can also be mycotransformed by such mechanisms as reduction, methylation and dealkylation.

Metalloids and some metals (e.g., As, Se, Hg, Sn, Pb) can be transformed by fungi into their methylmetal form which causes their volatilization in soil gazes and eventually in the atmosphere. In a greenhouse study, Giasson *et al.* (2006) suggested that phytoaccumulation of As and Se can slightly diminish because of phytovolatilization.

As showed by Zayed *et al.* (2000) and Giasson *et al.* (2006), Se may be lost in part by phytovolatilization in the dimethyl diselenide ($CH_3SeSeCH_3$) form. Dimethyl arsenic ($AsO(CH_3)_2(OH)$), methyl mercury (CH_3Hg^+) and tetramethyl lead ($Pb(CH_3)_4$) are the most common methylated forms of As, Hg and Pb that can also be phytovolatilized.

4.5 Mycorrhizoweathering of soil rocks and minerals

Bioavailability and toxicity of heavy metals in soils depend on their form rather than on total amounts. The availability of the eight metal fractions can be divided into three groups: (1) easily extractable and exchangeable, including water-soluble, exchangeable, and bound to reducible Fe and Mn oxides fractions; (2) potentially extractable and exchangeable, including strongly bound to minerals or weakly bound to organic matter (OM), strongly chelated by OM, bound to or occluded by carbonates, and bound to or occluded by sulphides fractions; and (3) nonextractable and nonexchangeable, found in residue fraction (Tessier *et al.*, 1979; Ma and Rao, 1997; Dinel *et al.*, 2000).

Heavy metals bound to or occluded by carbonates are more difficult to extract by vegetation. Carbonates can be the dominant heavy metal sink in a particular soil. Heavy metals may co-precipitate with carbonates incurporated in their structure, or may be sorbed by oxides (mainly Fe and Mn) that were precipitated onto the carbonates or other soil particles (Kabata-Pendias and Mukherjee, 2007). On the other hand, accumulation of heavy metals – Zn, Cd, As and Se – in plants can be enhanced by inoculation of roots by arbuscular mycorrhizal fungi (AMF) (Giasson *et al.*, 2006). Fungi produce protons, organic acids, phosphatases, and other metabolites for solubilization and complexation of metal cations (Singh, 2006).

Moreover, mycorrhizal fungi are able to acidify the rhizosphere by releasing organic acids like citric and oxalic acids (see Fig. 1) (Leyval and Joner, 2001). Oxalic acid is a leaching agent for a variety of metals, such as Al, Fe and Li, forming soluble metal oxalate complexes (Singh, 2006). The most important mechanisms for regulating heavy metal behavior by carbonates are related to variations in soil pH. Carboxylic acids released by AMF can solubilize heavy metals bound to carbonates and enhance their phytoaccumulation (Giasson *et al.*, 2005a).

Zinc and Cd speciation concentrations measurements from contaminated soil near a zinc smelter in Canada show that the metal fraction distribution is similar for Zn and Cd. In fact, the easily extractable and exchangeable fractions represent less than 27% for both Zn and Cd, which is not interesting for a phytoremediation technology. On the other hand, the two first metal fraction groups, consisting of easy and/or potentially extractable and exchangeable fractions including carbonate fraction, regroup around 86% of the metal total concentration for both Zn and Cd.

To determine if mycorrhizal fungi play a role in the speciation of heavy metals (biochemical weathering), Giasson *et al.* (2005a) used *in vitro* compartmented systems to study the mechanisms implicated in heavy metal (essential and non-essential) absorption by AMF colonized plant roots. The goal of their experiment was to determine whether mycorrhizal hyphae are directly involved in sequestration and uptake of essential Zn and non-essential Cd by plant roots, while these heavy metals were present in toxic concentrations in the Petri media. They wanted to verify the effects of endomycorrhizal (*Glomus intraradices*) hyphae on speciation of essential (Zn) and nonessential (Cd) heavy metals in order to change this water-insoluble carbonate form to a soluble and phytoavailable form.

Their results indicate that there is a solubilization of $ZnCO_3$ by hyphae and translocation to roots. Zinc saturation was reached in the *G. intraradices* colonized roots at approximately 400 µg/g, independently of initial $ZnCO_3$ concentrations. In the cadmium treatment, Cd saturation was not reached. In the lower Cd treatment, the plant to media metal ratio was 3:1, and in the higher treatment, the ratio was 1:1 (Giasson *et al.*, 2005a). In fact, mycorrhizal fungi are able to acidify the rhizosphere by releasing organic acids like citric and oxalic acids (Leyval and Joner, 2001). These organic acids can form coordination compounds or complexes with metals.

If the organic acids (e.g., citric and oxalic acids) contain two or more electron donor groups so that ring-like structures are formed, then the resulting complexes are metal chelates (Gadd, 2000). Berthelin *et al.* (2000) showed that releases of organic acids by ectomycorrhizae are efficient in weathering and solubilization of minerals by the following complexation dissolution processes:

$$M^+ (\text{Mineral})^- + HL \rightarrow H^+ (\text{Mineral})^- + ML \quad (1)$$

$$HL + LM \rightarrow L_2M + H^+ \quad (2)$$

where L = organic ligands and M^+ (Mineral)$^-$ are carbonates, phosphates, silicates and so on.

Because P availability is strongly controlled by dissolution of mineral P that can constitute a considerable portion of the available P, soil pH is a major factor in determining the relative importance of mycorrhizae in P uptake. Mineral phosphorous has greatest availability at slightly acid to near-neutral pH. At low pH, phosphorous solubility is limited by the low solubility of Fe and Al phosphates, whereas at alkaline pH phosphorous forms insoluble Ca and Mg phosphate minerals (Crowley and Alvey, 2002).

The availability of Cd from rock and mineral phosphates (apatite) can be enhanced with the release of organic acids such as tartaric acid by ectomycorrhizal fungi. *Suillus granulatus* was more efficient than *Pisolithus tinctorius* in that matter (Leyval and Joner, 2001). Mycorrhizoweathering of soil minerals (silicates, carbonates, phosphates) can enhance the availability of metals in the rhizosphere thereby enhancing plant uptake.

4.6 AMF and plant stress alleviation on mine sites

One of the main objectives in mine site reclamation is revegetation. This mining environment is characterized by poor physical and chemical conditions, poor nutrient (N, P) and organic matter contents, very low or very high pH, drought and high surface temperatures. Mycorrhizal colonization could improve vegetation establishment and survival particularly in such adverse conditions.

Young seedlings have to be protected from extremely high surface temperatures to prevent heat girdling of stems (Danielson, 1985). By colonizing the roots, the fungus enhances plant growth by making soil elements more accessible (George et al., 1992; Nadian et al., 1997; Gregory, 2006) and by improving water absorption (Sweat and Davis, 1984; Cui and Nobel, 1992). Accordingly, mycorrhizal colonization improves vegetation establishment and survival particularly in adverse conditions such as low fertility and arid soils (Jasper et al., 1989; Allen et al., 1996; Smith et al., 1998).

Mine spoils may be extremely acidic or alkaline. Acid mine drainage (AMD) is very frequent, especially in sulphide metal ore tailings, where rain water reacts with sulphide to form sulphuric acid (H_2SO_4). Leachate pH exiting from the tailings could be as low as 1. Plant roots can be colonized with mycorrhizae at pH values as low as 2.7, the critical pH for 95% maximum colonization of cassava roots varying with species from 4.4 to 4.8 (Ballen and Graham, 2002).

Hyphae of AMF may extend 8 cm from the root surface, but rhizomorphs of *Pisolithus* may extend 4 m into the soil, a result that suggests ectomycorrhizae are better adapted to long-distance transport than AMF (Danielson, 1985). Relatively few species of ectomycorrhizal symbionts have been identified as occurring on mine wastes, and of those, even fewer have been properly quantified with respect to their actual importance. To determine the degree of fungal symbiont adaptation to mine waste conditions, infection levels of each species must be quantified (Danielson, 1985). Ectomycorrhizae *Pisolithus tinctorius*, *Telephora terrestris*, and *Cenococcum geophilum* have been successfully field tested on spoils and tailings.

In their experiment, Chen et al. (2007) provided evidence for the potential use of local plant species in combination with AMF for ecological restoration of metalliferous mine tailings. It appears that considerable strain differences exist among AMF, and it would be profitable to screen isolates for adaptability to mine spoils. Old mine spoils with established vegetation may prove to be valuable sources of inoculum of adapted strains (Danielson, 1985).

5 CONCLUSION

Although usually considered important primarily for P uptake, AMF can improve assimilation of other non metallic nutrients such as N, K, S, B as well as of metallic nutrients (Zn, Cu, Mn, and others), particularly in unpolluted soils of low nutrient status. It has been suggested that mycorrhizae may benefit plant growth by increasing the availability of P from non-labile sources. The response to AMF colonization may vary among the different plant species. However, it should be considered to introduce mycorrhizae inoculums tolerant to metallic nutrients (e.g., Zn, Cu, Mn or others) into low-input agricultural soils in order to facilitate the recycling of organic, industrial and urban wastes on agricultural fields that would otherwise be extremely dangerous to agricultural ecosystems (Weissenhorn *et al.*, 1995c). For environmental considerations, mycorrhizal associations should be managed to attenuate the possibility of contaminating the soil and surface water (Jeffries *et al.*, 2003).

In order to exploit microbes as biofertilizers, biostimulants and bio-protectants against pathogens and heavy metals, ecological complexity of microbes in the mycorrhizosphere needs to be taken into consideration and optimization of rhizosphere/mycorrhizosphere systems need to be tailored (Khan, 2006). There is interspecific variation between AMF regarding translocation of metals to plants. As observed by Lasat (2002), effect of AMF associations on metal root uptake appears to be metal and plant specific. Greater root length densities, and presumably more hyphae, enable plants to explore a larger soil volume thus increasing access to cations (metals) not available to nonmycorrhizal plants (Mohammad *et al.*, 1995).

6 FUTURE RESEARCH

Arbuscular mycorrhizal fungi have great potential in the remediation of disturbed land and low fertility soil but the use of these mycorrhizae, and other beneficial microbial communities, by farmers in their fields is still lacking. Further experiments are needed to assess the ability of AMF to continue growing in the presence of multiple toxic metal or metalloid cations, either alone or in combination.

The understanding of interactions occurring between AMF and its biotic and abiotic environment is still in its infancy. The characterization of the composition of AMF exudates and the effects of these compounds on soil microbial community, plant nutrition, metal accumulation in plant shoots and shoot biomass production have implications for sustainable soil management and land rehabilitation.

REFERENCES

Al-Karaki, G.N., 2006, Nursery inoculation of tomato with arbuscular mycorrhizal fungi and subsequent performance under irrigation with saline water. *Sci. Hort.* **109**: 1–7.

Allen, M.F., Figueroa, C., Weinbaum, B.S, Barlow, S.B., and Allen, E.B., 1996, Differential production of oxalate by mycorrhizal fungi in arid ecosystems. *Biol. Fert. Soils* **22**: 287–292.

Almås, Å.R., Bakken, L.R., and Mulder, J., 2004, Changes in tolerance of soil microbial communities in Zn and Cd contaminated soils. *Soil Biol. Biochem.* **36**: 805–813.

Angle, J.S., Spiro, M.A., Heggo, A.M., El-Kherbawy, M., and Chaney, R.L., 1988, Soil microbial - legume interacts in heavy metal contaminated at Palmerton, PA., pp. 321–336. *Trace substances in the environment health*, 22nd Conference, St-Louis, MO, May 23–26.

Arines, J., Vilariño, A., and Sainz, M., 1990. Effect of vesicular-arbuscular mycorrhizal fungi on Mn uptake by red clover. *Agri. Ecosys. Environ.* **29**: 1–4.

Atkinson, D.J., Baddeley, A., Goicoechea, N., Green, J., Sanchez- Díaz, M., and Watson, C.A., 2002, Arbuscular mycorrhizal fungi in low input agriculture, pp. 211–222. *In* S. Gianinazzi, H. Schüepp, J.M. Barea, and K. Haselwandter (Eds.), *Mycorrhizal technology in agriculture: From genes to bioproducts*. Birkhäuser Verlag, Basel, Switzerland.

Audet, P., and Charest, C., 2006, Effects of AM colonization on "wild tobacco" plants grown in zinc-contaminated soil. *Mycorrhiza* **16**: 277–283.

Bai, J., Lin, X., Yin, R., Zhang, H., Junhua, W., Xueming, C., and Yongming, L., 2008, The influence of arbuscular mycorrhizal fungi on As and P uptake by maize (*Zea mays* L.) from AS-contaminated soils. *Appl. Soil Ecol.* **38**: 137–145.

Baker, A., Brooks, R., and Reeves, R., 1988, Growing for gold... and copper... and zinc. *New Sci.* **1603**: 44–48.

Ballen, K.G., and Graham, P.H., 2002, The role of acid pH in symbiosis between plants and soil organisms, pp. 383–404. *In* Z. Rengel (Ed.), *Handbook of plant growth – pH as the master variable*. Marcel Dekker, New York.

Barea, J.-M., and Jeffries, P., 1995, Arbuscular mycorrhizas in sustainable soil plant systems, pp. 521–560. *In* B. Hock and A. Varma (Eds.), *Mycorrhiza: Structure, function, molecular biology and biotechnology*. Springer, Berlin/Heidelberg, Germany.

Barea, J.-M., Azcón, R., and Azcón-Aguilar, C., 2002, Mycorrhizosphere interactions to improve plant fitness and soil quality. *Antonie van Leeuwenkoek* **81**: 343–351.

Berthelin, J., Leyval, C., and Mustin, C., 2000, Illustrations of the occurrence and diversity of mineral-microbe interactions involved in weathering of minerals, pp. 7–25. *In* J.D. Cotter-Howells, L.S. Campbell, E. Valsami-Jones, and M. Batchelder (Eds.), *Environmental mineralogy: Microbial interactions, anthropogenic influences, contaminated land and waste management*, Mineral Society Series 9. Mineral Society, London.

Bi, Y.L., Li, X.L. Christie, P., Hu, Z.Q., and Wong, M.H., 2003, Growth and nutrient uptake of arbuscular mycorrhizal maize in different depths of soil overlying coal fly ash, *Chemosphere* **50**: 863–869.

Biró, B., Posta, K., Füzy, A., Kadar, I., and Németh, T., 2005, Mycorrhizal functioning as part of the survival mechanisms of barley (*Hordeum vulgare* L.) at long-term heavy metal stress. *Acta Biol. Szegedien.* **49**: 65–67.

Boivin, M.-E.Y., Breure, A.M., Posthuma, L., and Rutgers, M., 2002, Determination of field effects of contaminants-significance of pollution-induced community tolerance. *Human Ecol. Risk Assess.* **8**: 1035–1055.

Boruvka L., and Drabek O., 2004, Heavy metal distribution between fractions of humic substances in heavy polluted soils. *Plant, Soil Environ.* **50**: 339–345.

Brady, N.C., and Weil, R.R., 2008, *The nature and properties of soils*. 14th Edition, Pearson Prentice Hall, Upper Saddle River, NJ.

Bradley, R., Burt, A.J., and Read, D.J., 1981, Mycorrhizal infection and resistance to heavy metal toxicity in *Calluna vulgaris*. *Nature* **292**: 335–337.

Brooks, R.R., 1995, *Biological systems in mineral exploration and processing*. Ellis Horwood, Toronto.

Brooks, R.R., 1998, *Plants that hyperaccumulate heavy metals: Their role in phytoremediation, microbiology, archaeology, mineral exploration and phytomining*, CAB International, New York.

Brundrett, M.C., and Abbott, L.K., 2002, Arbuscular mycorrhizas in plant communities, pp. 151–193. *In* K. Sivasithamparam, K. Dixon, and R.L. Barrett (Eds.), *Micro-organisms in plant conservation and biodiversity*. Kluwer Academic, Dordrecht, The Netherlands.

Burke, S.C., Angle, J.S., Chaney, R.L., and Cunningham, S.D., 2000, Arbuscular mycorrhizae effects on heavy metal uptake by corn. *Intern. J. Phytorem.* **2**: 23–29.

Bürkert, B., and Robson, A., 1994, ^{65}Zn uptake in subterranean clover (*Trifolium subterraneum* L.) by 3 vesicular arbuscular mycorrhizal fungi in a root-free sandy soil. *Soil Biol. Biochem.* **26**: 1117–1124.

Burleigh, S.H., Kristensen, B.K., and Bechmann, I.E., 2003, A plasma membrane zinc transporter from *Medicago truncatula* is up-regulated in roots by Zn fertilization, yet down-regulated by arbuscular mycorrhizal colonization. *Plant Mol. Biol.* **52**: 1077–1088.

Cardoso, I.M., and Kuyper, T.W., 2006, Mycorrhizas and tropical soil fertility. *Agri. Ecosys. Environ.* **116**: 72–84.

Cataldo, D.A., and Wildung, R.E., 1978, Soil and plant factors influencing the accumulation of heavy metals by plants. *Environ. Health Perspect.* **27**: 149–159.

Cavagnaro, T.R., 2008, The role of arbuscular mycorrhizas in improving plant zinc nutrition under low soil zinc concentrations: A review. *Plant Soil* **304**: 315–325.

Cavagnaro, T.R., Jackson, L.E., Scow, K.M., and Hristova, K.R., 2007, Effects of arbuscular mycorrhizas on ammonia oxidizing bacteria in an organic farm soil. *Microb. Ecol.* **54**: 618–626.

Chaudry, T.M., Hill, L., Khan, A.G., and Keuk, C., 1999, Colonization of iron and zinc-contaminated dumped filter cake waste by microbes, plants and associated mycorrhizae, pp. 275–283. *In* M.H. Wong and A.J.M. Baker (Eds.), *Remediation and management of degraded land*. CRC, Boca Raton, FL.

Chen, B.D., Li, X.L., Tao, H.Q., Christie, P., and Wong, M.H., 2003, The role of arbuscular mycorrhiza in zinc uptake by red clover growing in a calcareous soil spiked with various quantities of zinc. *Chemosphere* **50**: 839–846.

Chen, B.D., Zhu, Y.G., Duan, J., Xiao, X.Y., and Smith, S.E., 2007, Effects of the arbuscular mycorrhizal fungus *Glomus mosseae* on growth and metal uptake by four plant species in copper mine tailings. *Environ. Pollut.* **147**: 374–380.

Cheung, K.C., Zhang, J.Y., Deng, H.H., Ou, Y.K., Leung, H.M., Wu, S.C., and Wong, M.H., 2008, Interaction of higher plant (jute), electrofused bacteria and mycorrhiza on anthracene biodegradation. *Bioresour. Technol.* **99**: 2148–2155.

Christie, P., Li, X.L., and Chen, B.D., 2004, Arbuscular mycorrhizas can depress translocation of zinc to shoots of host plants in soils moderately polluted with zinc. *Plant Soil* **261**: 209–217.

Citterio, S., Prato, N., Fumagalli, P., Massa, N., Santagostino, A., Sgorbati, S., and Berta, G., 2005, The arbuscular mycorrhizal fungus *Glomus mosseae* induces growth and metal accumulation changes in *Cannabis sativa* L. *Chemosphere* **59**: 21–29.

Colpaert, J.V., 1998, Biological interactions: The significance of root-microbial symbioses for phytorestoration of metal-contaminated soils, pp. 75–91. *In* J. Vangronsveld and S. D. Cunningham (Eds.), *Metal-contaminated soils: In situ inactivation and phytorestoration*. Springer, New York.

Cooper, K.M., and Tinker, P.B., 1978, Translocation and transfer of nutrients in vesicular-arbuscular mycorrhizas. II. Uptake and translocation of phosphorus, zinc and sulfur. *New Phytol.* **81**: 43-52.

Courchesne, F., Séguin, V., and Dufresne, A., 2001, Solid phase fractionation of metals in the rhizosphere of forest soils, pp. 189–206. *In* G.R. Gobran, W.W. Wenzel, and E. Lombi (Eds.), *Trace elements in the rhizosphere*. CRC, Boca Raton, FL.

Crowley, D.E., and Alvey, S.A., 2002, Regulation of microbial processes by soil pH, pp. 351–382. *In* Z. Rengel (Ed.), *Handbook of plant growth – pH as the master variable*. Marcel Dekker, New York.

Cui, M., and Nobel, P.S., 1992, Nutrient status, water uptake and gas exchange for three desert succulents infected with mycorrhizal fungi. *New Phytol.* **122**: 643–649.

Dai, J., Becquer, T., Rouiller, J.H., Reversat, G., Bernhardt-Reversat, F., and Lavelle, P., 2004, Influence of heavy metals on C and N mineralization and microbial biomass in Zn-, Pb-, Cu-, and Cd-contaminated soils. *Appl. Soil Ecol.* **25**: 99–109.

Danielson, R.M., 1985, Mycorrhizae and reclamation of stressed terrestrial environments, pp. 173–201. *In* R.L. TateIII and D.A. Klein (Eds.), *Soil reclamation processes – microbiological analyses and applications*. Marcel Dekker, New York.

Davis, M.R.H., Zhao, F.J., and McGrath, S.P., 2004, Pollution induced community tolerance of soil microbes in response to a zinc gradient. *Environ. Toxi. Chem.* **23**: 2665–2672.

Davies, F.T., Puryear, J.D., Newton, R.J., Egilla, J.N., and Saraiva Grossi, J.A., 2001, Mycorrhizal fungi enhance accumulation and tolerance of chromium in sunflower (*Helianthus annuus*). *J. Plant Physiol.* **158**: 777–786.

Davies, F.T., Puryear, J.D., Newton, R.J., Egilla, J.N., and Saraiva Grossi, J.A., 2002, Mycorrhizal fungi increase chromium uptake by sunflower plants: Influence on tissue mineral concentration, growth, and gas exchange. *J. Plant Nutri.* **25**: 2389–2407.

Dehn, B., and Schüepp, H., 1989, Influence of VA mycorrhizae on the uptake and distribution of heavy metals in plants. *Agr. Ecosyst. Environ.* **29**: 79–83.

Del Val, C., Barea, J.M., and Azcón-Aguilar, C., 1999a, Assessing the tolerance to heavy metals of arbuscular mycorrhizal fungi isolates from sewage sludge-contaminated soils. *Appl. Soil Ecol.* **11**: 261–269.

Del Val, C., Barea, J.M., and Azcón-Aguilar, C., 1999b, Diversity of arbuscular mycorrhizal fungus populations in heavy-metal-contaminated soils. *Appl. Environ. Microbiol.* **65**: 718–723.

Deram, A., Languereau-Leman, F., Howsam, M., Petit, D., and Haluwyn, C.V., 2008, Seasonal patterns of cadmium accumulation in *Arrhenatherum elatius* (*Poaceae*): Influence of mycorrhizal and endophytic fungal colonisation. *Soil Biol. Biochem.* **40**: 845–848.

Díaz, G., and Honrubia, M., 1994, A mycorrhizal survey of plants growing on mine wastes in Southeast Spain. *Arid Soil Res. Rehab.* **8**: 59–68.

Dinel, H., Pare, T., Schnitzer, M., and Pelzer, N., 2000, Direct land application of cement kiln dust- and lime-sanitized biosolids: Extractability of trace metals and organic matter quality. *Geoderma* **96**: 307–320.

Dixon, R.K., 1988, The response of ectomycorrhizal *Quercus rubra* to soil cadmium, nickel and lead. *Soil Biol. Biochem.* **20**: 555–559.

Dixon R.K., and Buschena, C.A., 1988, Response of ectomycorrhizal *Pinus banksiana* and *Picea glauca* to heavy metals in soil. *Plant Soil* **105**: 265–271.

Douds, D.D. Jr., and Millner, P.D., 1999. Biodiversity of arbuscular mycorrhizal fungi in agroecosystems. Agric. Ecosys. Environ. **74**: 77–93.

Dueck, T.A., Visser, P., Ernst, W.H.O., and Schat, H., 1986, Vesicular-arbuscular mycorrhizae decrease zinc toxicity to grasses growing in zinc-polluted soil. *Soil Biol. Biochem.* **18**: 331–333.

Duffus, J.H., 2002, "Heavy Metals"– A Meaningless Term. *Pure Appl. Chem.* **74**: 793–807.

Ensley, B.D., 2000, Rationale for use of phytoremediation, pp. 3–12. *In* I. Raskin and B.D. Ensley (Eds.), *Phytoremediation of toxic metals. Using plants to clean up the environment.* John, Toronto.

Entry, J.A., Watrud, L.S., and Reeves, M., 1999, Accumulation of ^{137}Cs and ^{90}Sr from contaminated soil by three grass species inoculated with mycorrhizal fungi. *Environ. Pollut.* **104**: 449–457.

Fomina, M.A., Alexander, I.J., Colpaert, J.V., and Gadd, G.M., 2005, Solubilization of toxic metal minerals and metal tolerance of mycorrhizal fungi. *Soil Biol. Biochem.* **37**: 851–866.

Foster, J.W., and Hall, H.K., 1990, Adapative acification tolerance response of *Salmonella typhimurium*. *J. Bacteriol.* **172**: 771–778.

Frostegård, Å., Tunlid, A., and Bååth, E., 1993, Phospholipid fatty-acid composition, biomass, and activity of microbial communities from 2 soil types experimentally exposed to different heavy-metals. *Appl. Environ. Microbiol.* **59**: 3605–3617.

Gadd, G.M., 1990, Heavy metal accumulation by bacteria and other microorganisms. *Experientia* **46**: 834–840.

Gadd, G.M., 1993, Interactions of fungi with toxic metals. *New Phytol.* **124**: 25–60.

Gadd, G.M., 2000, Heterotrophic solubilization of metal-bearing minerals by fungi, pp. 57–75. *In* J.D. Cotter-Howells, L.S. Campbell, E. Valsami-Jones, and M. Batchelder (Eds.), *Environmental mineralogy: Microbial interactions, anthropogenic influences, contaminated land and waste management*, Mineral Society Series, 9. Mineral Society, London.

Gadd, G.M., 2005, Microorganisms in toxic metal-polluted soils, pp. 325–356. *In* F. Buscot and A. Varma (Eds.), *Microorganisms in soils: Roles in genesis and functions*. Part V. Book series: Soil biology, Vol. 3. Springer, Berlin/Heidelberg, Germany.

Gadd, G.M., and White, C., 1989, Heavy metal and radionuclide accumulation and toxicity in fungi and yeasts, pp. 19–38. *In* R.K. Poole and G.M. Gadd (Eds.), *Metal-microbe interactions*. Special publication of the Society for General Microbiology, Vol. 26. IRL Press/ Oxford University Press, New York.

Gaither, L.A., and Eide, D.J., 2001, Eukaryotic zinc transporters and their regulation. *Biometals* **14**: 251–270.

Galli, U., Schüepp, H., and Brunold, C., 1994, Heavy metal binding by mycorrhizal fungi. *Physiol. Plant.* **92**: 364–368.

George, E., Häussler, K.U., Vetterlein, K.U., Gorgus, E., and Marschner, H., 1992, Water and nutrient translocation by hyphae of *Glomus mosseae*. *Can. J. Botany*. **70**: 2130–2137.

Giasson, P., and Jaouich, A. 1998, La phytorestauration des sols contaminés au Québec. *Vecteur Environnement* **31**: 40–53.

Giasson, P., Jaouich, A., Gagné, S., and Moutoglis, P., 2005a, Endomycorrhizae involvement in Zn and Cd speciation change and phytoaccumulation. *Remediation* **15**: 75–81.

Giasson, P., Jaouich, A., Gagné, S., and Moutoglis, P., 2005b, Phytoremediation of zinc and cadmium: A study of arbuscular mycorrhizal hyphae. *Remediation* **15**: 113–122.

Giasson, P., Jaouich, A., Gagné, S., Massicotte, L., Cayer, P., and Moutoglis, P., 2006, Enhanced phytoremediation: A study of mycorrhizoremediation of heavy metal contaminated soil. *Remediation* **17**: 97–110.

Gildon, A., and Tinker, P.B., 1981, A heavy metal tolerant strain of a mycorrhizal fungus. *Trans. British Mycol. Soc.* **77**: 648–649.

Gildon, A., and Tinker, P.B., 1983a, Interactions of vesicular-arbuscular mycorrhizal infection and heavy metals in plants. 1. The effects of heavy metals on the development of vesicular-arbuscular mycorrhizas. *New Phytol.* **95**: 247–261.

Gildon, A., and Tinker, P.B., 1983b, Interactions of vesicular arbuscular mycorrhizal infection and heavy-metals in plants. 2. The effects of infection on uptake of copper. *New Phytol.* **95**: 263–268.

Giller, K.E., Witter, E., and McGrath, S., 1998, Toxicity of heavy metals to microorganisms and microbial processes in agricultural soils: A review. *Soil Biol. Biochem.* **30**: 1389–1414.

Göhre, V., and Paszkowski, U., 2006, Contribution of the arbuscular mycorrhizal symbiosis to heavy metal phytoremediation. *Planta* **223**: 1115–1122.

Gonzalez-Chavez, C., D'Haen, J., Vangronsveld, J., and Dodd, J.C., 2002a, Copper sorption and accumulation by the extraradical mycelium of different *Glomus* spp. (arbuscular mycorrhizal fungi) isolated from the same polluted soil. *Plant Soil* **240**: 287–297.

Gonzalez-Chavez, C., Harris, P.J., Dodd, J., and Meharg, A.A., 2002b, Arbuscular mycorrhizal fungi confer enhanced arsenate resistance on *Holcus lanatus*. *New Phytol.* **155**: 163–171.

Gonzalez-Chavez, M.C., Carrillo-Gonzalez, R., Wright, S.F., and Nichols, K.A., 2004, The role of glomalin, a protein produced by arbuscular mycorrhizal fungi, in sequestering potentially toxic elements. *Environ. Pollut.* **130**: 317–323.

Gregory, P.J., 2006, *Plant roots. growth, activity and interaction with soils*. Blackwell, Oxford.

Hamel, C., and Plenchette, C., 2007, *Mycorrhizae in crop production*. Haworth, Binghampton, NY.

Harrison, M.J., 2005, Signaling in the arbuscular mycorrhizal symbiosis. *Annu. Rev. Microbiol.* **59**: 19–42.

Harrison, M.J., Dewbre, G.R., and Liu, J., 2002, A phosphate transporter from *Medicago truncatula* involved in the acquisition of phosphate released by arbuscular mycorrhizal fungi. *Plant Cell* **14**: 2413–2429.

Hetrick, B.A.D., Wilson, G.W.T., and Figge, D.A.H., 1994, The influence of mycorrhizal symbiosis and fertilizer amendments on establishment of vegetation in heavy metal mine spoil. *Environ. Pollut.* **86**: 171–179.

Hildebrandt, U., Regvar, M., and Bothe, H., 2007, Arbuscular mycorrhiza and heavy metal tolerance. *Phytochem.* **68**: 139–146.

Hinojosa, M.B., Carreira, J.A., García-Ruíz, R., and Dick, R.P., 2005, Microbial response to heavy metal–polluted soils. *J. Environ. Qual.* **34**: 1789–1800.

Holtan-Hartvik, L., Bechman, H., Høyås, T.R., Linjordet, R., and Bakken, L.R., 2002, Heavy metals tolerance of soil denitrifying communities: N_2O dynamics. *Soil Biol. Biochem.* **34**: 1181–1190.

Homer, F.A., Reeves, R.D., and Brooks, R.R., 1997, The possible involvement of aminoacids in nickel chelation in some nickel-accumulating plants. *Curr. Top. Phytochem.* **14**: 31–33.

Hovsepyan A., and Greipsson, S., 2004, Effect of arbuscular mycorrhizal fungi on phyto-extraction by corn (*Zea mays*) of lead-contaminated soil. *Intern. J. Phytorem.* **6**: 305–321.

Hughes, M.N., and Poole, R.K., 1989, Metal mimicry and metal limitation in studies of metal – microbe interactions, pp. 1–17. In R.K. Poole and G.M. Gadd (Eds.), *Metal – microbe interactions*. Society for General Microbiology, IRL Press/Oxford University Press, New York.

Hutchinson, S.L., Schwab, A.P., and Banks, M.K., 2003, Biodegradation of petroleum hydrocarbons in the rhizosphere, pp. 355–386. *In* S.C. McCutcheon and J.L. Schnoor (Eds.), *Phytoremediation*. Wiley-Interscience, Hoboken, NJ.

Ietswaart, J.H., Griffioen, W.A.J., and Ernst, W.H.O., 1992, Seasonality of VAM infection in three populations of *Agrostis capillaries* (*Gramineae*) on soil with or without heavy metal enrichment. *Plant Soil* **139**: 67–73.

Jackson, A.P., and Alloway, B.J., 1992, The transfer of cadmium from agricultural soils to the human food chain, pp. 109–158. *In* D.C. Adriano (Ed.), *Biogeochemistry of trace metals*. Lewis, Boca Raton, FL.

Jackson, L.E., Burger, M., and Cavagnaro, T.R., 2008, Roots, nitrogen transformations, and ecosystem services. *Annu. Rev. Plant Biol.* **59**: 341–363.

Jamal, A., Ayub, N., Usman, M., and Khan, A.G., 2002, Arbuscular mycorrhizal fungi enhance zinc and nickel uptake from contaminated soil by soybean and lentil. *Intern. J. Phytorem.* **4**: 205–221.

Jansa, J., Smith, F.A., and Smith, S.E., 2008, Are there benefits of simultaneous root colonization by different arbuscular mycorrhizal fungi? *New Phytol.* **177**: 779–789.

Jasper, D.A., Abbott, L.K., and Robson, A.D., 1989, Hyphae of a vesicular-arbuscular mycorrhizal fungus maintain infectivity in dry soil, except when the soil is disturbed. *New Phytol.* **112**: 101–107.

Jastrow, J.D., Miller, R.M., and Lussenhop, J., 1998, Contributions of interacting biological mechanisms to soil aggregate stabilization in restored prairie. *Soil Biol. Biochem.* **30**: 905–916.

Jeffries, P., Gianinazzi, S., Perotto, S., Turnau, K., and Barea, J.M., 2003, The contribution of arbuscular mycorrhizal fungi in sustainable maintenance of plant health and soil fertility. *Biol. Fert. Soils* **37**: 1–16.

Joner, E.J., and Leyval, C., 1997, Uptake of ^{109}Cd by roots and hyphae of a *Glomus mosseae/Trifolium subterraneum* mycorrhiza from soil amended with high and low concentrations of cadmium. *New Phytol.* **135**: 353–360.

Joner, E.J., Briones, R., and Leyval, C., 2000, Metal-binding capacity of arbuscular mycorrhizal mycelium. *Plant Soil* **226**: 227–234.

Jun, J., Abubaker, J., Rehrer, C. Pfeffer, P.E., Shachar-Hill, Y., and Lammers, P.J., 2002, Expression in an arbuscular mycorrhizal fungus of genes putatively involved in metabolism, transport, the cytoskeleton and the cell cycle. *Plant Soil* **244**: 141–148.

Kabata-Pendias, A., 2001, *Trace elements in soils and plants*. 3rd Edition, CRC, Boca Raton, FL.

Kabata-Pendias, A., and Mukherjee, A.B., 2007, *Trace elements from soil to human*. Springer, Berlin/Heidelberg, Germany/New York.

Kapoor, R., and Viraraghavan, T., 1998, Biosorption of heavy metals on *Aspergillus niger*: Effect of pre-treatment. *Bioresour. Technol.* **63**: 109–113.

Karam, A., 2007, Métaux lourds et environnement du sol. Notes de cours. Département des sols et de génie agroalimentaire. Université Laval. Québec, Canada.

Karam, A., and De Coninck, A.S., 2007, Effect of turbot residue amendment on the sorption and desorption of cadmium in an acid loamy sand soil, pp. 384–385. *In* Abad Chabbi (Ed.), Proceedings of the International Symposium on Organic Matter Dynamics in Agro-Ecosystems, University of Poitiers, Les Presses de l'Imprimerie Oudin Poitiers, France. ISBN 978-2-7380-1245-6.

Karam, A., Côté, C., and Parent, L.É., 2003, Retention of copper in Cu-enriched organic soils, pp. 137–150. *In* L.-E. Parent and P. Ilnicki (Eds.), *Organic soils and peat materials for sustainable agriculture*. CRC LLC, Boca Raton, FL.

Karanika, E.D., Voulgari, O.K., Mamolos, A.P., Alifragis, D.A., and Veresoglou, D.S., 2008, Arbuscular mycorrhizal fungi in northern Greece and influence of soil resources on their colonization. *Pedobiologia*: 409–418.

Keller, C., McGrath, S.P., and Dunham, S.J., 2002, Trace metal leaching through a soil–grassland system after sewage sludge application. *J. Environ. Qual.* **31**: 1550–1560.

Khan, A.G., 2006, Mycorrhizoremediation – an enhanced form of phytoremediation. *J. Zhejiang Univ. Sci. B* **7**: 503–514.

Killham, K., and Firestone, M.K., 1983, Vesicular arbuscular mycorrhizal mediation of grass response to acidic and heavy metal deposition. *Plant Soil* **72**: 39–48.

Kistner, C., and Parniske, M., 2002, Evolution of signal transduction in intracellular symbiosis. *Trends Plant Sci.* **7**: 511–518.

Kothari, S.K., Marschner, H., and Römheld, V., 1990, Direct and indirect effects of VA mycorrhizal fungi and rhizosphere microorganisms on acquisition of mineral nutrients by maize (*Zea mays* L.) in a calcareous soil. *New Phytol.* **116**: 637–645.

Kucey, R.M.N., and Janzen, H.H., 1987, Effects of VAM and reduced nutrient availability on growth and phosphorus and micronutrient uptake of wheat and field beans under greenhouse conditions. *Plant Soil* **104**: 71–78.

Kumar, P.B.A.N., Dushenkov, V., Motto, H., and Raskin, I., 1995, Phytoextraction: The use of plants to remove heavy metals from soils. *Environ. Sci. Technol.* **29**: 1232–1238.

Laheurte, F., Leyval, C., and Berthelin, J., 1990, Root exudates of maize, pine and beech seedlings influenced by mycorrhizal and bacterial inoculation. *Symbiosis* **9**: 111–116.

Lambert, D.H., and Weidensaul, T.C., 1991, Element uptake by mycorrhizal soybean from sewage-sludge-treated soil. *Soil Sci. Soc. Am. J.* **55**: 393–398.

Lasat, M.M., 2002, Phytoextraction of toxic metals: A review of biological mechanisms. *J. Environ. Qual.* **31**: 109–120.

Lepp, N.W., 1992, Uptake and accumulation of metals in bacteria and fungi, pp. 277–298. *In* D.C. Adriano (Ed.), *Biogeochemistry of trace metals*. Lewis, Boca Raton, FL.

Leyval, C., and Joner, E.J., 2001, Bioavailability of heavy metals in the mycorrhizosphere, pp. 165–185. *In* G.R. Gobran, W.W. Wenzel, and E. Lombi (Eds.), *Trace elements in the rhizosphere*. CRC, Boca Raton, FL.

Leyval, C., Berthelin, J., Schontz, D., Weissenhorn, I., and Morel, J.L., 1991, Influence of endomycorrhizas on maize uptake of Pb, Cu, Zn, and Cd applied as mineral salts or sewage sludge, pp. 204–207. *In* J.G. Farmer (Ed.), *Heavy metals in the environment*, CEP Consultants, Edinburgh, UK.

Leyval, C. Turnau, K., and Haselwandter, K., 1997, Effect of heavy metal pollution on mycorrhizal colonization and function: Physiological, ecological and applied aspects. *Mycorrhiza* **7**: 139–153.

Li, X.L, Marschner, H., and George, E., 1991, Acquisition of phosphorus and copper by VA-mycorrhizal hyphae and root to shoot transport in white clover. *Plant Soil* **136**: 49–57.

Lingfei, Li., Anna, Y., and Zhiwei, Z., 2005, Seasonality of arbuscular mycorrhizal symbiosis and dark septate endophytes in a grassland site in southwest China. *FEMS Microbiol. Ecol.* **54**: 367–373.

Liu, W., and Lianfeng, D., 2008, Interactions between Bt transgenic crops and arbuscular mycorrhizal fungi : A new urgent issue of soil ecology in agroecosystems. *Acta Agri. Scandin. section B., Soil & Plant Science* **58**: 187–192.

Lombi, E., Wenzel, W.W., Gobran, G.R., and Adriano, D.C., 2001, Dependency of phyto-availability of metals on indigenous and induced rhizosphere processes: A review, pp. 3–24. *In* G.R. Gobran, W.W. Wenzel, and E. Lombi (Eds.), *Trace elements in the rhizosphere*. CRC, New York.

Lozet, J., and Mathieu, C., 1991, *Dictionary of soil science*. 2nd Edition, A.A. Balkema, Rotterdam, The Netherlands.

Lux, H.B., and Cumming, J.R., 2001, Mycorrhizae confer aluminium resistance to tulip-poplar seedlings. *Can. J. For. Res.* **31**: 694–702.

Ma, L.Q., and Rao, G.N., 1997, Heavy metals in the environment-chemical fractionation of cadmium, copper, nickel, and zinc in contaminated soils. *J. Environ. Qual.* **26**: 259–264.

Marie-Victorin, F. 1964, *Flore laurentienne*. Les Presses de l'Université de Montréal, Montreal, Canada.

Marschner, P., 2007, Plant-microbe interactions in the rhizosphere and nutrient cycling, pp. 159–182. *In* P. Marschner and Z. Rengel (Eds.), *Nutrient cycling in terrestrial ecosystems*. Part I. Book series: Soil biology, Vol. 10. Springer, Berlin/Heidelberg, Germany.

Marschner, P., Jentschke, G., and Godbold, D.L., 1998, Cation exchange capacity and lead sorption in ectomycorrhizal fungi. *Plant Soil* **205**: 93–98.

Martin, F., Perotto, S., and Bonfante, P., 2007, Mycorrhizal fungi: A fungal community at the interface between soil and roots, pp. 201–236. *In* R. Pinton, Z. Varanini, and P. Nannipieri (Eds.), *The rhizosphere: Biochemistry and organic substances at the soil-plant interface*. Marcel Dekker, New York.

McGrath, S.P., 1994, Effects of heavy metals from sewage sludge on soil microbes in agricultural ecosystems, pp. 247–274. *In* S.M. Ross (Ed.), *Toxic metals in soil-plant systems*, John , Chichester, UK.

McGrath, S.P., Chaudri, A.M., and Giller, K.E., 1995, Long-term effects of metals in sewage sludge on soils, microorganisms and plants. *J. Indus. Microbiol.* **14**: 94–104.

McIntyre, T., 2003, Phytoremediation of heavy metals from soils, pp. 887–904. *In* D.T. Tsao (Ed.), *Phytoremediation.*, Vol. 78. Advances in biochemical engineering biotechnology. Springer, New York.

Meharg, A.A., Bailey, J., Breadmore, K., and Macnair, M.R., 1994, Biomass allocation, phosphorus nutrition and vesicular-arbuscular mycorrhizal infection in clones of

Yorkshire Fog, *Holcus lanatus* L. (*Poaceae*) that differ in their phosphate uptake kinetics and tolerance to arsenate. *Plant Soil* **160**: 11–20.

Mench, M., and Martin, E., 1991, Mobilization of cadmium and other metals from two soils by root exudates by *Zea mays* L., *Nicotina tabacum* L., and *Nicotina rustica*. *Plant Soil* **132**: 187–196.

Mhatre, G.N., and Pankhurst, C.E., 1997, Bioindicators to detect contamination of soils with special reference to heavy metals, pp. 349–369. *In* C.E. Pankhurst, B.M. Doube, and V.V.S.R. Gupta (Eds.), *Biological indicators of soil health*. CAB International, New York.

Mohammad, M.J., Pan, W.L., and Kennedy, A.C., 1995, Wheat responses to vesicular-arbuscular mycorrhizal fungal inoculation of soils from eroded toposequence. *Soil Sci. Soc. Am. J.* **59**: 1086–1090.

Morley, G.F., and Gadd, G.M., 1995, Sorption of toxic metals by fungi and clay minerals. *Mycol. Res.* **99**: 1429–1438.

Morris, C., 1992, *Academic press dictionary of science and technology*. Academic, San Diego, CA.

Muchovej, R.M., 2001, *Importance of mycorrhizae for agriculture crops*. University of Florida Extension Service. Pamphlet SS-AGR-170, 5 pp. Available on line.

Mullen, M.D., Wolf, D.C., Beveridge, T.J., and Bailey, G.W., 1992, Sorption of heavy metals by the soil fungi *Aspergillus niger* and *Mucor rouxii*. *Soil Biol. Biochem.* **24**: 129–135.

Murphy, R.T., and Levy, J.F., 1983, Production of copper oxalate by some copper tolerant fungi. *Trans. British Mycol. Soc.* **81**: 165–168.

Nadian, H., Smith, S.E., Alston, A.M., and Murray, R.S., 1997, Effects of soil compaction on plant growth, phosphorus uptake and morphological charecteristics of vesicular-arbuscular mycorrhizal colonization of *Trifolium subterraneum*. *New Phytol.* **135**: 303–311.

Naidu, R. Oliver, D., and McConnel, S., 2003, Heavy metal phytotoxicity in soils, pp. 235–241. *In* A. Langley, M. Gilbey, and B. Kennedy (Eds), *Proceedings of the Fifth National Workshop on the Assesment of Site Contamination*. National Environment Protection Council (NEPC), Adelaide, Australia.

Niklińska, M., Chodak, M., and Laskowski, R., 2006, Pollution-induced community tolerance of microorganisms from forest soil organic layers polluted with Zn or Cu. *Appl. Soil Ecol.* **32**: 265–272.

Orlowska, E., Zubek, Sz, Jurkiewicz, A. Szarek-Lukaszewska, G., and Turnau, K., 2002, Influence of restoration on arbuscular mycorrhiza of *Biscutella laevigata* L. (*Brassicaceae*) and *Plantago lanceolata* L. (*Plantaginaceae*) from calamine spoil mounds. *Mycorrhiza* **12**: 153–160.

Ouziad, F., Hildebrandt, U., Schmelzer, E., and Bothe, H., 2005, Differential gene expressions in arbuscular mycorrhizal-colonized tomato grown under heavy metal stress. *J. Plant Physiol.* **162**: 634–649.

Pandolfini, T., Gremigni, P., and Gabbrielli, R., 1997, Biomonitoring of soil health by plants, pp. 325–347. *In* C.E. Pankhurst, B.M. Doube, and V.V.S.R. Gupta (Eds.), *Biological indicators of soil health*. CAB International, New York.

Paszkowski, U., 2006, A journey through signaling in arbuscular mycorrhizal symbioses 2006. *New Phytol.* **172**: 35–46.

Pawlowska, T.E., Blaszkowski, J., and Ruhling, A., 1996, The mycorrhizal status of plants colonizing a calamine spoil mound in southern Poland. *Mycorrhiza* **6**: 499–505.

Phipps, D.A., 1981, Chemistry and biochemistry of trace metals in biological systems. *In* N. W. Lepp (Ed.), *Effect of heavy metal pollution on plants*. Applied Science Publishers, Barking, UK.

Pichtel, J., and Salt, C.A., 1998, Vegetative growth and trace metal accumulation on metalliferous wastes. *J. Environ. Qual.* **27**: 618–624.

Piotrowski, J.S., Morford, S.L., and Rillig, M.C., 2008, Inhibition of colonization by a native arbuscular mycorrhizal fungal community via *Populus trichocarpa* litter, litter extract, and soluble phenolic compounds. *Soil Biol. Biochem.* **40**: 709–717.

Rand, G.M., Wells, P.G., and McCarty, L.S., 1995, Introduction to aquatic toxicology, pp. 3–67. *In* G.M. Rand (Ed.), *Fundamentals of aquatic toxicology*. Taylor & Francis, Washington, DC.

Rayner, M.H., and Sadler, P.J., 1989, Cadmium accumulation and resistance mechanisms in bacteria, pp. 39–47. *In* R.K. Poole and G.M. Gadd (Eds.), *Metal – microbe interactions*. Society for General Microbiology, IRL Press/Oxford University Press, New York.

Roesti, D., Ineichen,K., Braissant, O., Redecker, D., Wiemken, A., and Aragno, M., 2005, Bacteria associated with spores of the arbuscular mycorrhizal fungi *Glomus geosporum* and *Glomus constrictum*. *Appl. Environ. Microbiol.* **71**: 6673–6679.

Rufyikiri, G., Thiry, Y., Wang, L., Delvaux, B., and Declerck, S., 2002, Uranium uptake and translocation by the arbuscular mycorrhizal fungus, *Glomus intraradices*, under root-organ culture conditions. *New Phytol.* **156**: 275–281.

Rufyikiri, G., Thiry, Y., and Declerck, S., 2003, Contribution of hyphae and roots to uranium uptake and translocation by arbuscular mycorrhizal carrot roots under root-organ culture conditions. *New Phytol.* **158**: 391–399.

Sabrana, C., Avio, L., and Giovannetti, M., 1995, The occurrence of calcofluor and lectin-binding polysaccharides in the outer wall of arbuscular mycorrhizal fungal spores. *Mycol. Res.* **99**: 1249–1252.

Salido, A.L., Hasty, K.L., Lim, J.-M., and Butcher, D.J., 2003, Phytoremediation of arsenic and lead in contaminated soil using Chinese brake ferns (*Pteris vittata*) and Indian mustard (*Brassica juncea*). *Int. J. Phytoremed.* **5**: 89–103.

Salt, D.E., Prince, R.C., Baker, A.J.M., Raskin, I, and Pickering I.J., 1999, Zinc ligands in the metal hyperaccumulator *Thlaspi caerulescens* as determined using X-ray absorption spectroscopy. *Environ. Sci. Technol.* **33**: 713–717.

Schnoor, J.L. 1997, *Phytoremediation*. Ground-Water Remediation Technologies Analysis Center, Technology Evaluation Report TE-98-01, Pittsburgh, PA.

Schüßler, A., Schwarzott, D., and Walker, C., 2001, A new fungal phylum, the *Glomeromycota*: Phylogeny and evolution. *Mycol. Res.* **105**: 1413–1421.

Shetty, K.G., Hetrick, B.A.D., Figge, D.A.H., and Schwab, A.P., 1994, Effects of mycorrhizae and other soil microbes on revegetation of heavy metal contaminated mine spoil. *Environ. Pollut.* **86**: 181–188.

Siddiqui, Z.A., 2006, *PGPR: Biocontrol and Biofertilization*. Springer, The Netherlands.

Silver, S., Laddaga, R.A., and Misra, T.K., 1989, Plasmid-determined resistance to metal ions, pp. 49–63. *In* R.K. Poole and G.M. Gadd (Eds.), *Metal – microbe interactions*. Society for General Microbiology, IRL Press/Oxford University Press, New York.

Singh, H., 2006, Mycorrhizal fungi in rhizosphere bioremediation, pp. 533–572. *In* H. Singh (Ed.), *Mycoremediation: Fungal bioremediation*. John, New York.

Smith, M.R., Charvat, I., and Jacobson, R.L., 1998, Arbuscular mycorrhizae promote establishment of prairie species in a tallgrass prairie restoration. *Can. J. Bot.* **76**: 1947–1954.

Smith, R.A.H., and Bradshaw, A.D., 1979, The use of metal tolerant plant populations for the reclamation of metalliferous wastes. *J. Appl. Ecol.* **16**: 595–612.

Smith, S.E. and Read, D.J., 1997, *Mycorrhizal symbiosis*. 2nd Edition, Academic, London.

Sudová, R., and Vosátka, M., 2007, Differences in the effects of three arbuscular mycorrhizal fungal on P and Pb accumulation by maize plants. *Plant Soil* **296**: 77–83.

Sudová, R., Jurkiewicz, A., Turnau, K., and Vosátka, M., 2007, Persistence of heavy metal tolerance of the arbuscular mycorrhizal fungus *Glomus intraradices* under different cultivation regimes. *Symbiosis* **43**: 71–81.

Sweatt, M.R., and Davis, F.T. Jr., 1984, Mycorrhizae, water relations, growth, and nutrient uptake of geranium grown under moderately high phosphorus regimes. *J. Am. Soc. Horti. Sci.* **109**: 210–213.

Tessier, A., Campbell, P.G.C., and Bisson, M., 1979, Sequential extraction procedure for the speciation of particulate trace metals. *Analy. Chem.* **51**: 844–851.

Thomas, R.S., Franson, R.L., and Bethlenfalvay, G.J., 1993, Separation of vesicular-arbuscular mycorrhizal fungus and root effects on soil aggregation. *Soil Sci. Soc. Am. J.* **57**: 77–81.

Toler, H.D., Morton, J.B., and Cumming, J.R., 2005, Growth and metal accumulation of mycorrhizal sorghum exposed to elevated copper and zinc. *Water Air Soil Poll.* **164**: 155–172.

Toljander, J., 2006, *Interactions between soil bacteria and arbuscular mycorrhizal fungi.* Doctoral dissertation, Department of Forest Mycology and Pathology, Faculty of Natural Resources and Agricultural Sciences, SLU. Acta Universitatis Agriculturae Sueciae, Vol. 39.

Tosun, H., and Gönül, S.A., 2005, The effect of acid adaptation conditions on acid tolerance response of *Escherichia coli* O157:H7. *Turk. J. Biol.* **29**: 197–202.

Tullio, M., Pierandrei, F. Salerno, A., and Rea, E., 2003, Tolerance to cadmium of vesicular arbuscular mycorrhizae spores isolated from a cadmium-polluted and unpolluted soil. *Biol. Fert. Soils* **37**: 211–214.

Turnau, K., and Mesjasz-Przybylowicz, J., 2003, Arbuscular mycorrhizal of *Berkheya coddii* and other Ni-hyperaccumulating members of *Asteraceae* from ultramafic soils in South Africa. *Mycorrhiza* **13**: 185–190.

Turnau, K., Kottke, I., and Oberwinkler, F., 1993, Element localisation in mycorrhizal roots of *Pteridium aquilinum* (L.) Kuhn collected from experimental plots treated with cadmium dust. *New Phytol.* **123**: 313–324.

Varma, A., and Hock, B., 1999, Mycorrhiza: Structure, function, molecular biology, and biotechnology. 2nd Edition, Springer, New York.

Voegelin, A., Barmettler, K., and Kretzschmar, R. 2003, Heavy metal release from contaminated soils: Comparison of column leaching and batch extraction results. *J. Environ. Qual.* **32**: 865–875.

Wang, B., and Qiu, Y.L. 2006, Phylogenetic distribution and evolution of mycorrhizas in land plants. *Mycorrhiza* **16** : 299–363.

Wang, F., Lin, X., and Yin, R., 2005, Heavy metal uptake by arbuscular mycorrhizas of *Elsholtzia splendens* and the potential for phytoremediation of contaminated soil. *Plant Soil* **269**: 225–232.

Weissenhorn, I., and Leyval, C., 1995, Root colonization of maize by a Cd-sensitive and a Cd-tolerant Glomus mosseae and cadmium uptake in sand culture. *Plant Soil* **175**: 233–238.

Weissenhorn, I., Leyval, C., and Berthelin, J., 1993, Cd-tolerant arbuscular mycorrhizal (AM) fungi from heavy metal-polluted soils. *Plant Soil* **157**: 247–256.

Weissenhorn, I., Glashoff, A., Leyval, C., and Berthelin, J., 1994, Differential tolerance to Cd and Zn of arbuscular mycorrhizal (AM) fungal spores isolated from heavy metal polluted and unpolluted soils. *Plant Soil* **167**: 189–196.

Weissenhorn, I., Leyval, C., Belgy, G., and Berthelin, J., 1995a, Arbuscular mycorrhizal contribution to heavy metals uptake by maize (*Zea mays* L.) in pot culture with contaminated soil. *Mycorrhiza* **5**: 245–251.

Weissenhorn, I., Leyval, C., and Berthelin, J., 1995b, Bioavailablility of heavy metals and arbuscular mycorrhiza in a soil polluted by atmospheric deposition from a smelter. *Biol. Fert. Soils* **79**: 2228.

Weissenhorn, I., Mench, M., and Leyval, C., 1995c, Bioavailability of heavy metals and abundance of arbuscular mycorrhizas in a sewage sludge amended sandy soil. *Soil Biol. Biochem.* **27**: 287–296.

Wilkins, D.A., 1991, The influence of sheathing (ecto-) mycorrhizas of tree on the uptake and toxicity of metals. *Agricul. Ecosys. Environ.* **35**: 245–260.

Zayed, A., Pilon-Smits, E., Desouza, M., Lin, Z-Q., and Terry, N., 2000, Remediation of selenium-polluted soils and waters by phytovolatilization, pp. 61–83. *In* N. Terry and G. Banuelos (Eds.), *Phytoremediation of contaminated soil and water*. CRC LLC, New York.

Zhou, J.L., 1999, Zn biosorption by *Rhizopus arrhizus* and other fungi. *Appl. Microbiol. Biotechnol.* **51**: 686–693.

Zhu, Y.G., Christie., P., and Laidlaw, A.S., 2001, Uptake of Zn by arbuscular mycorrhizal white clover from Zn-contaminated soil. *Chemosphere* **42**: 193–199.

Chapter 5

ARBUSCULAR MYCORRHIZAL FUNGI COMMUNITIES IN MAJOR INTENSIVE NORTH AMERICAN GRAIN PRODUCTIONS

M.S. BEAUREGARD[1,2], C. HAMEL[2], AND M. ST.-ARNAUD[1]

[1]Institut de recherche en biologie végétale, Jardin botanique de Montréal, 4101, Sherbrooke St Est, Montréal, QC, H1X 2B2, Canada; [2]Semiarid Prairie Agricultural Research Centre, Agriculture and Agri-Food Canada, 1 Airport Road, Box 1030, Swift Current, SK, S9H 3X2, Canada

Abstract: With population increase, urban sprawl on some of the best agricultural soils and the interest for biofuels, serious pressures have been created on grain and oilseeds production in North America. Fertilizers are the main expense in intensive agricultural management practices. P fertilization is often closely related with soil degradation and contamination of surface water, causing eutrophication and accumulation of blue-green algae in certain locations of Canada. Arbuscular mycorrhizal (AM) symbioses have been shown to benefit plant growth in large part due to the very extensive hyphal network development in soil, exploiting nutrients more efficiently and improving plant uptake. AM symbiosis also increases resistance to stress and reduces disease incidence, representing a key solution in sustainable agriculture. Appropriate management of mycorrhizae in agriculture should allow a substantial reduction in chemical use and production costs. This chapter will review the effects of various fertilization practices on AMF community structure and crop productivity in major North American grain productions (i.e., corn, soybean, wheat, barley), and their reaction to other common management practices (i.e., tillage, rotation, pesticide use).

Keywords: Arbuscular mycorrhizal fungi; intensive agriculture; grain production.

1   INTRODUCTION

With population increase, urban sprawl on some of the best agricultural soils and growing interest for biofuels, serious pressures have been created on grain and oilseeds production in North America. However, agriculture still occupies an important part of North American territory and remains an important part of the economy (Table 1).

Many forms of crop management have been used systematically by farmers since many decades with the aim of rapidly increasing crop productivity. Among them mineral fertilizers now represent one of the main expenses in intensive agricultural management practices (Heffer and Prud'homme, 2006). Organic fertilizers such as manures and composts are easily available for most farmers but their nutrient content is often uneven and unpredictable from year to year. No matter which form of fertilizer is applied, conventional farming generates large nitrogen (N) and phosphorus (P) surplus, which can lead to both N and P leaching (Brady and Weil, 2002). Not only is there a cost for farmers associated to this loss, but the phenomenon has also been related to soil contamination, and can be a major threat for aquatic systems through surface and groundwater degradation (Kirchmann and Thorvaldsson, 2000). The role of P in anthropogenic eutrophication of water bodies is well known since many decades (Imboden, 1974). More recently, fertilizer runoff from agricultural fields was emphasized among the causes of excessive cyanobacteria growth and increasing of potentially harmful blooms leading to restricted access to lakes in certain locations of Canada and United States.

Table 1. North American cereal cropping importance according to the number of cultivated hectares (ha) and tonnes harvested in 2003. (FAO, 2004; Statistics Canada, 2007; USDA, 2007; http://www.fao.org/statistics/yearbook/vol_1_1/index.asp; http://www40.statcan.ca/l01/ cst01 /prim11a.htm; http://www.nass.usda.gov/About_NASS/index.asp)

Crops		Canada	Mexico	USA
Barley	ha	49,894,000	N.A.	2,000,000
	tonnes	12,327,600	1,109,424	6,011,080
Maize	ha	1,226,100	7,780,880	28,789,240
	tonnes	9,587,300	19,652,416	256,904,560
Soybean	ha	10,528,000	N.A.	29,000,000
	tonnes	2,268,300	75,686	65,795,340
Wheat	ha	10,467,400	626,517	21,383,410
	tonnes	23,552,000	3,000,000	63,589,820

N.A. = data not available

Arbuscular mycorrhizal fungi (AMF) are estimated to associate with over 90% of vascular plants, including most agricultural crops (Read *et al.*, 1976; Smith and Read, 1997). Their extensive soil hyphal network development has been largely studied. These obligatory symbionts have been demonstrated to benefit the growth of numerous plant species by improving their nutrient uptake as well as increasing their resistance to abiotic stresses and reducing damages caused by pathogenic microorganisms (Smith and Read, 1997; Clark and Zeto, 2000; Barea *et al.*, 2002; Schloter *et al.*, 2003; St-Arnaud and Vujanovic, 2007). A more appropriate management of mycorrhizae in agriculture is expected to allow a substantial reduction in the amount of minerals used without losses in productivity, whereas permitting a more sustainable production management. This chapter will review the AMF contribution to crop productivity and effects of various fertilization practices on AMF community structure in major North American grain productions (i.e., corn, soybean, wheat, barley), and their response to other common management practices (i.e., tillage, rotation, fallow).

2 ARBUSCULAR MYCORRHIZAE CONTRIBUTION TO GRAIN PRODUCTION

Several studies have been conducted to confirm AM symbiosis positive influence on major edible plants (Table 2). A survey published in 1988 reported that in 78 field trials, increased AMF colonization resulted in an average yield increase of 37% (McGonigle, 1988). A more recent meta-analysis of 290 field and greenhouse studies published between 1988 and 2003 confirmed this relation between colonization extent and crop productivity, and determined that increased colonization resulted in an overall 23% yield increase (Lekberg and Koide, 2005). It has been concluded that soybean, maize, barley and wheat yields were all increased by AMF colonization in greenhouse trials (Karagiannidis and Hadjisavvazinoviadi, 1998; Ilbas and Sahin, 2005; Lekberg and Koide, 2005; Nourinia et al., 2007).

Crop plants show variations in their dependence on mycorrhizae for nutrient uptake meaning that the ones with roots that cannot seek P efficiently receive the most benefit from mycorrhizal symbiosis. Other factors such as root surface area, root hair abundance and length, growth rate, response to soil conditions and exudations can be related to the plant dependency on AM symbiosis for nutrient uptake. Some crops are considered as facultative mycotrophs, while others are seen as obligate mycotrophs (Smith and Read, 1997). Various plants, such as leek or corn, are highly dependent on mycorrhizae to meet their basic P requirements, while others like wheat, barley and oat, benefit from the symbiosis but are less dependent (Plenchette, 1983; Ryan and Angus, 2003). Ryan et al. (2005) even noticed that high colonization by AMF was associated with reduced growth of winter wheat in low-P condition, strengthening the hypothesis that different plant species do not benefit equally from AMF because some of them acquire more nutrients from the symbiosis than others (Smith and Read, 1997). Screening of maize inbred lines to study their tolerance to low-P stress conditions has brought evidence that there are genetic variations in P uptake efficiency (Da Silva and

Table 2. Recent studies showing beneficial impacts of AMF on various edible plants cultivated in North America.

Type of plant	Conclusions	References
Citrus	• Enhanced drought tolerance of tangerine • Increased citrus P, Zn, and Cu contents • Increased root length of lemon	(Fidelibus et al., 2001; Ortas et al., 2002; Wu and Xia, 2006)
Cereals	• Enhanced growth of corn in compacted soil • Increased biomass production of corn in low P soil • Increased K, Ca, Mg uptake of corn • Enhanced growth of millet • Alleviated the adverse effects of chlorothalonil on rice • Exerted protective effects against toxicity of Cu, Zn, Pb, and Cd in contaminated soil on rice	(Bagayoko et al., 2000; Liu et al., 2002; Zhang et al., 2005, 2006; Miransari et al., 2007)
Legumes	• Reduced the development of pea root-rot • Increased growth and yield of peanut • Decreased incidence of peanut pod rot	(Abdalla and Abdel-Fattah, 2000; Thygesen et al., 2004; Quilambo et al., 2005)
Vegetables	• Enhanced development of pepper plants • Reduced *Phytophthora* blight in green pepper • Decreased *Fusarium* wilt incidence in cucumber	(Hao et al., 2005; Ozgonen and Erkilic, 2007)

Gabelman, 1992). Many experiments conducted on North America intensively grown crops have also showed that responsiveness to mycorrhizal colonisation changes with plant cultivars (Baon et al., 1993; Khalil et al., 1994; Zhu et al., 2001).

Studies have been conducted to investigate the effects of AMF on plant competition to support the hypothesis that the presence and abundance of such fungi can influence plant species dominance or mediate coexistence (Hamel et al., 1992; West, 1996; Marler et al., 1999; Hart et al., 2003; Yao et al., 2005). For example, Feldmann and Boyle (1998) found that the AMF benefits to maize yield come from maintaining a diverse weed cover crop.

A winter wheat cover crop was compared to a dandelion cover and found that dandelion produced higher AMF colonization, P-uptake and yield in the following maize crop showing that in some cases, weeds may provide an effective support to AMF between cropping periods (Kabir and Koide, 2000). Jordan et al. (2000) hypothesized that specific AMF could reduce the prevalence of non-host species in weed communities. Some studies have highlighted the capability of some AMF to strongly change the relative abundance of some important agricultural weeds. In fact, it has been reported that early growth rate of non-host weedy species were reduced in the presence of AMF (Francis and Read, 1994; Johnson, 1998). Conversely, there are also some indications that non-mycorrhizal plants may actively antagonistize AMF, e.g. via inhibitory compounds released into soil (Fontenla *et al.*, 1999). It is thus interesting to note that many common North American agricultural weeds belong to families that appear to be predominantly non-host (Francis and Read, 1994; Jordan *et al.*, 2000).

Data also support the idea that the influence of AM associations on plant competition is dependent not only on the presence but also on the identity of AMF (van der Heijden *et al.*, 2003; Vogelsang *et al.*, 2006). As the symbiosis between plants and AMF is non-specific, plant response to mycorrhizal colonization also varies according to the organism they are in symbiosis with. Therefore, biomass and P acquisition depend on the specific plant-AMF combination (Klironomos *et al.*, 2000; Smith *et al.*, 2004; Scheublin *et al.*, 2007). To illustrate this, a greenhouse study conducted on wheat demonstrated an increase in plant growth 42 days after inoculation with *Scutellospora calospora* but a significant decrease after inoculations with *Glomus* spp. or *Gigaspora decipiens*, under low-P condition (Graham and Abbott, 2000).

However, the whole dynamics of field soils in such plant production brings many more variables that should be taken into consideration. In fact, the degree to which AMF increase yields is greatly dependent on various factors such as soil type, nutrient status, crop, management practices, and soilborne microorganisms (Karagiannidis and Hadjisavvazinoviadi, 1998). Agricultural management practices, soil nature, abiotic stresses and other soil microorganisms are factors impacting plants and, thus, influencing AMF development and/or colonization, or *vice versa*.

3 ALLEVIATION OF MAJOR ABIOTIC STRESSES

AMF can adapt to a wide range of environments. They are found in soils with very different water regimes including very arid habitats. In these regions, low level of soil moisture can sometimes be compensated by increased root system area for water uptake through hyphal ramification in the soil (Khan *et al.*, 2003). It has been shown that mycorrhizal fungi can improve water use efficiency and sustain drought stress in wheat (Al-Karaki *et al.*, 2004), oat (Khan *et al.*, 2003), and corn (Subramanian and Charest, 1995, 1999; Subramanian *et al.*, 1995, 1997). Colonized soybeans had higher leaf water potential and relative water content than non-mycorrhizal plants under water stress conditions (Aliasgarzad *et al.*, 2006). AMF colonization also increased onion yield under water deficit condition (Bolandnazar *et al.*, 2007). In fact, mycorrhizal plants have, in general, higher water uptake due to hyphal extraction of soil water (Bethlenfalvay *et al.*, 1988; Ruiz-Lozano *et al.*, 1995; Al-karaki, 1998) and higher root hydraulic conductivity than non-mycorrhizal plants (Augé and Stodola, 1990). Some results also suggest that AM association enhances N assimilation by maize which enables the host plant to more efficiently withstand drought conditions and recover after stress is relieved (Subramanian and Charest, 1998).

There are also some evidences in the literature that AMF colonization can protect crops such as barley, cotton and lettuce against the negative effects of salt (Ruiz-Lozano and Azcón, 2000; Tian *et al.*, 2004; Nourinia *et* al., 2007). Moreover, in acidic soil conditions, growth and mineral acquisition of maize has been positively associated with AM root colonization (Clark, 1997; Alloush and Clark, 2001). Similarly, in alkaline soil, an experiment performed on durum wheat inoculated with *Glomus mosseae* also showed greater grain yields in mycorrhizal plants than in controls (Al-Karaki and Al-Omoush, 2002). Soil compaction is another important soil characteristic that greatly affects crop yields. Soil compaction can rapidly lead to reduced root density which can result in decreased water and nutrient uptake (Pardo *et al.*, 2000). Corn growth was increased by AM colonization in compacted soil, though the effectiveness of AMF-derived growth increase varied with the level of soil compaction, the AMF strain, and interaction with other soil microorganisms (Miransari *et al.*, 2007).

4 EFFECT OF AMF ON PLANT PATHOGENS

AMF may impact crop growth by affecting some soil microbial populations also present in the agroecosystems. Many researches focused on these relationships, showing that interactions between AMF and other soil microorganisms can be either detrimental or favourable to plant pathogens, other rhizosphere microbes, AMF or to mycorrhizal plants (Meyer and Linderman, 1986a, b; Paulitz and Linderman, 1989; Calvet *et al.*, 1992; St-Arnaud *et al.*, 1995; Rousseau *et al.*, 1996; Filion *et al.*, 1999; Vigo *et al.*, 2000; Elsen *et al.*, 2001; Talavera *et al.*, 2001; Gryndler *et al.*, 2002; St-Arnaud and Elsen, 2005). The effect of plants on soil biota may be related to the amount and quality of root exudates released in the soil. Ability to access and metabolize different nutrient sources vary from one microbial species to the other (Baudoin *et al.*, 2003). Plant root exudation pattern and therefore its impact on soil biological environment can be greatly modified by mycorrhizal colonization (Linderman, 1992). It has often been demonstrated that changes in the amount, quality or pattern of release of root exudates by mycorrhizal colonization could influence the other microorganisms present in the rhizosphere (Filion *et al.*, 1999; Graham, 2001; Sood, 2003; de Boer *et al.*, 2005; Lioussanne, 2007).

Several studies have investigated the changes occurring in the bacterial communities of mycorrhizal plants' rhizosphere. There are numerous reports of AMF influencing bacterial growth rate (Christensen and Jakobsen, 1993; Marschner and Crowley, 1996a, b; Marschner *et al.*, 1997). Experiments also demonstrated that mycorrhizal associations could be related to qualitative, quantitative and spatial shifts in those populations (Meyer and Linderman, 1986b; Linderman and Paulitz, 1990; Posta *et al.*, 1994; Andrade *et al.*, 1997). AM colonization can influence the species composition of the soil microbial community by increasing some groups and decreasing others (Christensen and Jakobsen, 1993; Vazquez *et al.*, 2000).

Such an important influence on soil microorganisms has rapidly raised interest on the possible role of AMF in bioprotection. Consequently, the number of research works conducted on the interactions between AMF and pathogens in diverse agricultural systems has exploded in the last decade. Even if mycorrhizal inoculation was sometimes associated with neutral effects (Bødker *et al.*, 2002) or enhanced disease symptoms, with some AMF isolates (Garmendia *et al.*, 2004), in most studies AMF inoculation reduced pathogen damages revealing the potential of AMF as a biological control agent (Selim *et al.*, 2005; Li *et al.*, 2007). It is also important to mention that although most studies have reported a decrease in fungal or nematode-induced root diseases severity in mycorrhizal treatments (Borowicz, 2001; Graham, 2001; Matsubara *et al.*, 2001; Castillo *et al.*, 2006; St-Arnaud and Vujanovic, 2007), there are only very few reports of such behaviour associated to most North American agronomic crop plants. For example, *Cochliobolus sativus*, currently one of the dominant pathogen of barley in Canada (Ghazvini and Tekauz, 2007), has been shown to be suppressed by various AMF independently of P availability and water stress (Boyetchko and Tewari, 1988, 1990; Rempel, 1989), as was transmission of *B. sorokiniana* in aerial parts of barley plants in Sweden (Sjöberg, 2005). Severity of take-all caused by *Gaeumannomyces graminis* var. *tritici* in wheat was also reduced by AMF inoculation in P-deficient soil but was not affected at a higher P level (Graham and Menge, 1982). Conversely, stem rust was more severe in mycorrhizal wheat plants inoculated with *Puccinia graminis* urediospores at the two-leaf stage, compared to non-mycorrhizal plantlets (Rempel, 1989). Despite its great potential to be used as a key component of sustainable agriculture, AMF effectiveness as disease control agents depends on many other factors, such as temperature, soil nutrient and water contents, time of mycorrhizal inoculation, amount of mycorrhizal inoculum, pathogen virulence, parameters which are all very difficult to study, not to mention control, in the field. On the other hand, management of soil microbial components, including AMF, definitely represents a promising direction toward the control of plant diseases.

5 EFFECT OF AMF ON PLANT BENEFICIAL MICROORGANISMS

Several types of microorganisms like N_2-fixing bacteria (de Varennes and Goss, 2007; Powell et al., 2007), P-solubilizing bacteria and fungi (Barea et al., 2002), antagonist of plant pathogens (Budi et al., 1999) and soil aggregating bacteria (Rillig et al., 2005) are associated with the rhizosphere of mycorrhizal plants. These organisms are generally grouped under the name 'plant growth promoting rhizosphere microorganisms' (PGPR) because they are able to exert beneficial effects on plant growth. Many leguminous plants have the ability to create a symbiosis with rhizobia, which can fix atmospheric nitrogen (N_2) and hence increase plant access to N sources. Various studies suggested that a specific interaction that influences both the nodulation and mycorrhizal colonization processes occurs between AMF and the N_2-fixing rhizobia in legumes (Ibijbijen et al., 1996; Saxena et al., 1997; Xavier and Germida, 2002, 2003). Growth and productivity of the legumes were always dependent on the combination of selected AMF and rhizobia, revealing that positive interactions between compatible symbionts could significantly increase growth and yields. Pot experiments done with soybean demonstrated that under controlled environment conditions, N_2-fixation in mycorrhizal plants is generally greater than in non-mycorrhizal plants, with more nodules and greater nodule dry weight (Goss and de Varennes, 2002). However, it seems that under field conditions N_2-fixation is not always promoted even if the tripartite symbiosis formed by indigenous arbuscular mycorrhizae, *Bradyrhizobium* and soybean is established (Antunes et al., 2006).

Synergistic effect of associative diazotroph bacteria on AMF activity has often been reported (Barea et al., 2002; Sala et al., 2007). Inoculation of barley with *Glomus mosseae* or *G. fasciculatum* together with *Azospirillum brasilense* produced a synergic effect on dry matter and grain yield, in a greenhouse study (Subba Rao, 1985). Biró et al. (2000) also noted a beneficial effect on soybean of co-inoculation with *Azospirillum brasilense*, *Rhizobium meliloti* and *Glomus fasciculatum*, while Russo et al. (2005) concluded after pot and field tests that an indirect effect of *Azospirillum* on mycorrhization can be assumed on corn and wheat plants as a consequence of the positive effect on root growth. Co-inoculation of wheat with strains of *Pseudomonas* and *Glomus clarum* have also shown a positive dry matter response (Walley and Germida, 1997). However, in this study, inoculation did not result in any increase in root dry weight or length. Another greenhouse trial conducted on maize inoculated with a biofertilizer containing *Glomus*, *Azotobacter* and *Bacillus* strains resulted in a significant increase of plant growth (Wu et al., 2005). Inoculation also improved soil properties, such as organic matter content and total N in soil as well as it increased the nutritional assimilation of plant (total N, P and K).

Higher nutrient assimilation can often be related to the beneficial effects of P- and K-solubilizing bacteria (Rodriguez and Fraga, 1999). They may indeed enhance mineral uptake by plants through solubilizing insoluble forms of P and K, and making them available in soil to plant roots. Many microorganisms are thus able to improve plant growth by solubilizing rock phosphate into plant available P form (Rodriguez and Fraga, 1999; Whitelaw, 2000; Reyes *et al.*, 2002). Among reports that showed synergistic interactions between P-solubilizing microorganisms and AMF (Villegas and Fortin, 2001, 2002; Hamel, 2004; Artursson *et al.*, 2006), a few studies have been conducted on wheat plants. In field trials performed in southern Egypt, the highest significant effect on *Triticum aestivum* L. yield and P content was observed when seeds were inoculated with a mixture of *Glomus constrictum* and two fungal isolates (*Aspergillus niger* and *Penicillium citrinum*), which are known as phosphate rock-solubilizing fungi (Omar, 1998). A recent work proposed by Babana and Antoun (2006) also showed that by inoculating wheat seeds with phosphate rock-solubilizing microorganisms and *Glomus intraradices* under field conditions, it is possible to obtain grain yields comparable to those produced by using diammonium phosphate fertilizer.

However, the application of microbial fertilizers has not resulted in constant effects. The mechanisms and interactions among crops, microbes and abiotic factors are still not well understood but there are great expectations regarding the fact that biofertilizers may complement mycorrhizal activities in sustainable agricultural systems.

6 IMPACT OF LAND USE ON AMF ABUNDANCE AND DIVERSITY

Crop management practices such as tillage, pesticide application, crop rotations and fertilization can impact the AM association, both directly, by damaging the AMF network, and indirectly, by modifying soil conditions essential to their survival and development. In general, agricultural practices affect the occurrence of AMF, with resulting effects on soil biological activity (Johnson *et al.*, 1992; Johnson and Pfleger, 1992; Helgason *et al.*, 1998; Menendez *et al.*, 2001). These impacts have raised a large interest in the scientific community.

A major effect of conventional crop management in field studies was the reduction of AMF biodiversity. To illustrate this, less than 10 different AMF species were identified in conventional agricultural soils (Talukdar and Germida, 1993; Cousins *et al.*, 2003), while more than 20 were found in grassland (Bever *et al.*, 1996) using trap cultures, both in Canada and USA. Oehl *et al.* (2005) recently supported this fact in Switzerland by demonstrating a significant difference between the number of species found in intensively managed maize fields and grasslands. Many studies have indeed indicated that AMF abundance and effectiveness are declining upon agricultural intensification (Douds and Millner, 1999; Oehl *et al.*, 2003; Gosling *et al.*, 2006). In major cropping systems, the diversity of host plant is by far lower than in an undisturbed ecosystem. Monoculture, which is very common in grain production, seems to create a selective pressure on AMF species leading to both a spore population decrease and a shift in the community (Johnson *et al.*, 1992; Rao *et al.*, 1995; Oehl *et al.*, 2003). For example, Bedini *et al.* (2007) reported that no more than six AMF spore morphotypes were detected in a maize monoculture. Only spores related to the genus *Glomus* were recovered, confirming data on its predominance in managed soils. Several studies reported prevalence of *Glomus* spp. in cropped soils, in contrast to rich AMF communities containing *Gigaspora* spp., *Scutellospora* spp. and Acaulospora spp. in uncultivated soils (Blaszkowski, 1993; Talukdar and Germida, 1993; Hamel *et al.*, 1994; Helgason *et al.*, 1998).

Long fallow periods or non-mycorrhizal crop plants have a profound effect on AMF activity and diversity. For example, a study conducted by Karasawa et al. (2002) indicated an increase in AMF colonization and growth of maize following a sunflower crop as compared to maize following mustard, a non-mycorrhizal crop. Earlier, Gavito and Miller (1998) observed a delay of more than 60 days in AMF colonization of corn following canola, a non-AMF host species, as compared to a previous crop of bromegrass and alfalfa (both mycorrhizal). However, similar AMF spore numbers were detected in wheat field whether the previous crop was corn or canola (Jansa *et al.*, 2002). Hamel *et al.* (2006a), also reported that cropping frequency did not influence AMF abundance in a wheat-based rotation in Canadian prairies, according to PLFA analysis.

During fallow periods, the viable AMF hyphal network decreases over time leading to a lower mineral uptake and growth of the subsequent mycorrhizal crop (Kabir *et al.*, 1999). It has been demonstrated that AM colonization and P uptake decrease with increasing length of a preceding fallow (Kabir *et al.*, 1999; Kabir and Koide, 2000). In fact, Kabir *et al.* (1999) have shown that a 90-day fallow in maize decreases AMF active hyphae by 57%, root colonization by 33%, and P, Zn, and Cu uptake by 19%, 54%, and 61% respectively. Another study conducted on maize and soybean has shown that AMF spore number increased after three years of continuous cropping. Under fallow, spore number declined during the first year, and then stabilised at a low level (Troeh and Loynachan, 2003). In some cases the adverse effect of fallow periods on AMF inoculum potential can be avoided by growing a cover crop. For example, Boswell *et al.* (1998)

found that growing winter wheat in comparison to fallow could increase AM inoculum potential and the growth and yield of maize. Authors have suggested that the absence of host plant during fallow negatively impacts AMF through energy source exhaustion. Moreover, in some part of North America, soil freezing and thawing during winter can directly disrupt extraradical hyphae (Boswell *et al.*, 1998).

More than a decade ago, it has been suggested that tillage disturb mycorrhizal activity in soil and therefore, plants nutrient uptake (Miller *et al.*, 1995; Boddington and Dodd, 2000; Mozafar *et al.*, 2000). Soil disturbance has often been shown to reduce the density of AMF spores, species richness and the length of extraradical mycelium of AMF relative to undisturbed soil (Boddington and Dodd, 2000). In a study conducted by Jansa *et al.* (2002) on wheat plants, it has been observed that AMF community composition was affected by tillage treatments. Fifteen AMF species were detected in the no-till fields, 14 in those under chisel treatments and 13 under conventional tillage conditions. Also, significantly more AMF spores were observed in soil from the no-tilled plots than from the tilled plots (when in rotation with canola). In this case, other factors, such as weed roots may have supported AMF development because canola is a non-mycorrhizal plant. Other research showing that soil disturbance by tillage in maize fields causes physical disruption of the fungus mycelium and therefore decreases the absorptive abilities of the mycorrhizae, have been summarized by Miller (2000). In reduced tillage systems, heavy P fertilization may not be as necessary as in heavily tilled systems because the intact mycorrhizal network increases the effective surface area for crop P uptake (Miller *et al.*, 1995; Miller, 2000).

7   IMPACT OF FERTILIZERS ON AMF

The addition of nutrients to the soil is a common practice in every intensive grain production in North America. In intensively managed agricultural systems, soil is often fertilized with N, P and K, and much of the plant biomass is harvested and not returned to the soil at the end of the growing season, which contributes to create very unique types of ecosystems. Various forms of fertilizer effectively increase crop yields (Schmidt *et al.*, 2001; Lithourgidis *et al.*, 2007). Mineral fertilizers are largely applied but several forms of more 'natural' fertilizers are also used. Mineral fertilizers are expensive and in some occasions, application of organic amendments to cultivated field reduces their necessity (Singer *et al.*, 2004). The nutrient content of organic amendments is not as consistent as the one of mineral fertilizers, which is less practical. Studies have been conducted with dozens of different types of manure or compost as well as mineral fertilizers on major agronomic plants. For example, swine bedding materials (Liebman *et al.*, 2004), crushed cotton gin compost (Tejada and Gonzalez, 2006), and urban refuse compost (Bazzoffi *et al.*, 1998) have been tried in corn production. In Canada, swine liquid compost application have been shown to benefit both barley and soybean in a crop rotation (Carter and Campbell, 2006). Both mineral and organic fertilizers bring to soil nutrients essential either to plants or soil microorganisms. It is a well known fact that AMF are important for the efficient uptake of nutrients, such as P and N (Smith and Read, 1997). Many authors have stressed increased P and N uptake by mycorrhizal plants but mostly under limiting availability conditions (Cruz *et al.*, 2004; Kanno *et al.*, 2006; Li *et al.*, 2006; Schreiner, 2007).

Fertilizer-AMF interactions are complex and difficult to predict. While in some cases manure addition leads to an increase of AM colonization (Tarkalson *et al.*, 1998), Ellis *et al.* (1992) found greater AM colonization in sorghum plants when neither manure nor fertilizer was added to soil. It appears from the literature that AMF directly affect N absorption and N assimilation (Barea *et al.*, 1987), particularly in neutral to slightly alkaline soils (Azcón *et al.*, 2001). The ability of mycorrhizal plants to better exploit soil N resources can occur directly through the uptake of organic molecules by AMF or, as stated before in this chapter, through different interactions with other soil microorganisms (i.e., competition with heterotrophic microbes for mineral N, enhancement of mineralization, better N_2-fixation, etc.). The arbuscular mycorrhizal symbiosis can both enhance decomposition of, and increase nitrogen capture from complex organic material in soil. Increased hyphal growth of the fungal partner was noted in the presence of the organic material, independently of the host plant (Hodge *et al.*, 2001). Feng *et al.* (2002) have also demonstrated that mycorrhizal fungi had significantly increased nitrogen uptake derived from soil in mycorrhizal cotton plants, while no significant influence on uptake of N derived from fertilizer was observed. Colonization also increased the amount of soil available N after fertilization treatment meaning that AMF may facilitate plant acquisition of nitrogen from sources which are otherwise not or less available to non-mycorrhizal plants. However this contribution of AMF to plant nutrition and growth would likely be more significant in organic farming systems or in unfertilized soil-plant systems than in highly N-fertilized agricultural fields.

Similarly, several reports have established the potential of arbuscular mycorrhizas to increase uptake of P by crops in otherwise P-deficient soils (Powell, 1981; Mohammad *et al.*, 2004). Plants with a moderate stress related to nutrient deficiency tend to release more soluble carbohydrate in their root exudates than unstressed plants (Schwab *et al.*, 1991). A selective pressure is then exerted on AMF strains that are more aggressively acquiring plant carbohydrates. But in North America, cereal crop plants are rarely under such stress as most growers are unwilling to risk low production and largely fertilize their fields.

Although it has been reported that the ability of AMF to improve plant P uptake is largest when the P source is organic (Feng et al., 2003), which could be explained by the fact that much of the P applied to the soil through mineral fertilizers is rapidly fixed into insoluble forms, demonstration has been made that AMF from fertilized soils produce fewer hyphae and arbuscules than fungi from unfertilized soils (Johnson, 1993). Moreover, AM colonization is generally severely limited by the high P-inputs used in vegetable production systems (Ryan and Graham, 2002). However, in a large-scale survey of 40 asparagus fields conducted in eastern Canada, PLFA 16:1ω5 in soil, used as an estimation of extraradical AM fungal development, was positively correlated with soil available P (Hamel et al., 2006b). This may be related to the higher plant growth and higher plant-derived C therefore available to sustain AMF hyphal network development. Nevertheless, at the moment in intensive crop productions, unless P-supply is balanced carefully with plant requirements, management practices favouring AM fungal activity may risk crop growth depression and profits reduction. Low-input agricultural systems have gained attention in many industrialized countries due to rising interest for the conservation of natural resources, reduction of environmental degradation, and escalating price of fertilizers. Conventional farming systems with lower application of fertilizers and pesticides have been developed (Mader et al., 2002). Under these conditions, plants are more dependent on an effective AMF symbiosis (Scullion et al., 1998; Galvez et al., 2001). AMF communities were generally impoverished in species composition in intensively managed agricultural lands (Johnson and Pfleger, 1992; Galvez et al., 2001; Jansa et al., 2002; Oehl et al., 2003, 2004), supporting the idea that organic farming could rely on a higher soil microbial biodiversity.

As P and N availability in agricultural systems is not as limiting to plant productivity as in other soil-plant systems, the main impacts of AMF may also be different. Hamel (2004) suggested that 'under these conditions, the major impact of AMF, which are root extensions and regulators of photosynthesis-derived C input to soil, could well be on microbial processes and soil quality.'

8 CONCLUSION

Since most intensively cultivated crops in North America are mycorrhizal, all possible interactions influencing plant growth must be considered. As conventional farming are now definitely high inputs agricultural systems generating large N and P surplus, the main benefit of AMF in the rhizosphere may not so much be related to nutrient uptake. AMF, by their interactions with soil particles and other organisms, represent an important component of soil quality. A better understanding of soil system would probably lead to a better management of AMF contribution to soil fertility and, may be, to a more sustainable agriculture, even in high yielding grain productions.

REFERENCES

Abdalla, M.E., and Abdel-Fattah, G.M., 2000, Influence of the endomycorrhizal fungus *Glomus mosseae* on the development of peanut pod rot disease in Egypt. *Mycorrhiza* **10**: 29–35.

Aliasgarzad, N., Neyshabouri, M.R., and Salimi, G., 2006, Effects of arbuscular mycorrhizal fungi and *Bradyrhizobium japonicum* on drought stress of soybean. *Biologia* **61**: S324–S328.

Al-Karaki, G., McMichael, B., and Zak, J., 2004, Field response of wheat to arbuscular mycorrhizal fungi and drought stress. *Mycorrhiza* **14**: 263–269.

Al-Karaki, G.N., 1998, Benefit, cost and water-use efficiency of arbuscular mycorrhizal durum wheat grown under drought stress. *Mycorrhiza* **8**: 41–45.

Al-Karaki, G.N., and Al-Omoush, M., 2002, Wheat response to phosphogypsum and mycorrhizal fungi in alkaline soil. *J. Plant Nutr.* **25**: 873–883.

Alloush, G.A., and Clark, R.B., 2001, Maize response to phosphate rock and arbuscular mycorrhizal fungi in acidic soil. *Com. Soil Sci. Plant Anal.* **32**: 231–254.

Andrade, G., Mihara, K.L., Linderman, R.G., and Bethlenfalvay, G.J., 1997, Bacteria from rhizosphere and hyphosphere soils of different arbuscular-mycorrhizal fungi. *Plant Soil* **192**: 71–79.

Antunes, P.M., de Varennes, A., Zhang, T., and Goss, M.J., 2006, The tripartite symbiosis formed by indigenous arbuscular mycorrhizal fungi, *Bradyrhizobium japonicum* and soya bean under field conditions. *J. Agr. Crop Sci.* **192**: 373–378.

Artursson, V., Finlay, R.D., and Jansson, J.K., 2006, Interactions between arbuscular mycorrhizal fungi and bacteria and their potential for stimulating plant growth. *Environ. Microbiol.* **8**: 1–10.

Augé, R.M., and Stodola, A.J.W., 1990, An apparent increase in symplastic water contributes to greater turgor in mycorrhizal roots of droughted Rosa plants. *New Phytol.* **115**: 285–295.

Azcón, R., Ruiz-Lozano, J., and Rodriguez, R., 2001, Differential contribution of arbuscular mycorrhizal fungi to plant nitrate uptake (N-15) under increasing N supply to the soil. *Can. J. Bot.* **79**: 1175–1180.

Babana, A.H., and Antoun, H., 2006, Effect of Tilemsi phosphate rock-solubilizing microorganisms on phosphorus uptake and yield of field-grown wheat (*Triticum aestivum* L.) in Mali. *Plant Soil* **287**: 51–58.

Bagayoko, M., George, E., Romheld, V., and Buerkert, A.B., 2000, Effects of mycorrhizae and phosphorus on growth and nutrient uptake of millet, cowpea and sorghum on a West African soil. *J. Agric. Sci.* **135**: 399–407.

Baon, J.B., Smith, S.E., and Alston, A.M., 1993, Mycorrhizal responses of barley cultivars differing in P-efficiency. *Plant Soil* **157**: 97–105.

Barea, J.M., Azcón-Aguilar, C., and Azcón, R., 1987, Vesicular-arbuscular improve both symbiotic N_2 fixation and N uptake from soil as assessed with a N-15 technique under field conditions. *New Phytol.* **106**: 717–726.

Barea, J.M., Azcón, R., and Azcón-Aguilar, C., 2002, Mycorrhizosphere interactions to improve plant fitness and soil quality. *Antonie van Leeuwenhoek* **81**: 343–351.

Baudoin, E., Benizri, E., and Guckert, A., 2003, Impact of artificial root exudates on the bacterial community structure in bulk soil and maize rhizosphere. *Soil Biol. Biochem.* **35**: 1183–1192.

Bazzoffi, P., Pellegrini, S., Rocchini, A., Morandi, M., and Grasselli, O., 1998, The effect of urban refuse compost and different tractors tires on soil physical properties, soil erosion and maize yield. *Soil Til. Res.* **48**: 275–286.

Bedini, S., Avio, L., Argese, E., and Giovannetti, M., 2007, Effects of long-term land use on arbuscular mycorrhizal fungi and glomalin-related soil protein. *Agric. Ecos. Environ.* **120**: 463–466.

Bethlenfalvay, G.J., Brown, M.S., Ames, R.N., and Thomas, R.S., 1988, Effects of drought on host and endophyte development in mycorrhizal soybeans in relation to water use and phosphate uptake. *Physiol. Plant.* **72**: 565–571.

Bever, J.D., Morton, J.B., Antonovics, J., and Schultz, P.A., 1996, Host-dependent sporulation and species diversity of arbuscular mycorrhizal fungi in a mown grassland. *J. Ecol.* **84**: 71–82.

Biró, B., Koves-Pechy, K., Voros, I., Takacs, T., Eggenberg, P., and Strasser, R.J., 2000, Interrelations between *Azospirillum* and *Rhizobium* nitrogen-fixers and arbuscular mycorrhizal fungi in the rhizosphere of alfalfa at sterile, AMF-free or normal soil conditions. *Appl. Soil Ecol.* **15**: 159–168.

Blaszkowski, J., 1993, Comparative studies of the occurrence of arbuscular fungi and mycorrhizae (*Glomales*) in cultivated and uncultivated soils of Poland. *Acta Mycol.* **28**: 93–140.

Boddington, C.L., and Dodd, J.C., 2000, The effect of agricultural practices on the development of indigenous arbuscular mycorrhizal fungi. I. Field studies in an Indonesian ultisol. *Plant Soil* **218**: 137–144.

Bødker, L., Kjøller, R., Kristensen, K., and Rosendahl, S., 2002, Interactions between indigenous arbuscular mycorrhizal fungi and *Aphanomyces euteiches* in field-grown pea. *Mycorrhiza* **12**: 7–12.

Bolandnazar, S., Aliasgarzad, N., Neishabury, M.R., and Chaparzadeh, N., 2007, Mycorrhizal colonization improves onion (*Allium cepa* L.) yield and water use efficiency under water deficit condition. *Sci. Hort.* **114**: 11–15.

Borowicz, V.A., 2001, Do arbuscular mycorrhizal fungi alter plant-pathogen relations? *Ecology* **82**: 3057–3068.

Boswell, E.P., Koide, R.T., Shumway, D.L., and Addy, H.D., 1998, Winter wheat cover cropping, VA mycorrhizal fungi and maize growth and yield. *Agric. Ecos. Environ.* **67**: 55–65.

Boyetchko, S.M., and Tewari, J.P., 1988, The effect of VA mycorrhizal fungi on infection by *Bipolaris sorokiniana* in barley. *Can. J. Plant Pathol.* **10**: 361.

Boyetchko, S.M., and Tewari, J.P., 1990, *Effect of phosphorus and VA mycorrhizal fungi on common root rot of barley. Innovation and integration*. Proceedings of the 8th North American Conference on Mycorrhizae, Sept. 5–8, Jackson, Wyoming.

Brady, N.C., and Weil, R.R., 2002. *The nature and properties of soils*, Prentice Hall, New Jersey, pp. 960.

Budi, S.W., van Tuinen, D., Martinotti, G., and Gianinazzi, S., 1999, Isolation from the *Sorghum bicolor* mycorrhizosphere of a bacterium compatible with arbuscular mycorrhiza development and antagonistic towards soilborne fungal pathogens. *Appl. Environ. Microbiol.* **65**: 5148–5150.

Calvet, C., Barea, J.M., and Pera, J., 1992, *In vitro* interactions between the vesicular-arbuscular mycorrhizal fungus *Glomus mosseae* and some saprophytic fungi isolated from organic substrates. *Soil Biol. Biochem.* **24**: 775–780.

Carter, M.R., and Campbell, A.J., 2006, Influence of tillage and liquid swine manure on productivity of a soybean-barley rotation and some properties of a fine sandy loam in Prince Edward Island. *Can. J. Soil Sci.* **86**: 741–748.

Castillo, C.G., Rubio, R., Rouanet, J.L., and Borie, F., 2006, Early effects of tillage and crop rotation on arbuscular mycorrhizal fungal propagules in an ultisol. *Biol. Fert. Soils* **43**: 83–92.

Christensen, H., and Jakobsen, I., 1993, Reduction of bacterial growth by a vesicular-arbuscular mycorrhizal fungus in the rhizosphere of cucumber (*Cucumis sativus* L). *Biol. Fert. Soils* **15**: 253–258.

Clark, R.B., 1997, Arbuscular mycorrhizal adaptation, spore germination, root colonization, and host plant growth and mineral acquisition at low pH. *Plant Soil* **192**: 15–22.

Clark, R.B., and Zeto, S.K., 2000, Mineral acquisition by arbuscular mycorrhizal plants. *J. Plant Nutr.* **23**: 867–902.

Cousins, J.R., Hope, D., Gries, C., and Stutz, J.C., 2003, Preliminary assessment of arbuscular mycorrhizal fungal diversity and community structure in an urban ecosystem. *Mycorrhiza* **13**: 319–326.

Cruz, C., Green, J.J., Watson, C.A., Wilson, F., and Martins-Loucao, M.A., 2004, Functional aspects of root architecture and mycorrhizal inoculation with respect to nutrient uptake capacity. *Mycorrhiza* **14**: 177–184.

Da Silva, A.E., and Gabelman, W.H., 1992, Screening maize inbred lines for tolerance to low-P stress condition. *Plant Soil* **146**: 181–187.

de Boer, W., Folman, L.B., Summerbell, R.C., and Boddy, L., 2005, Living in a fungal world: impact of fungi on soil bacterial niche development. *FEMS Microbiol. Rev.* **29**: 795–811.

de Varennes, A., and Goss, M.J., 2007, The tripartite symbiosis between legumes, rhizobia and indigenous mycorrhizal fungi is more efficient in undisturbed soil. *Soil Biol. Biochem.* **39**: 2603–2607.

Douds, D.D., and Millner, P., 1999, Biodiversity of arbuscular mycorrhizal fungi in agroecosystems. *Agric. Ecos. Environ.* **74**: 77–93.

Ellis, J.R., Roder, W., and Mason, S.C., 1992, Grain sorghum-soybean rotation and fertilization influence on vesicular-arbuscular mycorrhizal fungi. *Soil Sci. Soc. Amer. J.* **56**: 789–794.

Elsen, A., Declerck, S., and De Waele, D., 2001, Effects of *Glomus intraradices* on the reproduction of the burrowing nematode (*Radopholus similis*) in dixenic culture. *Mycorrhiza* **11**: 49–51.

Feldmann, F., and Boyle, C., 1998, Weed-mediated stability of arbuscular mycorrhizal fungi effectiveness in maize monocultures. *J. Appl. Bot.* **73**: 1–5.

Feng, G., Zhang, F.S., Li, X.L., Tian, C.Y., Tang, C.X., and Rengel, Z., 2002, Uptake of nitrogen from indigenous soil pool by cotton plant inoculated with arbuscular mycorrhizal fungi. *Com. Soil Sci. Plant Anal.* **33**: 3825–3836.

Feng, G., Song, Y.C., Li, X.L., and Christie, P., 2003, Contribution of arbuscular mycorrhizal fungi to utilization of organic sources of phosphorus by red clover in a calcareous soil. *Appl. Soil Ecol.* **22**: 139–148.

Fidelibus, M.W., Martin, C.A., and Stutz, J.C., 2001, Geographic isolates of *Glomus* increase root growth and whole-plant transpiration of citrus seedlings grown with high phosphorus. *Mycorrhiza* **10**: 231–236.

Filion, M., St-Arnaud, M., and Fortin, J.A., 1999, Direct interaction between the arbuscular mycorrhizal fungus *Glomus intraradices* and different rhizosphere microorganisms. *New Phytol.* **141**: 525–533.

Fontenla, S., Garcia-Romera, I., and Ocampo, J.A., 1999, Negative influence of non-host plants on the colonization of *Pisum sativum* by the arbuscular mycorrhizal fungus *Glomus mosseae*. *Soil Biol. Biochem.* **31**: 1591–1597.

Francis, R., and Read, D.J., 1994, The contributions of mycorrhizal fungi to the determination of plant community structure. *Plant Soil* **159**: 11–25.

Galvez, L., Douds, D.D., Drinkwater, L.E., and Wagoner, P., 2001, Effect of tillage and farming system upon VAM fungus populations and mycorrhizas and nutrient uptake of maize. *Plant Soil* **228**: 299–308.

Garmendia, I., Goicoechea, N., and Aguirreolea, J., 2004, Effectiveness of three *Glomus* species in protecting pepper (*Capsicum annuum* L.) against Verticillium wilt. *Biol. Contr.* **31**: 296–305.

Gavito, M.E., and Miller, M.H., 1998, Early phosphorus nutrition, mycorrhizae development, dry matter partitioning and yield of maize. *Plant Soil* **199**: 177–186.

Ghazvini, H., and Tekauz, A., 2007, Reactions of Iranian barley accessions to three predominant pathogens in Manitoba. *Can. J. Plant Pathol.* **29**: 69–78.

Gosling, P., Hodge, A., Goodlass, G., and Bending, G.D., 2006, Arbuscular mycorrhizal fungi and organic farming. *Agric. Ecos. Environ.* **113**: 17–35.

Goss, M.J., and de Varennes, A., 2002, Soil disturbance reduces the efficacy of mycorrhizal associations for early soybean growth and N_2 fixation. *Soil Biol. Biochem.* **34**: 1167–1173.

Graham, J.H., 2001, What do root pathogens see in mycorrhizas? *New Phytol.* **149**: 357–359.

Graham, J.H., and Abbott, L.K., 2000, Wheat responses to aggressive and non-aggressive arbuscular mycorrhizal fungi. *Plant Soil* **220**: 207–218.

Graham, J.H., and Menge, J.A., 1982, Influence of vesicular-arbuscular mycorrhizae and soil phosphorus on take-all disease of wheat. *Phytopathology* **72**: 95–98.

Gryndler, M., Vosatka, M., Hrselova, H., Catska, V., Chvatalova, I., and Jansa, J., 2002, Effect of dual inoculation with arbuscular mycorrhizal fungi and bacteria on growth and mineral nutrition of strawberry. *J. Plant Nutr.* **25**: 1341–1358.

Hamel, C., 2004, Impact of arbuscular mycorrhizal fungi on N and P cycling in the root zone. *Can. J. Soil Sci.* **84**: 383–395.

Hamel, C., Furlan, V., and Smith, D.L., 1992, Mycorrhizal effects on interspecific plant competition and nitrogen transfer in legume grass mixtures. *Crop Sci.* **32**: 991–996.

Hamel, C., Dalpé, Y., Lapierre, C., Simard, R.R., and Smith, D.L., 1994, Composition of the vesicular-arbuscular mycorrhizal fungi population in an old meadow as affected by pH, phosphorus and soil disturbance. *Agric. Ecos. Environ.* **49**: 223–231.

Hamel, C., Hanson, K., Selles, F., Cruz, A.F., Lemke, R., McConkey, B., and Zentner, R., 2006a, Seasonal and long-term resource-related variations in soil microbial communities in wheat-based rotations of the Canadian prairie. *Soil Biol. Biochem.* **38**: 2104–2116.

Hamel, C., Vujanovic, V., Jeannotte, R., Liu, A., Nakano, A., and St-Arnaud, M., 2006b, *Variation in arbuscular mycorrhizal fungi extraradicular biomass along a climatic gradient in an agricultural zone of Quebec, Canada.* 5th International Symposium on Society Congress, Aug. 4–10, Vienna, Austria.

Hao, Z.P., Christie, P., Qin, L., Wang, C.X., and Li, X.L., 2005, Control of fusarium wilt of cucumber seedlings by inoculation with an arbuscular mycorrhizal fungus. *J. Plant Nutr.* **28**: 1961–1974.

Hart, M.M., Reader, R.J., and Klironomos, J.N., 2003, Plant coexistence mediated by arbuscular mycorrhizal fungi. *Tren. Ecol. Evol.* **18**: 418–423.

Heffer, P., and Prud'homme, M. (2006). *Medium-term outlook for global Fertililizer demand, supply and trade, 2006-2010*, summary report presented at the 74th IFA Annual Conference Cape Town, Paris, France, International Fertilizer Industry Association.

Helgason, T., Daniell, T.J., Husband, R., Fitter, A.H., and Young, J.P.W., 1998, Ploughing up the wood-wide web. *Nature* **394**: 431.

Hodge, A., Campbell, C.D., and Fitter, A.H., 2001, An arbuscular mycorrhizal fungus accelerates decomposition and acquires nitrogen directly from organic material. *Nature* **413**: 297–299.

Ibijbijen, J., Urquiaga, S., Ismaili, M., Alves, B.J.R., and Boddey, R.M., 1996, Effect of arbuscular mycorrhizal fungi on growth, mineral nutrition and nitrogen fixation of three varieties of common beans (*Phaseolus vulgaris*). *New Phytol.* **134**: 353–360.

Ilbas, A.I., and Sahin, S., 2005, *Glomus fasciculatum* inoculation improves soybean production. *Acta Agri. Scand. S. B-Soil Plant Sci.* **55**: 287–292.

Imboden, D.M., 1974, Phosphorus model of lake eutrophication. *Limnol. Oceanogr.* **19**: 297–304.

Jansa, J., Mozafar, A., Anken, T., Ruh, R., Sanders, I.R., and Frossard, E., 2002, Diversity and structure of AMF communities as affected by tillage in a temperate soil. *Mycorrhiza* **12**: 225–234.

Johnson, N.C., 1993, Can fertilization of soil select less mutualistic mycorrhizae? *Ecol. Appl.* **3**: 749–757.

Johnson, N.C., 1998, Responses of *Salsola kali* and *Panicum virgatum* to mycorrhizal fungi, phosphorus and soil organic matter-implications for reclamation. *J. Appl. Ecol.* **35**: 86–94.

Johnson, N.C., and Pfleger, F.L., 1992. Vesicular-arbuscular mycorrhizae and cultural practices. In: *Mycorrhizae in sustainable agriculture.*, G. J. Bethlenfalvay and R. G. Linderman eds., ASA, CSSA, and SSSA, Madison, WI, 54, pp. 71–99.

Johnson, N.C., Tilman, D., and Wedin, D., 1992, Plant and soil controls on mycorrhizal fungal communities. *Ecology* **73**: 2034–2042.

Jordan, N.R., Zhang, J., and Huerd, S., 2000, Arbuscular-mycorrhizal fungi: potential roles in weed management. *Weed Res.* **40**: 397–410.

Kabir, Z., and Koide, R.T., 2000, The effect of dandelion or a cover crop on mycorrhiza inoculum potential, soil aggregation and yield of maize. *Agric. Ecos. Environ.* **78**: 167–174.

Kabir, Z., O'Halloran, I.P., and Hamel, C., 1999, Combined effects of soil disturbance and fallowing on plant and fungal components of mycorrhizal corn (*Zea mays* L.). *Soil Biol. Biochem.* **31**: 307–314.

Kanno, T., Saito, M., Ando, Y., Macedo, M.C.M., Nakamura, T., and Miranda, C.H.B., 2006, Importance of indigenous arbuscular mycorrhiza for growth and phosphorus uptake in tropical forage grasses growing on an acid, infertile soil from the brazilian savannas. *Trop. Grassl.* **40**: 94–101.

Karagiannidis, N., and Hadjisavvazinoviadi, S., 1998, The mycorrhizal fungus *Glomus mosseae* enhances growth, yield and chemical composition of a durum wheat variety in 10 different soils. *Nutr. Cycl. Agroecos.* **52**: 1–7.

Karasawa, T., Kasahara, Y., and Takebe, A., 2002, Differences in growth responses of maize to preceding cropping caused by fluctuation in the population of indigenous arbuscular mycorrhizal fungi. *Soil Biol. Biochem.* **34**: 851–857.

Khalil, S., Loynachan, T.E., and Tabatabai, M.A., 1994, Mycorrhizal dependency and nutrient uptake by improved and unimproved corn and soybean cultivars. *Agron. J.* **86**: 949–958.

Khan, I.A., Ahmad, S., and Ayub, N., 2003, Response of oat (*Avena sativa*) to inoculation with vesicular arbuscular mycorrhizae (VAM) in the presence of phosphorus. *Asian J. Plant Sci.* **2**: 371–373.

Kirchmann, H., and Thorvaldsson, G., 2000, Challenging targets for future agriculture. *Eur. J. Agro.* **12**: 145–161.

Klironomos, J.N., McCune, J., Hart, M., and Neville, J., 2000, The influence of arbuscular mycorrhizae on the relationship between plant diversity and productivity. *Ecol. Lett.* **3**: 137–141.

Lekberg, Y., and Koide, R.T., 2005, Is plant performance limited by abundance of arbuscular mycorrhizal fungi? A meta-analysis of studies published between 1988 and 2003. *New Phytol.* **168**: 189–204.

Li, B., Ravnskov, S., Xie, G.L., and Larsen, J., 2007, Biocontrol of *Pythium* damping-off in cucumber by arbuscular mycorrhiza-associated bacteria from the genus *Paenibacillus*. *Biocontr.* **52**: 863–875.

Li, H.Y., Smith, S.E., Holloway, R.E., Zhu, Y.G., and Smith, F.A., 2006, Arbuscular mycorrhizal fungi contribute to phosphorus uptake by wheat grown in a phosphorus-fixing soil even in the absence of positive growth responses. *New Phytol.* **172**: 536–543.

Liebman, M., Menalled, F.D., Buhler, D.D., Richard, T.L., Sundberg, D.N., Cambardella, C. A., and Kohler, K.A., 2004, Impacts of composted swine manure on weed and corn nutrient uptake, growth, and seed production. *Weed Sci.* **52**: 365–375.

Linderman, R.G., 1992, Vesicular-arbuscular mycorrhizae and soil microbial interactions. In: *Mycorrhizae in sustainable agriculture*, G. J. Bethlenfalvay and R. G. Linderman eds., American Society of Agriculture, Madison, WI, Special Publication No. 54, pp. 45–70.

Linderman, R.G., and Paulitz, T.C., 1990, Mycorrhizal-rhizobacterial interactions. In: *Biological control of soil-born plant pathogens*, D. Hornby, R. J. Cook, Y. Heniset al eds., CAB International, Wallingford, UK, pp. 261–283.

Lioussanne, L., 2007, Rôles des modifications de la microflore bactérienne et de l'exsudation racinaire de la tomate par la symbiose mycorhizienne dans le biocontrôle sur le *Phytophthora nicotianae*. Ph.D. thesis, Université de Montréal, pp. 264.

Lithourgidis, A.S., Matsi, T., Barbayiannis, N., and Dordas, C.A., 2007, Effect of liquid cattle manure on corn yield, composition, and soil properties. *Agron. J.* **99**: 1041–1047.

Liu, A., Hamel, C., Elmi, A., Costa, C., Ma, B., and Smith, D.L., 2002, Concentrations of K, Ca and Mg in maize colonized by arbuscular mycorrhizal fungi under field conditions. *Can. J. Soil Sci.* **82**: 271–278.

Mader, P., Fliessbach, A., Dubois, D., Gunst, L., Fried, P., and Niggli, U., 2002, Soil fertility and biodiversity in organic farming. *Science* **296**: 1694–1697.

Marler, M.J., Zabinski, C.A., and Callaway, R.M., 1999, Mycorrhizae indirectly enhance competitive effects of an invasive forb on a native bunchgrass. *Ecology* **80**: 1180–1186.

Marschner, P., and Crowley, D.E., 1996a, Physiological activity of a bioluminescent *Pseudomonas fluorescens* (strain 2-79) in the rhizosphere of mycorrhizal and non-mycorrhizal pepper (*Capsicum annuum* L). *Soil Biol. Biochem.* **28**: 869–876.

Marschner, P., and Crowley, D.E., 1996b, Root colonization of mycorrhizal and non-mycorrhizal pepper (*Capsicum annuum*) by *Pseudomonas fluorescens* 2-79RL. *New Phytol.* **134**: 115–122.

Marschner, P., Crowley, D.E., and Higashi, R.M., 1997, Root exudation and physiological status of a root-colonizing fluorescent pseudomonad in mycorrhizal and non-mycorrhizal pepper (*Capsicum annuum* L). *Plant Soil* **189**: 11–20.

Matsubara, Y., Ohba, N., and Fukui, H., 2001, Effect of arbuscular mycorrhizal fungus infection on the incidence of fusarium root rot in asparagus seedlings. *J. Jap. Soc. Hort. Sci.* **70**: 202–206.

McGonigle, T.P., 1988, A numerical analysis of published field trials with vesicular-arbuscular mycorrhizal fungi. *Func. Ecol.* **2**: 473–478.

Menendez, A.B., Scervino, J.M., and Godeas, A.M., 2001, Arbuscular mycorrhizal populations associated with natural and cultivated vegetation on a site of Buenos Aires province, Argentina. *Biol. Fert. Soils* **33**: 373–381.

Meyer, J.R., and Linderman, R.G., 1986a, Response of subterranean clover to dual inoculation with vesicular-arbuscular mycorrhizal fungi and a plant growth-promoting bacterium, *Pseudomonas putida*. *Soil Biol. Biochem.* **18**: 185–190.

Meyer, J.R., and Linderman, R.G., 1986b, Selective influence on populations of rhizosphere or rhizoplane bacteria and actinomycetes by mycorrhizas formed by *Glomus fasciculatum*. *Soil Biol. Biochem.* **18**: 191–196.

Miller, M.H., 2000, Arbuscular mycorrhizae and the phosphorus nutrition of maize: a review of Guelph studies. *Can. J. Plant Sci.* **80**: 47–52.

Miller, M.H., McGonigle, T.P., and Addy, H.D., 1995, Functional ecology of vesicular arbuscular mycorrhizas as influenced by phosphate fertilization and tillage in an agricultural ecosystem. *Crit. Rev. Biotechnol.* **15**: 241–255.

Miransari, M., Bahrami, H.A., Rejali, F., Malakouti, M.J., and Torabi, H., 2007, Using arbuscular mycorrhiza to reduce the stressful effects of soil compaction on corn (*Zea mays* L.) growth. *Soil Biol. Biochem.* **39**: 2014–2026.

Mohammad, A., Mitra, B., and Khan, A.G., 2004, Effects of sheared-root inoculum of *Glomus intraradices* on wheat grown at different phosphorus levels in the field. *Agric. Ecos. Environ.* **103**: 245–249.

Mozafar, A., Anken, T., Ruh, R., and Frossard, E., 2000, Tillage intensity, mycorrhizal and nonmycorrhizal fungi, and nutrient concentrations in maize, wheat, and canola. *Agron. J.* **92**: 1117–1124.

Nourinia, A.A., Faghani, E., Rejali, F., Safarnezhad, A., and Abbasi, M., 2007, Evaluation effects of symbiosis of mycorrhiza on yield components and some physiological parameters of barley genotypes under salinity stress. *Asian J. Plant Sci.* **6**: 1108–1112.

Oehl, F., Sieverding, E., Ineichen, K., Mader, P., Boller, T., and Wiemken, A., 2003, Impact of land use intensity on the species diversity of arbuscular mycorrhizal fungi in agroecosystems of Central Europe. *Appl. Environ. Microbiol.* **69**: 2816–2824.

Oehl, F., Sieverding, E., Mader, P., Dubois, D., Ineichen, K., Boller, T., and Wiemken, A., 2004, Impact of long-term conventional and organic farming on the diversity of arbuscular mycorrhizal fungi. *Oecol.* **138**: 574–583.

Oehl, F., Sieverding, E., Ineichen, K., Ris, E.A., Boller, T., and Wiemken, A., 2005, Community structure of arbuscular mycorrhizal fungi at different soil depths in extensively and intensively managed agroecosystems. *New Phytol.* **165**: 273–283.

Omar, S.A., 1998, The role of rock-phosphate-solubilizing fungi and vesicular-arbuscular-mycorrhiza (VAM) in growth of wheat plants fertilized with rock phosphate. *World J. Microbiol. Biotechnol.* **14**: 211–218.

Ortas, I., Ortakci, D., and Kaya, Z., 2002, Various mycorrhizal fungi propagated on different hosts have different effect on citrus growth and nutrient uptake. *Com. Soil Sci. Plant Anal.* **33**: 259–272.

Ozgonen, H., and Erkilic, A., 2007, Growth enhancement and Phytophthora blight (*Phytophthora capsici* Leonian) control by arbuscular mycorrhizal fungal inoculation in pepper. *Crop Protec.* **26**: 1682–1688.

Pardo, A., Amato, M., and Chiaranda, F.Q., 2000, Relationships between soil structure, root distribution and water uptake of chickpea (*Cicer arietinum* L.). Plant growth and water distribution. *Eur. J. Agro.* **13**: 39–45.

Paulitz, T.C., and Linderman, R.G., 1989, Interactions between fluorescent pseudomonads and VA mycorrhizal fungi. *New Phytol.* **113**: 37–45.

Plenchette, C., 1983, Growth responses of several plant species to mycorrhizae in a soil of moderate P fertility. *Plant Soil* **70**: 199–209.

Posta, K., Marschner, H., and Römheld, V., 1994, Manganese reduction in the rhizosphere of mycorrhizal and nonmycorrhizal maize. *Mycorrhiza* **5**: 119–124.

Powell, C.L., 1981, Inoculation of barley with efficient mycorrhizal fungi stimulates seed yield. *Plant Soil* **59**: 487–489.

Powell, J.R., Gulden, R.H., Hart, M.M., Campbell, R.G., Levy-Booth, D.J., Dunfield, K.E., Pauls, K.P., Swanton, C.J., Trevors, J.T., and Klironomos, J.N., 2007, Mycorrhizal and rhizobial colonization of genetically modified and conventional soybeans. *Appl. Environ. Microbiol.* **73**: 4365–4367.

Quilambo, O.A., Weissenhorn, I., Doddema, H., Kuiper, P.J.C., and Stulen, I., 2005, Arbuscular mycorrhizal inoculation of peanut in low-fertile tropical soil. II. Alleviation of drought stress. *J. Plant Nutr.* **28**: 1645–1662.

Rao, A.V., Tarafdar, J.C., Sharma, S.K., and Aggarwal, R.H., 1995, Influence of cropping systems on soil biochemical properties in an arid rainfed environment. *J. Arid Environ* **31**: 237–244.

Read, D.J., Koucheki, H.K., and Hodgson, J., 1976, Vesicular-arbuscular mycorrhiza in natural vegetation systems. *New Phytol.* **77**: 641–653.

Rempel, C.B., 1989, Interactions between vesicular-arbuscular mycorrhizae (VAM) and fungal pathogens in wheat. M.Sc. thesis, University of Manitoba, Winnipeg, Canada, pp. 134.

Reyes, I., Bernier, L., and Antoun, H., 2002, Rock phosphate solubilization and colonization of maize rhizosphere by wild and genetically modified strains of *Penicillium rugulosum*. *Microb. Ecol.* **44**: 39–48.

Rillig, M.C., Lutgen, E.R., Ramsey, P.W., Klironomos, J.N., and Gannon, J.E., 2005, Microbiota accompanying different arbuscular mycorrhizal fungal isolates influence soil aggregation. *Pedobiol.* **49**: 251–259.

Rodriguez, H., and Fraga, R., 1999, Phosphate solubilizing bacteria and their role in plant growth promotion. *Biotechnol. Adv.* **17**: 319–339.

Rousseau, A., Benhamou, N., Chet, I., and Piché, Y., 1996, Mycoparasitism of the extramatrical phase of *Glomus intraradices* by *Trichoderma harzianum*. *Phytopathology* **86**: 434–443.

Ruiz-Lozano, J.M., and Azcón, R., 2000, Symbiotic efficiency and infectivity of an autochthonous arbuscular mycorrhizal *Glomus* sp. from saline soils and *Glomus deserticola* under salinity. *Mycorrhiza* **10**: 137–143.

Ruiz-Lozano, J.M., Azcón, R., and Gomez, M., 1995, Effects of arbuscular-mycorrhizal *Glomus* species on drought tolerance: physiological and nutritional plant responses. *Appl. Environ. Microbiol.* **61**: 456–460.

Russo, A., Felici, C., Toffanin, A., Gotz, M., Collados, C., Barea, J.M., Moenne-Loccoz, Y., Smalla, K., Vanderleyden, J., and Nuti, M., 2005, Effect of *Azospirillum* inoculants on arbuscular mycorrhiza establishment in wheat and maize plants. *Biol. Fert. Soils* **41**: 301–309.

Ryan, M.H., and Angus, J.F., 2003, Arbuscular mycorrhizae in wheat and field pea crops on a low P soil: increased Zn-uptake but no increase in P-uptake or yield. *Plant Soil* **250**: 225–239.

Ryan, M.H., and Graham, J.H., 2002, Is there a role for arbuscular mycorrhizal fungi in production agriculture? *Plant Soil* **244**: 263–271.

Ryan, M.H., van Herwaarden, A.F., Angus, J.F., and Kirkegaard, J.A., 2005, Reduced growth of autumn-sown wheat in a low-P soil is associated with high colonisation by arbuscular mycorrhizal fungi. *Plant Soil* **270**: 275–286.

Sala, V.M.R., Freitas, S.D., and da Silveira, A.P.D., 2007, Interaction between arbuscular mycorrhizal fungi and diazotrophic bacterial in wheat plants. *Pesq. Agro. Bras.* **42**: 1593–1600.

Saxena, A.K., Rathi, S.K., and Tilak, K., 1997, Differential effect of various endomycorrhizal fungi on nodulating ability of green gram by *Bradyrhizobium* sp. (vigna) strains 24. *Biol. Fert. Soils* **24**: 175–178.

Scheublin, T.R., Van Logtestijn, R.S.P., and Van der Heijden, M.G.A., 2007, Presence and identity of arbuscular mycorrhizal fungi influence competitive interactions between plant species. *J. Ecol.* **95**: 631–638.

Schloter, M., Dilly, O., and Munch, J.C., 2003, Indicators for evaluating soil quality. *Agric. Ecos. Environ.* **98**: 255–262.

Schmidt, J.P., Lamb, J.A., Schmitt, M.A., Randall, G.W., Orf, J.H., and Gollany, H.T., 2001, Soybean varietal response to liquid swine manure application. *Agron. J.* **93**: 358–363.

Schreiner, R.P., 2007, Effects of native and nonnative arbuscular mycorrhizal fungi on growth and nutrient uptake of 'Pinot noir' (*Vitis vinifera* L.) in two soils with contrasting levels of phosphorus. *Appl. Soil Ecol.* **36**: 205–215.

Schwab, S.M., Menge, J.A., and Tinker, P.B., 1991, Regulation of nutrient transfer between host and fungus in vesicular-arbusculare mycorrhizas. *New Phytol.* **117**: 387–398.

Scullion, J., Eason, W.R., and Scott, E.P., 1998, The effectivity of arbuscular mycorrhizal fungi from high input conventional and organic grassland and grass-arable rotations. *Plant Soil* **204**: 243–254.

Selim, S., Negrel, J., Govaerts, C., Gianinazzi, S., and van Tuinen, D., 2005, Isolation and partial characterization of antagonistic peptides produced by *Paenibacillus* sp. strain B2 isolated from the sorghum mycorrhizosphere. *Appl. Environ. Microbiol.* **71**: 6501–6507.

Singer, J.W., Kohler, K.A., Liebman, M., Richard, T.L., Cambardella, C.A., and Buhler, D.D., 2004, Tillage and compost affect yield of corn, soybean, and wheat and soil fertility. *Agron. J.* **96**: 531–537.

Sjöberg, J., 2005, Arbuscular mycorrhizal fungi: occurrence in Sweden and interaction with a plant pathogenic fungus in barley. Ph.D. thesis, Swedish University of Agricultural Sciences, pp. 53.

Smith, S.E., and Read, D.J., 1997. *Mycorrhizal symbiosis*, 2nd edn., Academic, San Diego, CA/London, pp. 605.

Smith, S.E., Smith, F.A., and Jakobsen, I., 2004, Functional diversity in arbuscular mycorrhizal (AM) symbioses: the contribution of the mycorrhizal P uptake pathway is not correlated with mycorrhizal responses in growth or total P uptake. *New Phytol.* **162**: 511–524.

Sood, S.G., 2003, Chemotactic response of plant-growth-promoting bacteria towards roots of vesicular-arbuscular mycorrhizal tomato plants. *FEMS Microbiol. Ecol.* **45**: 219–227.

St-Arnaud, M., and Elsen, A., 2005. Interaction or arbuscular-mycorrhizal fungi with soil-borne pathogens and non-pathogenic rhizosphere micro-organisms. In: *In vitro culture of mycorrhizas*, S. Declerck, D.-G. Strullu and J. A. Fortin eds., Springer, Berlin/Heidelberg, Germany, pp. 217–231.

St-Arnaud, M., and Vujanovic, V., 2007, Effects of the arbuscular mycorrhizal symbiosis on plant diseases and pests. In: *in crop production*, C. Hamel and C. Plenchette eds., Haworth, New York, pp. 67–122.

St-Arnaud, M., Hamel, C., Vimard, B., Caron, M., and Fortin, J.A., 1995, Altered growth of *Fusarium oxysporum* f. sp. *chrysanthemi* in an *in vitro* dual culture system with the vesicular arbuscular mycorrhizal fungus *Glomus intraradices* growing on *Daucus carota* transformed roots. *Mycorrhiza* **5**: 431–438.

Subba Rao, N.S., 1985, Effect of combined inoculation of *Azospirillum brasilense* and vesicular arbuscular mycorrhiza on pearl millet (*Pennisetum americanum*). *Plant Soil* **81**: 283–286.

Subramanian, K.S., and Charest, C., 1995, Influence of arbuscular mycorrhizae on the metabolism of maize under drought stress. *Mycorrhiza* **5**: 273–278.

Subramanian, K.S., and Charest, C., 1998, Arbuscular mycorrhizae and nitrogen assimilation in maize after drought and recovery. *Physiol. Plant.* **102**: 285–296.

Subramanian, K.S., and Charest, C., 1999, Acquisition of N by external hyphae of an arbuscular mycorrhizal fungus and its impact on physiological responses in maize under drought-stressed and well-watered conditions. *Mycorrhiza* **9**: 69–75.

Subramanian, K.S., Charest, C., Dwyer, L.M., and Hamilton, R.I., 1995, Arbuscular mycorrhizas and water relations in maize under drought stress at tasselling. *New Phytol.* **129**: 643–650.

Subramanian, K.S., Charest, C., Dwyer, L.M., and Hamilton, R.I., 1997, Effects of arbuscular mycorrhizae on leaf water potential, sugar content, and P content during drought and recovery of maize. *Can. J. Bot.* **75**: 1582–1591.

Talavera, M., Itou, K., and Mizukubo, T., 2001, Reduction of nematode damage by root colonization with arbuscular mycorrhiza (*Glomus* spp.) in tomato-*Meloidogyne incognita* (Tylenchida: Meloidognidae) and carrot-*Pratylenchus penetrans* (Tylenchida: Pratylenchidae) pathosystems. *Appl. Entomol. Zool.* **36**: 387–392.

Talukdar, N.C., and Germida, J.J., 1993, Occurrence and isolation of vesicular-arbuscular mycorrhizae in cropped field soils of Saskatchewan, Canada. *Can. J. Microbiol.* **39**: 567–575.

Tarkalson, D.D., Jolley, V.D., Robbins, C.W., and Terry, R.E., 1998, Mycorrhizal colonization and nutrition of wheat and sweet corn grown in manure-treated and untreated topsoil and subsoil. *J. Plant Nutr.* **21**: 1985–1999.

Tejada, M., and Gonzalez, J.L., 2006, Crushed cotton gin compost on soil biological properties and rice yield. *Eur. J. Agro.* **25**: 22–29.

Thygesen, K., Larsen, J., and Bodker, L., 2004, Arbuscular mycorrhizal fungi reduce development of pea root-rot caused by *Aphanomyces euteiches* using oospores as pathogen inoculum. *Eur. J. Plant Pathol.* **110**: 411–419.

Tian, C.Y., Feng, G., Li, X.L., and Zhang, F.S., 2004, Different effects of arbuscular mycorrhizal fungal isolates from saline or non-saline soil on salinity tolerance of plants. *Appl. Soil Ecol.* **26**: 143–148.

Troeh, Z.I., and Loynachan, T.E., 2003, Endomycorrhizal fungal survival in continuous corn, soybean, and fallow. *Agron. J.* **95**: 224–230.

van der Heijden, M.G.A., Wiemken, A., and Sanders, I.R., 2003, Different arbuscular mycorrhizal fungi alter coexistence and resource distribution between co-occurring plant. *New Phytol.* **157**: 569–578.

Vazquez, M.M., Cesar, S., Azcón, R., and Barea, J.M., 2000, Interactions between arbuscular mycorrhizal fungi and other microbial inoculants (*Azospirillum, Pseudomonas, Trichoderma*) and their effects on microbial population and enzyme activities in the rhizosphere of maize plants. *Appl. Soil Ecol.* **15**: 261–272.

Vigo, C., Norman, J.R., and Hooker, J.E., 2000, Biocontrol of the pathogen *Phytophthora parasitica* by arbuscular mycorrhizal fungi is a consequence of effects on infection loci. *Plant Pathol.* **49**: 509–514.

Villegas, J., and Fortin, J.A., 2001, Phosphorus solubilization and pH changes as a result of the interactions between soil bacteria and arbuscular mycorrhizal fungi on a medium containing NH_4^+ as nitrogen source. *Can. J. Bot.* **79**: 865–870.

Villegas, J., and Fortin, J.A., 2002, Phosphorus solubilization and pH changes as a result of the interactions between soil bacteria and arbuscular mycorrhizal fungi on a medium containing NO_3^- as nitrogen source. *Can. J. Bot.* **80**: 571–576.

Vogelsang, K.M., Reynolds, H.L., and Bever, J.D., 2006, Mycorrhizal fungal identity and richness determine the diversity and productivity of a tallgrass prairie system. *New Phytol.* **172**: 554–562.

Walley, F.L., and Germida, J.J., 1997, Response of spring wheat (*Triticum aestivum*) to interactions between pseudomonas species and *Glomus clarum* NT4. *Biol. Fert. Soils* **24**: 365–371.

West, H.M., 1996, Influence of arbuscular mycorrhizal infection on competition between *Holcus lanatus* and *Dactylis glomerata*. *J. Ecol.* **84**: 429–438.

Whitelaw, M.A., 2000, Growth promotion of plants inoculated with phosphate-solubilizing fungi. *Adv. Agro.* **69**: 99–151.

Wu, S.C., Cao, Z.H., Li, Z.G., Cheung, K.C., and Wong, M.H., 2005, Effects of biofertilizer containing N-fixer, P and K solubilizers and AM fungi on maize growth: a greenhouse trial. *Geoderma* **125**: 155–166.

Wu, Q.S., and Xia, R.X., 2006, Arbuscular mycorrhizal fungi influence growth, osmotic adjustment and photosynthesis of citrus under well-watered and water stress conditions. *J. Plant Physiol.* **163**: 417–425.

Xavier, L.J.C., and Germida, J.J., 2002, Response of lentil under controlled conditions to co-inoculation with arbuscular mycorrhizal fungi and rhizobia varying in efficacy. *Soil Biol. Biochem.* **34**: 181–188.

Xavier, L.J.C., and Germida, J.J., 2003, Selective interactions between arbuscular mycorrhizal fungi and *Rhizobium leguminosarum* bv. *viceae* enhance pea yield and nutrition. *Biol. Fert. Soils* **37**: 261–267.

Yao, Q., Zhu, H.H., Chen, J.Z., and Christie, P., 2005, Influence of an arbuscular mycorrhizal fungus on competition for phosphorus between sweet orange and a leguminous herb. *J. Plant Nutr.* **28**: 2179–2192.

Zhang, X.H., Zhu, Y.G., Chen, B.D., Lin, A.J., Smith, S.E., and Smith, F.A., 2005, Arbuscular mycorrhizal fungi contribute to resistance of upland rice to combined metal contamination of soil. *J. Plant Nutr.* **28**: 2065–2077.

Zhang, X.H., Zhu, Y.G., Lin, A.J., Chen, B.D., Smith, S.E., and Smith, F.A., 2006, Arbuscular mycorrhizal fungi can alleviate the adverse effects of chlorothalonil on *Oryza sativa* L. *Chemosphere* **64**: 1627–1632.

Zhu, Y.G., Smith, S.E., Barritt, A.R., and Smith, F.A., 2001, Phosphorus (P) efficiencies and mycorrhizal responsiveness of old and modern wheat cultivars. *Plant Soil* **237**: 249–255.

Chapter 6

ARBUSCULAR MYCORRHIZAE: A DYNAMIC MICROSYMBIONT FOR SUSTAINABLE AGRICULTURE

JITENDRA PANWAR[1], R.S. YADAV[2], B.K. YADAV[3], AND J.C. TARAFDAR[3]
[1]*Biological Sciences Group, Birla Institute of Technology and Science, Pilani-333031,INDIA;*
[2]*Main Wheat Research Station, S. D. Agricultural University, Vijapur-382870, INDIA;*
[3]*Central Arid Zone Research Institute, Jodhpur- 342003, INDIA*

Abstract: Beneficial microorganisms associated with roots are of paramount importance and contributes for sustainable agriculture. Of the various microorganisms colonizing the rhizosphere, arbuscular mycorrhizal (AM) fungi occupy a unique ecological position as they are partly inside and partly outside the roots. They constitute a major portion of soil biomass and link the biotic and geochemical parts of the ecosystem. This symbiosis occurred in vast diversity of climate and soil-types. AM fungi enable plants to cope up with abiotic stresses by alleviating mineral deficiencies, overcoming the detrimental stresses of salinity and drought. Their association improves the adaptation of nursery raised and micropropagated seedlings to cope up with sudden stress due to change in environmental conditions and shift from *in vitro* to *in vivo* conditions. The contribution of AM fungi with respect to plant nutrition, water relations, soil stability, reclamation, rehabilitation and establishment of micropropagated plantlets is also discussed.

Keywords: Plant-microbe interactions; mycorrhizosphere; AM diversity; soil quality; micropropagation.

1 INTRODUCTION

The basic objective of agriculture is to produce food for the population. The ever-burgeoning human and animal population in the past decades not only found the food production inadequate but also degrade the whole agro-ecosystem. Further, the traditional agriculture system apparently sustainable at low productivity and at low population, pressure is breaking down under the onslaught of high human and animal population pressure. The intensification of agriculture as an inevitable consequence of the compulsion to produce more compounded with rapid and uncontrolled industrialization has put an enormous burden on the natural ecosystem. Although, the benefits of green revolution continue to accrue today but simultaneously the issues of sustainability once again has become increasingly prominent. Such concerns and problems posed by modern-day agriculture gave birth to new concepts in farming, such as organic farming, natural farming, biodynamic agriculture, do-nothing agriculture, eco-farming etc. The essential feature of such farming practices imply, i.e., back to nature. The problem has been further aggravated by shrinkage of the arable land due to soil degradation and deterioration of water quality which has now assumed global dimensions. Sustainable agriculture is the only proponent which encompasses soil and crop productivity by integration of agricultural management technology at the same time maintaining and enhancing the farm profitability and environmental quality. The production and productivity growth rate of major crops are also stagnated or even declined during this green revolution era (Table 1).

Table 1. Production and productivity growth rates (% per annum) of major crops.

Crops	Production			Productivity		
	1980–89	1990–99	2000–02	1980–89	1990–99	2000–02
Rice	3.62	1.09	(–)5.60	3.19	1.27	(–)0.72
Wheat	3.57	3.81	(–)0.28	3.10	2.11	0.73
Pulses	1.52	0.61	0.99	1.61	0.96	(–)1.84
Food grains	2.85	1.94	(–)3.73	2.74	1.52	(–)0.69
Oilseeds	5.20	2.13	(–)5.30	2.43	1.25	(–)3.83
Non-food grains	3.77	2.78	(–)2.21	2.31	1.04	(–)1.02
All major crops	3.19	2.28	(–)3.13	2.56	1.31	(–)0.87

Plant microbe interactions are the interesting events that contribute for the sustainable agriculture. Microbial world in particular the beneficial microbes associated with plant roots are of paramount importance in agriculture and crop productivity which include nitrogen fixers, 'P' solubilizers, growth enhancers, biocontrol agents, microbes important in industry and medicine, and mineral transporters besides helping in increasing soil binding and soil stability. Microbes have more capability to adapt to different environments like plants. Soil is a dynamic medium to nourish different microbial communities like bacteria, actinomycetes, fungi, algae, protozoan etc. which play significant role in cycling of plant nutrient elements, biological conversions, humus formation, ecosystem sustenance, geo-chemical cycling, and others besides supporting plant life and plant productivity. Among the microbial communities, arbuscular mycorrhizae are a mutualistic association between the roots of most plant species and fungi. Bidirectional movement of nutrients characterizes these symbionts where carbon flows to the fungus and inorganic nutrients get transported through mycorrhizal network to the plant.

2 THE MYCORRHIZOSPHERE AND MYCORRHIZAL ASSOCIATIONS

Traditionally, rhizosphere activity and mycorrhizal symbioses have been studied as if they represented separate ecosystem components, but they occur concurrently on many plant roots rather than as distinct entities. In all soil systems a growing plant will develop some sort of microbial plant interaction termed as a rhizosphere, similarly majority of higher plants are mycorrhizal. Thus, we cannot conclude that all rhizospheres are mycorrhizospheres, but it is also evident that most of the data collected from field studies of rhizosphere populations must have resulted from evolution of mycorrhizosphere. This conceptual oversight is gradually being corrected. From an environmental or soil community consideration the fungal-plant associations in mycorrhizal symbioses can be viewed on a gradient of increasing association of the plant and soil community (Table 2).

In evaluating the quantities of exudates and their chemical substituents, quantification of the intensity of the mycorrhizal associations is necessary. The phenomenon "Mycorrhiza" comprises all symbiotic associations of soil-borne fungi with roots or rhizoids of higher plants and was first introduced into the literature by Frank (1885). Allen (1991) described the fungal-plant interaction from a more neutral or microbially oriented aspect: "A mycorrhizae is a mutualistic symbiosis between plant and fungus localized in a root or root-like structure in which energy moves primarily from plant to fungus and inorganic resources move from fungus to plant". The group of fungi and plants, which are involved in the interaction, determines the type of mycorrhiza they form (Molina *et al.*, 1992). The physical association of mycorrhizal fungi with plant roots has been extensively described and serves as the primary basis for the classification of mycorrhizae. The different kinds of mycorrhiza are: Ectomycorrhizae; Arbuscular mycorrhizae; Ectendomycorrhizae; Arbutoid mycorrhizae; Monotroid mycorrhizae; Ericoid mycorrhizae and Orchid mycorrhizae. Arbutoid and monotropoid ectendomycorrhizae, as well as ericoid and orchidoid endomycorrhizae, can only be found in a few plant species and are, therefore, restricted to certain ecosystems. In contrast, ectomycorrhizas are established between a great variety mostly in Basidiomycota and the roots of many woody plants. This symbiosis is very common in forest ecosystems in the temperate zone and can be applied to plant production systems in tree nurseries (Munro et al., 1999). The most widespread type, however, is represented by the Arbuscular mycorrhizae, and this chapter concentrates on this type of symbioses, because the plant hosts include the most important crops used in agriculture and horticulture.

Table 2. Comparison of properties of non-rhizosphere, rhizosphere, and mycorrhizosphere soils.

Non-rhizosphere	Rhizosphere	Mycorrhizosphere
–All biogeochemical cycle rates limited by organic matter (humus) availability	–Maximum biogeochemical cycle	–All biogeochemical cycles are supported-controlled by both plant inputs and soil humus contents
–Nutrients may be leached to root tissue or ground waters –Nutrients may be directly incorporated directly into biomass –Microbial biomass limited by carbon and energy resources –Note that non-rhizosphere soil metabolic activity is not affected directly by rhizosphere interactions	–Nutrientsmineralized directly available for plant biomass as well as microbial biomass synthesis –Microbial biomass synthesis is controlled by plant productivity and rate of root exudates production	–Nutrients incorporated into fungal biomass and transported to plant tissue –Biomass controlled by plant productivity and transfer of photosynthates to the fungus. –Mycorrhizal fungi may enter in non-rhizosphere soil and catalyze biogeochemical processes

3 DIVERSITY OF ARBUSCULAR MYCORRHIZAL FUNGI

Arbuscular mycorrhizal (AM) fungi are ancient microorganisms appeared between 460 and 400 million years ago (Simon et al., 1993; Redecker et al., 2000) as plants started to colonize the land. AM fungi are of worldwide distribution. The AM root colonization is a dynamic process which is also influenced by several edaphic factors such as nutrient status of soil, season, AM strains, soil temperature, extent of soil pollution, soil pH, host cultivar susceptibility to AM colonization, feeder root condition at the time of sampling etc. The quality and type of AM propagules also affected the dynamics of root colonization, which are increased by increasing the age of plant (Chandra and Jamaluddin, 1999). These mycorrhizal associations are in the form of chlamydospores, zygospores and azygospores and have been recovered from the soils of different habitats e.g., nutrient deficient soils, sand dunes and deserts, industrial wastes, sodic soils, polluted sites, sewage, eroded sites and others like forests, open wastelands, scrubs, savanna, heaths, grasslands, coal waste etc. The fungus infects the root system of most cultivated crops and usually it invades several layers of the outer root cortex by penetration of hyphae in individual cells and form arbuscules within cells.

The AM fungal associations occur widely throughout the plant kingdom including most of the agricultural and horticultural crops (Gerdemann, 1968). Nowadays they are integral components of most terrestrial ecosystems and some 80% of terrestrial, vascular plant families act as hosts for the fungal endosymbionts (Harley and Harley, 1987; Brundrett, 2004). AM fungi were placed in Zygomycotina but recent studies have recognized the new fungal phylum namely Glomeromycota with a single class Glomeromycetes, containing 150 described species allover the world (Cavalier-Smith, 1998). Although, the separation of species within a genus is a difficult task but recently molecular techniques have been employed to examine intergeneric relationship by some workers (Kim *et al.*, 1999; Hiremath and Podila, 2000). The small number of different fungal species might suggest that their biodiversity is limited but the variation at the morphological, physiological, and genetic levels is rather high which results in a functional diversity that has an important impact for ecology and application in plant production systems (Table 3). Morton and Redecker (2001) based on concordant molecular and morphological characters discovered two new families, *Archaeosporaceae* and *Paraglomaceae* with two new genera *Archaeospora* and *Paraglomus* respectively.

4 ECOLOGY OF ARBUSCULAR MYCORRHIZA

Ecology is the study of the relationship of organisms to their environments. This environment includes all abiotic and biotic factors affecting the cell. Biotic properties of an ecosystem include not only those traits commonly classified as "natural" but also any consequences of anthropogenic interventions or interaction with the site. However, the nature, activity and future of any soil microbial community are determined by the capacity of the organisms to adapt to or modify negative soil properties.

Table 3. Variation among AM fungi (Summarizing genotype as well as phenotype variation and changes in biodiversity). (Franken and George, 2004.)

	Between species	Between isolates	Between nuclei
Genotype variation			
Genome size	15–1,000 Mb	Nd	Nd
rRNA	+	+	+
Genes encoding			
Chitin synthase	+	+	Nd
Translation elongation factor	+	Nd	–
Actin	+	Nd	+(In introns)
B-tubulin	+	Nd	Gene family
Glutathione S-transferase	Nd	Nd	Gene family
Binding protein	Nd	Nd	+
Phenotype variation			
Morphology			
Spore size, shape, and colour	+	–	
Presence of BLOs	+	–	
Germination	+	Nd	
Auxillary cells	+	–	
Arbuscule type	+	Nd	
Vesicles	+	–	
Branching promoting factor	+	Nd	
Nutrient exchange			
Phosphate	+	Nd	
Nitrogen	+	+	
Iron	+	Nd	
Carbohydrates	+	+	
Stress tolerance			
Heavy metals	+	+	
Drought	+	+	
Root pathogens	+	Nd	
Enzymatic parameters			
Various enzymes	+	Nd	
Superoxide dismutase	+	Nd	
RNA accumulation			
Fungal general pattern	+	Nd	
Fungal b-tubulins	+	Nd	
Plant general pattern	+	Nd	
Plant chitinase	Nd	+	
Plant glucanase	Nd	+	
Plant sucrose synthase	+	+	
Plant phosphate transporters	+	Nd	
Biodiversity in ecosystems			
Influenced by			
Plant species	+	+	
Soil management	+	+	

+ = differences detected; – = no difference detected; nd = not determined

Some organisms best equipped to cope with the ecological stresses are active, while more stringent organisms succumb. Although, the ultimate fate depends upon the genetic diversity of the organisms and the severity of the stress imposed as well as the time allowed for recovery.

The arbuscular mycorrhizal fungi are ubiquitous in nature, but little is known about the natural ecology of these fungal-plant association and the effects of certain soil amendments with natural waste products. The conventional agronomic practices may adversely affect the efficiency as well as the potential impact of these micro-symbionts (Table 4). AM fungi are also conditioned by soil factors. It is found that pH plays an important role in the distribution of AM fungi. Acidic to neutral soils have a large number of AM fungi. Hayman (1974) described the effect of light on AM fungi. Soil moisture exerts influence on mycorrhizal association (Redhead, 1977). Species and strains of AM fungi differ in their ranges to soil physico-chemical properties (Abbott and Robson, 1982). Soil factors not only influence soil fertility but affects mycorrhizal inoculum. The soil conditions like salinity, water logging, erosion, soil types, water holding capacity, soil porosity and fertility status, vegetations etc. appreciably influences the AM fungal association, distribution, composition and activity (Manoharachary, 2004). Therefore, in general one may conclude that conventional agricultural management prac-tices reduce AM fungal population while organically low- input system would be a viable proponent to increase their activity as well as for sustainable agricultural development. For a comprehensive review on mycorrhizae in natural ecosystems the reader is referred to Brundrett (1991), Fitter (1991) and Read (1991).

4.1   Plant nutrition and water relations

A substantial biomass component in many ecosystems is resultant influence of mycorrhizal associations (Allen, 1991). Mycorrhizae tend to be the largest component in the ecosystem primarily because both the fungi and the associated roots are turned over rapidly. Mycorrhizae as well dictate nutrient cycling rates and patterns by altering host plant resource acquisition and plant production. Odum and Biever (1994) have catalogued six main pathways in ecosystems through which the nutrients are recycled from plants, viz. grazing, seed consumption, feeding on nector, loss of soluble exudates, active extraction by parasitic and mutualistic organisms, and decomposition of plant structures. In this background, mycorrhizae play a vital role in last three categories in capturing nutrients (Jalali, 2001). Mycorrhizae therefore link the biotic and geochemical parts of the ecosystem. Their contribution in sustainable ecosystem is well recognized. With this recognition, the management of symbiotic fungal populations would become a potential tool for overall crop health and resultant sustainability.

Table 4. Summary of some of the potential effects of agricultural management practices on AM fungi in the field. (Atkinson et al., 2002).

Factors	Potential effect on			
	Effectivity of symbiosis	Host pressure	Spore populations and viability	Extra radical hyphae
Crop choice	–	✓	–	–
Variety choice	✓	–	–	✓
Sequence (rotation)	✓	✓	✓	✓
Tillage	–	✓	–	✓
Fallow period	–	✓	✓	–
Organic farming systems	✓	✓	✓	✓
Inoculants	✓	–	✓	–
Fumigants	✓	–	✓	–
pH changes	✓	–	✓	✓
Phosphorus	✓	–	–	–
Manures and crop residuse	✓	–	–	–

Mycorrhizal fungi are clearly instrumental in augmenting plant nutrient availability particularly in nutrient stressed ecosystems. A wide range of data is available demonstrating nutritional benefits to plant communities through mycorrhizal symbiotic associations. Most research has shown improved nutrient transfer to the plant tissue through the augmentation of the absorbing surface of roots by extension of the fungal mycelium into non-rhizosphere soil. Gains in phosphate, nitrogen, sulphur, potassium, calcium, zinc, iron, copper and water transfers are most commonly reported. The extensive literature documented these benefits has been reviewed by Allen (1991), Gupta (1991) and Mukerji et al. (1991). Besides this, mycorrhizal fungi may not only enhance soil-plant transfer of nutrient, but may also be instrumental in movement of nutrients between plants (Eason et al., 1991). Read et al. (1989) demonstrated through the use of $^{14}CO_2$ that carbon moves freely between plants connected by mycorrhizal mycelium. This plant-plant bridging by the fungal hyphae occurs between host plants of the same or different species. Arbuscular mycorrhizae were noted to connect the root systems of neighbors of different species and considered to have mediated the nutrient transfer. These workers suggested a competitive advantage of this bridging in plant groupings where mycorrhizal inoculum is limited. Panwar et al. (2007)

found that the hyphal front may advance at a rate of 2.0–4.3 cm d^{-1} in soil. Plants developing in a less desirable portion (e.g., a shaded site) of the ecosystem may gain benefits of the photosynthetic activity of the less stressed members of the community as well as the gains from nutrient production of the mycorrhizal function through an interconnected root system. Mycorrhizal inoculation enhances nodulation in legumes (Carling *et al.*, 1978). A combination of legumes, rhizobia and mycorrhizal fungus brings a significant improvement in plant growth through increased availability of phosphorus, sulphur and micronutrients with higher nitrogen fixation in soil. Thus, this combination may prove the cheapest way to enrich tropical soil with nitrogen.

Arbuscular mycorrhizal fungi play an important role also in the water economy of plants. These associations improve the unsaturated hydraulic conductivity of the roots either by modifying root morphology and root anatomy or indirectly by hormonal and structural changes in the host plant. These improvements are the factors contributing towards better uptake of the water by the plants. It has been suggested that the AM fungi help the plants in better absorption of water by the roots resulting on a better performance (Kehri and Chandra, 1990) and by exploring water in wider zones of soil (Safir *et al.*, 1971, 1972). It has been noted that the mycorrhizal plants show a better survival than non-mycorrhizal ones in extremely dry condition (Allen *et al.*, 1981). It appears that the most established benefits from mycorrhizal fungus to the host plant is through the widespread mycelial network which penetrates deeper and wider in the soil in search of water and nutrients thereby widening the zone of activity.

4.2 Improved soil quality

Plant health and productivity are rooted in the soil, and the quality of soil depends on the viability and diversity of its biota which determine the structures that support a stable and healthy agro-ecosystem (Doran and Linn, 1994). The importance of soil is now recognized not only as an agricultural resource base, but as a complex, fragile and yet dynamic system that must be protected as well as managed to ensure its long term stability and productivity. The goals of sustainability in agriculture could be viewed broadly as "maximum plant production with a minimum of soil loss". In this scenario of balanced agro-system inputs and outputs, the relevance of mycorrhizal endophytes has been described as that of a fundamental link between plant and soil (Miller and Jastrow, 1994). They have shown how the affinity between mycorrhizae and soil aggregates vary with root characteristics, with the intensity of root colonization, and with the quantum of soil mycelium associated with the root system.

4.3 Soil aggregation

The contribution of mycorrhizal associations in soil aggregate formation would be catalogued as:

- Hyphal growth into the soil matrix forms the skeletal structure that holds the primary soil particles together through physical entanglement (Miller and Jastrow, 1994).
- Roots and hyphae together create the physical and chemical environment to produce organic and amorphous materials (Tisdall, 1991) for the binding of particles.
- Hyphae and root enmesh microaggregates into macroaggregate structures, which accelerate the capacity for carbon and nutrient storage and provide microhabitats for soil microbes (Cambardella and Elliott, 1994).
- Secretion of glomalin that helps in aggregate stability by arbuscular mycorrhizal fungi (Wright and Upadhyaya, 1998).

Miller and Jastrow (1992) proposed that AM hyphae form and stabilize aggregates of soil through three distinct processes: (1) The AM hyphae physically entangle primary particles of soil; (2) roots and AM hyphae create conditions that enable microaggregates to form in soil; and (3) roots and AM hyphae enmesh and bind microaggregates and smaller macro-aggregates into larger macroaggregates. Therefore, mycorrhizal fungi are able to bind soil into semi-stable and stable aggregates, thus improving the structure of the soil. This improvement in soil structure has a direct impact on the indigenous microbial community through aeration and moisture infiltration, and an indirect effect via stimulation of plant root growth. This may make mycorrhizal plants particularly useful for reclamation of soils with problems of surface crusting and unstable structures like sand dunes as well as enhance host root development in disturbed soils. Clearly, it is well documented that this augmentation of root development increases the quantities of fixed carbon by reaching the soil microbial community. Because conceptually, the arbuscules presents a large area of close contact between the symbionts, and is assumed to be the interface over which carbon would be transferred and accumulated in the soil on sloughed off the hyphal network. Burns and Davis (1986) concluded that a holistic approach to the joint study of mycorrhizal host plant and soil responses would be vital if the microbial dynamics of rhizosphere and the resultant improvement of soil structure is an evolutionary mechanism that confers competitive advantage to plants.

Glomalin, a glycoprotein, produced in copious amounts by AM fungal hyphae plays a major role in aggregate stabilization (Wright and

Upadhyaya, 1998). A strong relationship between soil aggregate stability and glomalin as well as glomalin-related soil proteins (GRSPs) has been demonstrated (Wright and Anderson, 2000; Rillig, 2004). As a result of this relationship, glomalin has been credited with enhanced ecosystem productivity, improved soil aeration, drainage and microbial activity (Lovelock *et al.*, 2004). Because of its role in soil particle aggregation, glomalin is thought to significantly reduce organic matter degradation by protecting labile compounds with in soil aggregates thus enhancing carbon sequestration in soil ecosystems (Wright *et al.*, 2000; Rillig, 2004).

4.4 Soil microbial biomass

A variety of elegant studies of soil enzymes and respiratory activity in relationship to soil physical and chemical properties and biological interactions have conducted, but evaluation of the role of mycorrhizal fungi in these soil functions is rarely considered. This omission of a consideration of these important trophic interactions occurs in spite of the knowledge that mycorrhizal fungi constitute a major portion of soil biomass and that these fungi extend throughout the soil profile-far beyond the regions classified as rhizosphere soil. Fogel and Hunt (1983) in a study of Douglas fir *Pseudotsuga menziesii* stand found that mycorrhizae constituted 6% of the total standing crop. Furthermore, they found that roots and mycorrhizae contained larger reserves of nitrogen, phosphorus, potassium, and magnesium than did the forest floor or soil fungi. Fine roots and mycorrhizae contributed between 84% and 78% of the total tree organic matter to the soil. Read (1984) suggests that mycorrhizal fungal biomass may be the largest microbial biomass component of many forest soils. Allen (1991) states that "mycorrhizal fungi may be the single largest consumer group of net primary production in many, if not most terrestrial biomes." Arbuscular fungal hyphae have been quantified at densities of up to 38 m cm^{-3} (Allen and Allen, 1986).

From these observations, a number of benefits of mycorrhizal development to the total soil microbial community can be delineated. Foremost among the contribution to the soil community is the capacity to cycle nutrients- that are, mineralize accumulated biomass. Although the bulk of the research efforts have been directed at quantifying mineral nutrient transfer from fungal biomass or soil to plant tissue, it is reasonable to assume that a portion of the metabolic products of this fungal activity will be released to the soil microbial community. For example, mycorrhizal fungi have been shown to mineralize soil organic phosphate through synthesis of phosphatase (Dighton, 1983; Tarafdar and Marhsner, 1994).

4.5 Resilience of problematic soils

Since the advent of civilization, the fertile top soil has been degrading due to various environmental and ecological factors like saline and sodic soils, eroded soils, industrial waste lands, soils from mining regions, degraded forest lands, sand dunes and deserts etc. In these soils the native vegetation and animal communities scarred to endangered. Therefore, rehabilitation of these soils is a global problem and the afforestation and other agricultural activities are in progress but only limited success has been achieved. The importance of AM fungi in regeneration of these lands has been investigated by many workers (Danielson *et al.*, 1979; Pfeiffer and Bloss, 1988; Rao and Tak, 2001; Giri *et al.*, 2004). The greenhouse experiment conducted by Jasper *et al.* (1987) indicated that after four to five years of re-vegetation, the number of infective propagules appears to be restored to a level equivalent to that of undisturbed soil and they have given emphasis on investigating the possibility of improving re-vegetation by increasing the inoculum potential of disturbed sites. So far combined effects of mycorrhizal fungi on re-vegetation of disturbed sites have been reported by various workers. Also AM fungi had been found to grow up-to salinity of 12 dSm^{-1} electricalconductivity and increased the uptake of P, N, S and micronutrients in crops (Hirrel and Gerdemann, 1980). The AM infection counteracts adverse soil factors. They increase the tolerance of the crop to high acidity and temperature (Pond *et al.*, 1984; Poss *et al.*, 1985).

AM fungi isolates can decrease the heavy metal concentration in shoot or in root, or decrease translocation from root to shoot (Diaz *et al.*, 1996). The latter could be due to the high metal sorption capacity of these fungi, which could 'filter' metal ions during uptake (Joner *et al.*, 2000). The high concentration of heavy metals in the intracellular hyphae of a heavy metal tolerant AM fungi colonizing maize roots (Kaldorf *et al.*, 1999) strengthen the hypothesis of sequestration of metals by AM fungi structures. However, the competivity of such metal tolerant AM fungi in the field is often unknown and should be investigated. Potential of AM fungi for bioremediation of radionuclides (Ebbs *et al.*, 2000; Berreck and Hasel-wandter, 2001) and polycyclic aromatic hydrocarbons polluted soils (Binet *et al.*, 2000; Joner and Leyval, 2001) has well been reported.

5 ENVIRONMENTAL CONSIDERATIONS

The statement by Allen (1991) that "mycorrhizae represents one of the least understood, most widespread, and most important biological symbionts on earth" cannot be more appreciated than when application of our understanding of this fungal-plant interaction to ecosystem problems is considered. Although, commonly considered from an industrial polluted soil view point, reclamation management to improve soil quality is being more frequently related to "exhausted" agricultural soil system i.e. those cultivated in some cases for a long period by methods designed to maximum crop yields, at times at the sacrifice of maintaining soil quality. Traditionally, this soil management includes procedures that encourage development of a soil structure conducive to aboveground plant community development (Tate, 1987) and establishment of stable population of organisms involved in biogeochemical cycling (Tate, 1985). Due to their ability to improve longevity and productivity of aboveground plant communities, mycorrhizal associations are a critical component in soil reclamation management of even these "exhausted" agricultural soils.

Plant community gains from management of mycorrhizal associations for soil quality improvement are derived from both the soil structural enhancement resulting from the fungal contributions to soil aggregation and the improved availability of essential plant nutrients. Although, rapid and somewhat long-lasting benefits are accrued from amendment of degraded soil with a variety of organic matter sources and fertilizers but for minimization of anthropogenic intervention, a stable soil microbe-plant interactive community must be developed. Mycorrhizal associations are developed best under stressed conditions. Therefore, for rapid and enduring development of a degraded ecosystem, AM fungal propagules should be insured for reclamation of target site, for implementation of plant nutrient management procedures that will not prevent mycorrhizal development, and for successful reforestation to establish the aboveground plant community.

6 ESTABLISHMENT OF NURSERY AND MICRO-PROPAGATED SEEDLINGS

Mycorrhizal inoculations stimulate rooting and growth, and thereby transplant survival of cuttings and seedlings raised in sterilized nursery media, which is essential for the establishment of plant cover on mine soils, eroded soils, industrial waste, where the plant cover is difficult to establish (Hall and Armstrong, 1979; Hall, 1980). The inoculation with appropriate endomycorrhizal fungi has given excellent results in plants like Liquidamber, Ampelopsis, Yew, Lilac, Berberis, Chamaecyparis, Asparagus, Onion and Leek (Gianinazzi et al., 1989). Successful endomycorrhizal establishment has been reported in a variety of rooting media at nursery level containing sand, gravel, peat, expended clay, pumice, perlite, bark, sawdust, vermiculite or mixture of all these (Menge, 1983; Denhe and Backhaus, 1986).

Mycorrhizas, through their role in increasing the natural resistance of plants to abiotic and biotic stresses and in rendering their underground organs more efficient in exploiting soil resources, have opened interesting possibilities for the production of high-quality micropropagated plants with low inputs (Gianinazzi et al., 1990). AM fungi inoculation has been found to improve growth and nutrient uptake in a large variety of plants propagated *in-vitro* (Lovato et al., 1996; Sharma and Adholeya, 2004; Fortunato and Avato, 2008). There is a considerable interest in studying, not only the impact of AM fungi on plant grown but also that obtained by coupling AM fungi with other beneficial rhizospheric microorganisms in binary or multiple combinations (Cordier et al., 2000). The ability of the microorganisms to develop together without negative interactions opens the possibility of creating a given diversified microbial-rich environment around roots, beneficial for microplant development and therefore increasing the levels of tolerance of microplants for a large spectrum of biotic and abiotic stresses, linked to their transplantation into the field (Vestberg et al., 2002).

7 FUTURE LINES OF RESEARCH

Strengthening of identification and classification through molecular methods, multiplication and inoculum production of indigenous efficient AM fungal isolates, establishing germination of AM fungi on synthetic media, multiplication and commercialization of efficient fungi, identification of marker genes and genetic mechanisms for effective nutrient transport for increasing efficiency and adaptation to increase vigour and yield, establishing center for conservation of AM fungi, biotechnological application of AM fungi, technical and commercial point of view of mycorrhizal inoculation of micropropagated plants, development of protocol for INM and IPM for sustainable agriculture are some of the major lines of future research with respect to arbuscular mycorrhizal fungi.

As plants and microbial strains differ in the effectiveness of the symbiosis, the selection of adequate symbiotic consortia is bound to be of particular value, ensuring better chances for survival of plants. More research is needed to select those AM fungi strains which are most efficient and can survive the competition with spontaneously appearing or indigenous fungi. There is also a need to protect the diversity of AM fungi and vegetation, to ensure a robust plant community in abandoned agricultural lands.

REFERENCES

Abbott, A.K., and Robson, A.D., 1982, Infectivity of vesicular arbuscular mycorrhizal fungi in agricultural soils. *J. Agric. Res.* **33**: 1049–1059.

Allen, E.B., and Allen, M.F., 1986, Water relations of xeric grasses in the field: Interactions of mycorrhizae and competition. *New Phytol.* **104**: 559–571.

Allen, M.F., 1991, *The Ecology of Mycorrhizae*. Cambridge University Press, NewYork, pp. 184.

Allen, M.K., Smith, W.K., Moore, Jr., T.S., and Christensen, M., 1981, Comparative water relations and photosynthesis of mycorrhizal and non-mycorrhizal *Boutelova gracilis* H. B. K. Lag ex Steud. *New Phytol.* **88**: 683–693.

Atkinson, D., Baddeley, J.A., Goicoechea, N., Green, J., Sánchez-Diaz, M., and Watson, C.A., 2002, Arbuscular mycorrhizal in low input agriculture. In: Gianinazzi, S., Schiiepp, H., Barea, J. M., and Haselwandtar, K. (Eds.), *Mycorrhizal Technology in Agriculture: From Genes to Bioproducts*. Birkhäuser Verlag, Basel, pp. 211–222.

Berreck, M., and Haselwandter, K., 2001, Effect of arbuscular mycorrhizal symbiosis upon uptake of cesium and other cations by plants. *Mycorrhiza* **10**: 275–280.

Binet, P., Portal, J.M., and Leyval, C., 2000, Fate of polycyclic aromatic hydrocarbons (PAH) in the rhizosphere and mycorrhizosphere of ryegrass. *Plant Soil* **227**: 207–213.

Brundrett, M.C., 1991, Mycorrhizas in natural ecosystem. *Adv. Ecol. Res.* **21**: 171–313.

Brundrett, M.C., 2004, Diversity and classification of mycorrhizal fungi. *Biol. Rev.* **79**: 473–495.

Burns, R.G., and Davies, J.A., 1986, The microbiology of soil structure. *Biol. Agric. Hort.* **3**: 95–113.

Cambardella, C.A., and Elliott, E.T., 1994, Carbon and nitrogen dynamics of soil organic matter fractions from cultivated grassland soils. *Soil Sci. Soc. Am. J.* **58**: 123–130.

Carling, D.E., Riehle, W.G., Brown, M.F., and Tinker, P.B., 1978, Effects of vesicular-arbuscular mycorrhizal fungus on nitrate reductase and nitrogenase activities in nodulating and non-nodulating soybeans. *Phytopatholoslgy* **68**: 1590–1596.

Cavalier-Smith, T., 1998, A revised six kingdom system of life. Biol. Rev. 73: 203–260.

Chandra, K.K., and Jamuluddin,1999, Distribution of vesicular-arbuscular mycorrhizal fungi in coalmine overburden dumps. *Indian Phytopath.* **52**: 254–258.

Cordier, C., Lemoine, M.C., Lemanceau, P., Gianinazzi-Pearson, V., and Gianinazzi, S., 2000, The beneficial rhizosphere: A necessary strategy for microbial production. *Acta Hortic.* **530**: 259–268.

Danielson, R.M., Zak, J., and Parkinson, D., 1979, Plant growth and mycorrhizal development in amended coal spoil material. In: Wali, M. K. (Ed.), *Ecology and Coal Resource Development*. Pergamon, New York, pp. 912–919.

Dehne, H.W., and Backhaus, G. F., 1986, The use of vesicular-arbuscular mycorrhizal fungi in plant protection. I. Inoculum production. *Z. Pflkrankh. Pflschutz* **93**: 415–424.

Diaz, G., Azcon-Aguilar, C., and Honrubia, M., 1996, Influence of arbuscular mycorrhizae on heavy metal (Zn and Pb) uptake and growth of *Lygeum spartum* and *Anthyllis cytisoides*. *Plant Soil* **180**: 241–249.

Dighton, J., 1983, Phosphatase production by mycorrhizal fungi. *Plant Soil* **71**: 455–462.

Doran, J.W., and Linn, D.M., 1994, Microbial ecology of conservation management systems. In: Hatfield, J. L. and Stewart, B. A. (Eds.), *Soil Biology: Effects on Soil Quality*. Lewis, Boca Raton, FL, pp. 1–57.

Eason, W.R., Newman, E.I., and Chuba, P.N., 1991, Specificity of interplant cycling of phosphorus: The role of mycorrhizas. *Plant Soil* **137**: 267–274.

Ebbs, S., Kochian, L., Lasat, M., Pence, N., and Jiang, T., 2000, An integrated investigation of the phytoremediation of heavy metal and radionuclide contaminated soils: From the laboratory to the field. In: Wise, D. L., Trantolo, D. J., Cichon, E. J., Inyang, H. I., and

Stottmeister, U. (Eds.), *Bioremediation of Contaminated Soils*. Dekker, New York, pp. 745–769.

Fitter, A.H., 1991, Costs and benefits of mycorrhizas: Implications for functioning under natural conditions. *Experimen.* **47**: 35–355.

Fogel, R., and Hunt, G., 1983, Contribution of mycorrhizae and soil fungi to nutrient cycling in a Douglas-fir ecosystem. *Can. J. Forest. Res.* **13**: 219–232.

Fortunato, I.M., and Avato, P., 2008, Plant development and synthesis of essential oils in micropropagated and mycorrhiza inoculated plants of *Origanum vulgare* L. ssp. *hirtum* (Link) Ietswaart. *Plant Cell Tiss. Organ Cult.* **93**:139–149.

Frank, B., 1885, Ueber die auf Wurzelsybiose beruhende Ernährung gewisser Bäume durch unterirdische Pilze. *Bericht der deutschen Gesellschagt* **3**: 128–148.

Franken, P., and George, E., 2004, Diversity of arbuscular mycorrhizal fungi. In: Benckiser, G. and Schnell, S. (Eds.), *Biodiversity in Agricultural Production Systems*. Taylor & Francis Group, Boca Raton, FL/London, pp. 189–203.

Gerdemann, J.W., 1968, Vesicular-arbuscular mycorrhizae and plant growth. *Ann. Rev. Phytopathol.* **6**: 397–418.

Gianinazzi, S., Trouvelot, A., and Gianinazzi-Pearson, V., 1989, Conceptual approaches in agriculture for the rational use of VA-endomycorrhizae in agriculture: Possibilities and limitations. *Agric. Ecosys. Environ.* **29**: 153–161.

Gianinazzi, S., Trouvelot, A., and Gianinazzi-Pearson, V., 1990, Role and use of mycorrhizas in horticultural crop production. XXIII International Horticulture Congress Plenary lecture, Italy, pp. 25–30.

Giri,B., Kapoor, R., Agarwal, L., and Mukerji, K.G., 2004, preinoculation with arbuscular mycorrhizae helps *Acacia auriculiformis* grow in degraded Indian waste land soil. *Comm. Soil Sc. and Plant Anal.* **35**: 193–204.

Gupta, R.K., 1991, Drought response in fungi and mycorrhizal plants. In: Arora, D. K., Rai, B., Mukerji, K. G., and Kundson, G. R. (Eds.) *Handbook of Applied Microbiology. Vol. I, Soil and Plants*. Dekker, New York, pp. 55–75.

Hall, I.R.,1980, Growth of *Lotus pedunculatus* cav. in an eroded soil containing soil pellets infested with endomycorrhizal fungi. *New Zeal. J. Agric. Res.* **23**: 103–105.

Hall, I.R., and Armstrong, P., 1979, Effect of vesicular-arbuscular mycorrhizas on growth of white clover, lotus and ryegrass in some eroded soils. *New Zea. J. Agric. Res.* **22**: 558–608.

Harley, J.L., and Harley, E.L., 1987, A check list of mycorrhiza in British flora. *New Phytol.* **105**: 1–102.

Hayman, D.S., 1974, Plant growth responses to vesicular arbuscular mycorrhiza. VI. Effect of light and temperature. *New Phytol.* **73**: 71–80.

Hiremath, S.T., and Podila, G.K., 2000, Development of genetically engineered mycorrhizal fungi for biological control. In: Podila, G. K. and Douds, D. D. (Eds.), *Current Advances in Mycorrhizae Research*. APS, St. Paul, MN pp. 179–187.

Hirrel, M.C., and Gerdemann, J.W., 1980, Improved growth of onion and bellpepper in saline soil by two VAM fungi. *Soil Sc. Soc. Am. J.* **44**: 654–655.

Jalali, B.L., 2001, Mycorrhiza and plant health- need for paradigm shift. *Indian Phytopath.* **54**: 3–11.

Jasper, D.A., Robson, A.D., and Abbott, L.K., 1987, The effect of surface mining on the infectivity of vesicular-arbuscular mycorrhizal fungi. *Austr. J. Bot.* **35**: 641–652.

Joner, E.J., and Leyval, C., 2001, Arbuscular mycorrhizal influence on clover and ryegrass grown together in a soil spiked with polycyclic aromatic hydrocarbons. *Mycorrhiza* **10**: 155–159.

Joner, E.J., Briones, R., and Leyval, C., 2000, Metal binding capacity of arbuscular mycorrhizal fungi. *Plant Soil* **226**: 227–234.

Kaldorf, M., Kuhn, A.J., Schröder, W.H., Hildebrandt, U., and Bothe, H., 1999, Selective element deposits in maize colonized by a heavy metal tolerance conferring arbuscular mycorrhizal fungus. *J. Plant Physiol.* **154**: 718–728.
Kehri, H.K., and Chandra, S., 1990, Mycorrhizal association in crops under sewage farming. *J. Indian Bot. Soc.* **69**: 267–270.
Kim, S.J., Hiremath, S.T., and Podila, G.K., 1999, Cloning and identification of symbiosis-regulated genes from the ectomycorrhizal Laccaria bicolor. *Mycol. Res.* **103**: 168–172.
Lovato, P.E., Gianinazzi-Pearson, V., Trouvelot, A., and Gianinazzi, A., 1996, The state of art of mycorrhizas and micropropagation. *Adv. Horti. Sc.* **10**: 46–52.
Lovelock, C.E., Wright, S.F., Clark, D.A., and Ruess, R.W., 2004, Soil stocks of glomalin produced by arbuscular mycorrhizal fungi across a tropical rain forest landscape. *J. Ecol.* **92**: 278–287.
Manoharachary, C., 2004, Biodiversity, taxonomy, ecology, conservation and biotechnology of arbuscular mycorrhizal fungi. *Indian Phytopath.* **57**: 1–6.
Menge, G.A., 1983, Utilization of vesicular-arbuscular mycorrhizal fungi in agriculture. *Can. J. Bot.* **61**: 1015–1024.
Miller, R.M., and Jastrow, J.D., 1994, Vesicular-arbuscular mycorrhizae and biogeo-chemical cycling. In: Pfleger, F. L. and Lindermann, R. G. (Eds.), *Mycorrhizae and Plant Health.* APS, St. Paul, MN pp. 189–212.
Miller, R.M., and Jastrow, J.D., 1992, The role of mycorrhizal fungi in soil conservation. In: Bethlenfalvay, C. J. and Linderman, R. G. (Eds.), *Mycorrhizae in Sustainable Agriculture.* Crop Science Society and Soil Science Society of America, Madison, WI, pp. 29–44.
Molina, R., Massicotte, H., and Trappe, J.M., 1992, Specificity phenomena in mycorrhizal symbiosis: Community-ecological consequences and practical implecations. In: Allen, M. F. (Ed.), *Mycorrhizal Functioning: An Integrative Plant-Fungal Process.* Chapman & Hall, New York, pp. 357–423.
Morton, J.B., and Redecker, D., 2001, Two new families of Glomales, Archaesporaceae and Paraglomaceae, with two new genera Archaespora and Paraglomus, based on concordant molecular and morphological characters. *Mycologia* **93**: 181–195.
Mukerji, S., Mukerji, K.G., and Arora, D.K., 1991, Ectomycorrhizae. In: Arora, D. K., Rai, B., Mukerji, K. G., and Kundson, G. R. (Eds.), *Handbook of Applied Microbiology. Vol. I Soil and Plants.* Dekker, New York, pp. 187–215.
Munro, R.C., Wilson, J., Jefwa, K.W., and Mbuthia, K.W., 1999, A low-cost method of mycorrhizal inoculation improves growth of *Acacia tortilis* seedlings in the nursery. *For. Ecol. Managm.* **113**: 51–56.
Odum, E.P., and Biever, L.J., 1994, Resource quality, mutualism and energy partitioning in food chaons. *Am. Natur.* **124**: 360–376.
Panwar, J., Tarafdar, J.C., Yadav, R.S., Saini, V.K., Aseri, G.K., and Vyas, A., 2007, Technique for visual demonstration of germinating AM spore and their multiplication in pots. *J. Plant Nutr. Soil Sci.* **170**: 659–663.
Pfeiffer, C.M., and Bloss, H.E., 1988, Growth and nutrition of guayule (*Parthenium argentatum*) in a saline soil as influenced by vesicular-arbuscular mycorrhiza and phosphorus fertilization. *New Phytol.* **108**: 315–321.
Pond, E.C., Menge, J.A., and Jarrell, W.M., 1984, Improved growth of tomato in salinized soil by vesicular-arbuscular mycorrhizal fungi collected from saline soils. *Mycologia* **76**: 74–84.
Poss, J.A., Pond, E., Menge, J.A., and Jarrel, W.M., 1985, Effect of salinity on mycorrhizal onion and tomato in soil with and without additional phosphate. *Plant Soil* **88**: 307–319.
Rao, A.V., and Tak, R., 2001, Influence of mycorrhizal fungi on the growth of different tree species and their nutrient uptake in gypsum mine spil in India. *App. Soil Ecol.* **17**: 279–284.

Read, D.J., 1984, The structure and function of the vegetative mycelium of mycorrhizal roots. In: Jennings, D. H. and Rayner, A. D. M. (Eds.), *The Ecology and Physiology of the Fungal Mycelium*. Cambridge University Press, New York, pp. 215–240.

Read, D.J., 1991, Mycorrhizas in ecosystem. *Experientia* **47**: 376–391.

Read, D.J., Francis, R., and Finlay, R.D., 1989, Mycorrhizal mycelia and nutrient cycling in plant communities. In: Fitter, A. E. (Ed.), *Ecological Interactions in Soil: Plants, Microbes, and Animals*. Blackwell Scientific, Boston, MA, pp. 193–217.

Redecker, D., Kodner, R., and Graham, L.E., 2000, Glomalean fungi from the Ordovician. *Science* **289**: 1920–1921.

Redhead, J.F., 1977, Endotrophic mycorrhizas in Nigeria: Species of the Endogonaceae and their distribution. *Trans. Brit. Mycol. Soc.* **69**: 275–280.

Rillig, M.C., 2004, Arbuscular mycorrhizae, glomalin, and soil aggregation. *Can. J. Soil Sci.* **84**: 355–363.

Safir, G.R., Boyer, J.S., and Gerdemann, J.W., 1971, Mycorrhizal enhancement of water transport in soybean. *Science* **172**: 581–583.

Safir, G.R., Boyer, J.S., and Gerdemann, J.W., 1972, Nutrient status and mycorrhizal enhancement of water transport in soybean. *Plant Physiol.* **49**: 700–703.

Sharma, M.P., and Adholeya, A., 2004, Effect of arbuscular mycorrhizal fungi and phosphorus fertilization on the post vitro growth and yield of micropropagated strawberry grown in a sandy loam soil. *Can. J. Bot.* **2**: 322–328.

Simon, L., Bousquet, J., Levesque, R.C., and Lalonde, M., 1993, Origin and diversification of endomycorrhizal fungi and coincidence with vascular land plants. *Nature* **363**: 67–69.

Tarafdar, J.C., and Marschner, H., 1994, Phosphatase activity in the rhizosphere and hyphosphere of VA mycorrhizal wheat supplied with inorganic and organic phosphorus. *Soil Biol. Biochem.* **26**: 387–395.

Tate, R.L., 1987, *Soil Organic Matter: Biological and Ecological Effects*. John, New York, p. 291.

Tate, R.L., 1985, Microorganisms, ecosystem disturbance and soil formation processes. In: Tate, R. L. and Klein, D. A. (Eds.), *Soil Reclamation Processes: Microbiological Analyses and Applications*. Dekker, New York, pp. 1–33.

Tisdall, J.M., 1991, Fungal hyphae and structural stability of soil. *Aus. J. Soil Res.* **29**: 729–743.

Vestberg, M., Cassells, A.C., Schubert, A., Cordier, C., and Gianinazzi, S., 2002, Arbuscular mycorrhizal fungi and micropropagation of high value crops. In: Gianinazzi, S., Schüepp, H., Barea, J. M., and Haselwandter, K. (Eds.), *Mycorrhizal Technology in Agriculture from Gene to Bioproducts*. Birkhäuser Verlag, Basel, Switzerland, pp. 223–233.

Wright, S.F., and Anderson, R.L., 2000, Aggregate stability and glomalin in alternative croprotations for central Great Plains. *Biol. Fert. Soils* **31**: 249–253.

Wright, S.F., and Upadhyaya, A., 1998, A survey of soils for aggregate stability and glomalin, a glycoprotein produced by hyphae of arbuscular mycorrhizal fungi. *Plant Soil* **198**: 97–107.

Wright, S.F., Rillig, M.C., and Nichols, K.A., 2000, Glomalin: A soil protein important in carbon sequestration. In: *Proceedings of American Chemical Society Annual Meeting Symposium*, pp. 721–725.

Chapter 7

INDIRECT CONTRIBUTIONS OF AM FUNGI AND SOIL AGGREGATION TO PLANT GROWTH AND PROTECTION

KRISTINE A. NICHOLS
USDA-ARS-Northern Great Plains Research Laboratory, Mandan, ND 58554, USA

Abstract: Ecological and biological engineering contribute indirectly to the fitness of the soil environment and promote plant growth and protection. This engineering modifies soil physical, chemical, and biological attributes to enhance nutrient cycling, increase soil organic matter, and improve soil quality. Arbuscular mycorrhizal (AM) fungi, under most conditions, improve plant growth directly by providing greater and more efficient access via fungal hyphae for absorption of nutrients, especially P, and delivery of these nutrients to the plant. The AM symbiosis also augments disease resistance in host plants and suppresses the growth of non-mycorrhizal weeds. When plants moved from an aquatic to a terrestrial environment, mycorrhizal fungi were an integral part of their success by providing efficient nutrient absorption from the low organic matter mineral soil. In addition, AM fungi stabilize soil aggregates and promote the growth of other soil organisms by exuding photosynthetically-derived carbon into the mycorrhizosphere. Glomalin is a glycoprotein produced by AM fungi which probably originated as a protective coating on fungal hyphae to keep water and nutrients from being lost prior to reaching the plant host and to protect hyphae from decomposition and microbial attack. This substance also helps in stabilizing soil aggregates by forming a protective polymer-like lattice on the aggregate surface. AM fungal growth and biomolecules engineer well-structured soil where the distribution of water-stable aggregates and pore spaces provides resistance to wind and water erosion, greater air and water infiltration rates favorable for plant and microbial growth, nutrients in protect micro-sites near the plant roots, and protection to aggregate-occluded organic matter.

Keywords: Glomalin; soil aggregation; water-stable aggregates.

1 INTRODUCTION

The soil environment is a critical component to plant health because soils regulate root growth, water infiltration, aeration, the filtering and buffering of pollutants, and nutrient cycling and storage. To perform these functions, the physical, chemical, and biological components of soils need to interact on an intimate level. Soil organisms evolved mechanisms that result in the modification of their physiochemical environment to enhance plant growth (Rillig and Steinberg, 2002; Janos, 2007; Jordan et al., 2008). These mechanisms, such as the formation of water-stable soil aggregates, created habitats and a ready food supply for continued biological growth.

Arbuscular mycorrhizal (AM) fungi are arguably one of the most dominate and important organisms in the soil, comprising 5–50% of the total microbial biomass in soils (Olsson et al., 1999), and obligately-associated with the majority of vascular plants (Brundrett, 2002; Millner and Wright, 2002). These fungi receive carbon (about 12–27%) from the plant host in the form of simple hexose sugars, which are used for fungal growth and exuded into the mycorrhizosphere (Tinker et al., 1994). Root and mycorrhizal exudates attract soil organisms, which use these exudates to transform organic matter and soil minerals into plant-available nutrients.

The AM symbiosis played an integral role in helping plants to live when they moved from an aquatic to a terrestrial environment. Thread-like fungal hyphae fan out into the soil to scavenge even highly immobile nutrients, such as P. In addition to being able to grow out further into the soil than plant roots, the fine threads of hyphae have a much larger surface area to volume ratio than roots, and the fungal cell membrane is capable of concentrating solutes against a gradient (George et al., 1992). Rapid growth of fine, ephemeral hyphae occurs in microsites containing high concentrations of nutrients such as P, N, Fe, Cu and Zn (Clark and Zeto, 1996; Pawlowska et al., 2000) to deliver these nutrients to the plant for the lowest carbon 'cost'. In the environment of early Earth, mycorrhizal fungi assisted in the formation of soil aggregates. Soil aggregates are pellets of different shapes and sizes which contain a conglomeration of soil minerals (sand, silt, and clay); organic matter, such as plant debris; inorganic compounds like iron and aluminum oxides; roots; fungal hyphae; and other microbes (Chenu et al., 2000; Six et al., 2001). Roots and AM fungal hyphae act like a 'net' collecting soil minerals, organic matter, etc. Root and microbial exudates such as polysaccharides and glomalin (a glycoprotein produced on AM hyphae) provide the 'glue' to stick soil debris to the 'net' (Miller and Jastrow, 1990; Rillig and Mummey, 2006). In addition to helping to 'glue' aggregates together, glomalin appears to form a hydrophobic lattice around aggregates to keep them water-stable (Nichols and Wright, 2004). When present within aggregates, minerals and organic matter are less susceptible to wind and water erosion. Intra-aggregate organic matter is slowly decomposed by microbes and converted into plant-available nutrients. Soil aggregates create and maintain soil pores providing air and water infiltration rates favorable for plant growth.

Agricultural management influences aggregate formation and stability by its impacts on the physical destruction of soil aggregates and hyphal networks, and the allocation of photosynthetically-derived carbon belowground. Minimum or no tillage production systems reduce the physical impacts on aggregates while the use of continuous cover in the form living plants increases the amounts of photoassimulated carbon. Continuous cover may include removing fallow periods from annual cropping systems or using cover or perennial crops.

2 EVOLUTION OF AM SYMBIOSIS

About 500–600 Myear ago, algae and fungi were likely the first terrestrial associations between fungi and photosynthetic organisms (Redecker et al., 2000; Brundrett, 2002). These early land colonizers, present before the evolution of higher plants, developed mechanisms for foraging mineral nutrients in the harsh environment of early Earth and in the process would have 'engineered' soil by biomineralization (Brundrett, 2002; Rillig and Steinberg, 2002; Schüßler, 2002; Janos, 2007; Jordan et al., 2008). To efficiently forage nutrients, fungal hyphae formed fine threads to disperse widely throughout the soil, respond quickly to temporary, localized nutrient sources, compete successfully with other organisms, and produced enzymes to obtain mineral and organic nutrients (Brundrett, 2002). In this process, soil fungi would have increased their absorption of mineral nutrients beyond those required for immediate use as insurance against future shortages (Brundrett, 2002). Simultaneously, algal growth increased the atmospheric oxygen levels to the point where the ozone layer, which provides UV radiation protection, would have been formed (Pirozynski and Malloch, 1975). Under high CO_2 conditions and with little competition for sunlight, these protected algae quickly evolved more complex structures similar to modern-day bryophytes with few leaves or branches and underdeveloped roots (Pirozynski and Malloch, 1975; Schwartzman and Volk, 1989; Brundrett, 2002). These early plants exploited their photosynthetic capability and produced large amounts of carbon compounds to feed soil fungi which had abundant mineral nutrient resources. The mycorrhizal symbiosis subsequently evolved to where the symbiotic partners avoided conflicts over limited resources by utilizing materials each partner had in excess making the costs of production/acquisition balance against the benefits of mycorrhizal associations (Redecker et al., 2000; Brundrett, 2002).

In 1993, the BEG (European Bank of Glomeromycota) stated: "The study of plants without their mycorrhizas is the study of artifacts; the majority of plants, strictly speaking, do not have roots—they have mycorrhizas" (Schüßler et al., 2007). The fossil records show that the fungal structures found in the roots of Devonian plants, at least 400 million years ago, are almost exactly the same as the structures found in the roots of modern plants (Brundrett, 2002; Redecker *et al.*, 2000; Schüßler, 2002). The first exchange process probably began in the diffuse interface within the plant where certain cells of the endophytic fungus evolved to become more permeable and increase leakage of their contents (Brundrett, 2002; Redecker *et al.*, 2000; Schüßler, 2002). Selection pressures would have induced changes in membrane functions and wall structures by the host and fungus resulting in the formation of a specialized interface structure (i.e., the arbuscule) to provide a suitable habitat for AM fungi where carbon and nutrient exchange would be maximized to support more complex aboveground structures (Gensel and Andrews, 1987; Schwartzman and Volk, 1989; Brundrett, 2002; Schüßler, 2002).

By exploiting this mutually beneficial relationship, explosive innovations in plant development occurred during the Devonian Period, where the number of genera increased from 1 to 28 (Gensel and Andrews, 1987; Schwartzman and Volk, 1989) and plants grew from tiny creeping structures to a diverse array of 0.5–9 m high structures with leaves, roots, reproductive systems, and secondary growth (Gensel and Andrews, 1987). Roots, which began as subterranean stems, progressed from coarse dichotomous branches to roots with highly organized cell layers and branching to support more and larger above-ground structures, propagate the plant, serve as storage organs, and form conduits to distribute water and nutrients. Plants also developed mechanisms to control the extent of mycorrhizal formation by confining them in certain cell layers and controlling the timing of their formation and turnover. These control mechanisms resulted in slow evolutionary changes among mycorrhizal fungi (Redecker *et al.*, 2000; Brundrett, 2002). However, to continue to receive C from the plant, AM fungi had to constantly adapt to their growing environment. Selection pressures based on the soil environment probably stimulated more evolutionary changes in AM fungi, especially in the growth and function of extraradical hyphae. As an example, the 'engineering' of an environment favorable for plant and mycorrhizal growth may vary with different plant hosts, in different soil types and in response to different environmental conditions, but the process continued to evolve in the direction of maximizing the efficiency and productivity of the system. As a consequence, an active rhizosphere and mycorrhizosphere, containing a suite of biomolecules (such as polysaccharides, humic substances, and glomalin) possessing the requisite combination of hydrophilic, acidic, complexing, and sorptive properties, was formed (Johnson *et al.*, 2005; Rillig and Mummey, 2006; Schüßler *et al.*, 2007).

3 THE MYCORRHIZOSPHERE

On an annual basis, a major part of ecosystem C flux is the transfer of C from plants to fungi and from the fungi to the soil (Johnson et al., 2005; Schüßler et al., 2007). The result of this carbon transfer was the formation of a zone of intense microbial growth in the area surrounding AM colonized roots and extraradical hyphae (i.e. the mycorrhizosphere) greater than in the bulk soil (Andrade et al., 1998; Johnson et al., 2005; Rillig and Mummey, 2006) (Fig. 1). In the mycorrhizosphere, AM fungi influence bacterial, fungal, and microarthropod communities by providing them substrates in the forms of decomposing fine, ephemeral hyphae and the deposition of hyphal biomolecules, and by influencing soil structure (i.e. microbial habitats within aggregates and pore spaces) (Andrade et al., 1998; Rillig and Mummey, 2006). Abiotic and biotic factors (such as soil pH, soil chemical composition, root and hyphal exudation, and soil microflora) change in the mycorrhizosphere according to the interactions of the plant, fungus, soil, and soil microbes (Filion et al., 2003; Johnson et al., 2005; Rillig and Mummey, 2006). Hyphal morphology (i.e. wall thickness, width, branching patterns, and turnover), function (i.e. nutrient absorption, plant protection, and soil aggregate formation and stabilization), and longevity vary greatly across and within native and agricultural systems due to changes in the mycorrhizosphere (Rillig and Mummey, 2006).

Various microorganisms in the mycorrhizosphere may act as positive, mycorrhization helper organisms (i.e. organisms that promote AM root colonization or production of plant-available nutrients) or negative organisms (i.e. organisms that compete for resources or are fungal grazers or parasites) (Rillig, 2004). Positive organisms include Mn-oxidizing and reducing bacteria (i.e. *Streptomyces, Arthrobacter, Variovorax* and *Ralstonia*; Nogueira et al., 2007*)*, phosphate-solubilizing bacteria (such as *Enterobacter* sp. and *Bacillus subtilis*; Toro et al., 1997), nitrogen-transforming microbes (autotrophic ammonium oxidizers), and bacteria and fungi involved in soil aggregation. Also, the growth of the antagonistic bacterium *Paenibacillus* sp. strain B2 has resulted in the reduction of soilborne pathogens (including *Aphanomyces, Chalara, Fusarium, Phytophthora, Pythium, Rhizoctonia,* and *Verticillium)* (Budi et al., 1999; Filion et al., 2003; Selim et al., 2005). Negative organisms include soil mesofauna (such as *Collembola*; Johnson et al., 2005). The alterations of microbial communities have reduced plant disease susceptibility (Budi et al., 1999; Filion et al., 2003; Selim et al., 2005); increased the biodegradation of polycyclic aromatic hydrocarbons (Binet et al., 2000; Corgie et al., 2006); impacted the plant-availability and absorption of Mn and Fe (Nogueira et al., 2007); altered hyphal growth and morphology due to grazing by soil mesofauna (Johnson et al., 2005); and increased plant-available nutrients (Toro et al., 1997).

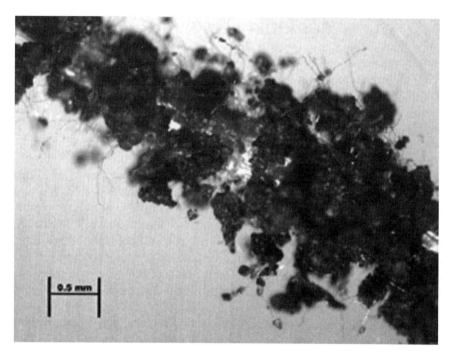

Fig. 1. Soil aggregates were 'bioengineered' in the mycorrhizosphere, containing AM fungal hyphae and colonized millet roots. The millet was planted as part of a cover crop mixture on July 7, 2007 following the harvest of a forage pea crop in the semi-arid northern Great Plains region of the United States. Roots were collected on August 30, 2007.

The interactions between the biological, physical, and chemical components of soil occur in the mycorrhizosphere (Toro *et al.*, 1997; Rillig and Mummey, 2006). In this zone, C is the energy source for a diverse array of microorganisms which work in complex symbiotic relationships (some of which were identified above) for their own growth (Rillig and Steinberg, 2002; Janos, 2007; Jordan *et al.*, 2008). A consequence of soil biological growth is the increased availability and acquisition of plant nutrients, better plant protection, and an 'engineered' environment, via the formation of soil aggregates which allows for better aeration, water infiltration, water retention, and plant root growth (Rillig *et al.*, 2007; Jordan *et al.*, 2008). Fungal hyphae and biomolecules, such as glomalin, are critically important to soil biological processes because of their interactions with the plant, soil, and soil microbes.

3.1 Extraradical fungal hyphae

In the soil, AM fungal hyphae are the thread-like filaments that comprise the body of mycorrhizal fungi and are the main functional organs in the mycorrhizal symbiosis. Without extraradical hyphae extending out (up to 8 cm) into the soil, the fungus would not be able to obtain the nutrients required to 'purchase' C from the plant (Rillig, 2004; Rillig and Mummey, 2006). Hyphal density ranges between 1 and 30 m g^{-1} of soil (Corgie *et al.*, 2006) and may contain 50–900 kg ha^{-1} of the soil carbon (Rillig, 2004). Olsson *et al.* (1999) found that AM fungal hyphae accounted for 5–50% of the microbial biomass, but more recently Cheng and Baumgartner (2006) estimated hyphal biomass at 20–30%. The role and importance of fungal hyphae in nutrient acquisition is well known, but the mechanisms of this function are not well understood. In studying differences in the architecture and wall thickness of extraradical hyphae, two types or morphologies are recognized: (i) the larger and thicker, more mellanized hyphae which is part of the 'permanent' fungal network and acts as 'conduit' or runner hyphae, and (ii) the thin-walled, very fine, ephemeral hyphae which is the 'absorptive' hyphae and will fan out into nutrient-rich microsites in the soil (Friese and Allen, 1991; Abbott *et al.*, 1992). The ephemeral hyphae are decomposed in days or weeks while the turnover of runner hyphae may be years (Rillig, 2004; Tresder *et al.*, 2007).

To support this resistance to decomposition and to keep nutrients from being lost to the soil environment prior to reaching the plant roots, a protective coating on hyphae is speculated. Glomalin (80% of which is bound tightly to the hyphal wall; Driver *et al.*, 2005) is speculated to provide this coating (Nichols and Wright, 2004; Purin and Rillig, 2008). The localization of glomalin in the hyphal wall indicates that this molecule may be important in mediating the interaction of the biotic and abiotic environment and in defending fungal hyphae against grazers by reducing hyphal palatability (Nichols and Wright, 2004; Driver *et al.*, 2005; Purin and Rillig, 2008).

Hyphal growth, and production of biomolecules, such as glomalin, is dependent upon how C received from the plant is allocated (Whitbeck, 2001; Tinker *et al.*, 1994). However, other growing conditions, such as soil texture, temperature, pH, or water-content, and signals from the plant, may cause differential fungal growth patterns (Bethlenfalvay *et al.*, 1999; Whitbeck, 2001; Steinberg and Rillig, 2003). Soil aggregates formed and stabilized by contributions from AM fungi help to buffer some of these adverse growing conditions.

3.2   Glomalin

Glomalin (a reddish-grown AM fungal glycoprotein) was first identified using immunological techniques with a monoclonal antibody (MAb32B11) (Wright *et al.*, 1996). Using indirect immunofluorescence with this antibody, glomalin was revealed on AM fungal hyphae, colonized roots, organic matter, soil particles, horticultural or nylon mesh, glass beads, and arbuscules within root cells (Wright *et al.*, 1996; Wright and Upadhyaya, 1999;

Wright, 2000). The gene (GiHsp 60) for glomalin production has been identified and shows genetic similarities to heat shock proteins indicating glomalin production may responded to stress (Gadkar and Rillig, 2006). Immunoelectron microscopy (Purin and Rillig, 2008), *in vitro* cultures (Driver *et al.*, 2005), and amino acid sequencing (Gadkar and Rillig, 2006) have indicated the majority (ca. 80%) of glomalin produced was localized in hyphal walls. Although glomalin is localized in the hyphal wall, studies indicate hyphal length and glomalin values are not related (Rillig and Steinberg, 2002; Bedini *et al.*, 2007). Therefore, it is theorized that glomalin is not present in the fine, ephemeral 'absorptive' hyphae (as was seen in immunofluorescence assays) making any relationship between glomalin and hyphal length difficult (Bedini *et al.*, 2007; Nichols and Wright, 2004).

Glomalin resists enzymatic and chemical decomposition in the soil and laboratory (Rillig and Steinberg, 2002; Steinberg and Rillig, 2003; Nichols and Wright, 2004) and has a turnover rate of 6–42 years (Rillig *et al.*, 2001; Harner *et al.*, 2007) and possibly up to 92 years (Preger *et al.*, 2007). The difficulties in solubilizing glomalin (requires repeated 1-hour intervals of autoclaving at 121°C in a neutral to slightly alkaline sodium salt solution; Wright *et al.*, 1996; Nichols and Wright, 2004) and the interactions of glomalin with metals, organic matter, and clay minerals (Nichols and Wright, 2004, 2005; Rosier *et al.*, 2007) are theorized to contribute to its stability. Atomic absorption analysis for iron indicated glomalin contained 0.2–8.8% with lower percentages in glomalin extracted from hyphae collected from a root-free zone in axenic sand:coal pot cultures (Wright and Upadhyaya, 1998; Nichols and Wright, 2005).

Glomalin extracted from soil is very similar to glomalin extracted from single-species pot cultures according to SDS-PAGE (Wright *et al.*, 1996; Wright and Upadhyaya, 1996; Rillig *et al.*, 2001), NMR (Rillig *et al.*, 2001), capillary electrophoresis (CE) (Wright *et al.*, 1998), crossed immuno/lectin affinity electrophoresis followed by LC/MS (Bolliger *et al.*, 2008), and C, H, N analysis by combustion (Rillig *et al.*, 2001). The C, N and H concentrations in glomalin extracted from soil averaged 36%, 3%, and 4%, respectively, while glomalin extracted from hyphae averaged 40%, 6%, and 7%, respectively (Rillig *et al.*, 2001; Lovelock *et al.*, 2004; Nichols and Wright, 2005; Tresder *et al.*, 2007). These values were used to calculate glomalin's contribution to soil organic C, which ranged from 1% to 31% with higher values from undisturbed, temperate soil where the mass value used was based on gravimetric weight of freeze-dried material (Nichols and Wright, 2005) and lower values from disturbed and undisturbed tropical and temperate soils where mass values were based on BRSP weight (Rillig *et al.*, 2001; Lovelock *et al.*, 2004; Tresder *et al.*, 2007). In addition, ^{1}H NMR spectra showed glomalin to be the most abundant and best characterized compound in crude extracts from soil and hyphal samples (Rillig *et al.*,

2001; Nichols and Wright, 2004). Peak location and size showed glomalin was most abundant in aliphatic groups with some carbohydrate and aromatic signatures. The low carbohydrate signatures also were found for glomalin samples by Schindler *et al.* (2007), but they found aromatic groups to be higher than aliphatic groups and much similarity between the glomalin samples and the IHSS Pahokee Peat standard of humic acid. No tannic acids peaks were found in ^{1}H NMR spectra of Rillig *et al.* (2001). Therefore, despite co-extraction of tannic material by the glomalin extraction procedure when samples are spiked with tannic acid (Halvorson and Gonzalez, 2008), there is no indication that glomalin extracted from soil and partially purified by acid precipitation contains a large portion of tannic acids (Rillig *et al.*, 2001).

Glomalin is, typically, quantified using the Bradford total protein assay and enzyme-linked immunosorbent assay (ELISA) using MAb32B11 (Wright *et al.*, 1996; Rillig, 2004; Nichols and Wright, 2004, 2005, 2006; Rillig *et al.*, 2007). Abundant amounts of glomalin (typically, 2–15 mg Bradford-reactive protein g^{-1} soil) have been measured in a wide range of soil environments (acidic, calcareous, grassland and cropland) (Wright and Upadhyaya, 1998; Wright *et al.*, 1999; Rillig *et al.*, 2001). Because this extraction procedure may co-extract other soil proteins (Rosier *et al.*, 2007), polyphenolics, such as tannic acids (Halvorson and Gonzalez, 2008), and/or other organic matter components, such as humic substances (Nichols and Wright, 2005, 2006), Rillig (2004) introduced new nomenclature for the soil extracts to be identified as glomalin-related soil protein (GRSP) and the Bradford total protein concentration as Bradford-reactive soil protein (BRSP). This change in nomenclature exhibits several problems in the area of glomalin research. However, despite the lack of information about the structure and function of glomalin and a definitive method for quantification, this methodology is still being used to measure glomalin in a wide variety of ecosystems (Nichols and Wright, 2004; Rillig, 2004; Rillig *et al.*, 2007). As these studies illustrate, research in this area continues because of the extremely important roles that have been theorized and demonstrated for glomalin (Nichols and Wright, 2004, 2005, 2006; Rillig, 2004; Rillig *et al.*, 2007).

Glomalin is theorized to (a) protect fungal hyphae from microbial attack, (b) provide a coating on runner or 'conduit' hyphae to prevent loss of water and nutrients, and (c) promote hyphal growth in the soil environment where turgor pressures vary with wet/dry cycles (Nichols and Wright, 2004; Rillig *et al.*, 2007). Saprophytic, pathogenic, and mutualistic, ectomycorrhizal fungi produce fungal proteins (called hydrophobins) which play similar functional roles to those identified for glomalin (Wessels, 1997; Whiteford and Spanu, 2002; Rillig *et al.*, 2007). These proteins have been well characterized and may serve as model proteins to develop methodology for further structural and functional analysis of glomalin. Although glomalin and hydrophobins do not appear to share structural similarities, these fungal proteins are similar in their location on fungal hyphae, high insolubility, speculated functionality, and multi-molecular formations (Wessels, 1997; Whiteford and Spanu, 2002; Nichols and Wright, 2004; Rillig *et al.*, 2007). In the growth media and on surfaces, hydrophobins, and possibly glomalin (as seen in immunofluorescence assays; Wright, 2000), self-assemble into amphipathic films comprised of rodlet-like formations at air-water or water-hydrophobic substance interfaces (Wessels, 1997, 1999). Scum formation has also been exhibited for hydrophobins, where N_2 gas is bubbled through the fungal culturing medium (Askolin *et al.*, 2001), and for glomalin (as stated above) by washing the potting medium with forced water, which introduced air into the water stream, creating air-water interface (Wright, 2000; Nichols and Wright, 2004). Similarly, Bradford-reactive soil protein has also been measured as part of tan-colored foam in rivers in the United States (Harner *et al.*, 2007).

Glomalin also shares similarities with other biomolecules, such as hydrophobic repellents and iron-accumulating transferrins. Repellents are similar to hydrophobins in that they contain a number of hydrophobic amino acids, but they do not contain the characteristic eight cysteine residues used to classify hydrophobins (like glomalin) and do not self-assemble into rodlets (unlike glomalin) (Kershaw and Talbot, 1998; Wright, 2000; Whiteford and Spanu, 2002; Rillig *et al.*, 2007). The similarities between transferrins and glomalin include resistance to proteolysis by trypsin or trypsin-like enzymes and high heat (up to ~70°C) (Iyer and Lonnerdal, 1993; Paulsson *et al.*, 1993) and hyper-accumulation of iron which produces multiple bands in SDS-PAGE depending on iron concentration and/or degree of degradation or deglycosylation (Bolliger *et al.*, 2008; Nagasako *et al.*, 1993; Rillig *et al.*, 2001).

Another functional role for glomalin is the formation and stabilization of soil aggregates. Glomalin may contribute to aggregate formation via the 'gluing' action of the oligosaccharides present in the glomalin molecule or the bridging of clay minerals and organic matter by iron or other polyvalent cations on the glomalin molecule (Wright and Upadhyaya, 1998; Nichols and Wright, 2004; Rillig, 2004). Also, the protective barrier glomalin provides to the fungal hyphae may protect soil aggregates from slaking (Nichols and Wright, 2004; Rillig *et al.*, 2007). Glomalin sloughed from growing hyphae or released during hyphal decomposition forms a lattice-like coating on the surface of aggregates (Wright, 2000; Nichols and Wright, 2004; Rillig *et al.*, 2007). This coating provides small openings for air, CO_2, and water exchange between the pores within the aggregate and the surrounding soil environment. When the glomalin barrier is not present, water rushes very quickly into the intra-aggregate pore spaces causing the air molecules to condense. Air-pressure in the pore space builds until the aggregate explodes.

4 SOIL AGGREGATION

Loss of topsoil, at a rate of 10 million hectares per year, is a serious problem in agroecosystems and has resulted in an estimated loss of nearly one-third of the world's arable land during the last 40 years (Pimentel *et al.*, 1995). Conservation agriculture has as its goals "to achieve high and sustainable productivity, quality and economic viability, while also respecting the environment ... [by] protecting soil and water" (Jones *et al.*, 2006). The formation and stabilization of soil aggregates appears to increase under conservation practices. Soil aggregates are important for: (1) maintaining soil porosity, which provides aeration and water infiltration rates favorable for plant and microbial growth, (2) increasing stability against wind and water erosion, and (3) storing carbon by protecting organic matter from microbial decomposition (Rillig *et al.*, 1999, 2007; Six *et al.*, 2001).

Soil aggregates are comprised of soil minerals (clay particles, fine sand and silt), plant roots, fungal hyphae, small plant or microbial debris, bacteria, free amorphous organic matter and organic matter strongly associated with clay coatings (Chenu *et al.*, 2000; Six *et al.*, 2001). The diversity of materials within aggregates makes both aggregate formation and stabilization complex processes (Andrade *et al.*, 1998; Miller and Jastrow, 1990). In the hierarchical formation of soil aggregates AM fungal hyphae may initiate aggregate formation by providing the framework, or 'net' upon which soil debris and microaggregates formed by the binding of primary particles collect (Miller and Jastrow, 1990; Rillig and Mummey, 2006). The efficiency of this process is enhanced by biochemical agents such as hydrophobins and glomalin, the alignment and grouping of particles due to wet/dry cycles, and the enrichment of bacterial communities in the mycorrhizosphere (Andrade *et al.*, 1998; Rillig and Mummey, 2006). Other fungi contribute to aggregate formation via the decomposition of stubble and mulch litter and the production of polysaccharides and mucigels (Chaney and Swift, 1986; Caesar-TonThat and Cochran, 2000).These polysaccharides will glue aggregates together quickly but do not to contribute to the long-term stability of aggregates, because they are water-soluble and easily decomposed (Chaney and Swift, 1986; Six *et al.*, 2001). Other fungal biomolecules, such as glomalin, hydrophobins, and repellents, may contribute to long-term stability by providing a hydrophobic barrier on the surface of aggregates (Wessels, 1997; Nichols and Wright, 2004). These aliphatic molecules work by increasing the contact angle for water penetration and by having very strong attractive forces between molecules which keeps them internally stable and induces the formation of rodlet-like layers or plaques (Degens, 1997; Chenu *et al.*, 2000; Nichols and Wright, 2004). When an aggregate has these layers on its surface, it is protected from disruption by rainfall due to mechanical dispersion by the kinetic

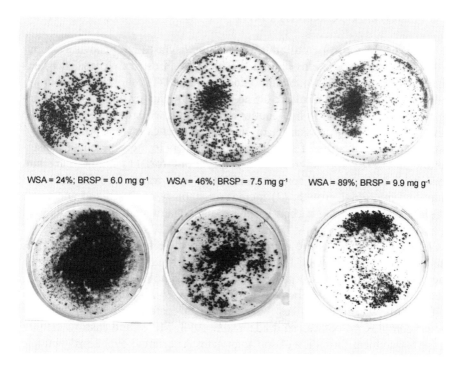

Fig. 2. Dry-sieved soil aggregates (1–2 mm) separated from the same soil type under the same climatic conditions in the semi-arid northern Great Plains region of the United States look about the same (A), but after the addition of water (such as during a rainfall event) (B), these aggregates do not act the same. Concentrations of water-stable aggregates (WSA) and glomalin (BRSP) change with management (from left to right): conventionally managed cropland, rangeland populated with tame grass species and heavily grazed, and rangeland populated with native grass species and rotationally grazed.

energy of raindrops, physiochemical dispersion, and the differential swelling of clays (Degens, 1997; Chenu et al., 2000). Figure 2 shows the interactions of glomalin and agroecosystem management on aggregate stability.

Other chemical (such as the interactions of clay minerals, organic matter, and polyvalent cations) and physical (such as drying and wetting actions, shrinking and swelling of clays, and freeze-thaw cycles) processes contribute to aggregate formation and stability (Chaney and Swift, 1986; Degens, 1997; Chenu et al., 2000). These chemical and physical process may be more important in microaggregates (<250 um) than in macroaggregates (≥250 um), but the exact mechanisms for the formation and stabilization of all aggregates are not well understood (Six et al., 2001; Rillig et al., 2007).

5 INDIRECT PLANT GROWTH AND PROTECTION

The interactions of the fungal hyphae in the mycorrhizosphere as well as the production of the fungal biomolecule, glomalin, have resulted in some indirect methods by which AM fungi may assist in the growth and protection of their host plants. In the mycorrhizosphere, a bacteria, *Paenibacillus* sp. strain B2, produces an antagonistic factor which will suppress a number of soilborne plant pathogens (see mycorrhizosphere section above). Other microorganisms also interact in the mycorrhizosphere to promote disease and weed suppression, as well as the production of plant-available nutrients to assist in the mycorrhizal symbiosis. Some mycorrhizosphere microbes along with fungal hyphae and glomalin may be used to 'engineer' soil aggregates to provide a stabilized habitat for further microbial growth (see soil aggregation section above). Soil aggregates improve soil structure, quality, and fertility. Finally, glomalin itself has been found to indirectly reduce the levels of potentially toxic metals, such as Cd, Pb, Mn, and Fe, in the plant host by molecularly binding these metals (Gonzalez-Chavez *et al.*, 2004; Chern *et al.*, 2007).

6 CONCLUSIONS

Modern agricultural practices have put new pressures on the plant-mycorrhizal symbiosis. Tillage practices physically disrupt soil aggregates and AM hyphal networks resulting in declining soil structure, fertility, and nutrient-cycling ability and forcing more C allocation by AM fungi to reestablishing these networks rather than to glomalin formation (Nichols and Wright, 2004). No tillage (NT) practices along with continuous cropping systems, using mycorrhizal host crops, and reducing synthetic inputs, especially P, enhance the plant-mycorrhizal symbiotic relationship (Preger *et al.*, 2007; Roldan *et al.*, 2007; Rillig *et al.*, 2007). These practices also increase the percentages of water-stable aggregates within the soil by increasing hyphal lengths, root and microbial exudates in the mycorrhizosphere, and allocating more C to glomalin production. In addition, higher levels of C sequestration are possible in these systems, since not only is C being allocated belowground to hyphal networks and the formation of the highly stable glomalin molecule, but organic matter occluded within aggregates appears to have a turnover time double free organic matter (Roldan *et al.*, 2007; Rillig, *et al.*, 2007; Six *et al.*, 2001).To maintain ecosystem function and a consistent food, feed, fiber, and energy supply to a growing global population, effective management of soil organisms, especially AM fungi, and agricultural systems needs to be developed and implemented.

REFERENCES

Abbott, L.K., Robson, A.D., Jasper, D.A., and Gazey, C., 1992, What is the role of VA mycorrhizal hyphae in soil? In: *Mycorrhizas in Ecosystems*. D.J. Read *et al.* (Eds.). CAB International, Wallingford, UK, pp. 37–41.

Andrade, G., Mihara, K.L., Linderman, R.G., and Bethlenfalvay, G.J., 1998, Soil aggregation status and rhizobacteria in the mycorrhizosphere. *Plant Soil*. **202**: 89–96.

Askolin, S., Nakari-Setala, T., and Tenkanen, M., 2001, Overproduction, purification, and characterization of the *Trichoderma reesei* hydrophobin HFBI. *Appl. Microbiol. Biotechnol.* **57**: 124–130.

Bedini, S., Avio, L., Argese, E., and Giovannetti, M., 2007, Effects of long-term land use on arbuscular mycorrhizal fungi and glomalin-related soil protein. *Agric. Ecosyst. Environ.* **120**: 463–466.

Bethlenfalvay, G.J., Cantrell, I.C., Mihara, K.L., and Schreiner, R.P., 1999, Relationships between soil aggregation and mycorrhizae as influenced by soil biota and nitrogen nutrition. *Biol. Fert. Soils* **28**: 356–363.

Binet, Ph., Portal, J.M., and Leyval, C., 2000, Fate of polycyclic aromatic hydrocarbons (PAH) in the rhizosphre and mycorrhizosphere of ryegrass. *Plant Soil*. **227**: 207–213.

Bolliger, A., Nalla, A., Magid, J., de Neergaard, A., Nalla, A.D., and Bog-Hansen, T.C., 2008, Re-examining the glomalin-purity of glomalin-related soil protein fractions through immunochemical, lectin-affinity and soil labeling experiments. *Soil Biol. Biochem.* **40**: 887–893.

Brundrett, M.C., 2002, Coevolution of roots and mycorrhizas of land plants. *New Phytol.* **154**: 275–304.

Budi, S.W., van Tuinen, D., Martinotti, G., and Gianinazzi, S., 1999, Isolation from *Sorghum bicolor* mycorrhizosphere of a bacterium compatible with arbuscular mycorrhiza development and antagonistic towards soilborne fungal pathogens. *Appl. Envion. Microbiol.* **65**: 5148–5150.

Caesar-TonThat, T.-C., and Cochran, V.L., 2000, Soil aggregate stabilization by a saprophytic lignin-decomposing basidiomycete fungus. I. Microbiological aspects. *Biol. Fert. Soils* **32**: 374–380.

Chaney, K., and Swift, R.S., 1986, Studies on aggregate stability. I. Reformation of soil aggregates. *J. Soil Sci.* **37**: 329–335.

Cheng, X., and Baumgartner, K., 2006, Effects of mycorrhizal roots and extraradical hyphae on 15N uptake from vineyard cover crop litter and the soil microbial community. *Soil Biol. Biochem.* **38**: 2665–2675.

Chenu, C., Le Bissonnais, Y., and Arrouays, D., 2000, Organic matter influence on clay wettability and soil aggregate stability. *Soil Sci. Soc. Am. J.* **64**: 1479–1486.

Chern, E.C., Tsai, D.W., and Ogunseitan, O.A., 2007, Deposition of glomalin-related soil protein and sequestered toxic metals into watersheds. *Environ. Sci. Technol.* **41**: 3566–3572.

Clark, R.B., and Zeto, S.K., 1996, Iron acquisition by mycorrhizal maize grown on alkaline soil. *J. Plant Nutrit.* **19**: 247–264.

Corgie, S.C., Fons, F., Beguiristain, T., and Leyval, C., 2006, Biodegradation of phenanthrene, spatial distribution of bacterial populations and dioxygenase expression in the mycorrhizosphere of *Lolium perenne* inoculated with *Glomus mossese*. *Mycorrhiza* **16**: 207–212.

Degens, B.P., 1997, Macro-aggregation of soils by biological bonding and binding mechanisms and the factors affecting these: a review. *Aust. J. Soil Res.* **35**: 431–459.

Driver, J.D., Holben, W.E., and Rillig, M.C., 2005, Characterization of glomalin as a hyphal wall component of arbuscular mycorrhizal fungi. *Soil Biol. Biochem.* **37**: 101–106.

Filion, M., St-Arnaud, M., and Jabaji-Hare, S.H., 2003, Quantification of *Fusarium solani* f. sp. *phaseoli* in mycorrhizal bean plants and surrounding mycorrhizosphere soil using real-time polymerase chain reaction and direct isolations on selective media. *Phytopathology* **93**: 229–235.

Friese, C.F., and Allen, M.F., 1991, The spread of VA mycorrhizal fungal hyphae in the soil: inoculum types and external hyphal architecture. *Mycol.* **83**: 409–418.

Gadkar, V., and Rillig, M.C., 2006, The arbuscular mycorrhizal fungal protein glomalin is a putative homolog of heat shock protein 60. *FEMS Microbiol. Lett.* **263**: 93–101.

Gensel, P.G., and Andrews, H.N., 1987, The evolution of early land plants. *Amer. Sci.* **75**: 478–489.

George, E., Haussler, K., Kothari, S.K., Ki, X.-L., and Marschner, H., 1992, Contribution of mycorrhizal hyphae to nutrient and water uptake by plants. In: *Mycorrhizas in Ecosystems.* D.J. Read *et al.* (Eds.). CAB International, Wallingford, UK.

Gonzalez-Chavez, M.C., Carillo-Gonzelez, R., Wright, S.E., and Nichols, K.A., 2004, The role of glomalin, a protein produced by arbuscular mycorrhizal fungi, in sequestering potentially toxic elements. *Environ. Poll.* **130**: 317–323.

Halvorson, J.J., and Gonzalez, J.M., 2008, Tannic acid reduces recovery of water-soluble carbon and nitrogen from soil and affects the composition of Bradford-reactive soil protein. *Soil Biol. Biochem.* **40**: 186–197.

Harner, M.J., Ramsey, P.W., and Rillig, M.C., 2007, Protein accumulation and distribution in floodplain soils and river foam. *Ecol. Lett.* **7**: 829–836.

Iyer, S., and Lonnerdal, B., 1993, Lactoferrin, lactoferrin receptors and iron metabolism. *Euro. J. Clin. Nutr.* **47**: 232–241.

Janos, D.P., 2007, Plant responsiveness to mycorrhizas differs from dependence upon mycorrhizas. *Mycorrhiza* **17**: 75–91.

Johnson, D., Krsek, M., Weillington, E.M., Stott, A.W., Cole, L., Bardgett, R.D., Read, D.J., and Leake, J.R., 2005, Soil invertebrates disrupt carbon flow through fungal networks. *Science* **309**: 1047.

Jones, C.A., Basch, G., Baylis, A.D., Bazzoni, D., Biggs, J., Bradbury, R.B., Chaney, K., Deeks, L.K., Field, R., Gómex, J.A., Jones, R.J.A., Jordan, V.W.L., Lane, M.C.G., Leake, A., Livermore, M., Owens, P.N., Ritz, K., Sturny, W.G., and Thoms, F., 2006, *Conservaion Agriculture in Europe – An Approach to Sustainable Crop Production by Protecting Soil and Water?* SOWAP, Jealott's Hill, Bracnell, UK.

Jordan, N.R., Larson, D.L., and Huerd, S.C., 2008, Soil modification by invasive plants: effects on native and invasive species of mixed-grass prairies. *Biol. Invasion.* **10**: 177–190.

Kershaw, M.J., and Talbot N.J., 1998, Hydrophobins and repellents: proteins with fundamental roles in fungal morphogenesis. *Fungal Gene. Biol.* **23**: 18–33.

Lovelock, C.E., Wright, S.F., Clark, D.A., and Ruess, R.W., 2004, Soil stocks of glomalin produced by arbuscular mycorrhizal fungi across a tropical rain forest landscape. *J. Ecol.* **92**: 278–287.

Miller, R.M., and Jastrow, J.D., 1990, Hierarchy of root and mycorrhizal fungal interactions with soil aggregation. *Soil Biol. Biochem.* **22**: 579–584.

Millner, P.D., and Wright, S.F., 2002, Tools for support of ecological research on arbuscular mycorrhizal fungi (Review article). *Symbiosis* **33**: 101–123.

Nagasako, Y., Saito, H., Tamura, Y., Shimamura, S., and Tomita, M., 1993, Iron-binding properties of bovine lactoferrin in iron-rich solution. *J. Dairy Sci.* **76**: 1876–1881.

Nichols, K.A., and Wright, S.F., 2004, Contributions of soil fungi to organic matter in agricultural soils. In: *Functions and Management of Soil Organic Matter in Agroecosystems.* F. Magdoff and R. Weil (Eds.). CRC, Washington, DC, pp. 179–198.

Nichols, K.A., and Wright, S.F., 2005, Comparison of Glomalin and Humic Acid In Eight Native U.S. Soils. *Soil Sci.* **170** : 985–997.

Nichols, K.A., and Wright, S.F., 2006, Carbon and nitrogen in operationally-defined soil organic matter pools. *Biol. Fert. Soils.* **43**: 215–220.

Nogueira, M.A., Nehls, U., Hampp, R., Poralla, K., and Cardoso, E.J.B.N., 2007, Mycorrhiza and soil bacteria influence extract-able iron and manganese in soil and uptake by soybean. *Plant Soil* **298**: 273–284.

Olsson, P.A., Thingstrup, I., Jakobsen, I., and Baath, E., 1999, Estimation of the biomass of arbuscular mycorrhizal fungi in linseed field. *Soil Biol. Biochem.* **31**: 1879–1887.

Paulsson, M.A., Svensson, U., Kishore, A.R., and Naidu, A.S., 1993, Thermal behavior of bovine lactoferrin in water and its relation to bacterial interaction and antibacterial activity. *J. Dairy Sci.* **76**: 3711–3720.

Pawlowska, T.E., Chaney, R.L., Chin, M., and Charvat, I., 2000, Effects of metal phyto-extraction practices on the indigenous community of arbuscular mycorrhizal fungi at a metal-contaminated landfill. *Appl. Environ. Microbiol.* **66**: 2526–2530.

Pimentel, D., Harvey, C., Resosudarmo, P., Sinclair, K., Kurz, D., McNair, M., Crist, S., Shpritz, L., Fitton, L., Saffouri, R., and Blair, R., 1995, Environmental and economic costs of soil erosion and conservation benefits. *Science* **267**: 1117–1123.

Pirozynski, K.A., and Malloch, D.W., 1975, The origin of land plants: a matter of myco-trophism. *BioSystems* **6**: 153–164.

Preger, A.C., Rillig, M.C., Johns, A.R., Du Preez, C.C., Lobe, I., and Amelung, W., 2007, Losses of glomalin-related soil protein under prolonged arable cropping: a chronosequence study in sandy soils of the South African Highveld. *Soil Biol. Biochem.* **39**: 445–453.

Purin, S., and Rillig, M.C., 2008, Immuno-cytolocalization of glomalin in the mycelium of arbuscular mycorrhizal fungus Glomus intraradices. *Soil Biol. Biochem.* **40**: 1000–1003.

Redecker, D., Morton, J.B., Bruns, T.D., 2000, Ancestral lineages of arbuscular mycorrhizal fungi (Glomales). *Mole. Phylo. Evol.* **14**: 276–284.

Rillig, M.C., 2004, Arbuscular mycorrhizae, glomalin, and soil aggregation. *Can. J. Soil Sci.* **84**: 355–363.

Rillig, M.C., Caldwell, B.A., Wosten, H.A.B., and Sollins, P., 2007, Role of protein in soil carbon and nitrogen storage: controls on persistence. *Biogeochem.* **85**: 25–44.

Rillig, M.C., and Mummey, D.L., 2006, Tansley review – mycorrhizas and soil structure. *New Phytol.* **171**: 41–53.

Rillig, M.C., and Steinberg, P.D., 2002, Glomalin production by an arbuscular mycorrhizal fungus: a mechanism of habitat modification? *Soil Biol. Biochem.* **34**: 1371–1374.

Rillig, M.C., Wright, S.F., Allen, M.F., and Field, C.B., 1999, Rise in carbon dioxide changes soil structure. *Nature* **400**: 628.

Rillig, M.C., Wright, S.F., Nichols, K.A., Schmidt, W.F., and Torn, M.S., 2001, Large contribution of arbuscular mycorrhizal fungi to soil carbon pools in tropical forest soils. *Plant Soil* **233**: 167–177.

Roldan, A., Salinas-Gracia, J.R., Alguacil, M.M., and Caravaca, F., 2007, Soil sustainability indicators following conservation tillage practices under subtropical maize and bean crops. *Soil Till. Res.* **93**: 273–282.

Rosier, C.L., Hoye, A.T., and Rillig, M.C., 2007, Glomalin-related soil protein: assessment of current detection and quantification tools. *Soil Biol. Biochem.* **38**: 2205–2211.

Schindler, F.A., Mercer, E.J., and Rice, J.A., 2007, Chemical characteristics of glomalin-related soil protein (GRSP) extracted from soils of varying organic matter content. *Soil Biol. Biochem.* **39**: 320–329.

Schüßler, A., 2002, Molecular phylogeny, taxonomy, and evolution of *Geosiphon pyriformis* and arbuscular mycorrhizal fungi. *Plant Soil* **244**: 75–83.

Schüßler, A., Martin, H., Cohen, D., Fitz, M., and Wipf, D., 2007, Addendum – arbuscular mycorrhiza-studies on the geosiphon symbiosis lead to the chacterization of the first glomeromycotan sugar transporter. *Plant Sign. Behav.* **2**: 314–317.

Schwartzman, D.W., and Volk, T., 1989, Biotic enhancement of weathering and the habitability of Earth. *Nature* **340**: 457–460.

Selim, S., Negrel, J., Govaerts, C., Gianinazzi, S., and van Tuinen, D., 2005, Isolation and partial characterization of antagonistic peptides produced by *Paenibacillus* sp. strain B2 isolated from the sorghum mycorrhizosphere. *Appl. Enivron. Microbiol.* **71**: 6501–6507.

Six, J., Carpenter, A., van Kessel, C., Merck, R., Harris, D., Horwath, W.R., and Lüscher, A., 2001, Impact of elevated CO_2 on soil organic matter dynamics as related to changes in aggregate turnover and residue quality. *Plant Soil* **234**: 27–36.

Steinberg, P.D., and Rillig, M.C., 2003, Differential decomposition of arbuscular mycorrhizal fungal hyphae and glomalin. *Soil Biol. Biochem.* **35**: 191–194.

Tinker, P.B., Durall, D.M., and Jones, M.D., 1994, Carbon use efficiency in mycorrhizas: theory and sample calculations. *New Phytol.* **128**: 115–122.

Toro, M., Azcón, R., and Barea, J.-M., 1997, Improvement of arbuscular mycorrhiza development by inoculation of soil with phosphate-solubilizing rhizobactera to improve rock phosphate bioavailability (^{32}P) and nutrient cycling. *Appl. Envion. Micobiol.* **63**: 4408–4412.

Treseder, K.K., Turner, K.M., and Mack, M.C., 2007, Mycorrhizal responses to nitrogen fertilization in boreal ecosystems: potential consequences for soil carbon storage. *Global Change Biol.* **13**: 78–88.

Wessels, J.G.H., 1997, Hydrophobins: proteins that change the nature of the fungal surface. *Adv. Microb. Physiol.* **38**: 1–45.

Wessels, J.G.H., 1999, Fungi in their own right. *Fungal Gene. Biol.* **27**: 134–145.

Whitbeck, J.L, 2001, Effects of light environment on vesicular-arbuscular mycorrhiza development in *Inga leiocalycina*, a tropical wet forest tree. *Biotropica* **33**: 303–311.

Whiteford, J.R., and Spanu, P.D., 2002, Hydrophobins and the interactions between fungi and plants. *Mole. Plant Pathol.* **3**: 391–400.

Wright, S.F., 2000, A fluorescent antibody assay for hyphae and glomalin from arbuscular mycorrhizal fungi. *Plant Soil* **226**: 171–177.

Wright, S.F., and Upadhyaya, A., 1996, Extraction of an abundant and unusual protein from soil and comparison with hyphal protein of arbuscular mycorrhizal fungi. *Soil Sci.* **161**: 575–585.

Wright, S.F., and Upadhyaya, A, 1998, A survey of soils for aggregate stability and glomalin, a glycoproteins produced by hyphae of arbuscular mycorrhizal fungi. *Plant Soil* **198**: 97–107.

Wright, S.F., and Upadhyaya, A., 1999, Quantification of arbuscular mycorrhizal fungi activity by the glomalin concentration on hyphal traps. *Mycorrhiza* **8**: 283–285.

Wright, S.F., Franke-Snyder, M., Morton, J.B., and Upadhyaya, A., 1996, Time-course study and partial characterization of a protein on hyphae of arbuscular mycorrhizal fungi during active colonization of roots. *Plant Soil* **181**: 193–203.

Wright, S.F., Upadhyaya,A., and Buyer, J. S., 1998, Comparison of N-linked oligosaccharides of glomalin from arbuscular mycorrhizal fungi and soils by capillary electrophoresis. *Soil Biol. Biochem.* **30**: 1853–1857.

Wright, S.F., Starr, J.L., and Paltineanu, I.C., 1999, Changes in aggregate stability and concentration of glomalin during tillage management transition. *Soil Sci. Soc. Am. J.* **63**: 1825–1829.

Chapter 8

ARBUSCULAR MYCORRHIZAE AND THEIR ROLE IN PLANT RESTORATION IN NATIVE ECOSYSTEMS

KRISH JAYACHANDRAN[1] AND JACK FISHER[2]
[1]*Environmental Studies Department and Southeast Environmental Research Center, Florida International University, 11200 SW 8 ST, Miami, Florida 33199, USA;* [2]*Fairchild Tropical Botanical Garden, 11935 Old Cutler Road, Coral Gables, Florida 33156, USA*

Abstract: There is high plant biodiversity in southern Florida, due to the floristic mixing of warm temperate Southeastern North America and tropical Caribbean. Arbuscular mycorrhizal (AM) fungi were found in the roots of native plants in the families Anacardiaceae, Arecaceae (Palmae), Cactaceae, Convolvulaceae, Cycadaceae, Euphorbiaceae, Fabaceae, Lauraceae, Rubiaceae, Simarubaceae and Smilacaeae that grow in the coastal maritime and inland hammocks of southern Florida. Seedlings of the following genera: *Amorpha, Coccothrinax, Gymnanthes, Hamelia, Jacquemontia, Licaria, Nectandra, Opuntia, Picramnia, Psychotria, Rhus, Sabal, Serenoa* and *Zamia* inoculated with AM fungi showed enhancement of growth and phosphorus uptake on local sandy, nutrient poor soils. Most native species were depend on AM fungi under natural conditions of poor or no soils, phosphorus limitations and often water stress. Restoration of endangered species of *Amorpha* (Fabaceae), *Jacquemontia* (Convolvulaceae), *Opuntia* (Cactaceae) and *Pseudophoenix* (Arecaceae) was considered using AM fungi. The symbiotic relationship between AM fungi and native plants is important in the low P ecosystem and also useful for restoration of native plants.

Keywords: Arbuscular mycorrhizae; native ecosystem; endangered plants; Everglades restoration; phosphorus.

1 INTRODUCTION

In southeastern Florida, much of the subtropical vegetation grows on coastal uplands (the Miami rock ridge), which are situated between the Atlantic Ocean on the east and the Everglades wetlands on the west. These upland sites have either of two subtropical forest types: pine rockland and hardwood hammock (Wunderlin and Hansen, 2000). Because of urbanization (metropolitan Miami) and agriculture, these two subtropical forest types are highly threatened by habitat loss with less than 5% remaining outside of Everglades National Park. Federal, state and local land managers are working to protect the few remaining fragments of these habitats. They seek to restore numerous endemic plants as part of a multi-species recovery plan for the greater Everglades region (U.S. Fish and Wildlife Service, 1999). Restoration of keystone species and introduction of rare species on degraded sites will require detailed biological information on the separate component species.

An important aspect of seedling establishment and survival on these shallow sandy soils is the relationship between their roots and soil microorganisms, particularly mycorrhizal fungi, which are ubiquitous and are significant biotic variables in many habitat restorations elsewhere (Smith and Read, 1997). However, surprisingly little information on mycorrhizal associations in the two South Florida vegetation types have been published, considering the national commitment of money and effort being placed on conservation and restoration issues in the Greater Everglades region. These upland plant communities bound and interdigitate the more widely known wetlands of the Everglades. Arbuscular mycorrhizal (AM) fungi were reported in wetland species by Aziz et al. (1995) and Jayachandran and Shetty (2003), and AM fungi in three pine rockland species (Fisher and Jayachandran, 1999, 2002; Fisher and Vovides, 2004). It is not surprising that AM fungi were found since the native flora consists of many widespread genera that have been verified as mycorrhizal elsewhere or are members of families that are known to be mycorrhizal.

Our goal was to answer the following questions: (1) Do native upland plant species (Table 1) form mycorrhizae and what types of mycorrhizae are present; (2) Does AM fungi affect seedling growth on native low nutrient soils; and (3) What levels of additional phosphorus (P) are equivalent to growth promoting affects of AM fungi? Our experimental results give land managers and native plant propagators information on the relative AM fungi dependency and the effectiveness of AM fungi compared to additions P fertilizer. Successful propagation of native plants for habitat restoration will require an understanding of the dependency of these plants to mycorrhizal fungi.

2 AM FUNGI COLONIZATION IN NATIVE PLANTS UNDER FIELD CONDITIONS

Majority of the native species tested showed AM fungi colonization. Some roots, especially those of palms, were difficult to clear and stain because of thick, lignified walls of epidermis and hypodermis. For these plant species,

AM fungi could only be observed reliably in thick transverse or oblique longitudinal sections that were processed like whole root fragments or after longitudinal splitting and dissection of the root cortex. Ultimate short fine roots of palms, *Zamia*, and several dicots were brittle and easily detached during digging and removal of soil. This was particularly apparent when roots of these plants were initially excavated in the field. Few fine roots possessing AM fungi were collected directly from naturally shallow, rocky soils. Occasionally, we found a proliferation of fine roots in a small pocket of humus or in deep crevices in the limestone substrate. In these sites, AM fungi were abundant but absent in roots of the same plant extracted elsewhere. For these reasons, we relied on pot (also called trap) cultures as the most reliable way to verify presence of AM fungi colonization. Since the ultimate feeder roots were difficult to extract from the limestone rock, we planted sterile seedlings in nurse culture pots, similar to those used for AMF inoculum, in order to examine the presence and type of AMF colonization. This use of seedlings was similar to the method of Brundrett and Abbott (1991).

Most seedlings growing in pots with AM fungi inoculum showed AM fungi colonization after eight weeks or longer. Each root sample was cleared in KOH, bleached with ammoniated H_2O_2, and stained with trypan blue or chlorazol black E in acidic glycerol (Brundrett *et al.*, 1996) to determine presence of AMF. The basic morphology of the AMF colonization was classified as *Acorus*- or *Paris*-type, following the scheme of Smith and Smith (1997).

2.1 Dicotyledons

The presence of root hairs was variable (Table 1). In most cases, features that are typical of *Acorus*-type (Smith and Smith, 1997) colonization were found. Arbuscules were mainly found in younger regions of roots, typically one per cortical cell. Non-septate hyphae were mostly found in the longitudinal intercellular spaces of the root cortex. Intercellular hyphae proliferated in deeper layers of cortex and not in the epidermis or peripheral layers that were adjacent to the region of hyphal penetration. Vesicles were found in older regions of roots.

Four dicot species had fungal coils in adjacent cortical cells typical of *Paris*-type (Smith and Smith, 1997). Only *Annona glabra* consistently had cortical parenchyma cells with multiple hyphal coils within each cell. The other three dicots had a mix types of AM fungi. One root of *Licaria triandra* was seen with *Paris*-like, highly coiled hyphae (several coils per cell) in the periphery of the cortex in one region, yet another region of the same root formed a longitudinal intercellular network with individual arbuscules per cell, typical of the *Acorus*-type. *Picramnia pentandra* similarly formed intracellular hyphal coils and single arbuscules per cortical cell. *Sophora tomentosa* var. *truncata* had features of both *Acorus*- and *Paris*-types in different regions.

Hyphal coils were also observed rarely and inconsistently in *Ocotea coriaceae*, *Psychotria nervosa*, *Simaruba glauca*, and *Tetrazygia bicolor*, but were not sufficiently common to classify them as *Paris*-type (Table 1).

Table 1. List of taxa with their families, plant community type, and observations on AM fungi colonization.

Taxon	Status[a]	AM fungi description					
		Cortical AM hyphae	Coils	Vesicles	Arbuscules	Root hairs	AM type
Acoelorrhaphe wrightii (Griseb. and H. Wendl.) H. Wendl. ex Becc.	T	+	*None*	+	+	*None*	*Acorus*
Annona glabra L.		+	+	None	None	None	*Paris*
Bourreria cassinifolia (A. Rich.) Griseb.	E	+	+ Few	+	+?	+ Short	*Acorus*
Coccothrinax argentata (Jacq.) L.H. Bailey	T	+	+ (Near epidermis)	None	+	None	*Acorus*
Consolea corallicola Small (*Opuntia spinosissima* Mill.)	E	+	None	+	None	+ Long (0.5 mm)	*Acorus* ?
Erithalis fruticosa L.	T	+	+ Rare	+	+	+	*Acorus*
Gymnanthes lucida Sw.		+	None	None	+	+	*Acorus*
Hamelia patens Jacq.		+	+ *Rare*	+	+	+	*Acorus*
Harrisia fragrans Small ex Britton and Rose	US	+ Rare	None	None	None	+ Long (0.8–1.2 mm)	*Acorus* ?
Ocotea (Nectandra) coriacea (Sw.) Griseb.		+ Rare	None	+	None	None	*Acorus*
Opuntia tricanthos (Willd.) Sweet	E	+?	None	None	None	? (+)	*Acorus* ?
Picramnia pentandra Sw.		+ Few intercellular	+	+ Within cell	+	+	*Acorus* + *Paris*

(continued)

Species	Status						AM Type
Polygala smallii R.R. Sm. and D.B. Ward	US	+	None	+	+	None	*Acorus*
Pseudophoenix sargentii H. Wendl. ex Sarg.	E	+	None	+	+	+ Rare	*Acorus*
		Dense network					
Psychotria nervosa Sw.		+	+ Rare	+	None	+ Short, rare	*Acorus*
Rhus copallinum L.		+	None	+	+	+	*Acorus*
Sabal palmetto (Walter) Lodd. ex Schult. and Schult. f.		+	+ Rare	+	+	None	*Acorus*
Serenoa repens (W. Bartram) Small		+	+ Near epidermis	+	+	None	*Acorus*
Smilax havanensis Jacq.	T	+	+ Multiple	None	+	+ Short	*Paris*
Sophora tomentosa L. var. *truncata* Torr. and A. Gray		+	+	+	+	+	*Acorus* + *Paris?*
Tephrosia angustissima var. *corallicola* (Small) Isely	E	+	None	+	+	+	*Acorus*
Tetrazygia bicolor (Mill.) Cogn.	T	+	+ Rare	+	+	+ Rare, long	*Acorus*
Thrinax morrisii H. Wendl.	E	+	None	+	+	None	*Acorus*
Trema micranthum (L.) Blume		+	None	+	+	+	*Acorus*
Zamia pumila L.		+	None	+	+	+	*Acorus*

[a]Endangered status: E – State of Florida endangered; T – State of Florida threatened (Coile, 2000); US – federally endangered (Fish and Wildlife Service, 1999)

2.2 Cacti

All species (*Consolea corallicola*, *Harrisia fragrans*, and *Opuntia tricanthos*) had noticeably long root hairs (0.5–1.2 mm long) that clung to sand particles and made a sand sheath around the roots. Although root surfaces had both septate and non-septate hyphae, only rarely were hyphae found in longitudinal intercellular spaces and appeared to be non-septate. We conclude that these species had questionable AM fungi when grown in pots. (Thus, the AM fungi type is noted with a question mark in Table 1) We were unable to collect and observe roots from wild plants.

2.3 Palms

All six species had lignified, rough, thick-walled epidermal cells that made staining and clearing difficult. They were best observed in thick transverse or longitudinal sections of ultimate roots. Root hairs were not found. All ultimate fine roots were brittle and easily detached during excavation and cleaning. *Coccothrinax argentata* had a dense network of intercellular hyphae in the cortex. Both intercellular and intracellular coils formed in the many layered hypodermis, and single arbuscules and vesicles were found. *Sabal palmetto* and *Serenoa repens* had dense intercellular hyphae in the cortex periphery with arbuscules and vesicles, but they also had coils in the epidermis and hypodermal cells. A more detailed description for *Serenoa* is given by Fisher and Jayachandran (1999). Coils were not observed in *Acoelorrhaphe wrightii*, *Pseudophoenix sargentii*, and *Thrinax morrisii*.

2.4 Other monocotyledons

Smilax havanensis had cortical cells filled with multiple hyphal coils without obvious intercellular hyphae. Cells with coils were connected by a single hypha that passed through the common wall. Occasionally, arbuscules were found one or two cells distant from coils, rarely adjacent to coils. These arbuscules were simple, with a single trunk hypha and were not definitely associated with the coils. *Smilax* displayed the classic *Paris*-type.

2.5 Cycad

Zamia pumila had typical *Acorus*-type AM fungi with longitudinal intercellular hyphae and arbuscules concentrated in the outer cortex. Vesicles formed in older roots, especially after secondary growth was present. The structure is described in greater detail by Fisher and Vovides (2004).

Of the 26 species exposed to AM fungi in pot culture and one (*Polygala*) only from the field, 24 formed clearly defined arbuscular mycorrhizae (AM) and three cacti (*Consolea*, *Harrisia*, *Optuntia*) had poorly developed or questionable AM. The lack of clear AM fungi structures in these cacti was unexpected since AM fungi colonization was reported in other cacti (Allen *et al.*, 1998; Barredo-Pool *et al.*, 1998; Carrillo-Garcia *et al.*, 1999) and also increased growth and P uptake (Pimienta-Barrios *et al.*, 2002; Rincón *et al.*, 1993). Although, Pimenta-Barrios *et al.* (2003) found that when a fungicide was applied to eliminate mycorrhizal colonization in natural plants of *Opuntia robusta*, physiological processes were unaffected. Interestingly, our three cacti had noticeably long root hairs that formed sand sheaths which required extra effort in freeing roots from soil particles. Other species had few or no root hairs (Table 1). This observation supports the loose relationship between AM and relatively thick feeder roots with short or no hairs, so-called magnolioid roots (Smith and Read, 1997).

Many of these 24 species were found with AM fungi in nature. We presume that those roots which did not have clear AM fungi in nature or had very low rates of colonization were, at least in part, artifacts of the difficulty in extracting fine feeder roots from the rocky substrate. Mycorrhizal status of other species growing in the same substrate, in which roots tend to proliferate in rock crevices and at great distances from the shoot, are best determined with trap cultures. We found that feeder roots containing AM fungi are easily lost during extraction or missed when roots proliferate at localized nutrient-rich or moist sites. Olsson et al. (2002) showed that humus-rich soil or organic matter promote or enhance mycorrhizal proliferation. This is the likely reason that the first survey of mycorrhizae in South Florida plants did not find AM fungi in many of these same species (Meador, 1977). Later investigations did find AM fungi in wetland plants of the Everglades (Aziz et al., 1995; Jayachandran and Shetty, 2003), plants of the pine rocklands (Fisher and Jayachandran, 2002; Fisher and Vovides, 2004), and plants of coastal dunes (Fisher and Jayachandran, 2002; Sylvia et al., 1993).

The two main structural types of AM in host roots were reviewed by Smith and Smith (1997): the *Arum*-type with intercellular hyphae in the root cortex; and the *Paris*-type with intracellular hyphal coils and no intercellular hyphae. Our survey found that 21 species formed the *Arum*-type and expands the categorization of AM according to colonization type as reviewed in Smith and Smith (1997). We observed that many of the species had some intracellular hyphal coils in the epidermis and hypodermis, but most hyphae occur in intercellular spaces deeper in the cortex. This variation has been reported widely in the literature and was classified as the *Arum*-type by Smith and Smith (1997) but may be the cause of some apparent conflicting reports of *Arum*- versus *Paris*-types depending upon interpretation of the intracellular hyphal coils that occur in the outer cortex. Within the inner cortex of the same roots, extensive networks of intercellular hyphae formed arbuscules and vesicles. Since Smith and Smith's review, Wubet et al. (2003) found *Arum*-type in all 11 indigenous trees in Ethiopia, although they report rare hyphal coils near the points of new infection. More typical *Arum*-type hyphae in the intercellular spaces of inner cortex may be poorly developed or not yet present in a particular root being observed. Among plants in a mangrove community, Sengupta and Chaudhuri (2002) reported 12 *Arum*-, 27 *Paris*-, and 13 with both-types.

Two species in our study (*Annona* and *Smilax*) formed only intracellular coils typical of the *Paris*-type colonization. We confirm the description of AM fungi in *Smilax* by Bedini et al. (2000) and Maremmani et al. (2003), where free arbuscules form in cells adjacent to cells filled with hyphal coils. Maremmani et al. (2003) found that two species of *Glomus*, which produce *Arum*-type AM in other plant species, form *Paris*-type in *Smilax*. Growth was also increased by AM fungi colonization. Their finding supports the general assumption that the host root mainly determines the type of AM structure, not the fungus (Smith and Smith, 1997).

Three species (*Licaria*, *Picramnia*, and *Sophora*) formed both *Arum*- and *Paris*-types of AM within roots of the same plant using the same mixed inoculum as with all the other species. However, we do not know how many different AM fungi are associated with the different types of colonization, nor if the same or different AM fungi species cause the mixed *Arum-Paris*-types in the same root system. We cannot state that these three species are "near-*Paris*" or "intermediate types" (Smith and Smith, 1997) because of the possibility that more than one AM fungi is involved in each symbiosis. It is generally assumed that the AM fungi structure is in great part regulated by the plant; each plant species has a particular type of colonizing fungal morphology, as seen in the findings of Bedini *et al.* (2000) described above. However, in tomato the AM fungi-type varied depending upon the AM fungi species, indicating fungal control of morphology (Cavagnaro *et al.*, 2001). Future inoculation with cultures of single AM species will clarify this point in the species we surveyed. If two or more AM fungi species are involved in these examples of mixed types, it will also be interesting to document whether they compliment one another in their benefit to the plant's nutrition as has been suggested (Sanders, 2002) and recently documented using two *Glomus* species in the same host (Drew *et al.*, 2003). The speed and amount of colonization of plant roots by AM fungi also varies with the fungal species and seems related to family classification of the fungus (Hart and Reader, 2002). Future research must clarify the identity of the AM fungi species involved in South Florida soils.

3 EFFECT OF AM FUNGI ON SEEDLING GROWTH

In 12 native species tested, seedlings inoculated with AM fungi grew more than those growing without AM fungi. Only *Serenoa* and one experiment with *Coccothrinax* showed no significant difference in total dry mass, although AM fungi promoted shoot growth in both of these experiments when shoot and root dry mass were analyzed separately. Most commonly, AM fungi increase the biomass of plants grown on nutrient poor soils, due mainly to enhanced uptake of P from soils in which available P is limiting (Smith and Read, 1997). We previously reported that AM fungi caused significantly more P uptake in an endangered Fabaceae (*Amorpha crenulata* Rydb.) of the pine rocklands and an endangered Convolvulaceae (*Jacquemontia reclinata* House ex Small) of the back dune-coastal hammock in South Florida (Fisher and Jayachandran, 2002). In the present study, we presume that enhancement of seedling dry mass by AM fungi inoculation is due to enhanced nutrient uptake. In most species, the promotion of growth by AM fungi was equivalent to additions of 10 or 20 mg/kg P. Only *Hamelia* showed no response to additions of P when AM fungi was absent, strongly indicating that *Hamelia* may be an obligate mycorrhizal plant. The non-AM plants had intensely red pigmented blades, a feature common in P deficient plants. Alternatively, perhaps P was not limiting growth, but rather uptake of another limiting nutrient was enhanced by AM fungi colonization. Cavagnaro *et al.* (2003) showed that the effect of AM fungi on the shoot and root mass of a liliaceous native plant was equal to additions of 8 and 24 mg/kg P, respectively. The native soil in their study had 6.6 mg/kg bicarbonate extractable P (Bray and Kurtz, 1949), similar to the soil in our study. The AM fungi in their study formed *Paris*-type AM and had an effect similar to those with the more widely studied *Arum*-types referred in their paper. In our experimental growth studies, only *Picramnia* formed the *Paris*-type (as well as *Arum*-type), and the results were similar.

Inoculation with AM fungi has promoted growth of seedlings or plantlets of other species that are related to the species in our study. Growth was increased in palms (Blal *et al.*, 1990; Morte and Honrubia, 2002; Ravolanirina *et al.*, 1989) and crop species related to wild species we surveyed, e.g., Anacardiaceae (Caravaca *et al.*, 2003), Fabaceae (Caravaca *et al.*, 2003), Rubiaceae (Vaast *et al.*, 1996; Kyllo *et al.*, 2003), and Lauraceae (Vidal *et al.*, 1992).

A complicating factor in tests for AM fungi effectiveness on growth is variation due to origin of the AM fungi and the host plant (Graham and Eissenstat, 1994; Henkel *et al.*, 1989; Schultz *et al.*, 2001). For coastal sea oats, Sylvia *et al.* (2003) found that effectiveness of the community pot cultures of AM fungi from widely separated sites varied in the same host plant genotype (or ecotype). Their results indicated a level of specificity between the local fungal community and the resident plant ecotype. Caravaca *et al.* (2003) showed that an inoculum of eight native AM fungi was more effective on growth and colonization than a single species inoculum. We used a mixture of native AM fungi derived from a single natural site. Some but not all seeds used in our research came from this location, although all seed sources grew in similar vegetation and soil types within 22 km. The limited colonization in our three cacti or degree of effectiveness in other species could be due to a sub-optimal combination of host plant and AM fungi inoculum.

Root/shoot (R/S) ratios of AM fungi inoculated plants were consistently smaller than or the same as non-AM fungi controls with or without additions of P. Such a decrease in R/S ratio in AM fungi plants has been shown previously in some species (Vaast *et al.*, 1996), although there was considerable variability (Allen, 1991; Corkidi and

Rincón, 1997; Janos et al., 2001). In species of *Psychotria*, Kyllo et al. (2003) found a complex relationship between R/S ratio (which they measured as root dry mass/leaf area) and light levels. They found that AM fungi colonization caused increased R/S ratio in light demanding species and decreased R/S ratio in shade tolerant species. Many of the species in our study thrive in open, high light habitats as adults, but as seedlings all appear to require shaded micro-sites to survive.

4 RELATIVE MYCORRHIZAL DEPENDENCY

Using the average dry mass of entire plants (root + shoot), the relative mycorrhizal dependency (RMD) of plants were calculated as percentage increase due to AM fungi. The results indicate highest dependency for *Hamelia* and wide variations between replicate experiments for *Psychotria*, *Coccothrinax* and *Tephrosia* (Table 2).

Table 2. Relative mycorrhizal dependency (RMD) of plants. Ranked from highest to lowest percentage increase in average dry mass of entire plant; (Treatment 2 − Treatment 1)/ Treatment 1. Each line = one experiment.

Taxon	RMD (%)
Hamelia patens	21,275
Hamelia patens	17,143
Psychotria nervosa	1,167
Tephrosia angustissima	697
Rhus copallinum	444
Coccothrinax argentata	253
Picramnia pentandra	196
Rhus copallinum	185
Gymnanthes lucida	172
Erithalis fruticosa	142
Ocotea coriacea	126
Sabal palmetto	109
Psychotria nervosa	104
Acoelorrhaphe wrightii	77
Sabal palmetto	77
Zamia pumila	77
Zamia pumila	45
Coccothrinax argentata	21
Serenoa repens	8
Tephrosia angustissima	4

5 SIGNIFICANCE FOR CONSERVATION AND RESTORATION ACTIVITIES

All the species examined (with some uncertainty for the three cacti) were colonized by AM fungi (Table 1), which was expected in these natural habitats with shallow, sandy, and nutrient poor soils. We found that growth of all 13 plant species tested, when grown on native soil, was enhanced by AM fungi colonization compared to seedlings without AM fungi. Koske and Gemma (1995) reported improved growth of cuttings and seedlings of endangered Hawaiian plants grown on various artificial soils mixes in the greenhouse. Thus, addition of AM fungi can improve the propagation of native tropical species and is an important factor to consider in conservation horticulture.

In South Florida, P is a limiting nutrient in native soils, as was shown in the native *Jacquemontia reclinata* and *Amorpha crenulata* (Fisher and Jayachandran, 2002), presumably because of the improved P uptake and growth promotion facilitated by AM fungi. Gemma *et al.* (2002) found that four species of Hawaiian plants responded similarly to AM fungi (a single *Glomus* species) when they were grown on native soils that were low in P. The Hawaiian soil used in their pot experiments had soil-solution P = 0.005 mg/L, but some field samples had <0.001 mg/L. We found similar low levels of soil-solution P in South Florida native soil = 0.002 mg/L and in our experimental pot soil = 0.003 mg/L as determined by the method of Olsen and Summers (1982). Gemma *et al.* (2002) suggested that this type of responsiveness to AM fungi under very low natural soil P levels should be referred to as "ecological mycorrhizal dependency" and we concur.

In other habitats, AM fungi inoculation of plants can aid in restorations, e.g. arid habitats (Requena *et al.*, 2001; Caravaca *et al.*, 2003), grasslands (Richter and Stutz, 2002), and coastal dunes (Gemma and Koske, 1997). In an experimental planting of two native grasses, root colonization by AM fungi of the two grasses were significantly different 14 weeks after sowing seeds but after 68 weeks were not significantly different (Salyards *et al.*, 2003). The results also indicated that fresh top soil (with numerous AM fungi species) was more effective than commercial inoculum (with only one AM fungus species) in the short term, but that with time all roots were colonized equally. These same two grasses showed differences in ability to facilitate AM fungi colonization and growth of a native *Salix* transplanted within the grasses (Salyards *et al.*, 2003).

In addition to promoting plant growth, other studies found that AM fungi enhanced competitive ability of native species against invasive species in potted plant experiments (Pendleton and Pendleton, 2003; Pendleton *et al.*, 2004) and promoted native species in field experiments (Smith *et al.*,

1998). AM fungi inoculation also promoted natural community development in a seeded tall grass prairie restoration (Smith *et al.*, 1998). All show that AM fungi should be considered as an important variable in native plant restoration and may improve restoration success.

We assume that natural seedling establishment depends upon AM fungi colonization. In certain situations in South Florida, where natural AM fungi inoculum could be absent or only present in low propagule numbers (e.g., cleared roadsides, sites where top soil was removed, reclaimed urban landscapes, or where soil from non-vegetated sources is added as in canal waste or coastal areas "enriched" with marine dredgings), the resulting soil environment would be similar to the classical low AM fungi habitats: mine tailings, strip mining disturbance, or volcanic eruption (Allen, 1991). In such cases, natural regeneration or active restoration could be limited or slowed by lack of natural AM fungi colonization.

Therefore, we suspect that restoration projects in highly disturbed sites of the Greater Everglades region of South Florida would benefit from introduced AM fungi, especially during nursery production of seedlings. The Florida Department of Forestry now routinely grows pine seedlings with ectomycorrhizae in its nursery as an aid to successful transplanting. Native Florida sea oats (*Uniola paniculata* L.) have benefited by pre-inoculation with AM fungi before outplanting in beach restoration (Sylvia, 1989; Sylvia *et al.*, 1993, 2003). Gemma and Koske (1997) showed that even nursery plants that already possessed AM fungi colonization benefited from addition of AM fungi inoculation at the time of planting in coastal dunes.

Because the Everglades hammocks and pine rocklands are naturally low P environments (Szulczewski *et al.*, 2008), any use of P fertilizer is a major concern for land managers (U.S. Fish and Wildlife Service, 1999). The horticultural use of AM fungi inoculum would promote native plant growth without the need of additional of P in situations where natural AM fungi are limiting in the field. Under nursery conditions, AM fungi inoculation of plants should promote plant growth without need of P fertilization and possible resulting P pollution in run off water. In the case of field restorations, we might wish to plant early pioneer plants which might not have high AM fungi dependency. However, all the lower dependent plants tested (Table 2) were slow growing palms and *Zamia*. The taller, sun tolerant woody plants, which are most useful for initial restorations (*Erithalis, Gymnanthes, Hamelia, Ocotea, Psychotria, Rhus*), have relatively high dependency. We must test other trees for AM fungi dependency in order to make sounder recommendations. Also, we must still document the potential benefits of AM fungi inoculation in native plant restorations and urban horticulture, namely: increased survivorship and long-term establishment of outplanted seedlings in field sites.

REFERENCES

Allen, E. B., Rincón, E., Allen, M. F., Pérez-Jimenez, A., and Huante, P., 1998, Disturbance and seasonal dynamics of mycorrhizae in a tropical deciduous forest in Mexico. *Biotropica* **30**: 261–274.
Allen, M. F., 1991, *The ecology of mycorrhizae*. Cambridge University Press, Cambridge.
Aziz, T., Sylvia, D. M., and Doren, R. F., 1995, Activity and species composition of arbuscular mycorrhizal fungi following soil removal. *Ecol. Appli.* **5**: 776–784.
Barredo-Pool, F., Varela, L., Arce-Montoya, M., and Orellana, R., 1998, Estudio de la asociación micorrízica en dos Cactáceas natives del Estado de Yucatán, México. In R. Zulueta Rodríguez, M. A. Escalona Aguilar, and D. Trejo Aguilar [eds.], *Avances de la investigación micorrízica en México*, pp. 69–76. Universidad Veracruzana, Xalapa, Mexico.
Bedini, S., Maremmani, A., and Giovannetti, M., 2000, *Paris*-type mycorrhizas in *Smilax aspera* L. growing in a Mediterranean sclerophyllous wood. *Mycorrhiza* **10**: 9–13.
Blal, B., Morel, C., Gianinazzi-Pearson, V., Fardeau, J. C., and Gianinazzi, S., 1990, Influence of vesicular-arbuscular mycorrhizae on phosphate fertilizer efficiency in two tropical acid soils planted with micropropagated oil palm (*Elaeis guineensis* Jacq.). *Biol. Fert. Soils* **9**: 43–48.
Bray, R. H., and Kurtz, L. T., 1949, Determination of total, organic and available form of phosphorus in soil. *Soil Sci.* **59**: 39–45.
Brundrett, M. C., and Abbott, L. K., 1991, Roots of jarrah forest plants. I. Mycorrhizal associations of shrubs and herbaceous plants. *Austr. J. Bot.* **39**: 445–457.
Brundrett, M., Bougher, N., Dell, B., Grove, T., and Malajczuk, N., 1996, Working with mycorrhizas in forestry and agriculture. *Austr. Centre Int. Agri. Res. Monogr.* **32**: 1–374.
Caravaca, F., Barea, J. M., Palenzuela, J., Figueroa, D., Alguacil, M. M., and Roldán, A., 2003, Establishment of shrub species in a degraded semiarid site after inoculation with native or allochthonous arbuscular mycorrhizal fungi. *Appl. Soil Ecol.* **22**: 103–111.
Carrillo-Garcia, A., León de la Luz, J.-L., Bashan, Y., and Bethlenfalvay, G. J., 1999, Nurse plants, mycorrhizae, and plant establishment in a disturbed area of the Sonoran Desert. *Restor. Ecol.* **7**: 321–335.
Cavagnaro, T. R., Gao, L.-L., Smith, F. A., and Smith, S. E., 2001, Morphology of arbuscular mycorrhizas is influenced by fungal identity. *New Phytol.* **151**: 469–475.
Cavagnaro, T. R., Smith, F. A., Ayling, S. M., and Smith, S. E., 2003, Growth and phosphorus nutrition of a *Paris*-type arbuscular mycorrhizal symbiosis. *New Phytol.* **157**: 127–134.
Coile, N.C.,2000. Notes on Florida's endangered and threatened plants. Florida Dept. of Agriculture and consumer services, 3rd edition. Botany section contribution No.38.
Corkidi, L., and Rincón, E., 1997, Arbuscular mycorrhizae in a tropical sand dune ecosystem on the Gulf of Mexico. II Effects of arbuscular mycorrhizal fungi on the growth of species distributed in different early successional stages. *Mycorrhiza* **7**: 17–23.
Drew, E. A., Murray, R. S., Smith, S. E., and Jakobsen, I., 2003, Beyond the rhizosphere: growth and function of arbuscular mycorrhizal external hyphae in sands of varying pore size. *Plant Soil* **251**: 105–114.
Fisher, J. B., and Jayachandran, K., 1999, Root structure and arbuscular mycorrhizal colonization of the palm *Serenoa repens* under field conditions. *Plant Soil* **217**: 229–241.
Fisher, J. B., and Jayachandran, K., 2002, Arbuscular mycorrhizal fungi enhance seedling growth in two endangered plant species from south Florida. *Intern. J. Plant Sci.* **163**: 559–566.
Fisher, J. B., and Vovides, A. P., 2004, Mycorrhizae are present in cycad roots. *Bot. Rev.* **70**: 16–23.

Gemma, J. N., and Koske, R. E., 1997, Arbuscular mycorrhizae in sand dune plants of the North Atlantic coast of the U.S.: field and greenhouse inoculation and presence of mycorrhizae in planting stock. *J. Environ. Manag.* **50**: 251–264.

Gemma, J. N., Koske, R. E., and Habte, H., 2002, Mycorrhizal dependency of some endemic and endangered Hawaiian plant species. *Amer. J. Bot.* **89**: 337–345.

Graham, J. H., and Eissenstat, D. M., 1994, Host genotype and the formation and function of VA mycorrhizae. *Plant Soil* **159**: 170–185.

Hart, M. M., and Reader, R. J., 2002, Taxonomic basis for variation in the colonization strategy of arbuscular mycorrhizal fungi. *New Phytol.* **153**: 335–344.

Henkel, T. W., Smith, W. K., and Christensen, M., 1989, Infectivity and effectivity of indigenous vesicular-arbuscular mycorrhizal fungi from contiguous soils in southwestern Wyoming. *New Phytol.* **112**: 205–214.

Janos, D. P., Schroeder, M. S., Schaffer, B., and Crane, J. H., 2001, Inoculation with arbuscular mycorrhizal fungi enhances growth of *Litchi chinensis* Sonn. trees after propagation by air-layering. *Plant Soil* **233**: 85–94.

Jayachandran, K., and Shetty, K. G., 2003, Growth response and phosphorus uptake by arbuscular mycorrhizae of wet prairie sawgrass. *Aquat. Bot.* **76**: 281–290.

Koske, R. E., and Gemma, J. N., 1995, Vesicular-arbuscular mycorrhizal inoculation of Hawaiian plants: a conservation technique for endangered tropical species. *Pacific Sci.* **49**: 181–191.

Kyllo, D. A., Velez, V., and Tyree, M. T., 2003, Combined effects of arbuscular mycorrhizas and light on water uptake of the neotropical understory shrubs, *Piper* and *Psychotria*. *New Phytol.* **160**: 443–454.

Maremmani, A., Bedini, S., Matoševic, I., Tomai, P. E., and Giovannetti, M., 2003, Type of mycorrhizal associations in two coastal nature reserves of the Mediterranean basin. *Mycorrhiza* **13**: 33–40.

Meador, R. E., 1977, The role of mycorrhizae in influencing succession on abandoned Everglades farmland. MS thesis, University of Florida, Gainesville, FL, 98 pp.

Morte, A., and Honrubia, M., 2002, Growth response of *Phoenix canariensis* to inoculation with arbuscular mycorrhizal fungi. *Palms* **46**: 76–80.

Olsen, S. R., and Summers, L. E., 1982, Phosphorus. *In* A. L. Page, R. H. Miller, and D. R. Keeney [eds.] *Methods of soil analysis, part 2 – chemical and microbiological properties, agronomy No 9 Part 2*. American Society of Agronomy, Soil Science Society America, Madison, WI.

Olsson, P. A., Jakobsen, I., and Wallander, H., 2002, Foraging and resource allocation strategies of mycorrhizal fungi in a patchy environment. *In* M. G. A. van der Heijden and I Sanders [eds.] *Mycorrhizal ecology*, pp. 93–115. Springer, Berlin.

Pendleton, R. L., and Pendleton, B. K., 2003, Soil microorganisms affect survival and growth of shrubs grown in competition with cheatgrass (New Mexico). *Ecol. Restor.* **21**: 215–216.

Pendleton, R. L., Pendleton, B. K., Howard, G. L., and Warren, S. D., 2004, Response of Lewis flax seedlings to inoculation with arbuscular mycorrhizal fungi and cyanobacteria. *In* A. L. Hild, N. L. Shaw, E. E. Meyer, D. T. Booth, and E. D. McArthur [comps.] *Seed and soil dynamics in shrubland ecosystems*, Proceedings RMRS-P-31, pp. 64–68. US Department of Agriculture, Forest Service, Rocky Mountain Research Station, Albuquerque, New Mexico.

Pimienta-Barrios, E., Pimienta-Barrios, En., Salas-Galván, M. E., Zañudo-Hernandez, J., and Nobel, P. S., 2002, Growth and reproductive characteristics of the columnar cactus *Stenocereus queretaroensis* and their relationships with environmental factors and colonization by arbuscular mycorrhizae. *Tree Physiol.* **22**: 667–674.

Pimienta-Barrios, E., Gonzalez del Castillo-Aranda, M. E., Muñoz-Urias, A., and Nobel, P. S., 2003, Effects of Benomyl and drought on the mycorrhizal development and daily net CO_2 uptake of a wild platyopuntia in a rocky semi-arid environment. *Ann. Bot.* **92**: 239–245.

Ravolanirina, F., Blal, B., Gianinazzi, S., and Gianinazzi-Pearson, V., 1989, Mise au point d'une méthode rapide d'endomycorhization de vitroplants. *Fruits* **44**: 165–170.

Requena, N., Perez-Solis, E., Azcón-Aguilar, C., Jeffries, P., and Barea, J., 2001, Management of indigenous plant-microbe symbioses aids restoration of desertified ecosystems. *Appl. Environ. Microbiol.* **67**: 495–498.

Richter, B. S., and Stutz, J. C., 2002, Mycorrhizal inoculation of big sacaton: implications for grassland restoration of abandoned agricultural fields. *Restor. Ecol.* **10**: 607–616.

Rincón, E., Huante, P., and Ramírez, Y., 1993, Influence of vesicular-arbuscular mycorrhizae on biomass production by the cactus *Pachycereus pectin-aboriginum*. *Mycorrhiza* **3**: 79–81.

Salyards, J. R., Evans, R. Y., and Berry, A. M., 2003, Mycorrhizal development and plant growth in inoculated and non-inoculated plots of California native grasses and shrubs. *Native Plants* (Fall 2003): 143–149.

Sanders, I. R., 2002, Specificity in the arbuscular mycorrhizal symbiosis. *In* M. G. A. van der Heijden and I. Sanders [eds.] *Mycorrhizal ecology*, pp. 415–437. Springer, Berlin.

Schultz, P. A., Miller, R. M., Jastrow, J. D., Rivetta, C. V., and Bever, J. D., 2001, Evidence of a mycorrhizal mechanism for the adaptation of *Andropogon gerardii* (Poaceae) to high- and low-nutrient parairies. *Amer. J. Bot.* **88**: 1650–1656.

Sengupta, A., and Chaudhuri, S., 2002, Arbuscular mycorrhizal relationships of mangrove plant community at the Ganges River estuary in India. *Mycorrhiza* **12**: 169–174.

Smith, F. A., and Smith, S. E., 1997, Tansley Review No. 96. Structural diversity in (vesicular)-arbuscular mycorrhizal symbioses. *New Phytol.* **137**: 373–388.

Smith, S. E., and Read, D. J., 1997, *Mycorrhizal symbiosis*. Second Edition. Academic, San Diego, CA.

Smith, M. R., Charvat, I., and Jacobson, R. L., 1998, Arbuscular mycorrhizae promote establishment of prairie species in a tall grass prairie restoration. *Can. J. Bot.* **76**: 1947–1954.

Sylvia, D. M., 1989, Nursery inoculation of sea oats with vesicular-arbuscular mycorrhizal fungi and out-planting performance on Florida beaches. *J. Coastal Res.* **5**: 747–754.

Sylvia, D. M., Jarstfer, A. G., and Vostátka, M., 1993, Comparisons of vesicular-arbuscular mycorrhizal species and inocula formulations in a commercial nursery and on diverse Florida beaches. *Biol. Fert. Soils* **16**: 139–144.

Sylvia, D. M., Alagely, A. K., Kane, M. E., and Philman, N. L., 2003, Compatible host – mycorrhizal fungus combinations for micropropagated sea oats. *Mycorrhiza* **13**: 177–183.

Szulczewski, M. D., Li, Y., Zhou, M., and Savabi, M. R., 2008, Phosphorus fractions in calcareous from soils the southern Everglades and nearby farmlands. *Soil Sci. Soci. J.* (in press).

U.S. Fish and Wildlife Service, 1999, *South Florida multi-species recovery plan*. Southeast Region, U.S. Fish and Wildlife Service. Atlanta, GA.

Vaast, P., Zasoski, R. J., and Bledsoe, C. S., 1996, Effects of vesicular-arbuscular mycorrhizal inoculation at different soil P availabilities on growth and nutrient uptake of in vitro propagated coffee (*Coffea arabica* L.) plants. *Mycorrhiza* **6**: 493–497.

Vidal, M. T., Azcon-Aguilar, C., Barea, J. M., and Pliegoalfaro, F., 1992, Mycorrhizal inoculation enhances growth and development of micropropagated plants of avocado. *Hort. Sci.* **1**: 25–30.

Wubet, T., Kottke, I., Teketay, D., and Oberwinkler, F., 2003, Mycorrhizal status of indigenous trees in dry Afromontane forests of Ethiopia. *For. Ecol. Manag.* **179**: 387–399.

Wunderlin, R. P., and Hansen, B. F., 2000, *Flora of Florida*. Vol. 1. University Press of Florida, Gainesville, FL.

Chapter 9

EFFECTS OF INTERACTIONS OF ARBUSCULAR MYCORRHIZAL FUNGI AND BENEFICIAL SAPROPHYTIC MYCOFLORA ON PLANT GROWTH AND DISEASE PROTECTION

M.G.B. SALDAJENO[1], W.A. CHANDANIE[1], M. KUBOTA[2], AND M. HYAKUMACHI[2]

[1]*United Graduate School of Agricultural Science, Gifu University, Gifu, Japan;* [2]*Laboratory of Plant Pathology, Faculty of Applied Biological Sciences, Gifu University, Gifu, Japan*

Abstract: Arbuscular mycorrhizal (AM) fungi and beneficial saprophytic mycoflora like plant growth promoting fungi (PGPF) are capable of promoting plant growth and may suppress several plant diseases. The interaction of these microorganisms in the plant rhizosphere may affect plant growth and microbial community composition. Mixtures of these microorganisms generally increase the genetic diversity in the rhizosphere microorganisms that may persist longer and utilize a wider array of mechanisms to increase plant growth. In particular, combinations of AM fungi and PGPF may provide protection at different times, under different conditions, and occupy different or complementary niches. In this chapter, the consequences of co-inoculation of the AM fungi and beneficial saprophytic mycoflora in terms of plant growth promotion, root colonization and disease suppression are discussed and its implication to sustainable agriculture is considered.

Keywords: Disease control; interaction; mycorrhizae; plant growth; saprophytic fungi.

1 INTRODUCTION

Most higher plants are known to form one of the most intricate fungal-root associations with a special group of microorganisms known as arbuscular mycorrhizal (AM) fungi. AM fungi are included in the phylum Zygomycota, order Glomales (Redecker *et al.*, 2000) but now they are placed in the phylum 'Glomeromycota' (Schussler et al., 2001). The association of AM fungi with plants is intimately beneficial for both partners. Significant alterations in the root physiology occur (Linderman, 1992) when plants become mycorrhizal and this association also alters root exudation. An altered root exudation may effect the composition of rhizosphere microorganisms (Linderman, 1988, 1992).

AM fungi interact with almost all organisms in the mycorrhizosphere including beneficial, plant pathogenic, saprophytic and even predatory microfauna (Bagyaraj, 1984). Much efforts have been geared to elucidate microbial interactions between AM fungi and other soil microorganisms of the rhizosphere but researches are mainly focused on the relationship of AM fungi with plant growth promoting rhizobacteria (Andrade et al., 1997; Siddiqui, 2006) and soil-borne pathogens (Jalali and Jalali, 1991; Siddiqui and Mahmood, 1995). Few studies were done on its interaction with beneficial saprophytic fungi (Calvet et al., 1992, 1993; Green et al., 1999). Saprophytic fungi live on the rhizoplane (McAllister et al., 1996; Garcia-Romera et al., 1998) and mycorrhizosphere of plants; generally procure their nutritional requirements from organic matter and other elements in the soil (Garcia-Romera et al., 1998). Saprophytic fungi form the largest group of fungi. These fungi have enzymes that work to digest the cellulose and lignin found in the organic matter, with the lignin being an important source of carbon for many organisms. Out of saprophytes, plant growth promoting fungi (PGPF) are non-pathogenic soil inhabitants (Hyakumachi, 1994; Chandanie et al., 2006b) and known to promote growth of several plants including cucumber (Meera et al., 1994, 1995; Shivanna et al., 2005), wheat (Shivanna et al., 1994, 1996), and soybean (Shivanna et al., 1996). These fungi can control several plant diseases like cucumber anthracnose (Chandanie et al., 2005a, 2006a, b), Fusarium crown and root rot in tomatoes (Horinouchi et al., 2007), take-all disease of wheat, Pythium and Rhizoctonia damping off, Pythium foliar blight, Sclerotium blight, Fusarium wilt, and brown patch diseases (Hyakumachi, 1994; Hyakumachi and Kubota, 2004a) and may be used in the management of plant diseases. Interaction of these two groups of microorganisms may be beneficial for both plant growth and plant disease control.

2 INTERACTION BETWEEN AM FUNGI AND BENEFICIAL SAPROPHYTIC MYCOFLORA

A number of studies on the interaction of AM fungi with wide variety of soil microorganisms (Bagyaraj, 1984; Linderman and Paulitz, 1990; Linderman, 1992; Gryndler, 2000) exist under various conditions. Effects of these interactions may be exploited for the benefit of sustainable agriculture. Since both are beneficial microorganisms, their synergistic or additive effect could be more beneficial for increasing growth and yield and also for the control of various plant diseases. Effects of interaction between AM fungi and beneficial saprophytic fungi is reviewed and presented in Table 1. Since the soil contains extremely rich pool of microbial entities with highly diversified and complex relationships, this characteristic of soil may sometimes contribute difficulty to reproduce similar results (Gryndler, 2000). Therefore, interactive effects of these microorganisms should be studied under different soil types and in various environmental conditions before their use for plant growth promotion and disease control.

2.1 *In vitro* interactions

The interactions of AM and saprophytic fungi *in vitro* experiments generally have stimulatory, inhibitory or no effect on the germination of spores, conidia and growth of hyphae of one or both species. For instance, Calvet *et al.* (1992) in the interaction study of *G. mosseae* and some saprophytic fungi isolated from organic substrates found that germination rate of *G. mosseae* was hastened and development of mycelia from germinated spores were enhanced in the presence of *Trichoderma* spp. Conversely, the presence of *Penicillium decumbens* and *Aspergillus fumigatus* inhibited spore germination of *G. mosseae*. They suspected that these two fungi might have produced antibiotic-like substances affecting spore germination of *G. mosseae* under axenic condition. Likewise, McAllister *et al.* (1996) observed that soluble and volatile substances produced by *Alternaria alternata* and *Fusarium equiseti* inhibit the spore germination of *G. mosseae*. On the other hand, Calvet *et al.* (1989) found that germination of *G. mosseae* resting spores on water agar was not affected by the presence of *Trichoderma* spp. and the inoculation of any of the isolates strongly enhanced the production of vegetative spores. Filion *et al.* (1999) noted that conidial germination of *Trichoderma harzianum* was stimulated in the presence of the AM fungal extract while germination of *F. oxysporum f. sp. chrysanthemi* conidia was reduced. The measured effects were directly correlated with extract concentration.

Martinez *et al.* (2004) *in vitro* experiments paired several strains of *Trichoderma pseudokoningii* with spores of *G. mosseae* and *Gigantea rosea*. Some strains of *T. pseudokoningii* resulted in the inhibition of spore germination of both AM fungi while others had no effect on germination. The soluble exudates and volatile substances produced by the saprophytes may have resulted in the inhibition of spore germination of AM fungi. The effect of genus *Trichoderma* on AM spore germination may differ with the species used (Rousseau *et al.*, 1996; Siddiqui and Mahmood, 1996; Green *et al.*,

1999). The interaction of AM fungi with other beneficial microorganisms apart from saprophytic fungi tested *in vitro* had almost similar effects. For example, Fracchia *et al.* (2003) demonstrated that the presence of the yeast *Rhodotorula mucilaginosa* increased the hyphal length of *G. mosseae* and *G. rosea* spores. Saprophytic fungi mainly influence AM fungi when the latter are in the presymbiotic phase of the symbiosis development (McAllister *et al.*, 1994; Garcia-Romera *et al.*, 1998) and affected the spore germination of AM fungi both positively and negatively (Martinez *et al.*, 2004).

Table 1. Interaction effects of arbuscular mycorrhizal (AM) fungi with beneficial saprophytic mycoflora.

AM fungus	PGPF	Interaction effects	Reference
Gigaspora rosea	*Rhodotorula mucilaginosa*	Increased AMF root colonization; SDW and RDW increased when PGPF inoculated earlier than AMF on red clover	Fracchia *et al.*, 2003
Gigaspora rosea	*Trichoderma pseudokoningii* (several strains)	Root colonization by AMF decreased; no effect on the CFU of PGPF; root and shoot dry weights of soybean were decreased	Martinez *et al.*, 2004
Glomus coronatum	*F. oxysporum*	Enhanced AMF colonization; no effect on the growth of tomato; no effect on nematode control	Diedhiou *et al.*, 2003
G. deserticola	*F. oxysporum*	Increased AMF colonization; no effect on CFU of PGPF; increased SDW and RDW of pea	Fracchia *et al.*, 2000
G. deserticola	*T. harzianum*	No effect on AMF or PGPF colonization; no effect on the growth of maize	Mar Vázquez *et al.*, 2000
G. etunicatum	*Gliocladium virens*	Effects of PGPF on AMF colonization were variable on cucumber depending on medium used	Paulitz and Linderman, 1991
G. intraradices	*A. niger*	No effect on AMF colonization; increased shoot biomass of lettuce	Kohler *et al.*, 2007
G. intraradices	*Clonostachys rosea*	Reduced AMF root colonization; reduced CFU of PGPF; increased growth of tomato	Ravnskov *et al.*, 2006
G. intraradices	*T. harzianum*	Adverse effect on SDW; (+) effect on RDW; reduced root colonization by AM fungus	Green *et al.*, 1999
G. intraradices	*T. harzianum*	Stimulated conidial germination of PGPF in axenic culture	Filion *et al.*, 1999
G. intraradices	*T. harzianum*	Significant decrease in severity and incidence of disease on tomato	Datnoff *et al.*, 1995
G. mosseae	*T. harzianum*	Negative or no effect on root colonization of AMF; increased SDW and height of pigeon pea; reduced nematode population and wilting index	Siddiqui and Mahmood, 1996
G. mosseae	*V. chlamydosporium*		

G. mosseae	T. harzianum	Enhanced AMF growth in dual cultures; increased PGPF population in rhizosphere soil; increased growth of Geranium; reduced root-rot	Haggag and Abd-El latif, 2001
G. mosseae	P. oxalicum		
G. mosseae	T. harzianum	No effect on AMF or PGPF colonization; no effect on maize growth	Mar Vázquez et al., 2000
G. mosseae	G. virens	Reduced AMF colonization on cucumber when peat moss-Czapek inoculum was used	Paulitz and Linderman, 1991
G. mosseae	T. koningii	Variable effects on AMF colonization; PGPF population in soil and plant dependent on time of inoculation on lettuce and maize	McAllister et al., 1994
G. mosseae	F. solani		
G. mosseae	T. pseudokoningii (several strains)	Variable effect on soybean growth, AMF spore germination and root colonization ; no effect on CFU of PGPF	Martinez et al., 2004
G. mosseae	T. aureoviride	Increased AMF root colonization on marigold ; enhanced dry weight and foliar area	Calvet et al., 1993
G. mosseae	T. aureoviride	T. aureoviride and T. harzianum enhanced germination rate of AMF spores on axenic culture while no effect on AMF or PGPF colonization and plant growth; A. fumigatus and P. decumbens inhibited spore germination of AMF	Calvet et al., 1992
G. mosseae	T. harzianum		
G. mosseae	A. fumigatus		
G. mosseae	P. decumbens		
G. mosseae	T. harzianum	No effect on germination of AMF resting spores; enhanced production of AMF vegetative spores in axenic culture	Calvet et al., 1989
G. mosseae	T. aureoviride		
G. mosseae	F. concolor	No effect on AMF colonization and soybean growth except when F. oxysporum, F. solani were co–inoculated with G. mosseae increased AMF root colonization and SDW	Garcia-Romero et al., 1998
G. mosseae	F. equiseti		
G. mosseae	F. graminearum		
G. mosseae	F. lateritium		
G. mosseae	F. moniliforme		
G. mosseae	F. oxysporum		
G. mosseae	F. solani		
G. mosseae	F. stilboide		
G. mosseae	F. equiseti	Reduced AMF spore germination; varied effect on maize growth and CFU of PGPF depending on time of inoculation	McAllister et al., 1996
G. mosseae	A. alternata		
G. mosseae	F. oxysporum	Increased AMF colonization; no effect on CFU of PGPF; increased SDW and RDW of pea	Fracchia et al., 2000
G. mosseae	Rhodotorula mucilaginosa	Increased AMF root colonization; Increased SDW and RDW of soybean when PGPF inoculated earlier than AMF	Fracchia et al., 2003

(continued)

G. mosseae	Phoma sp	No effect on AMF colonization; reduced isolation frequency of PGPF in roots; increased SDW of cucumber; reduced level of disease protection compared to AMF alone	Chandanie et al., 2005a, b, 2006a
G. mosseae	P. simplicissimum	No effect on AMF or PGPF colonization and growth of cucumber; induced resistance against anthracnose and damping-off	Chandanie et al., 2006a, b

SDW = shoot dry weight; RDW = root dry weight; AMF = arbuscular mycorrhizal fungi; CFU = colony forming unit; PGPF = plant growth promoting fungi

2.2 Effect of co-inoculation on AM colonization

The effects of saprophytic fungi on AM colonization differ widely as in case of *in vitro* experiments. Inoculation of *G. mosseae* or *G. deserticola* with *F. oxysporum* resulted in an increased colonization of the roots by AM fungi (Fracchia *et al.*, 2000). Similarly, Haggag and Abd-El latif (2001) observed an increased root colonization of *G. mosseae* in geranium when inoculated with *T. harzianum* and *P. oxalicum*. Diedhiou *et al.* (2003) reported that combined application of *G. coronatum* and the non-pathogenic *Fusarium oxysporum* enhanced mycorrhization of tomato roots while dual inoculation with *G. mosseae* and strains of *Fusarium* sp. led to increase in AM colonization of soybean (Garcia-Romera *et al.*, 1998). However, Mar Vázquez *et al.* (2000) found that none of the microorganisms used showed negative effects on AM establishment but mycorrhizal colonization induced qualitative changes in the bacterial population depending on the combination of inoculants involved.

Chandanie *et al.* (2005b, 2006b) demonstrated that the percent root length colonized by *G. mosseae* was not adversely affected by the existence of *Phoma* or *Penicillium* while presence of *Trichoderma* enhanced mycorrhizal colonization. Siddiqui and Mahmood (1996) reported that *T. harzianum* has an adverse effect on root colonization by *G. mosseae* while Ravnskov *et al.* (2006) found that *Clonostachys rosea* and *G. intraradices* were mutually inhibitory, although their combination promoted plant growth. In addition, *C. rosea* reduced root colonization caused by *G. intraradices*. Similarly, *G. intraradices* adversely affected *C. rosea* population in the soil.

2.3 Effect of co-inoculation on saprophytic fungi

The effect AM fungi on root colonization of the saprophytic co-inoculants was measured by counting the population of the co-inoculants in the rhizosphere soil. Under natural condition, Haggag and Abd-El latif (2001) noted that in the presence of *G. mosseae*, population and survival of *T. harzianum* and *P. oxalicum* increased even after 90 days after inoculation. Conversely, Garcia-Romera *et al.* (1998) demonstrated that the dual inoculation with *G. mosseae* and some strains of *Fusarium* sp. did not influence the colony forming units (CFU) of *Fusarium* in the rhizosphere soil, while Ravnskov *et al.* (2006) found that inoculation of *G. intraradices* decreased CFU of *C. rosea* by 78% and 58% suggesting that *C. rosea* and *G. intraradices* were mutually inhibitory.

Chandanie *et al.* (2005b, 2006b) observed that root colonization ability of *Phoma* sp. was significantly reduced by the inoculation of *G. mosseae* in cucumber plants. Root colonization by *T. harzianum* was also slightly reduced in the presence of *G. mosseae*. On the contrary, AM fungus had no significant effect on colonization of *Penicillium simplicissimum* both in the rhizosphere and roots of cucumber. The interaction effects between individual saprophytic fungi and AM fungi differ according to the species of AM fungi or saprophytic fungi involved. For instance, AM fungi adversely affected the population development of some saprophytic fungi in the roots; it has slight or no effect on populations of other saprophytic fungi. On one hand, the presence of some saprophytic fungi in the soil exerted no influence on AM colonization in roots while others when combined with *G. mosseae* seemed to promote AM formation in the host roots.

The interaction of AM fungi and saprophytic fungi may be contradictory between species of the same genus of the saprophytic fungus and even within the strains of the same species of the AM fungus and saprophytic fungus (Martinez *et al.*, 2004). For example, McAllister *et al.* (1994) reported a synergistic interaction between *G. mosseae* and *T. aureoviridae* while Calvet *et al.* (1993) observed the antagonistic interaction between *G. mosseae* and *T. koningii*. Green *et al.* (1999) and Rousseau *et al.* (1996) observed antagonistic interaction between *G. intraradices* and *T. harzianum* while synergistic reaction was observed by Datnoff *et al.* (1995).

Mycorrhizal plants are capable of producing compounds which can interfere with rhizosphere microorganisms and modify microbial community around the mycorrhizal roots (Linderman, 1988; Linderman and Paulitz, 1990). In addition, although not well documented, the extraradical mycelium of AM fungi might also impact strongly on the microbial population around the mycorrhizal roots (Filion *et al.*, 1999). For instance, it was observed that substances released by the extraradical mycelia of *G. intraradices* either stimulated or inhibited conidial germination of non-pathogenic *T. harzianum* and root pathogen *F. oxysporum chrysanthemi*. The reduction in root colonization of *Phoma* sp. and *T. harzianum* might be related to the exudates secreted by the mycorrhizal roots and/or the extraradical mycelia of *G. mosseae* (Chandanie *et al.*, 2006b). On the other hand, the strong sporulation ability of *P. simplicissimum* under natural conditions and the inherent characteristics related to the species may explain the tolerance of the fungus to such exudates.

2.4 Effect of interaction on plant growth

The interaction effect of AM fungi and other beneficial soil microorganisms on plant growth have been demonstrated. Fracchia *et al.* (2000) found that dual inoculation of *G. mosseae* or *G. deserticola* and *F. oxysporum* led to enhanced growth of plants. Similarly, combined inoculation of *Trichoderma aureoviride* and *G. mosseae* had a synergistic effect on the growth of marigold plants (Calvet *et al.*, 1993). Haggag and Abd-El latif (2001) found that combined inoculation of *G. mosseae* and *T. harzianum* or *P. oxalicum* enhanced growth of geranium plants. A field study of Diedhiou *et al.* (2003) showed that combined application of *G. coronatum* and the non-pathogenic strain of *Fusarium oxysporum* did not increase plant growth. Garcia-Romera *et al.* (1998) noted that dual inoculation with *G. mosseae* and some strains of *Fusarium* sp. led to enhanced growth of soybean plants. The shoot dry weight of soybean plants cultivated in non-sterilized soils or soils inoculated with *G. mosseae* were not negatively affected by the presence of *Fusarium*. Ravnskov *et al.* (2006) found that *Clonostachys rosea* and *G. intraradices* were mutually inhibitory, but promoted plant growth with some alteration in soil microbial communities.

Chandanie *et al.* (2005b, 2006b) noted that plant growth was stimulated when *Trichoderma* was combined with AM fungus; no stimulation was observed when *Penicillium* was combined with AM fungus. However, use of *Phoma* with AM fungus was found inhibitory to growth but stimulated growth when applied alone. They also noted that plants treated with *G. mosseae* alone did not increase shoot dry weight or root dry weight compared to the controls. PGPF generally increase plant growth through mineralization, suppression of deleterious microorganisms, and hormone production (Hyakumachi and Kubota, 2004b). Hyakumachi (2000) has demonstrated that amendment of soil with PGPF-infested barley grains showed increased production of NH_4-N and NO_3-N. Moreover, it was pointed out that the total amount of nitrogen in PGPF infected-barley grains remains the same despite different PGPF isolates were used, but NH_4-N amount varies depending on the isolate. Hyakumachi (2000) illustrated correlations between reduction of barley grain weight and cellulase activity, degradation activity of starch, and the dry weight of bent grass. Results suggest that the plant-growth promoting effect of PGPF is related to the mineralization of organic substrates because PGPF provide necessary mineral nutrients to plant in easily assimilated form.

Plant growth promotion by PGPF was also attributed to the suppression of indigenous pathogenic *Pythium* spp. in the soil (Hyakumachi, 1994) which resulted to a notable growth promotion of field-grown cucumbers. In addition, some fungal species are capable of producing growth hormone in axenic culture (Ram, 1959). Some strains of *Phoma* sp. for instance have been found to produce abscisic acid which is reported to promote plant growth (Hyakumachi and Kubota, 2004b). The disease inhibition by AM fungi might be related to the increase in phosphorus content but increase in phosphorus may not be a sole reason of disease inhibition. In addition to changes in nutrient uptake in the root system, a mycorrhizosphere effect and activation of plant defense mechanisms are thought to be responsible for disease inhibition by AM fungi (Linderman, 1994; Demir and Akkopru, 2005). Moreover, the use of *Glomus* sp. is also reported to increase phenylalanine and serine in tomato roots (Suresh, 1980); these amino acids have an inhibitory effect on plant pathogens (Reddy, 1974).

2.5 Effect of interaction on air-borne diseases

Studies on the effects of AM fungi and beneficial soil microorganisms on disease suppression is scanty, as most studies were done on the direct interaction of AM fungi with the pathogen itself (Krishna and Bagyaraj, 1983; Caron *et al.*, 1986; Kaye *et al.*, 1984; Trotta *et al.*, 1996; Garcia-Garrido and Ocampo, 1988; Bødker *et al.*, 2002; Karagiannidis *et al.*, 2002; Rosendahl and Rosendahl, 1990). Chandanie *et al.* (2005b, 2006b) observed that inoculation of PGPF (*Phoma* sp., *P. simplicissimum*, or *T. harzianum*) into the root system of cucumber provided considerable protection against the anthracnose pathogen *Colletotrichum orbiculare*. Although the treatment of *G. mosseae* had no significant effect on the disease development, combined inoculations of *G. mosseae* with *Phoma* sp. reduced the level of disease. However, the level of protection induced by *P. simplicissimum* or *T. harzianum* was not altered by combining it with *G. mosseae*.

The resistance against the *C. orbiculare* was achieved when there was no direct contact between the pathogen and the inducer within the plant. The induction of systemic resistance is implicated as the mechanism of disease suppression (Chandanie *et al.*, 2006b).This is in agreement with the results of Meera *et al.* (1994) and Hyakumachi and Kubota (2004a). PGPF-mediated ISR has been demonstrated by increased lignin deposition at the point of penetration by the pathogen *C. orbiculare* in the epidermal tissues of cucumber hypocotyls (Hyakumachi and Kubota, 2004a) and also conspicuous superoxide generation by culture filtrates of respective PGPF isolates (Koike *et al.*, 2001). Biochemical analysis have revealed systemic accumulation of salicylic acid and increased activities of chitinase, ß-1,3-glucanases and peroxidase in cucumber plants induced by PGPF (Hyakumachi and Kubota, 2004a; Yedidia et al., 1999). Hossain et al. (2007) hypothesized that multiple defense mechanisms are involved in P. simplicissimum-mediated ISR in Arabidopsis plants. Additional studies on the expression of pathogenesis-related genes and the signaling pathways involved in PGPF-mediated ISR are required.

Plants treated with G. mosseae showed a tendency to intensify development of leaf disease symptoms and this has been correlated to improved nutrition and higher physiological activities compared to non-AM plants (Dehne, 1982). However, Chandanie et al. (2006b) did not observed any significant intensification or reduction of disease development in plants inoculated with AM fungus compared to uninoculated cucumber plants. The result suggests that G. mosseae in cucumber roots could not induce positive or negative systemic effect against C. orbiculare. This finding corroborates the results observed that AM symbiosis formed by G. intraradices in cucumber plants had no systemic influence on development of powdery mildew colonies in the shoot portion caused by the fungus Podosphaera xanthii (Larsen and Yohalem, 2004).

2.6 Effect of interaction on soil-borne diseases

Using the commercial formulations of G. intraradices and T. harzianum on tomatoes, Datnoff et al. (1995) demonstrated that G. intraradices with T. harzianum reduce both the incidence and severity of Fusarium crown and root rot under field conditions. Similarly, Haggag and Abd-El latif (2001) noted that the application of G. mosseae with T. harzianum or P. oxalicum reduced Geranium root rot caused by Fusarium solani and Macrophomina phaseolina both in artificially and naturally-infested soils.

Chandanie et al. (2005b, 2006b) tested P. simplicissimum and T. harzianum against Rhizoctonia damping-off. They pre-treated cucumber seedlings for 7 and 12 days with AM fungus and Trichoderma or Penicillium and transplanted plants into soil infested with pathogen R. solani. Prior inoculation of P. simplicissimum, T. harzianum or G. mosseae to the rhizosphere and/or roots of cucumber seedlings protected plants from the damping off disease caused by R. solani. Combined inoculations of a PGPF isolate with the AM fungi were highly effective for the control of R. solani compared to single inoculation of each PGPF species. The degree of protection provided by seven days pre-treatment of seedlings with G. mosseae increased when the duration of treatment was increased to 12 days, but such an increase was not found with P. simplicissimum or its combined treatment with G. mosseae. The levels of protection were dependent on the pathogen inoculum potential. Treatments were less effective at high population density of the pathogen. The degree of protection achieved by G. mosseae was highly dependent on the duration of pre-inoculation time before pathogen introduction. The enhanced disease protection observed with the longer pre-inoculation time may be due to pre-establishment of the AM fungi in or on roots. Moreover, unlike the PGPF, *G. mosseae* was ineffective when supplied simultaneously to roots with the pathogen. Since AM fungi are biotrophic, their mode of establishment and competition with root pathogens seems to differ from saprophytic PGPF (Chandanie *et al.*, 2006b). Only a well established AM fungi symbiosis could reduce damage caused by root pathogens (Cordier *et al.*, 1996; Kloepper *et al.*, 2004;) but simultaneous addition of AM fungi with pathogen could reduce severity of some root diseases as well (Rosendahl and Rosendahl, 1990). Reason for discrepancies in results includes differences in AM fungal inoculum potential. If the inoculum contains lots of healthy viable non-dormant spores which germinate well and establish faster, it may be effective in the control of already existing pathogen in the soil. Moreover, the pathogen inoculum potential and its virulence may also have an impact on the results. The protective capability of pre-inoculated *G. mosseae* could be a result of combined mechanisms including competetion, altered root exudation, anatomical and morphological changes in the root system and activation of plant defense mechanism (Azcón-Aguilar and Barea, 1996; Pozo *et al.*, 1998, 1999, 2002).

The disease suppression capability of AM fungi and the PGPF have been discussed and reviewed in detail (Dehne, 1982; Bagyaraj, 1984; Paulitz and Linderman, 1991; Hooker *et al.*, 1994; Hyakumachi, 1994; Linderman, 1994; Azcón-Aguilar and Barea, 1996; Xavier and Boyetchko, 2002; Hyakumachi and Kubota, 2004b) and has been thought to be due to several mechanisms, although much of these mechanisms are still poorly understood. It is generally accepted that antagonistic capability includes the actions of hyperparasitism, antibiosis and competition although PGPF isolated from zoysia grass (Chandanie *et al.*, 2005b, 2006b) did not show hyperparasitism and antibiosis against other fungi (Hyakumachi and Kubota, 2004b). Another important mechanism is the induction of systemic resistance (Hyakumachi and Kubota, 2004b).

3 CONCLUSIONS AND RELEVANCE TO SUSTAINABLE AGRICULTURE

The effects of saprophytic fungi and AM fungi may vary depending on the inherent characteristic of saprophytic and AM fungi. Results may be contradictory within species of the same genus and even within strains of the same species. Biocontrol of plant pathogens is considered as a major practice in sustainable agriculture and is regarded as a directed and accurate management of common ecosystems components to protect plants against pathogens (Azcón-Aguilar and Barea, 1996). Since interactions of biocontrol agents with beneficial organisms in the rhizosphere might stimulate biocontrol in an agro ecosystem (Calvet et al., 1993; Paulitz and Linderman, 1991; Brimmer and Boland, 2003), it is important to study interactions of biocontrol agents with non-target beneficial organisms in the soil to ensure the successful development, commercialization and usage of biocontrol strategies. There is very limited knowledge about interactions between AM fungi and biocontrol agents (Paulitz and Linderman, 1991).

AM fungi are key components of sustainable plant-soil ecosystems (Jeffries and Barea, 2001). Most plants of agricultural importance are vastly benefited from AM associations to overcome biotic and abiotic stresses (Linderman, 1994; Nelson, 1987; Tisdall, 1994), and AM fungi are very significant among groups of beneficial microorganisms. Additionally, these fungi are ubiquitous components of both natural and agricultural ecosystems (Smith and Read, 1997) and their associations are effective in reducing root diseases caused by various soil-borne pathogens (Dehne, 1982; Hooker et al., 1994; Azcón-Aguilar and Barea, 1996). Since AM symbiosis is known to alter microbial population composition in the rhizosphere (Linderman, 1988; Linderman and Paulitz, 1990), testing the interaction of AM fungi and saprophytic fungi like the PGPF is useful to understand the possible additive or synergistic effects. A thorough understanding of the AM fungi-beneficial saprophytic fungi interactions is indispensable for their successful utilization for biocontrol and for increasing growth and yields of crops without inorganic fertilizers and pesticides.

REFERENCES

Andrade, G., Mihara, K.L., Linderman, R.G., and Bethlnfalvay, G.J., 1997, Bacteria from rhizosphere and hyphosphere soils of different arbuscular-mycorrhizal fungi. *Plant Soil* **192**: 71–79.

Azcón-Aguilar, C., and Barea, J.M., 1996, Arbuscular mycorrhizas and biological control of soil-borne plant pathogens - an overview of the mechanisms involved. *Mycorrhiza* **6**: 457–464.

Bagyaraj, D.J., 1984, Biological interactions with VA mycorrhizal fungi. In: Powell CL, Bagyaraj DJ (eds), *VA Mycorrhiza*. CRC, Boca Raton, FL, pp. 132–153.

Bødker, L., Kjøller, R., Kristensen, K., and Rosendahl, S., 2002, Interactions between indigenous arbuscular mycorrhizal fungi and *Aphanomyces euteiches* in field-grown pea. *Mycorrhiza* **12**: 7–12.

Brimmer, T.A., and Boland, G.J., 2003, A review of the non-target effects of fungi used to biologically control plant diseases. *Agri. Ecos. Environ.* **100**: 3–16.

Calvet, C., Pera, J., and Barea, J.M., 1989, Interactions of *Trichoderma* spp. with *Glomus mosseae* and two wilt pathogenic fungi. *Agri. Ecos. Environ.* **29**: 59–65.

Calvet, C., Barea, J.M., and Pera, J., 1992, *In vitro* interactions between the vesicular-arbuscular mycorrhizal fungus *Glomus mosseae* and some saprophytic fungi isolated from organic substrates. *Soil Biol. Biochem.* **24**: 775–780.

Calvet, C., Pera, J., and Barea, J.M., 1993, Growth response of marigold (*Tagetes erecta* L.) to inoculation with *Glomus mosseae, Trichoderma aureoviride* and *Pythium ultimum* in a peat-perlite mixture. *Plant Soil* **148**: 1–6.
Caron, M., Fortin, J.A., and Richard, C., 1986, Effect of inoculation sequence on the interaction between *Glomus intraradices* and *Fusarium oxysporum* f. sp. *radicis-lycopersici* in tomatoes. *Can. J. Plant Pathol.* **8**: 12–16.
Chandanie, W.A., Ito, M., Kubota, M., and Hyakumachi, M., 2005a, Interaction between arbuscular mycorrhizal fungus *Glomus mosseae* and plant growth promoting fungus *Phoma* sp. on their root colonization and disease suppression of cucumber. *Ann. Rep. Interdis. Res. Inst. Environ. Sci.* **24**: 91–102.
Chandanie, W.A., Kubota, M., and Hyakumachi, M., 2005b, Interaction between arbuscular mycorrhizal fungus *Glomus mosseae* and plant growth promoting fungus *Phoma* sp. on their root colonization and growth promotion of cucumber. *Mycosci.* **46**: 201–204.
Chandanie, W.A., Kubota, M., and Hyakumachi, M., 2006a, Effect of combined inoculation of plant growth promoting fungus *Penicillium simplicissimum* and arbuscular mycorrhizal fungus *Glomus mosseae* on their colonization, plant growth promotion and disease suppression of cucumber. *Ann. Rep. Interdis. Res. Inst. Environ. Sci.* **25**: 109–119.
Chandanie, W.A., Kubota, M., and Hyakumachi, M., 2006b, Interactions between plant growth promoting fungi and arbuscular mycorrhizal fungus *Glomus mosseae* and induction of systemic resistance to anthracnose disease in cucumber. *Plant Soil* **286**: 209–217.
Cordier, C., Gianinazzi, S., and Gianinazzi-Pearson, V., 1996, Colonisation patterns of root tissues by *Phytophthora nicotianae* var. *parasitica* related to reduced disease in mycorrhizal tomato. *Plant Soil* **185**: 223–232.
Datnoff, L.E., Nemec, S., and Pernezny, K., 1995, Biological control of Fusarium crown and root rot of tomato in Florida using *Trichoderma harzianum* and *Glomus intraradices*. *Biological Cont.* **5**: 427–431.
Dehne, H.W., 1982, Interaction between vesicular-arbuscular mycorrhizal fungi and plant pathogens. *Phytopathology* **72**: 1115–1119.
Demir, S., and Akkopru, A., 2005, Using of arbuscular mycorrhizal fungi (AMF) for biocontrol of soil-borne fungal pathogens. In: Chincholkar SB, Mukerji KG (eds), *Biological Control of Plant Diseases: Current concepts*. Howarth, New York.
Diedhiou, P.M., Hallmann, J., Oerke, E.C., and Dehne, H.W., 2003, Effects of arbuscular mycorrhizal fungi and a non-pathogenic *Fusarium oxysporum* on *Meloidogyne incognita* infestation of tomato. *Mycorrhiza* **13**: 199–204.
Filion, M., St-Arnaud, M., and Fortin, J.A., 1999, Direct interaction between the arbuscular mycorrhizal fungus *Glomus intraradices* and different rhizosphere microorganisms. *New Phytol.* **141**: 525–533.
Fracchia, S., García-Romera, I., Godeas, A., and Ocampo, J.A., 2000, Effect of the saprophytic fungus *Fusarium oxysporum* on arbuscular mycorrhizal colonization and growth of plants in greenhouse and field trials. *Plant Soil* **223**: 175–184.
Fracchia, S., Godeas, A., Scervino, J.M., Sampedro, I., Ocampo, J.A., and García-Romera, I., 2003, Interaction between the soil yeast *Rhodotorula mucilaginosa* and the arbuscular mycorrhizal fungi *Glomus mosseae* and *Gigaspora rosea*. *Soil Biol. Biochem.* **35**: 701–707.
Garcia-Garrido, J.M., and Ocampo, J.A., 1988, Interaction between *Glomus mosseae* and *Erwinia carotovora* and its effects on the growth of tomato plants. *New Phytol.* **110**: 551–555.
García-Romera, I., Garcia-Garrido, J.M., Martin, J., Fracchia, S., Mujica, M.T., Godeas, A., and Ocampo, J.A., 1998, Interaction between saprophytic *Fusarium* strains and arbuscular mycorrhizas of soybean plants. *Symbiosis* **24**: 235–246.
Green, H., Larsen, J., Olsson, P.A., Jensen, D.F., and Jakobsen, I., 1999, Suppression of the biocontrol agent *Trichoderma harzianum* by mycelium of the arbuscular mycorrhizal fungus *Glomus intraradices* in root-free soil. *Appl. Environ. Microbiol.* **65**: 1428–1434.

Gryndler, M., 2000, Interactions of arbuscular mycorrhizal fungi with other soil organisms. In: Kapulnik Y, Douds Jr DD (eds), *Arbuscular Mycorrhizas: Physiology and Function*. Kluwer Academic, The Netherlands, pp. 239–262.

Haggag, W.M., and Abd-El latif, F.M., 2001, Interaction between vesicular arbuscular mycorrhizae and antagonistic biocontrol microorganisms on controlling root rot disease incidence of geranium plants. OnLine *J. Biol. Sci.* **1**: 1147–1153.

Hooker, J.E., Jaizme-Vega, M., and Atkinson, D., 1994, Biocontrol of plant pathogens using arbuscular mycorrhizal fungi. In: Gianinazzi S, Schuepp H (eds), *Impact of Arbuscular Mycorrhizas on Sustainable Agriculture and Natural Ecosystems*. Birkhauser, Basel, Switzerland, pp. 191–200.

Horinouchi, H., Muslim, A., Suzuki, T., and Hyakumachi, M., 2007, *Fusarium equiseti* GF191 as an effective biocontrol agent against Fusarium crown and root rot of tomato in rock wool systems. *Crop Prot.* **26**: 1514–1523.

Hossain, M.M., Sultana, F., Kubota, M., Koyama, H., and Hyakumachi, M., 2007, The plant growth-promoting fungus *Penicillium simplicissimum* GP17-2 induces resistance in *Arabidopsis thaliana* by activation of multiple defense signals. *Plant Cell Physiol.* **48**: 1724–1736.

Hyakumachi, M., 1994, Plant growth promoting fungi from turfgrass rhizosphere with potential for disease suppression. *Soil Microorgan.* **44**: 53–68.

Hyakumachi, M., 2000, Studies on biological control of soilborne plant pathogens. *J. Gen. Plant Pathol.* **66**: 272–274.

Hyakumachi, M., and Kubota, M., 2004a, Biological control of plant diseases by plant growth promoting fungi. *Proceedings of the International Seminar on Biological Control Soilborne Plant Diseases*, Japan-Argentina Joint Study, pp. 87–123.

Hyakumachi, M., and Kubota, M., 2004b, Fungi as plant growth promoter and disease suppressor. In: Arora DK (ed), *Fungal Biotechnology in Agricultural, Food, and Environmental Applications*. Marcel Dekker, New York, pp. 101–110.

Jalali, B.L., and Jalali, I., 1991, Mycorrhiza in plant disease control. In: Arora DK, Rai B, Mukerji KG, Knudsen GR (eds), *Handbook of Applied Mycology, Vol. 1, Soil and Plants*. Marcel Dekker, New York, pp. 131–154.

Jeffries, P., and Barea, J.M., 2001, Arbuscular mycorrhiza: a key component of sustainable plant-soil ecosystems. In: Hock (ed), *The Mycota: Fungal Associations*, Vol. IX, pp. 95–113.

Karagiannidis, N., Bletsos, F., and Stavropoulos, N., 2002, Effect of Verticillium wilt (*Verticillium dahliae* Kleb.) and mycorrhiza *(Glomus mosseae)* on root colonization, growth and nutrient uptake in tomato and eggplant seedlings. *Sci. Horti.* **94**: 145–156.

Kaye, J.W., Pfleger, F.L., and Stewart, E.L., 1984, Interaction of *Glomus fasciculatum* and *Pythium ultimum* on greenhouse-grown poinsettia. *Can. J. Bot.* **62**: 1575–1579.

Kloepper, J.W., Ryu, C.M., and Zhang, S., 2004, Induced systemic resistance and promotion of plant growth by *Bacillus* spp. *Phytopathology* **94**: 1259–1266.

Kohler, J., Caravaca, F., Carrasco, L., and Roldan, A., 2007, Interactions between a plant growth-promoting rhizobacterium, an AM fungus and a phosphate-solubilising fungus in the rhizosphere of *Lactuca sativa*. *Appl. Soil. Ecol.* **35**: 480–487.

Koike, N., Hyakumachi, M., Kageyama, K., Tsuyumu, S., and Doke, N., 2001, Induction of systemic resistance in cucumber against several diseases by plant growth promoting fungi: lignification and superoxide generation. *Eur. J. Plant Pathol.* **107**: 523–533.

Krishna, K.R., and Bagyaraj, D.J., 1983, Interaction between *Glomus fasciculatum* and *Sclerotium rolfsii* in peanut. *Can. J. Bot.* **61**: 2349–2351.

Larsen, J., and Yohalem, D., 2004, Interactions between mycorrhiza and powdery mildew of cucumber. *Mycol. Progr.* **3**: 123–128.

Linderman, R.G., 1988, Mycorrhizal interactions with the rhizosphere microflora: the mycorrhizosphere effect. *Phytopathology* **78**: 366–371.

Linderman, R.G., 1992, Vesicular-arbuscular mycorrhizae and soil microbial interactions. In: Bethlenfalvay GJ, Linderman RG (eds), *Mycorrhizae in Sustainable Agriculture*. ASA Special Publication No. 54, Madison, WI, pp. 45–70.

Linderman, R.G., 1994, Role of VAM fungi in biocontrol. In: Pfleger FL, Linderman RG (eds), *Mycorrhizae and Plant Health*. APS, St Paul, MN, pp. 1–26.

Linderman, R.G., and Paulitz, T.C., 1990, Mycorrhizal-rhizobacterial interactions. In: Hornby D, Cook RJ, Henis Y, Ko WH, Rovira AD, Schippers B, Scott PR (eds), *Biological Control of Soil-Borne Plant Pathogens*. CAB International, Wallingford, UK, pp. 261–283.

Martinez, A., Obertello, M., Pardo, A., Ocampo, J.A., and Godeas, A., 2004, Interactions between *Trichoderma pseudokoningii* strains and the arbuscular mycorrhizal fungi *Glomus mosseae* and *Gigaspora rosea*. *Mycorrhiza* **14**: 79–84.

Mar Vázquez, M., César, S., Azcón, R., and Barea, J.M., 2000, Interactions between arbuscular mycorrhizal fungi and other microbial inoculants (*Azospirillum*, *Pseudomonas*, *Trichoderma*) and their effects on microbial population and enzyme activities in the rhizosphere of maize plants. *Appl Soil Ecol*. **15**: 261–272.

McAllister, C.B., García-Romera, I., Godeas, A., and Ocampo, J.A., 1994, Interactions between *Trichoderma koningii*, *Fusarium solani* and *Glomus mosseae*: effects on plant growth, arbuscular mycorrhizas and the saprophyte inoculants. *Soil Biol. Biochem*. **26**: 1363–1367.

McAllister, C.B., Garcia-Garrido, J.M., García-Romera, I., Godeas, A., and Ocampo, J.A., 1996, In vitro interactions between *Alternaria alternata*, *Fusarium equiseti* and *Glomus mosseae*. *Symbiosis* **20**: 163–174.

Meera, M.S., Shivanna, M.B., Kageyama, K., and Hyakumachi, M., 1994, Plant growth promoting fungi from zoysiagrass rhizosphere as potential inducers of systemic resistance in cucumbers. *Phytopathology* **84**: 1399–1406.

Meera, M.S., Shivanna, M.B., Kageyama, K., and Hyakumachi, M., 1995, Responses of cucumber cultivars to induction of systemic resistance against anthracnose by plant growth promoting fungi. *Eur. J. Plant Pathol*. **101**: 421–430.

Nelson, C.E., 1987, The water relations of vesicular-arbuscular mycorrhizal systems. In: Safir G (ed), *Ecophysiology of VA Mycorrhizal Plants*. CRC, Boca Raton, FL, pp. 71–91.

Paulitz, T.C., and Linderman, R.G., 1991, Lack of antagonism between the biocontrol agent *Gliocladium virens* and vesicular arbuscular mycorrhizal fungi. *New Phytol*. **117**: 303–308.

Pozo, M.J., Azcón-Aguilar, C., Dumas-Gaudot, E., and Barea, J.M., 1998, Chitosanase and chitinase activities in tomato roots during interactions with arbuscular mycorrhizal fungi or *Phytophthora parasitica*. *J. Expt. Bot*. **49**: 1729–1739.

Pozo, M.J., Azcón-Aguilar, C., Dumas-Gaudot, E., and Barea, J.M., 1999, β-1,3-glucanase activities in tomato roots inoculated with arbuscular mycorrhizal fungi and/or *Phytophthora parasitica* and their possible involvement in bioprotection. *Plant Sci*. **141**: 149–157.

Pozo, M.J., Cordier, C., Dumas-Gaudot, E., Gianinazzi, S., Barea, J.M., and Azcón-Aguilar, C., 2002, Localized versus systemic effect of arbuscular mycorrhizal fungi on defence responses to *Phytophthora* infection in tomato plants. *J. Exp. Bot*. **53**: 525–534.

Ram, C.S.V., 1959, Production of growth-promoting substances by *Fusarium* and their action on root elongation in *Oryzae sativa* L. *Proc. Ind. Acad. Sci*. **49**: 167–182.

Ravnskov, S., Jensen, B., Knudsen, I.M.B., Bødker, L., Jensen, D.F., Karliński, L., and Larsen, J., 2006, Soil inoculation with the biocontrol agent *Clonostachys rosea* and the mycorrhizal fungus *Glomus intraradices* results in mutual inhibition, plant growth promotion and alteration of soil microbial communities. *Soil Biol. Biochem*. **38**: 3453–3462.

Reddy, P.P., 1974, Studies on the action of amino acids on the root-knot nematode *Meloidogyne incognita*. Ph.D. dissertation, University of Agricultural Sciences, Banglore, India.

Redecker, D., Morton, J.B., and Bruns, T.D., 2000, Ancestral lineages of arbuscular mycorrhizal fungi (*Glomales*). *Mol. Phylogen. Evol*. **14**: 276–284.

Rosendahl, C.N., and Rosendahl, S., 1990, The role of vesicular-arbuscular mycorrhiza in controlling damping-off and growth reduction in cucumber caused by *Pythium ultimum*. *Symbiosis* **9**: 363–366.

Rousseau, A., Benhamou, N., Chet, I., and Piché, Y., 1996, Mycoparasitism of the extramatrical phase of *Glomus intraradices* by *Trichoderma harzianum*. *Phytopathology* **86**: 434–443.

Schussler, A., Schwarzott, D., and Walker, C., 2001, A new fungal phylum, the Glomeromycota, phylogeny and evolution. *Mycol. Res.* **105**: 1413–1421.

Shivanna, M.B., Meera, M.S., and Hyakumachi, M., 1994, Sterile fungi from zoysiagrass rhizosphere as plant growth promoters in spring wheat. *Can. J. Microbiol.* **40**: 637–644.

Shivanna, M.B., Meera, M.S., Kageyama, K., and Hyakumachi, M., 1996, Growth promotion ability of zoysiagrass rhizosphere fungi in consecutive plantings of wheat and soybean. *Mycosci.* **37**: 163–168.

Shivanna, M.B., Meera, M.S., Kubota, M., and Hyakumachi, M., 2005, Promotion of growth and yield in cucumber by zoysiagrass rhizosphere fungi. *Microbes Environ.* **20**: 34–40.

Siddiqui, Z.A., 2006, PGPR: prospective biocontrol agents of plant pathogens. In: Siddiqui ZA (ed), *PGPR: Biocontrol and Biofertilization*. Springer, Dordrecht, The Netherlands, pp. 111–142.

Siddiqui, Z.A., and Mahmood, I., 1995, Role of plant symbionts in nematode management: a review. *Bioresour. Technol.* **54**: 217–226.

Siddiqui, Z.A., and Mahmood, I., 1996, Biological control of *Heterodera cajani* and *Fusarium udum* on pigeonpea by *Glomus mosseae*, *Trichoderma harzianum*, and *Verticillium chlamydosporium*. *Isr. J. Plant Sci.* **44**: 49–56.

Smith, S.E., and Read, D.J., 1997, *Mycorrhizal Symbiosis*. 2nd edn. Academic, UK.

Suresh, C.K., 1980, Interaction between vesicular arbuscular mycorrhizal and root knot nematode in tomato. M.Sc. (Agric.) thesis, University Agricultural Sciences, Banglore, India.

Tisdall, J.M., 1994, Possible role of soil microorganisms in aggregation in soils. *Plant Soil* **159**: 115–121.

Trotta, A., Varese, G.C., Gnavi, E., Fusconi, A., Sampò, S., and Berta, G., 1996, Interactions between the soilborne root pathogen *Phytophthora nicotianae* var. *parasitica* and the arbuscular mycorrhizal fungus *Glomus mosseae* in tomato plants. *Plant Soil* **185**: 199–209.

Xavier, L.J.C., and Boyetchko, S.M., 2002, Mycorrhizae as biocontrol agents. In: Mukerji KG, Manoharachary C, Chamola BP (eds), *Techniques in Mycorrhizal Studies*. Kluwer Academic, Dordrecht, The Netherlands, pp. 493–536.

Yedidia, I., Benhamou, N., and Chet, I., 1999, Induction of defense responses in cucumber plants (*Cucumis sativus* L.) by the biocontrol agent *Trichoderma harzianum*. *Appl. Envir. Microbiol.* **65**: 1061–1070.

Chapter 10

THE MYCORRHIZOSPHERE EFFECT: A MULTITROPHIC INTERACTION COMPLEX IMPROVES MYCORRHIZAL SYMBIOSIS AND PLANT GROWTH

R. DUPONNOIS[1], A. GALIANA[2], AND Y. PRIN[2]

[1]IRD, UMR 113 CIRAD/INRA/IRD/SUPAGRO/UM2, Laboratoire des Symbioses Tropicales et Méditerranéennes (LSTM), Montpellier, France; [2]CIRAD, UMR 113 CIRAD/INRA/IRD/SUPAGRO/UM2, Laboratoire des Symbioses Tropicales et Méditerranéennes (LSTM), Montpellier, France

Abstract: Mycorrhizal fungi are essential components of sustainable soil–plant systems. Hyphae of arbuscular mycorrhizal (AM) fungi play important role in the formation and stability of soil aggregates and contribute to the composition of plant community structures. Mycorrhizal symbiosis generally increases root exudation and influences rhizosphere microbial communities. Mycorrhizal hyphae exude chemical compounds that have a selective effect on the microbial communities in the rhizosphere and in the soil. These microbial compartments are commonly named "mycorrhizosphere" and there has been increasing evidence that the mycorrhizosphere communities have an important role in plant growth and soil fertility. For instance, it has been demonstrated that mycorrhizal fungi had a selective effect on bacteria potentially beneficial to the symbiosis and to the plants. Hence, mycorrhizal symbiosis provides a microbial complex regulated by multitrophic interactions. This fungal symbiosis had an indirect effect on plant growth through its selective pressure on mycorrhizosphere communities in addition to its classical direct effect. This chapter presents highlights on multitrophic interactions and its importance in sustainable agriculture, especially in tropical and mediterranean countries.

Keywords: Bacteria; mycorrhizosphere effect; functional diversity; pseudomonads; soil fertility.

1 INTRODUCTION

Symbiotic mycorrhizal fungi such as arbuscular mycorrhizal (AM) fungi or ectomycorrhizal (EC) fungi are ubiquitous component of most ecosystems throughout the world and form a key component of soil microbiota influencing plant growth and uptake of nutrients (Bethlenfalvay and Linderman, 1992; Van der Heijden et al., 1998). The mycorrhizal symbiosis mobilizes and transports nutrients to roots (Smith and Read, 1997). It is well known that AM fungi improved nutrient uptake, especially nitrogen and phosphorus, by increasing the abilities of the host plants to explore a larger volume of soil than roots alone and to mobilize phosphate from a greater surface area (Jakobsen et al., 1992; Joner et al., 2000). It also reduces water stress (Augé, 2001) and improves soil aggregation in eroded soils (Caravaca et al., 2002). Mycorrhizal fungi affect the diversity of plant communities (van der Heijden et al., 1998; Klironomos et al., 2000; O'Connor et al., 2002) and influence relationships between plants (West, 1996; Marler et al., 1999; van der Heijden et al., 2003). Mycorrhizal plants transfer more assimilates to the roots than non-mycorrhizal ones. These fungi effect mainly results of the carbon demand of the fungal symbiont which may assimilate 10% of the carbon allocated to the roots (Fitter, 1991) and of the higher respiration rate of mycorrhizal roots compared with non mycorrhizal roots (Kucey and Paul, 1982). Moreover, mycorrhizal fungi alter root exudation both quantitatively and qualitatively (Rambelli, 1973; Leyval and Berthelin, 1993), as they catabolise some of the root exudates and modify root metabolic functions. The microbial communities of the soil surrounding mycorrhizal roots and extrametrical mycelium are different from those of the rhizosphere of non mycorrhizal plants and the bulk soil (Katznelson et al., 1962; Garbaye and Bowen, 1987, 1989; Garbaye, 1991). Hence, the rhizosphere concept has been widened to associate this fungal effect, resulting in the introduction of terms "mycorrhizosphere" and "hyphosphere" (Rambelli, 1973; Linderman, 1988). The mycorrhizosphere named the zone influenced by both the root and the mycorrhizal fungus whereas the hyphosphere is the zone surrounding individual fungal hyphae (Linderman, 1988). Specific relationships occur between mycorrhizal fungi and mycorrhizosphere microbiota and there is abundant literature attesting that mycorrhizal symbiosis is largely influenced by soil microorganisms (Rambelli, 1973; Bowen, 1980; De Oliveira, 1988; De Oliveira and Garbaye, 1989). However, these interactions have been mainly focused on the effects of mycorrhizosphere microbial communities on the mycorrhizal formation (extent of mycorrhizal colonization) and on the mycorrhizal efficiency on the host plant growth. Recently, it has been proposed that mycorrhizal symbiosis is a component of a microbial complex regulated by multitrophic interactions (Frey-Klett et al., 2005). In addition to their known direct effect on plant growth, mycorrhizal symbionts could positively act on host plant development through a selective effect on bacterial communities involved in soil functioning and soil fertility. This new concept of the mycorrhizal symbiosis is of particular importance in tropical and mediterranean areas subjected to desertification. Desertification process usually results from degradation of natural plant communities (population structure, succession pattern and species diversity) and of physico-chemical and biological soil properties (nutrient availability, microbial activity, soil structure, etc.) (Garcia et al., 1997; Requena et al., 2001). In addition, the loss or reduction of the activity of mycorrhizal fungi was often detected (Bethlenfalvay and Schüepp, 1994). Hence, the management of soil mycorrhizal potential in tropical and mediterranean environments is of great importance since mycorrhizal symbiosis determines plant biodiversity, ecosystem variability and productivity directly from its influence on plant mineral nutrition but also indirectly from its impact on soil microbial functioning. The natural role of mycorrhizosphere microorganisms has been marginalized in intensive agriculture and forest management but, due to the increased environmental awareness, particular interest has been done on low-input cropping systems. In low-input, sustainable agrosystem production, natural activities of microbes contribute to the biocontrol of pathogens and improve supply of nutrients.

This chapter will focus on the interactions between mycorrhizal fungi and specific groups of microorganisms potentially beneficial to the plant growth (Rhizobia, Plant Growth Promoting Rhizobacteria) and the influence of the mycorrhizal symbiosis on the functioning of soil microbial communities. The review will highlight the aspects of the interactions in the mycorrhizosphere which may have practical applications in afforestation programs in tropical and mediterranean soils.

2 EFFECTS OF MYCORRHIZAL FUNGI ON NODULATION AND N_2 FIXATION BY LEGUMES

Mycorrhiza formation is known to enhance nodulation and N_2 fixation by legumes (Reddell and Warren, 1986; Amora-Lazcano et al., 1998; André et al., 2005). Mycorrhizal and rhizobial symbioses often act synergistically on infection rate, mineral nutrition and plant growth (Amora-Lazcano et al., 1998). The positive fungal effect on plant P uptake is beneficial for the functioning of the nitrogenase enzyme of the rhizobial symbiont leading to a higher N_2 fixation and, consequently to a better root growth and mycorrhizal development (Johansson et al., 2004). The fungal effect on rhizobial development is dependant to the mycorrhizal extent along the root systems but also to the fungal symbiont. Testing the effect of the ectomycorrhizal fungus,

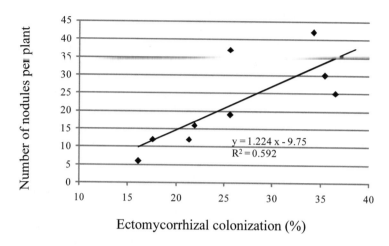

Fig. 1. Correlation between ectomycorrhizal colonization of *Acacia holosericea* seedlings with *Pisolithus albus* and the number of rhizobial nodules per plant after eight month's culture in a not disinfected sandy soil.

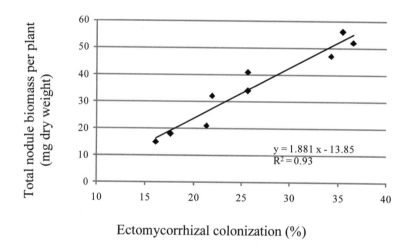

Fig. 2. Correlation between ectomycorrhizal colonization of *Acacia holosericea* seedlings with *Pisolithus albus* and the total nodule biomass per plant after eight month's culture in a not disinfected sandy soil (Duponnois, unpublished data, 2008)

Pisolithus albus strain IR100, on the growth of *Acacia holosericea* (a fast growing Australian Acacia species) and on the nodulation formation, it has been found that the number of nodules per plant and their total biomass were significantly correlated with the ectomycorrhizal root colonization (Fig. 1 and 2). This fungal effect was also dependent of the mycorrhizal symbiont (Fig. 3 and 4) (Duponnois and Plenchette, 2003). However, the influence of the mycorrhizal symbiosis on nodule development is not limited to a quantitative effect and the fungi can modify the structure of rhizobial bacteria along the root system. For instance, the arbuscular mycorrhizal fungus, *Glomus intraradices*, induced different dynamics of two rhizobia, *Sinorhizobium terangae* strain ORS 1009 and *Mesorhizobium plurifarium* strain ORS 1096, co-inoculated to *Acacia tortilis* ssp. *raddiana* (André *et al.*, 2003) (Fig. 4). Mycorrhizal infection increases the competitiveness of ORS1009 (Fig. 4). This result suggests that more specific relationships could occur during the development of the tripartite symbiosis, at physiological and molecular level (Van Rhijn *et al.*, 1997; Blilou *et al.*, 1999). More recently, it has been demonstrated that below-ground diversity of AM fungi was a major factor contributing to the maintenance of plant diversity and to ecosystem functioning (van der Heijden *et al.*, 1998). This fungal diversity

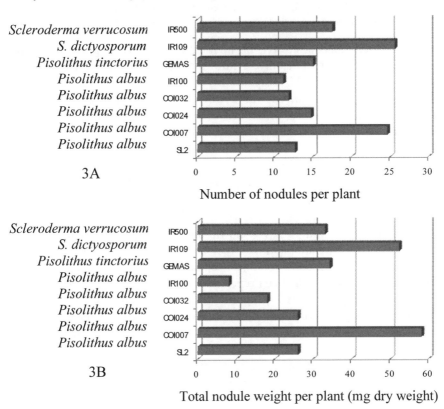

Fig. 3. Influence of different ectomycorrhizal fungal strains on rhizobial development (A: number of nodules per plant; B: total nodule weight par plant) with Acacia holosericea after four month's culture under glasshouse conditions.

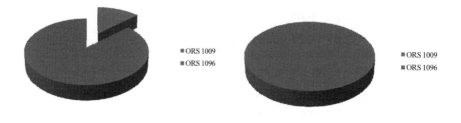

With *Glomus intraradices*   With out *Glomus intraradices*

Fig. 4. Identification of rhizobial strains (*Sinorhizobium terangae* strain ORS 1009 and *Mesorhizobium plurifarium* strain ORS 1096) from nodules collected along *Acacia tortilis* root system inoculated with both rhizobial strains with or without the arbuscular mycorrhizal fungus *Glomus intraradices* after four month's culture in a disinfected sand (From André et et al., 2003). Results are expressed as percentages of each rhizobial strain from each treatment.

has also a beneficial effect on the nodulation process. In greenhouse, the influence of six different ectomycorrhizal fungi isolates and of a combination of these six ectomycorrhizal symbionts was measured on the nodulation of *Acacia mangium* seedlings with *Bradyrhizobium* sp. isolate AUST 13C. After four month's culture, the results show that the number of nodules per plant was linked with the number of inoculated fungal strains (Fig. 5).

3   FLUORESCENT PSEUDOMONAD FUNCTIONAL DIVERSITY

Numerous studies have shown that the mycorrhizosphere effect exerted a significant stimulating effect on the populations of fluorescent pseudomonads in the soil. For instance, this quantitative influence of the mycorrhizal symbiosis has been reported with several plant species such as Douglas fir (Frey et al., 1997), hybrid larch, Sitka spruce (Grayston et al., 1994), hazel trees mycorrhized with truffles (Mamoun and Olivier, 1989), a fast growing Australian Acacia species, *Acacia holosericea* (Founoune et al., 2002). More recently, Ramanankierana et al. (2006) compared the abundance of fluorescent pseudomonads and their functional diversity in different compartments (rhizosphere, mycorrhizosphere, hyphosphere and bulk soil) resulting from the ectomycorrhization of *Uapaca bojeri*, an endemic Euphorbiaceae of Madagascar. Their results showed that the number of fluorescent pseudomonads was significantly higher in the hyphosphere soil than in the rhizosphere and mycorrhizosphere soil (Fig. 6). The lowest abundance

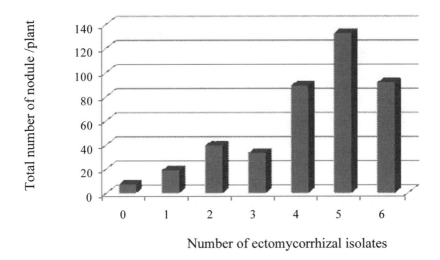

Fig. 5. Effect of ectomycorrhizal fungus – species richness on the number of nodules per *Acacia mangium* seedlings after four month's culture in glasshouse conditions.

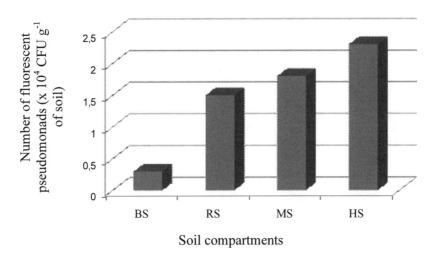

Fig. 6. Number of fluorescent pseudomonads in soil compartments (Bulk soil, BS; Rhizosphere soil, RS; Mycorrhizosphere soil, MS; Hyphosphere soil, HS) resulting from the ectomycorrhization of *Uapaca bojeri* in glasshouse conditions. Columns indexed by the same letters expressed data that are not significantly different according to the Newman-Keul's test ($p < 0.05$).

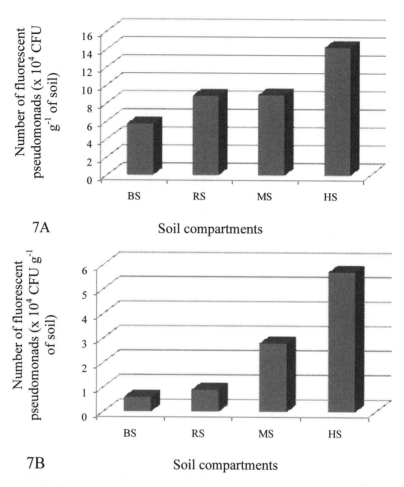

Fig. 7. Distribution of the fluorescent pseudomonad according to their functional abilities (A: phosphate solubilizing fluorescent pseudomonads; B: lipase producing fluorescent pseudomonads) in soil compartments (Bulk soil, BS; Rhizosphere soil, RS; Mycorrhizosphere soil, MS; Hyphosphere soil, HS) resulting from the ectomycorrhization of *Uapaca bojeri* in glasshouse conditions. Columns indexed by the same letters expressed data that are not significantly different according to the Newman-Keul's test ($p < 0.05$).

was recorded in the bulk soil (Fig. 6). Functional abilities (lipasic and phosphate solubilizing activities) of fluorescent pseudomonads have been also determined in each soil compartment. It has been found that the percentages of phosphate solubilizing and lipase producing fluorescent pseudomonads were significantly higher in the hyphosphere soil than in the other soil compartments (Fig. 7). These results showed that the ectomycorrhizal mycelium increased the multiplication of fluorescent pseudomonads differently than that recorded from the ectomycorrhizosphere effect. The hyphae of ectomycorrhizal fungi could be the sources of carbon to the soil microbial communities from fungal exudates (Sun *et al.*, 1999) and/or from following senescence of hyphae (Bending and Read, 1995) that could be used by fluorescent pseudomonads. In addition to this quantitative effect, the ectomycorrhizal symbiosis has also modified the distribution of phosphate solubilizing and lipase producing fluorescent pseudomonads, especially in the hyphosphere soil compartment. It is known that extramatrical mycelium can absorb and then translocate to the host plant, soluble phosphorus from mineral and organic matter, trough the excretion of organic acids and phosphatase, respectively (Landeweert *et al.*, 2001). In addition, most of the fluorescent pseudomonads isolated from the hyphosphere compartment were able to solubilize inorganic phosphate (tricalcium orthophosphate) compared to those from the bulk soil (Fig. 7). These results corroborate those of Frey-Klett *et al.* (2005) who demonstrated that phosphate solubilizing fluorescent pseudomonads were more abundant in the hyphosphere than in the bulk soil. The mycorrhizal effect in not limited to the phosphorus solubilization process but can also interest organic matter degradation. Lipases are a group of enzymes that catalyse the hydrolysis of triacylglycerols to diacylglycerols, monoacylglycerols, fatty acids and glycerol (Thompson *et al.*, 1999). Lipase activity is also involved in the soil humification processes (Lähdesmäki and Piispanen, 1988).

4 THE MYCORRHIZAL EFFECT AND SOIL MICROBIAL FUNCTIONALITIES

Culture-independent methods such as fatty acid extraction (Cavigelli *et al.*, 1995; Ibekwe and Kennedy, 1999) and PCR-DGGE (Ferris *et al.*, 1996; Muyzer and Smalla, 1998; Assigbetse *et al.*, 2005) are increasingly used for the analysis of soil microbial community structure. In contrast little is known of the importance of the functional diversity of soil microbial communities (Pankhurst *et al.*, 1996) resulting from a limited access to suitable techniques. The functional diversity of microbial communities includes the range and relative expression of activities involved in decomposition, nutrient transformation, plant growth promotion, etc. (Giller *et al.*, 1997). The diversity of decomposition functions performed by heterotrophic microbes is one component of microbial functional diversity. Hence to directly measure the diversity of decomposition functions, an assay has been developed to provide a measure of a component of the catabolic functional diversity in soil. This assay gives catabolic response profiles (patterns of *in situ* catabolic potential, ISCP) by measuring the short-term utilization of a range of readily available substrates added to soils (Degens and Vojvodic-Vukovic, 1999). Patterns of

ISCP provide a real time measure of microbial functional diversity since they give a direct measurement of substrate catabolism by microbial communities in soils without prior culturing of microorganisms. This methodological approach has been widely used to describe the mycorrhizal effect on the functional diversity of soil microbial communities.

Table 1 summarizes the effects of mycorrhizal inoculation on the catabolic evenness of soil microbial communities following mycorrhizal inoculation in different experimental and environmental conditions. The catabolic evenness, E, representing the variability of a substrate among the range of tested substrates, was calculated using the Simpson–Yule index, $E = 1/p^2i$ with p_i = respiration response to individual substrates/total respiration activity induced by all substrates for a soil treatment (Magurran, 1988). These results show that mycorrhizal inoculation significantly enhances the catabolic evenness of soil microbial communities. It has been hypothesized that increases in the microbial catabolic diversity will enhance the resistance of soils to stress or disturbance (Giller et al., 1997). Hence, it shows that the extent of mycorrhizal plant and soil colonization is an important component in soil functioning more particularly in the context of global change.

Table 1. Effect of mycorrhizal inoculation on catabolic evenness in different experimental and environmental conditions.

Plant species	Fungal symbiont	E	References
AM fungal inoculation			
Cupressus atlantica	Control	12.0 a [1]	Ouahmane et al., 2007
	Glomus intraradices	12.2 b	
Sorghum bicolor	Control	4.3 a	Dabire et al., 2007
	Glomus intraradices	6.5 b	
Eucalyptus camaldulensis	Control	17.8 a	Kisa et al., 2007
	Glomus intraradices	20.7 b	
Acacia holosericea	Control	12.9 a	Duponnois, R. unpublished data, 2008
	Glomus intraradices	16.5 b	
Ectomycorrhizal inoculation			
Acacia holosericea	Control	15.2 a	Duponnois, R. unpublished data, 2008
	Scleroderma dictyosporum	18.3 b	
Uapaca bojeri	Control	4.7 a	Ramankierana et al, 2007
	Scleroderma sp.	6.9 b	

For each reference, data in the same column followed by the same letters are not significantly different according to the Newman-Keul's test ($p < 0.05$).

5 CONCLUSION

All these results show the importance of the mycorrhizal symbiosis in soil biofunctioning and in the development of some specific groups of microorganisms known to play a key role in soil fertility and ecosystem productivity. In addition several studies have underlined the importance of mycorrhizal fungus diversity to the maintenance of plant diversity and to ecosystem functioning. These results emphasize the need to protect mycorrhizal fungi and to consider these symbiotic fungi in ecosystem and agrosystem management practices. Recent results on the influence of plant nurses on ecosystem functions have to be taken in account in the future more particularly in the rehabilitation practices of mediterranean and tropical ecosystems.

REFERENCES

Amora-Lazcano, E., Vazquez, M.M., and Azcon, R., 1998, Response of nitrogen-transforming microorganisms to arbuscular mycorrhizal fungi. *Biol. Fert. Soils* **27**: 65–70.

André, S., Neyra, M., and Duponnois, R., 2003, Arbuscular mycorrhizal symbiosis changes the colonization pattern of *Acacia tortilis* ssp. *raddiana* rhizosphere by two strains of rhizobia. *Microb. Ecol.* **45**: 137–144.

André, S., Galiana, A., Le Roux, C., Prin, Y., Neyra, M., and Duponnois, R., 2005, Ectomycorrhizal symbiosis enhanced the efficiency of inoculation with two *Bradyrhizobium* strains and *Acacia holosericea* growth. *Mycorrhiza* **15**: 357–364.

Assigbetse, K., Gueye, M., Thioulouse, J., and Duponnois, R., 2005, Soil bacterial diversity responses to root colonization by an ectomycorrhizal fungus are not root-growth dependent. *Microb. Ecol.* **50**: 350–359.

Augé, R.M., 2001, Water relations, drought and vesicular arbuscular mycorrhizal symbiosis. *Mycorrhiza* **11**: 3–42.

Bending, G.D., and Read, D.J., 1995, The structure and function of the vegetative mycelium of ectomycorrhizal plants. V. foraging behavior and translocation of nutrients from exploited litter. *New Phytol.* **130**: 401–409.

Bethlenfalvay, G.J., and Linderman, R.G., 1992, *Mycorrhizae in sustainable agriculture*. ASA Special Publication No. 54, Madison, WI.

Bethlenfalvay, G.J., and Schüepp, H., 1994, Arbuscular mycorrhizas and agrosystem stability. In: Gianinazzi, S., Schüepp, H. (Eds), *Impact of Arbuscular Mycorrhizas on Sustainable Agriculture and Natural Ecosystems*. Birkhäuser Verlag, Basel, Switzerland, pp. 117–131.

Blilou, I., Ocampo, J.A., and Gracia-Garrido, J.M., 1999, Resistance of pea roots to endomycorrhizal fungus or *Rhizobium* correlates with enhanced levels of endogenous salicylic acid. *J. Exp. Bot.* **50**: 1663–1668.

Bowen, G.D., 1980, Misconceptions, concepts and approaches in rhizosphere biology. In: Ellwood, D.C., Hedger, J.N., Latham, M.J., Lynch, J.M., Slater, J.M. (Eds), *Contempary Microbial Ecology*. Academic, London, pp. 283–304.

Caravaca, F., Barea, J.M., Figueroa, D., and Roldan, A., 2002, Assessing the effectiveness of mycorrhizal inoculation and soil compost addition for reafforestation with *Olea europaea* subsp. *sylvestris* through changes in soil biological and physical parameters. *Appl. Soil Ecol.* **20**: 107–118.

Cavigelli, M.A., Robetson, G.P., and Klug, M.J., 1995, Fatty acid methyl ester (FAME) profiles as measures of soil microbial community structure. *Plant Soil* **170**: 99–113.

Dabire, A.P., Hien, V., Kisa, M., Bilgo, A., Sangare, K.S., Plenchette, C., Galiana, A., Prin, Y., and Duponnois, R., 2007, Responses of soil microbial catabolic diversity to arbuscular mycorrhizal inoculation and soil disinfection. *Mycorrhiza* **17**: 537–545.

De Oliveira, V.L., 1988, Interactions chez les microorganisms du sol et l'établissement de la symbiose ectomycorhizienne chez le hêtre (*Fagus silvatica* L.) avec *Hebeloma crustuliniforme* et *Paxillus involutus*. Ph. D. thesis, University of Nancy 1. 118 p.

De Oliveira, V.L., and Garbaye, J., 1989, Les microorganismes auxiliaires de l'établissement des symbioses ectomycorhiziennes. *Eur. J. Pathol.* **19**: 54–64.

Degens, B.P., and Vojvodic-Vukovic, M., 1999, A sampling strategy to assess the effects of land use on microbial functional diversity in soils. *Austr. J. Soil Res.* **37**: 593–601.

Duponnois, R., and Plenchette, C., 2003, A mycorrhiza helper bacterium enhances ectomycorrhizal and endomycorrhizal symbiosis of Australian *Acacia* species. *Mycorrhiza* **13**: 85–91.

Ferris, M.J., Muyzer, G., and Ward, D.M., 1996, Denaturing gradient gel electrophoresis profiles of 16S rRNA-defined populations inhabiting a hot spring microbial mat community. *Appl. Envir. Microbiol.* **62**: 340–346.

Fitter, A.H., 1991, Costs and benefits of mycorrhizas: implications for functioning under natural conditions. *Experimentia* **47**: 350–355.

Founoune, H., Duponnois, R., Meyer, J.M., Thioulouse, J., Masse, D., Chotte, J.L., and Neyra, M., 2002, Interactions between ectomycorrhizal symbiosis and fluorescent pseudomonads on *Acacia holosericea*: isolation of mycorrhiza helper bacteria (MHB) from a soudano-Sahelian soil. *FEMS Microbiol. Ecol.* **4**: 37–46.

Frey, P., Frey-Klett, P., Garbaye, J., Berge, O., and Heulin, T., 1997, Metabolic and genotypic fingerprinting of fluorescent pseudomonads associated with the Douglas Fir - *Laccaria bicolor* mycorrhizosphere. *Appl. Envir. Microbiol.* **63**: 1852–1860.

Frey-Klett, P., Chavatte, M., Clausse, M.L., Courrier, S., Le Roux, C., Raaijmakers, J., Martinotti, M.G., Pierrat, J.C., and Garbaye, J., 2005, Ectomycorrhizal symbiosis affects functional diversity of rhizosphere fluorescent pseudomonads. *New Phytol.* **165**: 317–328.

Garbaye, J., 1991, Biological interactions in the mycorrhizosphere. *Experientia* **47**: 370–375.

Garbaye, J., and Bowen, G.D., 1987, Effect of different microflora on the success of mycorrhizal inoculation of *Pinus radiata*. *Can. J. Forest Res.* **17**: 941–943.

Garbaye, J., and Bowen, G.D., 1989, Ectomycorrhizal infection of *Pinus radiata* by some microorganisms associated with the mantle of ectomycorrhizas. *New Phytol.* **112**: 383–388.

Garcia, C., Hernandez, T., Roldan, A., and Albaladejo, L., 1997, Biological and biochemical quality of a semiarid soil after induced revegetation. *J. Envir. Qual.* **26**: 1116–1122.

Giller, K.E., Beare, M.H., Lavelle, P., Izac, A.-M.N., and Swift, M.J., 1997, Agricultural intensification, soil biodiversity and agroeco-system function. *Appl. Soil Ecol.* **6**: 3–16.

Grayston, S.J., Campell, C.D., and Vaughan, D., 1994, Microbial diversity in the rhizospheres of different tree species. In: Pankhurst, C.E. (Ed), *Soil Biota: Management in Sustainable Farming Systems*. CSIRO, Adelaide, pp. 155–157.

Ibekwe, A.M., and Kennedy, A.C., 1999, Fatty acid methyl ester (FAME) profiles as tool to investigate community structure of two agricultural soils. *Plant Soil* **206**: 151–161.

Jakobsen, I., Abbott, L.K., and Robson, A.D., 1992, External hyphae of vesicular-arbuscular mycorrhizal fungi associated with *Trifolium subterraneum*. I. Spread of hyphae and phosphorus inflow into roots. *New Phytol.* **120**: 371–380.

Johansson, J.F., Paul, L.R., and Finlay, R.D., 2004, Microbial interactions in the mycorrhizosphere and their significance for sustainable agriculture. *FEMS Microbiol. Ecol.* **48**: 1–13.

Joner, E.J., Aarle, I.M., and Vosatka, M., 2000, Phosphatase activity of extra-radical arbuscular mycorrhizal hyphae: a review. *Plant Soil* **226**: 199–210.
Katznelson, H., Rouatt, J.W., and Peterson, E.A., 1962, The rhizosphere effect of mycorrhizal and nonmycorrhizal roots of yellow birch seedlings. *Can. J. Bot.* **40**: 377–382.
Kisa, M., Sanon, A., Thioulouse, J., Assigbetse, K., Sylla, S., Spichiger, R., Dieng, L., Berthelin, J., Prin, Y., Galiana, A., Lepage, M., and Duponnois, R., 2007, Arbuscular mycorrhizal symbiosis can counterbalance the negative influence of the exotic tree species *Eucalyptus camaldulensis* on the structure and functioning of soil microbial communities in a sahelian soil. *FEMS Microbiol. Ecol.* **62**: 32–44.
Klironomos, J.N., McCune, J., Hart, M., and Neville, J., 2000, The influence of arbuscular mycorrhizae on the relationship between plant diversity and productivity. *Ecol. Lett.* **3**: 137–141.
Kucey, R.M.N., and Paul, E.A., 1982, Carbon flow photosynthesis, and N_2 fixation in mycorrhizal and nodulated faba beans (*Vicia faba* L.). *Soil Biol. Biochem.* **14**: 407–412.
Lähdesmäki, P., and Piispanen, R., 1988, Degradation products and the hydrolytic enzyme activities in the soil humification processes. *Soil Biol. Biochem.* **20**: 287–292.
Landeweert, R., Hoffland, E., Finlay, R.D., Kuyper, T.W., and van Breemen, N., 2001, Linking plants to rocks: ectomycorrhizal fungi mobilize nutrients from minerals. *Trend. Ecol. Evol.* **16**: 248–254.
Leyval, C., and Berthelin, J., 1993, Rhizodeposition and net release of soluble organic compounds of pine and beech seedlings inoculated with rhizobacteria and ectomycorrhizal fungi. *Biol. Fert. Soils* **15**: 259–267.
Linderman, R.G., 1988, Mycorrhizal interactions with the rhizosphere: the mycorrhizosphere effect. *Phytopathology* **78**: 366–371.
Magurran, A.E., 1988, *Ecological Diversity and Its Measurement*. Croom Helm, London.
Mamoun, M., and Olivier, J.M., 1989, Dynamique des populations fongiques et bactériennes de la rhizosphère des noisetiers truffiers. II. Chélateur du fer et répartition taxonomique chez les Pseudomonas fluorescents. *Agron.* **9** : 345–351.
Marler, M.J., Zabinski, C.A., and Callaway, R.M., 1999, Mycorrhizae indirectly enhance competitive effects of an invasive forb on a native bunchgrass. *Ecology* **80**: 1180–1186.
Muyzer, G., and Smalla, K., 1998, Application of denaturing gradient gel electrophoresis (DGGE) and temperature gradient gel electrophoresis (TGGE) in microbial ecology. *Antonie van Leeuwenhoek* **73**: 127–141.
O'Connor, P.J., Smith, S.E., and Smith, F.A., 2002, Arbuscular mycorrhizas influence plant diversity and community structure in a semiarid herbland. *New Phytol.* **154**: 209–218.
Ouahmane, L., Thioulouse, J., Hafidi, M., Prin, Y., Ducousso, M., Galiana, A., Plenchette, C.,Kisa, M., and Duponnois, R., 2007, Soil functional diversity and P solubilization from rock phosphate after inoculation with native or exotic arbuscular mycorrhizal fungi. *Forest Ecol. Manag.* **241**: 200–208.
Pankhurst, C.E., Ophel-Keller, K., Doube, B.M., and Gupta, V.V.S.R., 1996, Biodiversity of soil microbial communities in agricultural systems. *Biodiv. Conser.* **5**: 197–209.
Ramanankierana, N., Rakotoarimanga, N., Thioulouse, J., Kisa, M., Randrianjohany, E., Ramoroson, L., and Duponnois, R., 2006, The ectomycorrhizosphere effect influences functional diversity of soil microflora. *Inter. J. Soil Sci.***1**: 8–19.
Ramanankierana, N., Ducousso, M., Rakotoarimanga, N., Prin, Y., Thioulouse, J., Randrianjohany, E., Ramaroson, L., Kisa, M., Galiana, A., and Duponnois, R., 2007, Arbuscular mycorrhizas and ectomycorrhizas of *Uapaca bojeri* L. (Euphorbiaceae): sporophore diversity, patterns of root colonization and effects on seedling growth and soil microbial catabolic diversity. *Mycorrhiza* **17**: 195–208.
Rambelli, A., 1973, The rhizosphere of mycorrhizae. In: Marks, G.C., Kozlowski, T.T. (Eds), *Ectomycorrhizae: Their Ecology and Physiology*. Academic, New York, pp. 229–343.

Reddell, P., and Warren, R., 1986, Inoculation of Acacia with mycorrhizal fungi: potential benefits. In: Turnbull, J.W. (Ed), *Australian Acacia in Developing Countries*. ACIAR, Canberra, pp. 50–53.

Requena, N., Perez-Solis, E., Azcon-Aguilar, C., Jeffries, P., and Barea, J.M., 2001, Management of indigenous plant–microbe symbioses aids restoration of desertified ecosystems. *Appl. Envir. Microbiol.* **67**: 495–498.

Smith, S.E., and Read, D.J., 1997, *Mycorrhizal Symbiosis*. 2nd edn. Academic, UK.

Sun, Y.P., Unestam, T., Lucas, S.D., Johanson, K.J., Kenne, L., and Finlay, R.D., 1999, Exudation reabsorption in mycorrhizal fungi, the dynamic interface for interaction with soil and other microorganisms. *Mycorrhiza* **9**: 137–144.

Thompson, C.A., Delaquis, P.J., and Mazza, G., 1999, Detection and measurement of microbial activity: a review. *Crit. Rev. Food Sci.* **39**: 165–187.

van der Heijden, M.G.A., Klironomos, J.N., Ursic, M., Poutoglis, P., Streitwolf-Engel, R., Boller, T., Wiemken, A., and Sanders, I.R., 1998, Mycorrhizal fungal diversity determines plant biodiversity, ecosystem variability and productivity. *Nature* **396**: 69–72.

van der Heijden, M.G.A., Wiemken, A., and Sanders, I.R., 2003, Different arbuscular mycorrhizal fungi alter coexistence and resource distribution between co-occurring plant. *New Phytol.* **157**: 569–578.

Van Rhijn, P., Fang, Y., Galili, S., Shaul, O., Atzon, N., Wininger, S., Eshed, Y., Lum, M., Li, Y., To, V., Fujishige, N., Kapulik, Y., and Hirsch, A.M., 1997, Expression of early nodulin genes in alfalfa mycorrhizae indicates that signal transduction pathways used in forming arbuscular mycorrhizae and Rhizobium-induced nodules may be conserved. *Plant Biol.* **94**: 5467–5472.

West, H.M., 1996, Influence of arbuscular mycorrhizal infection on competition between *Holcus lanatus* and *Dactylis glomerata*. *J. Ecol.* **84**: 429–438.

Chapter 11

ECTOMYCORRHIZAE AND THEIR IMPORTANCE IN FOREST ECOSYSTEMS

KAZUYOSHI FUTAI, TAKESHI TANIGUCHI AND RYOTA KATAOKA
Graduate School of Agriculture, Kyoto University, Sakyo-ku 606-8502, Kyoto, JAPAN

Abstract: Ectomycorrhizal (ECM) associations involve the most diverse category of myocrrhizae. The diversity derives from the fungal partners; more than 5,000 species of fungi, mainly Basidiomycetes, with a limited number of Ascomycetes and Zygomycetes, make the relationship very diverse. On the contrary, however, relatively few families of plants such as Fagaceae, Pinaceae, Betulaceae, and Dipterocarpaceae are involved in the ECM associations. These plants, however, are distributed over wide areas of temperate and boreal forests, and are therefore economically important. ECM fungi make associations with plants by forming a sheath (mantle) around fine root tips with hyphae that grow inward between root cells of the cortex and make Hartig net, and emanate outward through the soil, increasing the surface area to absorb nutrients and water. Thus, the mycorrhizal fungi gain photosynthates and other essential substances from the plant and in return help the plant take up water and minerals. Pine wilt disease (PWD) is a globally serious forest disease, and also shows the importance of ectomycorrhizal relationships. Pine trees planted on a mountain slope were killed by PWD, but some trees survived at the top of the slope, where mycorrhizal associations developed far better than on lower slopes. ECM associations, beside fertilization, also increase the supply of water to the pines, and elevate host resistance against disease and parasites. Moreover, inoculation of pine seedlings with ECM fungi under laboratory conditions confirmed the increase in their resistance to PWD. Pine seedlings can tolerate the adverse effects of environmental stress such as acid mist when infected with ECM fungi. These fungi can also make a significant contribution to forest ecosystems by increasing biomass and creating a network among trees through which nutrients may transported. ECM fungi also improve the growth of host plants at the seedling stage. Many pioneer plants in wastelands are facilitated in their establishment by ECM. This association has been successfully applied to reforestation programs in tropical forests by inoculating mycorrhizae on to nursery seedlings.

Keywords: Basidiomycetes; forest ecosystem; networks; succession.

1 INTRODUCTION

Ectomycorrhizal symbiosis is the most diverse of all mycorrhizal associations. The diversity arises primarily from the fungal partners including about 5,000 to 6,000 species, mostly Basidiomycetes, some Ascomycetes, and a few Zygomycetes. The roots of ectomycorrhizal trees and shrubs including Pinaceae, Cupressaceae, Fagaceae, Betulaceae, Salicaceae, Dipterocarpaceae, and Myrtaceae support a great species richness of fungal symbionts. This limited group of plants constitutes a large component of temperate forest dominant trees; therefore, ectomycorrhizal symbiosis is economically important.

Ectomycorrhizal fungi form a symbiotic relationship with a plant by forming a sheath around its root tip. The fungus then penetrates the root along the middle lamellae between cell walls by inward growth of hyphae, thereby form a Hartig net, a complex network of fungal hyphae that is the site of nutrient exchange between the fungus and the host plant. The fungi and the plant essentially fuse walls, and nutrient exchange appears to take place across these walls. The fungus gains carbon and other essential organic substances from the tree and in return helps the trees take up water, mineral salts and metabolites with increased surface area of hyphae emanating through the soil. Ectomycorrhizal fungi also protect host trees from attack by parasites, predators, nematodes and other soil pathogens. Thus, most forest trees are highly dependent on their fungal partners and could possibly not exist without them in areas of poor soil quality.

2 SPECIFICITY AND DIVERSITY OF ECTOMYCORRIZAE

2.1 Specificity

Molina and Trappe (1982) in early specificity experiments examined the specificity of 27 fungi in ECM formation with seven Pacific Northwest conifers, and indicated that the fungi varied widely to form mycorrhizae with the various conifers. These can be classified into three groups:

- Fungi with wide ECM host potential, low specificity, and sporocarps usually associated with diverse hosts in the field
- Fungi with intermediate host potential yet specific or limited in sporocarp-host associations, and
- Fungi with narrow host potential, only form ECM with a specific host species or species within a genus and likewise limited in their sporocarp association

The fruiting body assessments and long-term fungal community collections suggest a range of specificity patterns from generalist to specialist for both fungal species and vascular plants. In mixed spruce and hardwood forest communities in the northeastern United States, hardwoods and spruce shared only 8 of 54 fungal species while 19 were associated only with spruce (Bills *et al.*, 1986). In greenhouse experiments, Molina and Trappe (1994) examined host specificity between fungal and plant partners, and also studied the influence of neighboring plants on ECM development using seedlings of 6 coniferous trees grown in monoculture and dual culture inoculated with spore slurries of 15 species of ECM hypogeous fungi (11 *Rhizopogon* species, and each of 4 other genera). None of the fungal species had broad host range affinities. A variety of specificity responses were exhibited by the different fungal taxa, ranging from genus-restricted to intermediate host range. In dual culture, 9 of the 11 *Rhizopogon* species examined formed abundant ECM on *Pinus ponderosa*, and formed some ECM on secondary hosts such as *Abies grandis, Tsuga heterophylla, Pseudotsuga menziesii* and *Picea sitchensis*. None of the fungi tested, however, developed ECM on these secondary hosts in monoculture, which suggests potential interplant linkages and community dynamics.

The specificity found under laboratory culture conditions is not always consistent with that in nature, because field conditions alter specificity patterns indicated in culture experiments. To determine plant species associating with a fungus, there is the need to trace single hyphae through the soil; however, this procedure is almost impossible due to the fragile nature of individual hyphae. The use of molecular methods has enabled researchers to identify accurately *in situ* both fungi and plants that form ECM in the field, thereby facilitating the investigation of ECM specificity patterns in mixed-tree-species forests. Many studies have been conducted to assess ECM specificity patterns in the field using molecular methods. For instance, Horton and Bruns (1998) investigated ECM associations in a mixed stand of Douglas fir (*Pseudotsuga menziesii*) and bishop pine (*Pinus muricata*). They identified fungi directly from field-collected ECM root tips using PCR-based methods and found sixteen species of fungi, out of which twelve were associated with both hosts. *Rhizopogon parksii* was specific to Douglas fir; three other species colonized only one of the hosts, but were too infrequent to draw conclusions about specificity. By evaluating the biomass of ECM root tips sampled in the stand, the authors concluded that multiple-host fungi dominated on mycorrhizal roots and colonized the roots of competing plant hosts.

Horton *et al.* (1999), using molecular methods, assessed patterns of ECM between an angiosperm and a conifer and concluded that sharing of ECM fungi by *Arctostaphylos* sp. and *P. menziesii* facilitated the establishment of the conifer in sites dominated by the angiosperm. They further confirmed the results of vegetation surveys and seedling survival assays which suggested that *Pseudotsuga* establishes only in *Arctostaphylos*. Because of the importance of ECM specificity to ecosystem function, and the conflicting results of laboratory and field experiments, using molecular methods is the best way to ascertain specificity in the field.

Because ECM associations are essential for many plants for their growth and survival, ECM specificity is thought to be a crucial determinant of ecosystem function and which benefits both plant and fungal partners. Cullings *et al.* (2000) suggested that ECM specificity may limit the ability of some plants to migrate and become established, and thus influence the rates and directions of ecosystem change. Therefore, assessment of ECM specificity patterns is critical to ecosystem function.

However, there are great differences both between and within fungal species in terms of forming mycorrhiza and promoting growth of the host plant. Lamhamedi *et al.* (1990) examined the ability of 28 monokaryons and 78 reconstituted dikaryons of *Pisolithus tinctorius* to form ECM on *Pinus pinaster* and *Pinus banksiana* and found a marked difference in the ability to form mycorrhiza and promote growth of *P. pinaster* both between and within monkaryons and dikaryons. Some monokaryons and dikaryons failed to form ECM. Monokaryons formed fewer ECM on *P. pinaster* than di-karyons. The heterokaryotic state was necessary for the full expression of ECM forming ability. Growth of *P. pinaster* was more strongly correlated with ECM formation by dikaryons than by monokaryons.

In a review of the current state of knowledge of interactions between *Pisolithus tinctorius* and its hosts, Cairney and Chambers (1997) demonstrated that this ECM fungus displays much intraspecific heterogeneity of host specificity, physiology, and the benefits the fungus can impart upon the host plant. It is not clear at present how far such heterogeneity reflects systematic segregation within *P. tinctorius*.

The variation within ribosomal DNA (rDNA) genes of 19 isolates of *Pisolithus* from different geographic origins and hosts was examined by PCR-RFLP analysis. Cluster analysis based on the restriction fragments grouped the isolates into three distinct groups: group I contained isolates collected in the northern hemisphere, except Pt 1, group II contained those collected in Brazil and group III contained isolate Pt 1. Additional analysis of other rDNA regions, IGS, 17S and 25S rDNA, resulted in similar groupings (Gomes *et al.*, 2004).

As mentioned above, recent molecular methods allowed for the examination of ECM diversity on plant seedlings. For example, ECM fungi on root tips of introduced *Eucalyptus robusta* and *Pinus caribea* as well as the endemic *Vateriopsis seychellarum* and indigenous *Intsia bijuga* in the Seychelles were identified by anatomotyping and rDNA sequencing (Tedersoo *et al.*, 2007). Sequencing revealed 30 species of ECM fungi on root tips of

V. seychellarum and *I. bijuga*, with three species overlapping. *Eucalyptus robusta* shared five of these taxa, whereas *P. caribea* hosted three unique species of ECM fungi that were likely co-introduced with containerized seedlings. The low diversity of native ECM fungi is attributed to deforestation and the long-term isolation of the Seychelles. Native ECM fungi associate with exotic eucalypts, whereas co-introduced ECM fungi persist in pine plantations for decades.

Facilitation of seedling establishment by ECM appears to be determined by the affinity between plants and ECM species, which reflects long-term relationships. A great diversity of plants and fungi engage in mycorrhizal associations. In natural habitats, and in an ecologically meaningful time span, these associations have evolved to improve the health of both plant and fungal symbionts. In systems managed by humans, mycorrhizal associations often improve plant productivity, but this is not always the case. Mycorrhizal fungi might be considered to be parasitic on plants when the net cost of the symbiosis exceeds net benefits. Parasitism can be developmentally induced, environmentally induced, or possibly genotypically induced (Johnson *et al.*, 1997).

2.2 Diversity

In a field survey of a Swedish boreal forest, between 60,000 and 1.2 million ectomycorrhizae were found in one square meter of forest soil and 95% of the root tips examined formed ectomycorrhizae (Jonsson, 1998). Bruns (1995) reported that 13 to 35 species exist in about 0.1 ha and the ECM fungal diversity is very high. Individual ECM fungal species were reported to possess different physiological features (Hung and Trappe, 1983; Abuzinadah and Read, 1986; Samson and Fortin, 1986) and functional roles to their host trees (Cairney, 1999; Koide *et al.*, 2007). High ECM diversity suggests that there is a potential for significant community-level effects of these associations on host plant performance. Jonsson *et al.* (2001) reported that biomass production of birch seedlings (*Betula pendula*) was greater when inoculated with eight ECM fungal species than with single species under low fertility conditions, but not under high fertility. Baxter and Dighton (2001) reported that ECM diversity per seedling was a better determinant of improved nutrient status of birch (*B. populifolia*) than species composition or colonization rates. The ECM diversity increases plant productivity and improves nutrient uptake of the host plant to a greater degree under nutrient limiting conditions.

Baxter and Dighton (2005) examined the effect of ECM diversity on *P. rigida* in unsterilised field soils. After one growing season, growth and nutrient uptake of *P. rigida* seedlings increased with increasing ECM diversity on tree root systems, and this effect was not considered to be due to fungal species composition. This result suggests that multiple inoculations of ECM fungi into host plants may achieve a successful outcome in afforestation efforts. In order to use this multiple inoculation successfully, further trials to select the number and composition of ECM fungal species and to develop methods of multiple inoculation of ECM species are needed.

3   THE ECTOMYCORRHIZAL RHIZOSPHERE

The rhizosphere is characterized by increased microbial activity stimulated by leakage and exudation of organic substances from the root (Grayston et al., 1998). Root exudates have been regarded as messengers that communicate and initiate biological and physical interactions between roots and soil-born organisms (Walker et al., 2003) and roots themselves are now understood to be rhizosphere ambassadors that facilitate communication between participants (Bais et al., 2006). On encountering a challenge, roots typically respond by secreting two classes of compounds. Small molecules such as amino acids, organic acids, sugars, phenolics, and other secondary metabolites account for much of the diversity of root exudates, whereas high-molecular weight compounds such as mucilage (polysaccharides) and proteins are less diverse but often comprise a larger proportion of the root exudates (Stintzi and Browse, 2000; Stotz et al., 2000; Bais et al., 2006). Plant roots seem to communicate with soil-borne organisms, although some can be positive (symbiotic) and others can be negative (parasitic or pathogenic) to the plant.

In positive associations root secretions may play symbiotic roles, depending on the other elements involved in the association (Walker et al., 2003). For instance, flavonoids in root exudates of legumes are well known to play an important role in activating *Rhizobium meliloti* genes responsible for the nodulation process (Peters et al., 1986). Akiyama et al. (2005) identified strigolactone, 5-deoxy-strigol, a group of sesquiterpenes from *Lotus japonicus* root exudates as an activating factor which triggers hyphal branching in dormant mycorrhizal fungi, *Gigaspora margarita*.

The growth rate of ectomycorrhizal fungi is promoted by pine root exudates, though different fungal species often react differently (Melin, 1963). Both palmitic acid and a cytokinin, isopentenylaminopurine are able to function as growth promoting factors (Sun and Fries, 1992). Horan and Chilvers (1990) investigated the ability of mycorrhizal fungi to penetrate membranes of plant roots, and suggested that there is a selective chemotropic attraction of these mycorrhizal fungi to substances diffusing from compatible host root apices. Such chemotropism could provide the signal that initiates the ectomycorrhizal infection process.

In negative plant-soil borne organism associations root exudates may function to defend plant roots. To survive continual attack by pathogenic and/or parasitic organisms, the delicate and physically unprotected root cells must depend on secretion of phytoalexins, defense proteins etc. (Flores et al., 1999). For instance, when elicited by fungal cell wall extracts from *Phytophthora cinnamomi*, or by in situ challenge with *Pythium ultimum*, basil roots exude rosmarinic acid that has antimicrobial activity against an array of soil microorganisms (Bais et al., 2002).

4 WATER AND NUTRIENT SUPPLY

4.1 Water supply

Hypotheses to explain mycorrhizal enhancement of root hydraulic conductivity are based on work with arbuscular mycorrhizae (AM), and water use has been found to be greater for AM plants than for non-mycorrhizal plants. AM and ectomycorrhizae are different in many respects, so they may alter host plant water uptake via different mechanisms. Coleman et al. (1990) examined hydraulic conductivity of Douglas fir (*Pseudotsuga menziesii*) seedlings inoculated with *Laccaria bicolor* or *Hebeloma crustuliniforme*, and non-inoculated seedlings infected naturally with *Thelephora* that were grown under three low levels of P fertilization (1, 10 and 100 µM P). Seedling morphology, tissue P levels, hydraulic conductivity and plant growth substance levels in xylem sap were measured after 9 months growth. Increased tissue P and decreased root/shoot ratio correlated with increased hydraulic conductivity in each of the mycorrhizal treatments. When adjusted for the effect of these two factors, hydraulic conductivity of *Laccaria* and *Hebeloma* seedlings was still lower than that for the *Thelephora* seedlings. In a subsequent experiment the hydraulic conductivity of seedlings with *Hebeloma* and *Rhizopogon vinicolor* mycorrhizae was compared to that of non-mycorrhizal seedlings (grown at 100 mM P) and no differences were found among treatments. The lack of an ectomycorrhizal effect on hydraulic conductivity is quite different from the enhancement of host hydraulic conductivity by AM fungi.

Nardini et al. (2000) investigated the physiological impact of ectomycorrhizal infection on the association between *Tuber melanosporum* and *Quercus ilex*. They compared a number of physiological parameters on 2-year-old seedlings inoculated for 22 months to those of non-inoculated plants. Inoculated seedlings had a 100% infection rate in root tips compared to a 25% infection rate in root tips of non-inoculated seedlings. Inoculated seedlings had higher values of net assimilation and stomatal conductance than non-inoculated seedlings. Root hydraulic conductance per unit root surface area of inoculated seedlings was reduced to 0.44% that of non-inoculated seedlings but had 2.5 times more fine root surface area than non-inoculated seedlings. When root conductance was scaled by leaf area, the inoculated seedlings had 1.27 times more the root conductance per unit leaf area compared to non-inoculated seedlings. Inoculated seedlings also had significantly higher hydraulic conductance of shoots with leaves, of shoots without leaves and lower leaf blade hydraulic resistances. Thus, the seedlings of *Quercus ilex* clearly suffered a disadvantage of lower hydraulic conductance of roots per unit root surface area due to the infection of *Tuber melanosporum*. This, however, seemed to be compensated by the increase in the amount of root (mass of fine roots and surface area) to provide a sufficient water supply to shoots.

4.2 Water stress

Bogeat-Triboulot *et al.* (2004) inoculated *Pinus pinaster* seedlings grown in a sandy dune soil with *Hebeloma cylindrosporum* and left others to natural colonisation. Six months later, they subjected half of the seedlings of both treatments to 3-weeks moderate drought. Root colonisation analysis showed that root tips were colonised to almost 100% independent of the inoculation. DNA determination of the ectomycorrhizal morphotypes showed that inoculated seedlings were extensively colonised by *H. cylindrosporum* (more than 75%) whereas non-inoculated seedlings were colonised by the exotic species *Thelephora terrestris* (50%) and *Laccaria bicolor* (30%) and to a lesser extent by *H. cylindrosporum* (20%). Drought did not affect these frequencies. Total plant biomass was not affected by the mycorrhizal status or by drought but the root/shoot biomass ratio as well as the root/leaf surface area ratio were much lower in seedlings extensively colonised by *H. cylindrosporum*. Root hydraulic conductivity was higher in plants mainly colonised by *H. cylindrosporum*, showing that this fungus improved the water uptake capacity of the root system as compared to *T. terrestris* and/or *L. bicolor*. This positive effect was also found, to a lesser extent, under drought conditions.

When inoculated with reconstituted dikaryons of *Pisolithus* sp. growth parameters (shoot length, shoot/root ratio and leaf area), nutrition and physiological indicators (transpiration rate, stomatal conductance and xylem water potential) of maritime pine (*Pinus pinaster*) seedlings were influenced during drought and in recovery from drought (Lamhamedi *et al.*, 1992). Seedlings colonized with certain dikaryons were more sensitive to water stress and showed less mycorrhiza formation under water stress than seedlings colonized with other dikaryons. Non-inoculated seedlings were significantly smaller than those inoculated with dikaryons. Transpiration rate, stomatal conductance and xylem water potential varied among mycorrhizal treatments during the water stress and recovery periods. After rewatering, the controls and seedlings inoculated with dikaryon 34 × 20 had a weaker recovery of transpiration rate, stomatal conductance and xylem water potential than the other treatments and experienced damage due to the water stress. Concentrations of various nutrient elements differed in shoots of *Pinus pinaster* colonized by the various dikaryons. Based on their results, Lamhamedi et al. (1992) expected that breeding of ectomycorhizal fungi could constitute a new tool for improving reforestation success in arid and semi-arid zones. Their results also suggest that the effects on water relation of host trees provided by ECM must be different not only between different fungal species but also between different dikaryons of the same species.

Pine wilt disease (PWD) is a serious forest epidemic which is caused by a nematode *Bursaphelenchus xylophilus* that is carried from dead pines to healthy pines by a sawyer beetle, *Monochamus* species. This disease has been spreading from one forest to another. It has been found, however, when a stand located on slopes is devastated by PWD, some pine trees survive on the ridge. Pines seem to survive better on the upper part of a slope than on the lower part. To compare growth conditions between various provenances of Japanese black pine, *Pinus thunbergii* and Japanese red pine, *Pinus densiflora*, approx. 4,000 pine seedlings of 23 families were planted on a slope in 1973. The area of the stand was ca. 1.4 ha, the slope is at an incline of 25 degrees, and the height of the slope is about 50 m. Since 1979, PWD spread into this stand and by the end of 1993, pine wilt damage became severe, with more than 70% of the trees killed (Fig. 1a). However, some of the provenances, even at the end of 1993 such as provenance No. 236 and 241 shown as the framed area in Fig. 1a, survived in a higher ratio. As shown in Table 1, the survival ratio of either provenance was apparently higher at the upper part, followed by middle and lowest at the lower part of the slope.

The quantity of mineral nutrients and water content are generally poor at the upper part of a slope compared with the lower part. Water stress seems more severe at upper parts than at lower parts of a slope. Similarly

Table 1. Survival ratio (%) of pine trees on a slope.

	No. 241	No. 236
Upper part	74	65
Middle part	54	39
Lower part	35	17

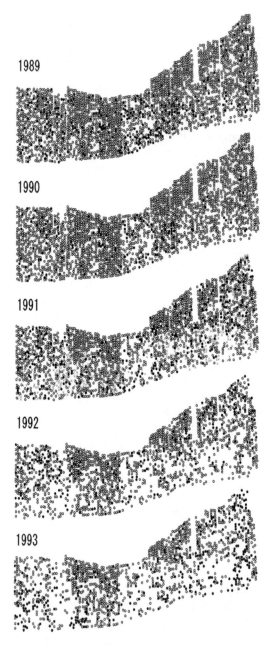

Fig. 1a. Decrease in surviving pine trees since 1989 to 1993 on a slope area in a red square served for mycorrhizal ratio measurement (Courtesy of Dr. Nakai, I.).

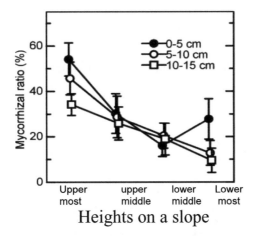

Heights on a slope

Fig. 1b. Mycorrhizal ratio in three depths of soil collected from different sites at the slope.

pine wilt is typically more severe at upper slopes, because pine wilt is exaggerated by drought. However, findings here are contrary due to mycorrhizal symbiosis. Species of *Pinus* are well-known ectomycorrhizal plants which obtain a proportion of their mineral nutrients and water from fungal symbionts. Under conditions of low nutrients and water, mycorrhizal relationships mitigate the deficiency of nutrients and drought stress. The mycorrhizal density was examined at four different heights of the slope (Akema and Futai, 2005). As shown in Fig. 1b, mycorrhizal ratios were higher at the higher part of the slope than at the lower part. The pines on the upper slope are continually under combined stresses of drought and nutrient deficiency. These plants mitigated their stress to drought and nutrient deficiency by well-developed mycorrhizal symbiosis, even when they were exposed to summer drought which damaged other pine trees grown at lower sites of the slope.

4.3 Nutrient supply

Marschner and Dell (1994) examined nutrient uptake in mycorrhizal symbiosis for three groups of mycorrhizae; the ectomycorrhizae (ECM), the ericoid mycorrhizae (EM), and the arbuscular mycorrhizae (AM). They argued that mycorrhizal infection may affect the mineral nutrition of the host plant directly by enhancing plant growth through nutrient acquisition by the fungus, or indirectly by modifying transpiration rates and the composition of rhizosphere microflora. ECM capacity for the external hyphae to take up and deliver nutrients to the plant has been demonstrated for P, NH_4^+, NO_3^- and K, but not for Ca, SO_4^{2-}, Cu, Zn and Fe. Knowledge of the role of ECM in the uptake of nutrients other than P and N is limited (Table 3). ECM fungi produce ectoenzymes that provide host plants with the organic N and P that are normally unavailable to AM fungi or non-mycorrhizal roots.

Table 2. The effect of ECM fungi on water stress tolerance of plants.

ECM fungi	Host plant	References
Culture experiment		
Boletus edulis, Cenococcum geophilum, Hebeloma crustuliniforme, Laccaria bicolor, Laccaria laccata, Lactarius controversus, Lactarius rufus, Leccinum auruntiacum, Rhizopogon vinicolor, Suillus albidipes, Suillus brevipes, Suillus caerulescens, Suillus granulatus, Suillus lakei, Suillus luteus, Suillus ponderosus, Suillus sibiricus, Tricholoma focale	Culture experiments examined the ability of each ECM fungus to tolerate imposed water stress with polyethylene glycol etc. in pure culture, and host plant was not used in the experiments.	Coleman and Bledsoe, 1989
Cenococcum graniforme, Gomphidius viscidus, Suillus luteus		Bingyun and Nioh, 1997
Cenococcum graniforme, Suillus luteus, Thelephora terrestris		Mexal and Reid, 1973
Effect of pre-inoculation of ECM fungi into plants in field		
Pisolithus albus, Scleroderma dictyosporum	*Acacia holosericea*	Duponnois et al., 2005
Pisolithus tinctorius	*Quercus velutina*	Dixon et al., 1983
Scleroderma verrucosum	*Cistus albidus, Quercus coccifera*	Caravaca et al., 2005
Effect of ECM fungi on plants in laboratory		
Hebeloma crustuliniforme, Laccaria laccata, Rhizopogon vinicolor	*Pseudotsuga menziesii*	Dosskey et al., 1991
Hebeloma longicaudum, Laccaria laccata, Paxillus involutus, Pisolithus tinctorius	*Picea mariana, Pinus banksiana*	Boyle and Hellenbrand, 1991
Gomphidius viscidus, Suillus luteus	*Pinus tabulaeformis*	Bingyun and Nioh, 1997
Laccaria laccata, Rhizopogon vinicolor, Pisolithus tinctorius	*Pseudotsuga menziesii*	Parke et al., 1983
Pisolithus tinctorius	*Pinus taeda*	Svenson et al., 1991
Rhizopogon occidentalis, Rhizopogon salebrosus, Rhizopogon vulgaris	*Pinus muricata*	Kennedy and Peay, 2007
Suillus mediterraneensis	*Pinus halepensis*	Morte et al., 2001

Plant nutrients, with the exception of nitrogen, are ultimately derived from weathering of primary minerals. Traditional theories about the role of ectomycorrhizal fungi in plant nutrition have emphasized quantitative effects on uptake and transport of dissolved nutrients. Qualitative effects of the symbiosis on the ability of plants to access organic nitrogen and phosphorus sources have also become increasingly apparent. Recent research suggests that ectomycorrhizal fungi mobilize other essential plant nutrients directly from minerals through excretion of organic acids. This enables ectomycorrhizal plants to utilize essential nutrients from insoluble mineral sources and affects nutrient cycling in forest systems (Landeweert *et al.*, 2001).

Wallander *et al.* (2004) inoculated pine (*Pinus sylvestris*) seedlings with indigenous ectomycorrhizal fungi using forest soil with four levels of wood ash addition (0, 1, 3 and 6 t ha^{-1}), and estimated the demand for P and K by seedlings grown in the different soils by measuring the uptake of ^{32}P and ^{86}Rb in a root. Utilisation of P from apatite was also tested in a laboratory where uptake by the ectomycorrhizal mycelium was separated from uptake by roots. The demand for P and K in the seedlings was similar regardless of the ash treatment. Uptake of P from apatite was on average 23% of total seedling P and was not related to the fungal biomass (ergosterol) in roots. The improved P uptake from apatite by ECM fungi found in earlier studies is probably not a general phenomenon among ECM fungi.

Martin *et al.* (1987) reviewed the metabolism of carbon and nitrogen compounds in ectomycorrhizal associations of trees. The absorption and translocation of mineral ions by mycelia require an energy source and a reducing agent which are both supplied by respiratory catabolism of carbohydrates produced by the host plant. Photosynthates are also required to generate the carbon skeletons for amino acid and carbohydrate syntheses during the growth of the mycelia. Competition for photosynthates occurs between the fungal cells and the various vegetative sinks in the host tree. The nature of carbon compounds involved in these processes, their routes of metabolism, the mechanisms of control and the partitioning of metabolites between the various sites of utilization are poorly understood.

The ability of ectomycorrhizal fungi to utilize organic nitrogen sources has been intensively investigated (Chalot and Brun, 1998). The fate of soil proteins, peptides and amino acids has been studied from a number of perspectives. Exocellular hydrolytic enzymes have been detected and characterized in a number of ectomycorrhizal and ericoid fungi. Studies on amino acid transport through the plasma membrane have demonstrated the ability of ectomycorrhizal fungi to take up the products of proteolytic activities. Investigations on intracellular metabolism of amino acids have allowed the identification of the metabolic pathways involved. Further translocation of amino acids in symbiotic tissue has been established by experiments using isotopic analysis. For instance, alanine metabolism in the ectomycorrhizal fungus *Paxillus involutus* was investigated using [^{15}N] alanine (Chalot et al., 1995). Short-term exposure of mycelial discs to [^{15}N] alanine showed that the greatest flow of ^{15}N was to glutamate and to aspartate. Levels of enrichment were as high as 15–20% for glutamate and 13–18% for aspartate, whereas that of alanine reached 30%. Radiolabel was also detected in the amino-N of glutamine and in serine and glycine, although at lower levels. Pre-incubation of mycelia with amino oxy acetate, an inhibitor of transamination reactions, resulted in complete inhibition of the flow of the label to glutamate, aspartate, and amino-N of glutamine, whereas [^{15}N] alanine rapidly accumulated. This evidence indicates the direct involvement of alanine aminotransferase for translocation of ^{15}N from alanine to glutamate. Alanine may be a convenient reservoir of both nitrogen and carbon.

Utilization of organic nitrogen by ectomycorrhizal fungi was examined by Tibbett et al. (1998), who grew arctic and temperate strains of *Hebeloma* spp. in axenic culture on glutamic acid, alanine, lysine and NH_4^+ as sole sources of nitrogen (N), with excess carbon (C) or deficient C (supplied as glucose). All strains tested had the capacity to assimilate amino acids and generally utilized alanine and glutamic acid more readily than NH_4^+. Some strains were able to utilize amino C when starved of glucose C, and could mineralize amino-N to NH_3-N. Arctic strains, in particular, appeared to be pre-adapted to the utilization of seed protein N and glutamic acid N, which is often liberated in high concentrations after soil freezing.

In relation to nitrogen uptake, the effects of Collembola grazing activities on ectomycorrhizal symbiosis were studied by Ek et al. (1994). Using laboratory microcosms, *Pinus contorta* seedlings in association with *Paxillus involutus* were grown in sandy soil and the collembola *Onychiurus armatus* was added in different densities. To study effects on nutrient uptake by the extramatrical mycorrhizal mycelium, cups containing $^{15}NH_4^+$ – and phytin-amended soil were evenly distributed in the microcosms. These cups were covered with a net that allowed the mycelium to penetrate but not collembola or plant roots. Extramatrical hyphal growth was impeded at a high density of *O. armatus*, while low densities of *O. armatus* increased extramatrical hyphal growth, colonization rate of side plants, and the biomass of *P. involutus*. However, the amount of *P. involutus* on/in the plant roots was not affected. Thus, low densities of collembolans induced a shift towards a larger proportion of *P. involutus* growing extramatrically. The presence of *O. armatus* in low numbers enhanced uptake and transfer of ^{15}N by *P. involvus* to the plants by up to 76%.

Table 3. ECM fungi increase the nutrient supply for plants.

ECM fungi	Host plant	Nutrient	References
Paxillus involutus, Suillus luteus, Suillus bovinus, Thelephora terrestris	Pinus sylvestris	P	Colpaert et al., 1999
Cenococcum geophilum, Pisolithus tinctorius	Pinus taeda	P	Rousseau et al., 1994
Laccaria bicolor, Pisolithus tinctorius, Paxillus involutus	Pinus rigida	P	Cumming, 1996
Descolea maculata, Pisolithus tinctorius, Laccaria laccata	Eticalyptus diversicolor	P	Bougher et al., 1990
Paxulus muelleri, Cortinarius globuliformis, Thaxterogaster sp., Hysterangium inflatum, Hydnangiurn carneum, Hymenogaster viscidus, Hymenogaster zeylanicus, Setchelliogaster sp., Laccaria laccata, Scleroderma verrucusom Amanita xanthocephala Descolea maculata	Eucalyptus globulus	P	Burgess et al., 1993
S. variegatus and unkonown ECM	Pinus sylvestris	P	Wallander, 2000
Indigenous ectomycorrhizal fungi	Quercus robur, Betula pendula	N, P, K	Newton and Pigott, 1990
Indigenous ectomycorrhizal fungi	Pinus sylvestris	P, K	Wallander et al., 2005
Paxilus involtus	Picea abies Betua pendula	P, Ca	Andersson et al., 1996
Laccaria laccata	Picea mariana	N	Quoreshi and Timmer, 2000
A. rubescens, L. deterrimus	Pinus sylvestris	N	Taylor et al., 2004
Hebeloma crustuliniforme, Amanita muscaria, Paxillus incolutus	Betula pendula	N	Abuzinadah and Read, 1986
P. tinctorius	Pinus resinosa	N	Wu et al., 2003
Rhizopogon roseolus, Suillus bovinus, Pisolithus tinctorius, Paxillus involutus	Pinus sylvestris	N	Finlay et al., 1988
T. terrestris, unidentified ECM	Pinus contorta	N	Finlay and Söderström, 1992
Indigenous ectomycorrhizal fungi	Picea engelmannii	N	Grenon et al., 2004
Indigenous ectomycorrhizal fungi	Fagus sylvatica	N	Geßler et al., 2005
Pisolithus arhizus	Pinus sylvestris	N	Högberg, 1989
Piloderma croceum, Piloderma spp., unidentified fungus	Picea abies	N, P, K, Ca	Mahmood et al., 2003

The organic soil horizons of heathland and temperate forest ecosystems are characteristically rich in phenolics, which present barriers to organic N availability to soil microflora. The abilities of ectomycorrhizal,, ericoid mycorrhizal and wood decomposing saprotrophic fungi to degrade model compounds representing the insoluble phenolic lignin, and soluble phenolics, which provide physical and chemical barriers respectively to organic N availability, were compared (Bending and Read, 1997). No clear relationship was found between ability to degrade lignin and soluble phenolics. The presumptive assays indicated that most mycorrhizal fungi have only low abilities to degrade these compounds relative to the wood decomposing fungi.

5 FACILITATION OF SEEDLING ESTABLISHMENT AND GROWTH PROMOTION

To develop an effective production system of mycorrhizal seedlings for afforestation or reforestation, significant efforts have been undertaken. There have been several problems, however, including inoculation techniques, survival of early developmental stages, specificity between host-ECM fungi, and environmental factors. Each of these issues is discussed below.

5.1 Inoculation technique

Pisolithus tinctorius has been used in forestry inoculation programmes (Cairney and Chambers, 1997), and a number of inoculation protocols were developed for this fungus. Marx *et al.* (1982) examined the effectiveness of *P. tinctorius* in forming ectomycorrhizae on container-grown seedlings of ten pine species, Douglas fir, western hemlock, and bur oak. Their results were as follows.

- Inocula of *P. tinctorius* was mixed into rooting media before sowing of seed. A medium of vermiculite and 5–10% by volume peat moss with nutrients was best for growing mycelial inoculum. Peat moss, which contains humic acids, was used for keeping pH of the inoculum below 6.0 at which range it was the most effective. Inoculum was also most effective after leaching with water to remove nutrients.
- No single inoculum characteristic, such as number of *P. tinctorius* propagules in large and small particles, microbial contamination, residual glucose, bulk density, and moisture content, as well as results of a fast assay for ectomycorrhizal development on loblolly pine seedlings, was consistently correlated with effectiveness of

inoculum in forming *P. tinctorius* ectomycorrhizae on seedlings in containers.
- A captan drench after seeding significantly improved effectiveness of inoculum that was initially low in effectiveness. (captan is a fungicide used until the end of 1980s in the United States).
- Seedling growth was correlated with *P. tinctorius* ectomycorrhizal development in only a few tests. The probable cause for lack of growth stimulation of seedlings by *P. tinctorius* ectomycorrhizae is photosynthate drain on the juvenile seedlings by *P. tinctorius*.
- Jack pine seedlings grown in a medium containing high levels of N, P, and K formed about half as many *P. tinctorius* ectomycorrhizae as similarly treated seedlings grown at about half this level of fertility. The seedlings grown at high fertility status were, however, larger regardless of ectomycorrhizal treatment.

In the basic pure-culture technique, Chilvers *et al.* (1986) grew ectomycorrhizal fungi on stiff absorbent paper over Potato Dextrose Agar. The paper was subsequently removed aseptically and laid, fungus-side down, on to a two-dimensional array of roots of seedlings grown axenically on filter paper over a mineral salts agar medium. With the high inoculum potential achieved by this technique, ectomycorrhizae formed within 1 week of inoculating *Pisolithus tinctorius* or *Paxillus involutus* on to *Eucalyptus globulins* sub.sp. *bicostata*, and over 50% of root apices were colonized by ectomycorrhizae within 2 weeks. The paper-based inoculum has also been applied successfully to convert peripheral roots of seedlings grown in peat moss within clay pots.

To prevent secondary infection with native fungi, Teste *et al.* (2006) conducted greenhouse experiments using Douglas fir (*Pseudotsuga menziesii* var. glauca) seedlings where chemical methods (fungicides) or physical methods (mesh barriers) were used. They partly succeeded in reducing ectomycorrhizal colonization by approximately 55% with the application of tiazole fungicide. Meshes with pore sizes of 0.2 and 1 μm were effective in preventing the formation of mycorrhizae via hyphal growth across the mesh barriers.

To compare the suitability of substrate for mycorrhizal Norway spruce seedlings, Repáč (2007) examined *Sphagnum* peat, spruce bark compost, peat + perlite (1:1, v:v) and compost + perlite (1:1, v:v) by inoculating seed with a vegetative alginate-bead inoculum of *Hebeloma crustuliniforme*, *Hygrophorus agathosmus* or *Paxillus involutus* or left uninoculated prior to seed addition. Growth and percentage of mycorrhization of bare root seedlings cultivated in the greenhouse were evaluated after the first growing season. Seedlings grown in peat-based substrates had significantly larger aboveground and total dry weight, but significantly lower mycorrhization percentage than those grown in compost-based substrates. There were no significant differences between fungal treatments (including control) for both the percentage of mycorrhization and growth of seedlings. Growth parameters were negatively correlated with the extent of mycorrhization, indicating allocation of host photosynthates to the fungi. Based on these results the authors concluded that the artificially introduced fungi were not efficient in mycorrhizal formation, because naturally occurring fungi were common in all treatments. Operational inoculation of specific tree seedlings with the test fungi was proposed.

5.2 Early development of seedlings

To compare the symbiotic function of ECM fungus with amendment formulations, Walker (2001) examined shoot growth and root growth in terms of mass and total length, of containerized sugar pine (*Pinus lambertiana*) and Jeffrey pine (*Pinus jeffreyi*) seedlings treated with two nutrient formulations, and those with ECM inoculation with pelletized basidiospores of *Pisolithus tinctorius*. Overall, results of experiments indicate that the high rate of amendment formulation produced the most favorable array of attributes in both sugar and Jeffrey pine, but that *P. tinctorius* is likely a more promising mycobiont for inoculation of the latter species than the former.

Pisolithus tinctorius has increased plantation survival and growth in Nevada and in the southeastern United States. When bare-root stock of Douglas fir, lodgepole pine, white fir, and grand fir, was inoculated with *P. tinctorius* and handled by standard procedures, performance was no better than for stock which was naturally infected with indigenous flora in a nursery in southwestern Oregon (Castellano and Trappe, 1991). Climate, planting sites and nursery practices in the Pacific Northwest differ drastically from those in the southeastern United States. Thus, the reaction of each ECM fungus to each phase of the nursery and planting process must be carefully analyzed before any ECM fungus can be introduced for nursery inoculation.

If we select the incorrect ECM fungus and/or use the incorrect procedure for host plants without careful analysis, the inoculation may decrease the survival ratio of the seedlings. For instance, when plants were inoculated with three ECM fungi, *Hebeloma crustuliniforme, Paxillus involutus* and *Thelephora terrestris* in a peat/vermiculite mixture during the transition period from in vitro to glasshouse conditions, survival of inoculated plants was generally lower than that of uninoculated controls. Lowest survival (60%) was observed in the presence of *T. terrestris*. This fungus, however, gave the highest frequency of root infections, a significant increase in shoot height and a doubling of shoot dry weight. The content of N, P and K in plants with ECM was higher than for uninfected control plants (Heslin and Douglas, 1986).

Nutrient uptake rates in mycorrhizal roots may influence the growth, vigor and survivability of the seedling. Under controlled conditions in a semi-hydroponic culture system with quartz sand as substrate and a percolating nutrient solution, Eltrop and Marschner (1996) compared growth, nitrogen uptake and mineral nutrient concentrations in plant tissue between mycorrhizal and non-mycorrhizal Norway spruce (*Picea abies*) seedlings. The culture system allowed the determination of nutrient uptake rates in mycorrhizal root systems with an intact extramatrical mycelium. Among three ECM fungi examined, the infection rate of the roots by *P. tinctorius* and *Laccaria laccata* was high but the infection rate by *Paxillus involutus* was low.

When nitrogen was supplied with ammonium nitrate, dry weight of roots and shoots was significantly lower in mycorrhizal than in non-mycorrhizal plants. Depletion of ammonium in the external solution was more rapid than was depletion of nitrate. When plants were supplied with ammonium but not nitrate as the N source, dry weight was lower in mycorrhizal plants infected with *P. tinctorius* than it was in non-mycorrhizal plants. Therefore, N uptake rates were increased in mycorrhizal plants with *P. tinctorius* only when they were supplied with ammonium but not with nitrate.

Uptake rates of N, P, K, Ca and Mg was not significantly different between non-mycorrhizal and mycorrhizal plants. This finding indicates that the extramatrical mycelium may play an important role in nutrient uptake only when spatial nutrient availability is limited. The decreased growth of mycorrhizal plants is attributed to the demand of the mycorrhizal fungus for photosynthates, i.e., source limitation. Eltrop and Marschner (1996) examined the possible reasons for this growth depression in mycorrhizal plants. Based on the results of several experiments, they concluded that increased root respiration was mainly responsible for the growth reduction in mycorrhizal compared with non-mycorrhizal plants, whereas the production of fungal biomass in the extramatrical mycelium of mycorrhizal plants was of minor importance.

By placing a layer of pine or oak litter on the surface of the nursery bed soil to mimic natural litter cover, Aučina *et al.* (2007) studied the effects of pine and oak litter on species composition and diversity of mycorrhizal fungi colonizing 2-year-old *Pinus sylvestris* seedlings grown in a bare-root nursery in Lithuania. Oak litter appeared to be most favorable to seedling survival, with a 73% survival rate in comparison to the untreated mineral bed-soil (44%). This field experiment provides preliminary evidence that changes in the supply of organic matter through litter manipulation may have far-reaching effects on the chemistry of soil, thus influencing the growth and survival of *Pinus sylvestris* seedlings and their mycorrhizal communities.

5.3 Environmental factors

Mycorrhizal associations generally benefit vascular plants in nutrient-poor conditions because symbiotic fungi increase the absorptive surface of roots and offer to their host plants better access to soil mineral nutrients (Koide, 1991). Some mycorrhizal fungi are also able to utilize soil organic nitrogen and phosphorus pools. However, mycorrhiza also impart a considerable cost to the host plant, as mycorrhizal roots may consume a 2.5-fold greater amount of host carbon compared with non-mycorrhizal roots (Jones et al., 1991). Therefore, non-mycorrhizal plants are common in high arctic and alpine areas, which are often poor in nitrogen and phosphorus. The relative proportion of mycorrhizal plants has been found to decrease along with increasing altitude, suggesting that the advantage of mycorrhizal symbiosis may change along an altitudinal gradient. This effect may be related to the environmental factors that possibly constrain the amount of photosynthesized carbon to be shared with mycorrhizal fungi (Ruotsalainen et al., 2002).

The widespread decline in dominant tree species may occur world wide by the introduction of invasive insects and pathogens, and lead to cascading effects on other tree species and microorganisms including ECM fungi (Lewis et al., 2004). In the eastern USA, for example, eastern hemlock (*Tsuga canadensis*) is declining because of infestation by the hemlock woolly adelgid (*Adelges tsugae*). Northern red oak (*Quercus rubra*) is a common replacement species in declining hemlock stands, but reduced mycorrhizal inoculum potential in infested hemlock stands may cause oak to grow more slowly compared with oak growing in oak stands. We grew red oak seedlings for one growing season in declining hemlock-dominated stands infested with hemlock woolly adelgid (HWA) and in adjacent oak-dominated stands. Ectomycorrhizal root tip density and morphotype richness in soil cores were 63% and 27% less, respectively, in declining hemlock stands than in oak stands. Similarly, ectomycorrhizal percent colonization and morphotype richness on oak seedlings were 33% and 30% less, respectively, in declining hemlock stands than in oak stands. In addition, oak seedlings in the hemlock stands had 29% less dry mass than did seedlings in oak stands. Analysis of covariance indicated that morphotype richness could account for differences in oak seedling dry mass between declining hemlock stands and oak stands. Additionally, oak seedling dry mass in declining hemlock stands significantly decreased with decreasing ectomycorrhizal percent colonization and morphotype richness. These results suggest that oak seedling growth in declining hemlock stands is affected by reduced ectomycorrhizal inoculum potential. Further, the rate of forest recovery following hemlock decline associated with HWA infestation may be slowed by indirect effects of HWA on the growth of replacement species, through effects on ectomycorrhizal colonization and morphotype richness.

Mycorrhizal fungi might be considered to be parasitic on plants when net cost of the symbiosis exceeds net benefits (Johnson et al., 1997). Parasitism can be developmentally induced, environmentally induced, or possibly genotypically induced. Morphological, phenological, and physiological characteristics of the symbionts influence the functioning of mycorrhizae at an individual scale. Biotic and abiotic factors at the rhizosphere, community, and ecosystem scales further mediate mycorrhizal functioning. Despite the complexity of mycorrhizal associations, it might be possible to construct predictive models of mycorrhizal functioning. These models will need to incorporate variables and parameters that account for differences in plant responses to, and control of, mycorrhizal fungi, and differences in fungal effects on, and responses to, the plant. Developing and testing quantitative models of mycorrhizal functioning in the field requires creative experimental manipulations and measurements. Such work will be facilitated by recent advances in molecular and biochemical techniques. A greater understanding of how mycorrhizae function in complex natural systems is a prerequisite to managing them in agriculture, forestry, and site restoration.

6 DISEASE RESISTANCE AND RESPONSE TO ATMOSPHERIC POLLUTANTS

6.1 Disease resistance

Disease protection by mycorrhizal fungi has been demonstrated to be important in the field (Table 4); however, mechanism(s) by which ECM fungi inhibit diseases requires attention. Whipps (2004) suggested four major modes of actions by which mycorrhizal fungi increase the disease resistance of host species: (1) direct competition or inhibition, (2) enhanced or altered plant growth, nutrition and morphology, (3) biochemical changes associated with plant defence mechanisms and induced resistance, and (4) development of an antagonistic microbiota. Mode (1) includes competition for nutrients, particularly carbon compounds derived from phytosysthesis. This phenomenon has been studied as a potential mode of action for pathogen control in arbuscular mycorrhizal (AM) plants, but has thus far been neglected in ECM plants, and further studies are needed (Whipps, 2004). In addition, mode (1) includes the production and release of antibiotics and the physical sheathing of the mantle of ectomycorrhiza (Marx, 1969; Duchesne et al., 1988a, b; Branzanti et al., 1999). Oxalic and other organic acids have been reported to function as antibiotics (Duchesne et al., 1988b, 1989; Rasanayagam and

Jefferies, 1992). An example of mode (2) may be the mycorrhizal resistance effect for pine trees to infection by the pinewood nematode, *Bursaphelenchus xylophilus*, a pathogen involved in Pine Wilt Disease. The inoculation of Japanese red pine *Pinus densiflora*, with ectomycorrrhizal fungi *Suillus lubens* and *Rhizopogon rubescens* improved growth of seedlings, thereby decreasing seedling mortality caused by pinewood nematode (Kikuchi et al., 1991). Mode (3) occurs during plant resistance responses including production of phenolics and phytoalexins; formation of structural barriers to prevent pathogen ingress; production of numerous pathogenesis related (PR) proteins; production of enzymes that degrade the cell wall, such as chitinases; and production of enzymes associated with production of phenolics, phytoalexins, and structural barriers, including phenylalanine amminia lyase, chalcone synthase, chalcone isomerase, and superoxide dismutase (Guenoune et al., 2001; Guillon et al., 2002). The induced resistance created by inoculation with ECM fungi prior to challenges by pathogens was studied, and it was revealed that infusion of phenolic compounds and production of terpenes led to plant resistance to the pathogens (Krupa and Fries, 1971; Strobel and Sinclair, 1991a, b). Mode (4) seems to be realized by the improvement of mycorrrhizal formation and enhancement of plant growth by several bacteria (Bending et al., 2002; Duponnois and Garbaye, 1991). In addition, rhizosphere bacteria directly inhibit the growth and sporulation of pathogenic fungi thus inhibiting disease (Schelkle and Peterson, 1996; Pedersen and Chakravarty, 1999).

A high rate of fertilizers (N/P/K) and fungicides were applied to maintain seedlings in forest nursery (Lilja et al., 1997); however, these treatments generally reduced the ectomycorrhiza-host symbiosis (Tammi et al., 2001). Sen (2001) examined the suppressive effects of forest-adapted mycorrhizal fungi and nursery-adapted ones on *Rhizoctonia* sp. under conditions of low nitrogen. *Wilcoxina mikorae*, which is adapted for nursery growth, did not suppress the disease caused by *Rhizoctonia* sp., while *Suillus bovines*, which is adapted to forest ecosystems, suppressed the disease. Moreover, *Pseudomonas* sp. and *Bacillus* sp. were isolated from the surrounding soil of *Suillus bovines*-mycorrhiza. Sen (2001) suggested that interactions between these bacteria and the suppressive effect of mycorrhizal fungi may be occurring.

Interactions between mycorrhizal fungi and other microorganisms are important for control of plant disease under natural conditions. *Pseudomonas fluorescens*, which can suppress pathogenic fungi such as *Rhizoctonia* spp., *Fusarium* spp., *Phytophthora* spp., and *Heterobasidion* spp. had a higher population density in the mycorrhizosphere soil of Douglas fir-*L. bicolor* than in other soil (Frey-Klett et al., 2005). Moreover, some strains have been

Table 4. ECM fungi reduce severity of various fungal diseases of plants.

ECM fungi	Host plant	Pathogenic fungi	References
Clitocybe claviceps, Laccaria bicolor, Paxillus involutus	Picea glauca, Pinus contorta,	Fusarium moniliforme, F. oxysporum	Chakravarty et al., 1999
Hebeloma crustiliniforme, Laccaria laccata, Pisolithus tinctorius	Pinus sylvestris	F. moniliforme, Rhizoctonia solani	Chakravarty and Unestam, 1987b
Hebeloma crustiliniforme, Hebeloma sinapizans, Laccaria laccata, Paxillus involutus	Castanea sativa	Phytophthora cambivora, P. cinnamomi	Branzanti et al., 1999
Laccaria bicolor, Laccaria laccata	Picea abies, Pseudotsuga menziesii	Fusarium oxysporum	Sampangi et al., 1986
Laccaria bicolor	Pseudotsuga menziesii	Fusarium oxysporum	Strobel and Sinclair, 1991b
Laccaria laccata, Pisolithus tinctorius	Pinus sylvestris	Cylindrocarpon destructans	Chakravarty and Unestam, 1987a
Laccaria laccata	Pseudotsuga menziesii	Fusarium oxysporum	Sylvia, 1983
Paxillus involutus	Picea mariana	Cylindrocladium floridanum	Morin et al., 1999
Paxillus involutus	Pinus banksiana	Fusarium moniliforme	Hwang et al., 1995
Paxillus involutus	Pinus contorta	F. moniliforme, F. oxysporum	Pedersen et al., 1999
Paxillus involutus	Pinus resinosa	F. moniliforme, F. oxysporum	Chakravarty et al., 1991
Paxillus involutus	Pinus resinosa	Fusarium oxysporum	Duchesne et al., 1988b
Paxillus involutus	Pinus resinosa	Fusarium oxysporum	Duchesne et al., 1989
Paxillus involutus	Pinus resinosa	Fusarium oxysporum	Farquhar and Peterson, 1991
Paxillus involutus, Suillus bovines, Wilcoxina mikolae	Pinus sylvestris	Rhizoctonia sp.	Sen, 2001

reported as mycorrhiza helper bacteria; *Bacillus subtilis* MB3, *Pseudomonas* spp. (BBc6 and Bc13) (Schelkle and Peterson, 1996) and *Streptomyces* spp. (Becker et al., 1999; Maier et al., 2004; Schrey et al., 2005) are reported to have suppressed plant pathogens. *Streptomyces* AcH505 particularly showed fungal specificity, which suppressed the growth of the plant pathogen while it enhanced the growth of mycorrhizal fungi (Schrey et al., 2005). Thereafter, three compounds (Auxofuran, WS-5995 B and C) induced by *Streptomyces* AcH505 were identified, and it was revealed that auxofuran enhanced the growth of mycorrhizal fungi and WS-5995 B and C suppressed the growth of the plant pathogen (Riedlinger et al., 2006).

6.2 Response to atmospheric pollution

Dighton and Jansen (1991) reviewed the effects of pollutants on mycorrhizae, primarily ectomycorrhizae. The authors had to conclude that the effects of pollutants on mycorrhizae are not clear. For instance, some damage to roots and mycorrhizal fungi by pollutants result from the toxicity of increased Al availability and imbalances in Ca: Al ratios. But quantitative information on the degree to which the functioning of the mycorrhizal association is impaired had been lacking. The effects of pollutants on the rate of photosynthesis and subsequent allocation of carbohydrates to maintenance of mycorrhizae are also possible mechanisms for damage due to pollution. Thus, the methods used in experiments, levels of pollutants used and relationships between studies on seedlings to those of a mature forest do not suggest consistent models of pollution effects.

Stankevičienė and Pečiulytė (2004) studied ectomycorrhizae (ECM) and soil microfungi in soil cores obtained from seven unequally polluted forest plots spaced at different distances from a fertilizer factory in Lithuania. To evaluate the ECM state they chose the following three criteria: (1) number of ectomycorrhizal root tips in 100 cm of soil, (2) length of ectomycorrhizal roots in 100 cm^3 of soil, and (3) diversity of ectomycorrhizal morphotypes in the soil. Abundance of ECM roots and soil microfungi was visibly different in separate investigation plots. Average numbers of ECM root tips during the investigation period (2000–2002) in different forests ranged from 134 to 1,017/100 cm^3 of soil and the length of ECM roots ranged from 12.2 to 79.8 cm/100 cm^3. Total numbers of viable soil fungi varied from 1.5 to 566.6 thousands CFU/g dry weight soil. The forest farthest from the factory exhibited the highest ECM abundance and diversity of ECM morphotypes, while the abundance of soil microfungi was lowest. The lowest diversity of ECM morphotypes was detected in forests characterized by the highest concentration of heavy metals and lowest concentrations of nutrients (N, P), and the highest microfungal abundance was in forests with the highest nutrient concentrations. It was concluded that microfungi and ECM can act as important evaluation criteria in soil monitoring for afforestation purposes due to their significant reaction to pollution-induced chemical soil changes.

To determine the effect of acid precipitation on Japanese black pine (*Pinus thunbergii*) with and without mycorrhizae (*Pisolithus tinctorius*),

1-year-old seedlings were exposed to simulated acid mist (SAM), pH 3.0, for 10 min per day twice a week for 3 or 4 months. To estimate the effect of SAM and that of ECM, the following criteria were measured: (1) fresh weight of roots, shoots, and whole seedlings, (2) chlorophyll a and b contents, (3) transpiration rate, (4) extractable phosphorus content in shoots or roots, and (4) percentage of mycorrhizal root tips and dichotomous root tips. SAM adversely affected the weight of whole seedlings and the transpiration rate, and decreased the extractable phosphorus content of seedlings. These adverse effects were partly compensated with ECM inoculation. SAM, however, retarded mycorrhiza formation (Maehara et al., 1993).

7 ECTOMYCORRHIZAL FUNGI AS BIOREMEDIATION AGENTS

Persistent organic pollutants (POPs) such as DDT (dichlorodiphenyl trichloroethane) and polycyclic aromatic hydrocarbons (PAHs) have historically been serious environmental and public health problems, because of their long distance mobility, high organism accumulation and persistence. ECM fungi may have the ability to grow in, and possibly decompose such compounds because these fungi, associated with the host plant, are distributed throughout the soil and provided with a long-term supply of photosynthetic carbon from their plant hosts. In addition, ECM fungi may enhance the activites of indigenous soil microorganisms. To date, the ability of ECM fungi to degrade lignin and soluble phenolics (Bending and Read, 1997), has given rise to the possibility of applying ECM to bioremediation of aromatic pollutants (Donnelly et al., 1994). Out of the 42 species of ECM fungi screened thus far, 33 have been shown to degrade one or more classes of chemicals (Meharg and Cairney, 2000). Moreover, Braun-Lullemann et al. (1999) isolated 16 genera (27 strains) via screening of those strains superior for degradation of PAHs. *Amanita excelsa, Leccinum versipelle, Suillus grevillei, S. luteus*, and *S. variegates* could remove up to 50% of PAHs, including very recalcitrant compounds such as benzo[α]pyrene from solution culture. Meharg et al. (1997) demonstrated that isolates of *Suillus variegafus* and *Paxillus involutus* mineralize 2, 4-dichlorophenol both in axenic culture and in symbiosis with *Pinus sylvstris*.

Ectomycorrhizal fungi impart a protective effect to plants via prevention of translocation of heavy metals into the host (Galli et al., 1993; Tam, 1995). For example, Turnau et al. (1994) revealed that heavy metals like Cd, Cu and Fe accumulated mainly in electron opaque granules and in the outer pigmented layer of the cell wall, both characterized by the presence of polysaccharides and cystein-rich proteins. On the other hand, some studies report that ectomycorrhizal fungi do not limit heavy metal concentration in their hosts (Godbold *et al.*, 1998; Bucking and Heyser, 1994). Sell *et al.* (2005) showed that *P. involutus* significantly enhanced total Cd extraction by *P. canadensis*. Several fungal species reduced Zn accumulation in pine seedlings; *Thelephora terrestris*, however, increased Zn concentration in its host plants (Colpaert and Van assche, 1992).

Organic acids can bind heavy metals, and thus serve as detoxification agents. Ahonen-Jonnarth *et al.* (2000) studied the action of low molecular weight organic acids produced by mycorrhizal *Pinus sylvestris* exposed to elevated aluminium and heavy metal concentrations. The production of acid-bound substances led to enhanced absorption of heavy metals which combined with phosphate in soil (Fomina *et al.*, 2006).

Ectomycorrhizal fungi may indirectly influence degradation or uptake of soil contaminants in the rhizosphere via mycorrhizosphere effects (Meharg and Cairney, 2000). There may be several advantages to using a combined plant-mycorrhiza-bacteria system for biodegradation of soil pollutants. Interactions between ECM fungi and other soil microbes are complex and are currently poorly understood; however, it is known that bacterial communities can be markedly altered in the mycorrhizosphere compared to the rhizosphere of non-mycorrhizal roots (Rambelli, 1973). Cellular interactions and catabolic activities of mycorrhizal root associated non-sporulating bacteria were investigated in a simplified phytoremediation simulation involving a woody plant species (Sarand *et al.*, 1998). The tolerance and degradation of m-toluate by *Pinus sylvestris*, a symbiotic *Suillus bovinus* and *Pseudomonas fluorescens* strains was determined (Sarand *et al.*, 1999). It may be difficult, however, to separate the role and/or interaction of each organism.

8 VEGETATION SUCCESSION AND ECTOMYCORRHIZAL FUNGI

The development of a biological community by the action of vegetation on the environment leading to the establishment of new species is termed succession. Primary succession is the term used when this phenomenon occurs on a new, sterile area, such as that uncovered by a retreating glacier or created by an erupting volcano. Most successional processes are secondary successions, i.e., the recovery of disturbed sites by fire, disease, or human-induced land clearing processes (Clements, 1916; Begon *et al.*, 1996). Succession has been mainly explained by changes in environments and resource competition between plants (Connell and Slatyer, 1977; Tilman, 1985). However, it has been revealed that mycorrhizal fungi also play an important role in vegetative succession (Smith and Read, 1997; van der Heijden *et al.*, 1998).

In the early stages of primary succession at a severely disturbed site, non-mycorrhizal and facultative mycorrhizal plants tend to occur and dominate (Allen *et al.*, 1987; Allen, 1988, 1991). These are followed by obligately arbuscular mycorrhizal plants, and later by ectomycorrhizal and ericoid mycorrhizal species (Read, 1989, 1992).

In early stages of succession, ECM fungi inoculum is limited. In an early succession volcanic desert, ECM colonization of *Salix reinii* seedlings strongly depended on nearby established *S. reinii* shrubs. Growth and nitrogen content of seedlings increased significantly with number of associated ECM fungal species and ECM root tips (Nara and Hogetsu, 2004). In addition, ECM fungi on *S. reinii* contributed to tree succession from *S. reinii* to *Betula ermanii* and *Larix kaempferi* (Nara, 2006b). These results suggest that infection by ECM fungi is important in early forest succession, where ECM fungal inoculum is limited.

In Australia an introduced tree, *P. radiate*, did not regenerate naturally because native fungal partners were absent (Tommerup *et al.*, 1987). Inoculation of ECM fungi improved the establishment of pine seedlings. In natural tropical forests, about 95% of tree species are endomycorrhizal (Le Tacon *et al.*, 1987) and the seedlings of introduced non-native tree species (pines, eucalypts and casuarinas) were stunted and chlorotic when ECM fungi was not inoculated. However, inoculation of ECM fungi into the seedlings improved growth.

In secondary succession, nitrogen is a limiting nutrient and a key plant growth factor (Odum, 1960; Golley, 1965; Tilman, 1987). The nitrogen status of the ecosystem changes from which inorganic nitrogen predominates, to the condition in which plant residues sequester nitrogen largely in organic form. Abuzinadah and Read (1986) reported that the ability of ECM fungal species to utilize peptides and proteins differed. *Laccaria laccata* and *Pisolithus tinctorius* had little ability to grow on peptides or proteins, but *Suillus bovinus* and *Rhizopogon roseolus* grew vigorously on both compounds. *L. laccata* and *P. tinctorius* were frequently observed in environments in which inorganic nitrogen predominated, whereas *S. bovinus* and *R. roseolus* were frequently observed in environments in which organic nitrogen predominated. ECM colonization with the fungi, which are adaptive and can acquire nitrogen, may increase the nitrogen uptake and adaptation of the host plants.

Some works reveal that nitrogen is saturated in polluted forests by anthropogenic factors such as deposition of atmospheric pollutants (Aber *et al.*, 1989; Harrison *et al.*, 1995) and presence of exotic invasive species (Taniguchi *et al.*, 2007). Deposition of nitrogen in forests may induce limitations or imbalances of other nutrients such as phosphorus and potassium (Mohren et al., 1986; Aber et al., 1989; Harrison et al., 1995). Taniguchi et al. (2008) reported that phosphatase activity of dominant ECM fungal species was higher in a nitrogen-rich forest than in a nitrogen-poor forest. Therefore, the ability of ECM fungal species to take up phosphorus or potassium may be significant in nitrogen-saturated forests.

Table 5. ECM fungi act as bioremediation agents.

ECM fungi	Host Plant	Pollutants	References
Thelephora terrestris, Laccaria laccata	Pinus sylvestris	PAHs	Genney et al., 2004
Suillus bovinus, Hebeloma crustuliniforme	Pinus sylvestris	PAHs	Joner et al., 2006
Paxillus involutus, Suillus bovinus	Pinus sylvestris	benzoic acid, 4-hydroxybenzoic acid	Dittmann et al., 2002
Paxillus involutus, Suillus variegatus	Pinus sylvestris	2,4-dichlorophenol	Meharg et al., 1997
Suillus bovinus	Pinus sylvestris	m-toluene	Sarand et al., 1998
Paxillus involutus, Pisolithus tinctorius	Salix viminalis, Populus canadensis	Cd	Sell et al., 2005
Rhizopogon roseolus	Pinus sylvestris	Cd, Al	Turnau et al., 1996
Paxillus involutus, Suillus variegatus	Pinus sylvestris	Cd, Zn	Hartley-Whitaker et al., 2000
Suillus bovinus	Pinus sylvestris	Zn	Bucking and Heyser, 1994
Thelephora terrestris, Paxillus involtus, Amanita muscaria	Pinus sylvestris	Zn	Colpaert and Van assche, 1992
Suillus variegatus, Rhizopogon roseolus, Paxillus involtus	Pinus sylvestris	Al, Cd, Cu, Ni	Ahonen-Jonnarth et al., 2000
Laccarius thiogalus, Lactarius rufus, Paxillus involtus	Picea abies	Al	Hentschel et al., 1993 Jentschke et al., 1991a, b
Laccaria bicolor	Pinus sylvestris	Ni, Cd	Ahonen-Jonnarth and Finlay, 2001
Lactarius rufus, Lactarius hibbardae, Laccaria proxima, Scleroderma flavidum	Betula papyrifera	Ni	Jones and Hutchinson, 1986
Thelephora terrestris, Paxillus involtus, Amanita muscaria	Betula papyrifera	Zn	Brown and Wilkins, 1985
Paxillus involutus	Pinus sylvestris	Zn	Fomina et al., 2006
Suillus bovinus	Pinus sylvestris	Zn	Adriaensen et al., 2003

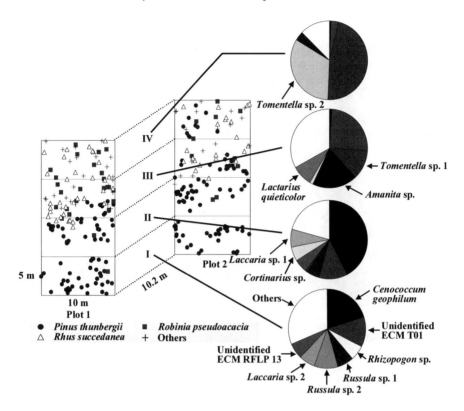

Fig. 2. ECM community on pine seedlings in subplot I (*P. thunbergii*-dominated subplot), II (*P. thunbergii*-dominated subplot mixed with a few *R. pseudoacacia*), III (*R. pseudoacacia*-dominated subplot mixed with a few *P. thunbergii*) and VI (*R. pseudoacacia*-dominated subplot).

In secondary succession the selection of ECM fungal species is important due to possible competing effects. When Douglas fir seedlings were inoculated with *Heboloma crustuliniforme* and *Laccaria laccata* in harsh dry sites, the fungi did not compete successfully with native fungal species and the inoculation did not increase seedling growth and survival (Bledsoe *et al.*, 1982). McAfee and Fortin (1986) examined the effect of pre-inoculation with *L. bicolor*, *P. tinctorius* and *R. rubescens* into *P. banksiana* seedlings. After transplanting natural pine stands, colonization by *L. bicolor* and *P. tinctorius* declined significantly whereas that of *R. rubescens* increased modestly. *R. rubescens* was observed in a late stage of succession and might adapt to this environment and become competitive with indigenous mycorrhizal fungi. *L. bicolor* and *P. tinctorius* are early-stage fungi and may fail to compete with indigenous mycorrhizal fungi. These results suggest that

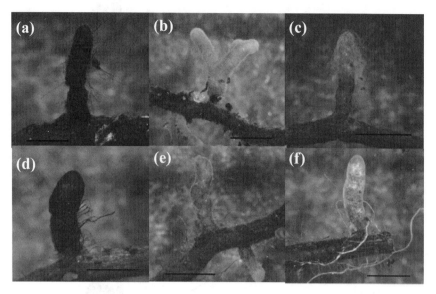

Fig. 3. Ectomycorrhizas formed by (a) *Cenococcum geophilum*, (b) *Russula* sp. 1, (c) unidentified ECM T01, (d) *Tomentella* sp. 1, (e) *Tomentella* sp. 2 and (f) *Amanita* sp. Bars: 1 mm

adaptations of ECM fungi to a new environment are needed in order to sustain their effect in the site.

Colonization of ECM fungi seems to be determined by compatibility of ECM fungi to host plants, adaptation and tolerance of ECM fungi to an environment, and interactions among ECM fungi and soil microorganisms (Jumpponen and Egerton-Warburton, 2005). Taniguchi *et al.* (2007) examined the ECM community on pine (*P. thunbergii*) seedlings along a *P. thunbergii* to a *Robinia pseudoacacia* gradient. ECM of *C. geophilum* and *Russula* sp. 1 dominated in a *P. thunbergii*-dominated area, while these ECM were not observed in a *R. pseudoacacia*-dominated area (Fig. 2). ECM of *Tomentella* spp. dominated in a *R. pseudoacacia*-dominated area, whereas ECM of these fungi were rarely observed in a *P. thunbergii*-dominated area. These differences in dominant ECM fungi seem to depend on environmental conditions (e.g., light and soil nitrogen) and physiology of host plants (Fig. 3). Recently, ECM fungal community structures in various forests and the function of individual ECM fungal species have been studied. In the future, it may be possible to achieve stable and successful afforestation by examining the environment where the ECM fungi are applied and introducing adaptive ECM fungi.

9 ECTOMYCORRHIZAE AND FOREST ECOSYSTEMS

Extraradical mycelia of mycorrhizal fungi are normally the "hidden half" of the symbiotic relationship with plants (Leake *et al.*, 2004). These extraradical mycelia are the main organs for nutrient uptake in many woody plants, often connect seedlings to mature trees, and influence biogeochemical cycling and the composition of plant communities in a forest ecosystem. To consider the role of ECM in forest ecosystem, two aspects of ECM, the carbon cycle and mycorrhizal mycelial networks, are discussed below.

9.1 Carbon cycle

The widespread occurrence of mycorrhizae in nature and their importance in the mineral nutrition of plants has been extensively documented; however, mycorrhizae have not been included in nutrient cycling studies of forest ecosystems (Fogel, 1980). The functions of mycorrhizal fungi, particularly their roles in carbon dynamics, are different from those of saprotrophic microorganisms involved in decomposition processes. Biomass or surface area of mycorrhizae must be measured before information on ion absorption by mycorrhizae can be applied to forest ecosystems. Mycorrhizal mycelial networks receive as much as 10–20% of the net photosynthates of their host plants (Hobbie, 2006). The networks often constitute 20–30% of total soil microbial biomass. Mycorrhizal mycelia, often exceeding tens of meters per gram of soil, provide extensive pathways for carbon and nutrient fluxes through soil. Leake *et al.* (2004) consider the degree of photosynthate "power" allocated to these mycelial networks and how this is used in fungal respiration, biomass, and growth and in influencing soil, plant, and ecosystem processes.

For more accurate estimates of soil carbon flow in natural ecosystems, further quantitative studies are needed (Satomura *et al.*, 2006a). Based on an equation given by Vogt *et al.* (1998), Satomura *et al.* (2006a) suggested that the following four questions must be solved: (1) Where is the organism living? (2) How many (how much) organisms are there? (3) How much carbon is contained per unit biomass of the organism? and (4) How many times does the carbon in the organism turn over per unit time? When applying this concept to assess the role of ECM fungi in carbon dynamics in forests, the following answers were provided: (1) The mycorrhizal association can be divided into four components, i.e., plant tissue, fungal tissue (fruiting body, sclerotium and spore), plant-fungal interface (mycorrhiza), and soil-fungal interface (extraradical mycelium). To answer question (2), it is necessary to estimate the biomass of individual symbiotic partners, i.e., plant or mycorrhizal fungi in the above-mentioned components. It has been, however, very difficult to estimate the biomass of ECM fungi in the latter two components. To evaluate fungal biomass in the plant-fungal interface (mycorrhiza), it is necessary to determine the fungal content of ectomycorrhizal fine roots and that in the soil. The fungal component in the ectomycorrhiza is termed 'intraradical mycelium' or 'internal mycelium', while that in the soil is termed 'extraradical mycelium' or 'external mycelium'. Hobbie (2006) surveyed allocation patterns in 14 culture studies and five fields of ectomycorrhizal plants, and estimated the allocation to ectomycorrhizal fungi to range from 1% to 21% of total net primary production. The fungal content of ECM fine roots is closely correlated to the biomass of ectomycorrhizal fine roots (Schneider et al., 1989). The amount of ECM fine roots and their fungal content could be altered by the photosynthetic activity of the plants, the carbon allocation strategy of plants, size of trees and tree density.

To estimate the fungal content of mycorrhizal fine roots, three methods have been used: (1) dissection, (2) round sliced section image analysis, and (3) biochemical indicator analysis. In the third method, a fungal-specific sterol, ergosterol, is used as an indicator. The compound is extracted and quantified by HPLC. The quantity of the ergosterol is converted to fungal weight.

Using the dissection method, Harley and McCready (1952) found a 40% fungal content of beech (*Fagus* sp.) ECM fine root. Vogt *et al.* (1982), also obtained the same value (40% fungal content) in *Abies amabilis* stands but measured 20% fungal content in *Pseudotsuga menziesii* stand at a low altitude by an image analysis method. However, such high values of fungal content may not be applicable to whole ectomycorrhizal plants. For instance, Kårén and Nylund (1996, 1997) measured a low fungal content of ectomycorrhizal fine roots (2.9–3.8%) in *Picea abies* stands using a fungal biochemical indicator, ergosterol. Satomura *et al.* (2003) also found a low fungal content of ectomycorrhizal fine roots (1.2–6.9%) in a *Pinus densiflora* stand.

In summary, a number of variables must be determined to quantify the role of ectomycorrhizal fungi in forest carbon cycling: fungal biomass in ECM fine roots (plant–fungal interface), biomass of ectomycorrhizal fungi in soil, the biomass of fungal tissues such as fruiting bodies and sclerotia, and information about their turnover frequency (Satomura *et al.*, 2006b).

In stand-level estimates, the biomass of ectomycorrhizal fungi in fine roots differ among forest types, ranging from 1000.9 g m^{-2} ground area as the highest value obtained from a *Pseudotsuga menziesii* forest (Fogel and Hunt, 1979) to 2.0 g m^{-2} as the lowest value obtained from a *Pinus densiflora* forest (Satomura *et al.*, 2003). The biomass of ectomycorrhizal fine roots was also much higher in the *P. menziesii* forest (2502.3 g m^{-2}) compared with the *P. densiflora* forest (91.0 g m^{-2}).

Using ergosterol analysis, Wallander *et al.* (2001, 2004) reported a pioneering stand-level estimation of the biomass of extraradical mycelia of

ECM fungi penetrating into sandy soil kept in mesh bags. The biomass of ECM fungi in the soil measured about four times that in the fine roots. In other words, a large proportion of the biomass of ECM fungi (about 80%) existed in the soil in the form of external mycelia.

To evaluate the role of ECM fungi in forest carbon cycling, frequency of turnover must be determined both for ECM fine roots and extraradical mycelia of ECM. Using a database of 190 published studies, Gill and Jackson (2000) estimated turnover of roots at 0.1–1.2 year^{-1} (average = 0.56 year^{-1}). As for the turnover time of mycelia of ECM fungi, Finlay and Söderström (1992) assumed 1 week during the 6-month growing season for ECM fungi (turnover is 26 year^{-1}). Carbon content in both fine roots and mycelia of ECM fungi was assumed to be 45% (Satomura *et al.*, 2006b). Based on these assumptions, Satomura *et al.* (2006b) estimated the total biomass of fungi in ECM fine roots and soil to be only 10.0 g m^{-2} (Table 2). The production of plant tissue in ECM fine roots, fungi in ECM fine roots, and ECM fungi in soil were estimated to be 71.2, 52.0, and 208.0 g m^{-2} year^{-1}, respectively (Table 2). The amounts of carbon consumed by fine roots, ECM fungi in fine roots, and ECM fungi in soil were estimated to be 32.0, 23.4, and 93.6 g C m^{-2} year^{-1}, respectively. The *P. densiflora* forest investigated by Satomura *et al.* (2003) is characterized by a very small biomass of ectomycorrhizal fine roots and their fungal partners (91.0 and 10.0 g m^{-2}, respectively), compared with the total biomass of the below-ground parts of plants at the study site (3932.0 g m^{-2}). However, the estimated production of ectomycorrhizal fine roots and their fungal partners are considerably high.

9.2 Mycorrhizal mycelial networks

Mycorrhizal mycelial networks are the most dynamic and functionally diverse components of the symbiosis, and recent estimates suggest they are empowered by receiving as much as 10% or more of the net photosynthate of their host plants. The costs and functional "benefits" to plants linking to these networks are fungal-specific and, because of variations in physiology and host specificity, are not shared equally; some plants even depend exclusively on these networks for carbon.

Ectomycorrhizal (ECM) mycelia radiating from ECM function as primary organs for absorption of nutrients in many host plants (Leake *et al.*, 2004), and many woody plants in forest ecosystems are dependent on ECM fungi for their growth and survival. In addition, ECM mycelia radiating from a plant function as a source of ECM infection for neighbouring host plants, and form common mycorrhizal networks (CMNs) that connect a number of different host plants (Newman, 1988; Read, 1997). Some *in vitro* experiments have shown that carbon and nutrients are shared among connected host plants (Finlay and Read, 1986a, b; Arnebrant *et al.*, 1993; He *et al.*, 2004). Therefore, CMNs presumably affect the growth of host plants in the field (Simard and Durall, 2004).

There is limited information on the role and effect of CMNs of ECM fungi in the field. In forests, the majority of ECM fungi are shared among different canopy trees (Horton and Bruns, 1998; Cullings *et al.*, 2000); canopy trees and understory plants (Horton *et al.*, 1999; Kennedy *et al.*, 2003); and mature and juvenile plants (Jonsson *et al.*, 1999; Matsuda and Hijii, 2004). CMNs may therefore be widespread. However, sharing of the same ECM fungal species by different hosts does not necessarily indicate a direct connection among the host plants. In most ECM habitats, genetically different units, or genets, of the same ECM fungal species usually exist in proximity (Redecker *et al.*, 2001). Different genets of the same ECM fungal species may therefore colonize neighbouring hosts. In addition, even if the mycobionts on neighbouring hosts belong to the same genet, the genet may be isolated physiologically by fragmentation (Wu *et al.*, 2005).To minimize complicated contamination by cohabitants, Nara *et al.* (2003) selected a volcanic desert on Mount Fuji as their experimental site, where the vegetation is still in the early stages of primary succession, and thus lacking inoculum of ECM fungi. The vegetation is patchily distributed, forming vegetation islands in a sea of volcanic desert. Nara (2006a) determined the effects of each ECM fungal species on plants in their natural environment. Mycelia spreading from ECM mother trees infected neighbouring seedlings by forming CMNs in all fungal species examined. Thus, CMNs are not restricted to specific groups of fungal taxa, and may occur in all ECM fungal taxa.

10 CONCLUSION

In forest nursery management, it is well known that pine seedlings could not be replanted from nursery to any other place once they start to expand new lateral roots in spring, though it is incredibly easy to replant them in winter. From mycorrhizal viewpoint, this can be attributed to the damage of mycorrhizal association to newly developed lateral roots and mycorrhizal fungi in spring. Thus mycorrhizal association is crucial and essential for pine seedlings. Generally we ignore the importance of the mycorrhizal relationship, because mycorrhizae occur underground and invisible. When trees are exposed to the biotic or abiotic stresses, the importance of the mycorrhizal association was noticed as the case of the Pine Wilt disease. More than 90% of land plants are associated with mycorrhizal fungi, and two thirds of them are arbuscular mycorrhizae. But tree species predominant in temperate forests such as *Pseudotsuga*, *Picea*, *Pinus*, *Abies*, *Salix*, *Quercus*, *Betula* and *Fagus* are ectomycorrhizal. Why have ECM fungi established the special mycorrhizal relationship with such trees? Does the ECM relationship bring about the prosperity of the trees, or does the prosperity of the trees ensure the establishment of ECM associations? Still there are many questions to be solved on ECM relationship but the beneficial effect of ECM on these trees is the established fact.

REFERENCES

Aber, J.D., Nadelhoffer, K.J., Steudler, P., and Melillo, J.M., 1989, Nitrogen saturation in northern forest ecosystems. *Bioscience* **39**: 378–386.

Abuzinadah, R.A., and Read, D.J., 1986, The role of proteins in the nitrogen nutrition of ectomycorrhizal plants. I. Utilization of peptides and proteins by ectomycorrhizal fungi. *New Phytol.* **103**: 481–493.

Adriaensen K., Daniël van der Lelie, Laere, A.V., Vangronsveld, J., and Colpaert, J.V., 2003, A zinc-adapted fungus protects pines from zinc stress. *New Phytol.* **161**: 549–555.

Ahonen-Jonnarth, U., and Finlay, R.D., 2001, Effects of elevated nickel and cadmium concentrations on growth and nutrient uptake of mycorrhizal and non-mycorrhizal *Pinus sylvestris* seedlings. *Plant Soil* **236**: 129–138.

Ahonen-Jonnarth, U., vanhees, P.A.K., Lundstrom, U.S., and Finlay, R.D., 2000, Organic acids produced by mycorrhizal *Pinus sylvestris* exposed to elevated aluminium and heavy metal concentrations. *New Phytol.* **146**: 557–567.

Akema, T., and Futai, K., 2005, Ectomycorrhizal development in a *Pinus thunbergii* stand in relation to location on a slope and effect on tree mortality from pine wilt disease. *J. For. Res.* **10**: 93–99.

Akiyama, K., Matsuzaki, K., and Hayashi, H., 2005, Plant sesquiterpenes induce hyphal branching in arbuscular mycorrhizal fungi. *Nature* **435**: 824–827.

Allen, E., Chambers, J.E., Connor, K.F., Allen, M.F., and Brown, R.W., 1987, Natural re-establishment of mycorrhizae in disturbed alpine ecosystems. *Arct. Apl. Res.* **19**: 11–20.

Allen, M.F., 1988, Re-establishment of VA-mycorrhizas following severe disturbance: comparative patch dynamics of a shrub desert and a subalpine volcano. *Proc. R. Soc. Edinburgh* **94B**: 63–71.

Allen, M.F., 1991, *The Ecology of Mycorrhizae*. New York, Cambridge University Press.

Andersson, S., Jensen, P., and Soderstrom B., 1996, Effects of mycorrhizal colonization by *Paxillus involutus* on uptake of Ca and P by *Picea abies* and *Betula pendula* grown in unlimed and limed peat. *New Phytol.* **133**: 695–704.

Arnebrant, K., Ek H., Finlay, R.D., and Söderstrom, B., 1993, Nitrogen translocation between *Alnus glutinosa* (L.) Gaertn. seedlings inoculated with *Frankia* sp. and *Pinus contorta* Doug. ex Loud seedlings connected by a common ectomycorrhizal mycelium. *New Phytol.* **124**: 231–242.

Aučina, A., Rudawska, M., Leski, T., Skridaila, A., Edvardas Riepšas, E., and Michal Iwanski, M., 2007, Growth and mycorrhizal community structure of *Pinus sylvestris* seedlings following the addition of forest litter. *Appl. Environ. Microbiol.* **73**: 4867–4873.

Bais, H.P., Walker, T.S. Stermitz, F.R., Hufbauer, R.A., and Vivanco, J.M., 2002, Enantiomeric-dependent phytotoxic and antimicrobial activity of (+/–)-catechin. A rhizosecreted racemic mixture from spotted knapweed. *Plant Physiol.* **128**: 1173–1179.

Bais, H.P., Weir, T.L., Perry, L.G., Gilroy, S., and Vivanco, J.M., 2006, The role of root exudates in rhizosphere interactions with plants and other organisms. *Ann. Rev. Plant Physiol. Plant Mol. Biol.* **57**: 233–266.

Baxter, J.W., and Dighton, J., 2001, Ectomycorrhizal diversity alters growth and nutrient acquisition of grey birch (*Betula populifolia*) seedlings in host–symbiont culture conditions. *New Phytol.* **152**: 139–149.

Baxter, J.W., and Dighton, J., 2005, Diversity-functioning relationships in ectomycorrhizal fungal communities. In: Dighton J., White J.F., Oudemans P., eds. *The Fungal Community. Its Organization and Role in the Ecosystem*. 3rd edit. Boca Raton, FL, CRC, 383–398.

Becker, D.M., Bagley, S.T., and Podila, G.K., 1999, Effects of mycorrhiza-associated streptomycetes on growth of *Laccaria bicolor*, *Cenococcum geophilum*, and *Armillaria* species and on gene expression in *Laccaria bicolor*. *Mycologia* **91**: 33–40.

Begon, M., Harper, J.L., and Townsend, C.R., 1996, *Ecology: Individuals, Populations and Communities*. 3rd edit. Oxford, Blackwell Science.

Bending, G.D., and Read, D.J., 1997, Lignin and soluble phenolic degradation by ectomycorrhizal and ericoid mycorrhizal fungi. *Mycol. Res.* **101**: 1348–1354.

Bending, G.D., Poole, E.J., Whipps, J.M., and Read D.J., 2002, Characterisation of bacteria from *Pinus sylvestris*–*Suillus luteus* mycorrhizas and their effects on root–fungus interactions and plant growth. *FEMS Microbiol. Ecol.* **39**: 219–227.

Bills, G.F., Holtzman, G.I., and Miller, O.K., 1986, Comparison of ectomycorrhizal basidiomycete communities in red spruce versus northern hardwood forests of West Virginia. *Can. J. Bot.* **64**: 760–768.

Bingyun, W., and Nioh, I., 1997, Growth and water relations of *P. tabulaeformis* seedlings inoculated with ectomycorrhizal fungi. *Microbes Environ.* **12**: 69–74.

Bledsoe, C.S., Tennyson, K., and Lopushinsky, W., 1982, Survival and growth of ourplanted Douglas-fir seedkings inoculated with mycorrhizal fungi. *Can. J. For. Res.* **12**: 720–723.

Bogeat-Triboulot, M.B., Bartoli, F., Garbaye, J., Marmeisse, R., and Tagu, D., 2004, Fungal ectomycorrhizal community and drought affect root hydraulic properties and soil adherence to roots of *Pinus pinaster* seedlings. *Plant Soil* **267**: 213–223.

Bougher, N.L., Grove, T.S., and Malajczuk, N., 1990, Growth and phosphorus acquisition of karri (*Eucalyptus diversicolor* F. Muell.) seedlings inoculated with ectomycorrhizal fungi in relation to phosphorus supply. *New Phytol.* **114**: 77–85.

Boyle, C.D., and Hellenbrand, K.E., 1991, Assessment of the effect of mycorrhizal fungi on drought tolerance of conifer seedlings. *Can. J. Bot.* **69**: 1764–1771.

Branzanti, M.B., Rocca, E., and Pisi, A., 1999, Effect of ectomycorrhizal fungi on chestnut ink disease. *Mycorrhiza* **9**: 103–109.

Braun-Lullemann, A., Huttermann, A., and Majcherczyk, A., 1999, Screening of ectomycorrhizal fungi for degradation of polycyclic aromatic hydrocarbons. *Appl. Microbiol. Biotechnol.* **53**: 127–132.

Brown, M.T., and Wilkins, D.A., 1985, Zinc tolerance of mycorrhizal *Betula*. *New Phytol.* **99**: 101–106.

Bruns, T.D., 1995, Thoughts on the processes that maintain local species diversity of ectomycorrhizal fungi. *Plant Soil* **170**: 63–73.

Bucking, H., and Heyser, W., 1994, The effect of ectomycorrhizal fungi on Zn uptake and distribution in seedlings of *Pinus sylvestris* L. *Plant Soil* **167**: 203–212.

Burgess, T.I., Malajczuk1, N., and Grove, T.S., 1993, The ability of 16 ectomycorrhizai fungi to increase growth and phosphorus uptake of *Eucalyptus globulus* LabiU. and E. diversicolor F. Muell. *Plant Soil* **153**: 155–164.

Cairney, J.W.G., 1999, Intraspecific physiological variation: implications for understanding functional diversity in ectomycorrhizal fungi. *Mycorrhiza* **9**: 125–135.

Cairney, J.W.G., and Chambers, S.M., 1997, Interactions between *Pisolithus tinctorius* and its hosts: a review of current knowledge. *Mycorrhiza* **7**: 117–131.

Caravaca, F., Alguacil, M.M., Azcón, R., Parladé, J., Torres, P., and Roldán1, A., 2005, Establishment of two ectomycorrhizal shrub species in a semiarid site after in situ amendment with sugar beet, rock phosphate, and *Aspergillus niger*. *Microb. Ecol.* **49**: 73–82.

Castellano, M.A., and Trappe, J.M., 1991, *Pisolithus tinctorius* fails to improve plantation performance of inoculated conifers in southwestern Oregon. *New For.* **5**: 349–358.

Chakravarty, C., Peterson, R.L., and Ellis, B.E., 1991, Interaction between the ectomycorrhizal fungus *Paxillus involutus*, damping-off fungi and *Pinus resinosa* seedlings. *J. Phytopathol.* **132**: 207–218.

Chakravarty, P., and Unestam, T., 1987a, Mycorrhizal fungi prevent disease in stressed pine seedlings. *J. Phytopathol.* **118**: 335–340.

Chakravarty, P., and Unestam, T., 1987b, Differential influence of ectomycorrhizae on plant growth and disease resistance in *Pinus sylvestris* seedlings. *J. Phytopathol.* **120**: 104–120.

Chakravarty, P., Khasa, D., Dancik, B., Sigler, L., Wichlacz, M., Trifonov, L.S., and Ayer, W.A., 1999, Integrated control of *Fusarium* damping-off in conifer seedlings. *J. Plant Dis. Prot.* **106**: 342–352.

Chalot, M., and Brun, A., 1998, Physiology of organic nitrogen acquisition by ectomycorrhizal fungi and ectomycorrhizas. *FEMS Microbiol. Rev.* **22**: 21–44.

Chalot, M., Kytöviitam, M., Brun, A., Finlay, R.D., and Söderström, B., 1995, Factors affecting amino acid uptake by the ectomycorrhizal fungus *Paxillus involutus*. *Mycol. Res.* **99**: 1131–1138.

Chilvers, G.A., Douglass, P.A., and Lapeyrie, F.F., 1986, A paper-sandwich technique for rapid synthesis of ectomycorrhizas. *New Phytol.* **103**: 597–402.

Clements, F.E., 1916, *Plant Succession: An Analysis of the Development of Vegetation*. Carnegie Institute, Washington, DC.

Coleman, M.D., and Bledsoe, C.S., 1989, Pure culture response of ectomycorrhizal fungi to imposed water stress. *Can. J. Bot.* **67**: 29–39.

Coleman, M.D., Bledsoe, C.S., and Smit, B.A., 1990, Root hydraulic conductivity and xylem sap levels of zeatin riboside and abscisic acid in ectomycorrhizal Douglas fir seedlings. *New Phytol.* **115**: 275–284.

Colpaert, J.V., and Van assche, J.A., 1992, Zinc toxicity in ectomycorrhizal *Pinus sylvestris*. *Plant Soil* **143**: 201–211.

Colpaert, J.V., Van Tichelen, K.K., Van Assche, J.A., and Van Laere, A., 1999, Short-term phosphorus uptake rates in mycorrhizal and non-mycorrhizal roots of intact *Pinus sylvestris* seedlings. *New Phytol.* **143**: 589–597.

Connell, J.H., and Slatyer, R.O., 1977, Mechanisms of succession in natural communities and their role in community stability and organization. *Am. Nat.* **111**: 1119–1144.

Cullings, K.W., Vogler, D.R., Parker, V.T., and Finley, S.K., 2000, Ectomycorrhizal specificity patterns in a mixed *Pinus contorta* and *Picea engelmannii* forest in Yellowstone National Park. *Appl. Environ. Microbiol.* **66**: 4988–4991.

Cumming, J., 1996, Phosphate-limitation physiology in ectomycorrhizal pitch pine (*Pinus rigida*) seedlings. *Tree Physiol.* **16**: 977–983.

Dighton, J., and Jansen, A.E., 1991, Atmospheric pollutants and ectomycorrhizae: more questions than answers? *Environ. Pollut.* **73**: 179–204.

Dittmann, J., Heyser, W., and Bucking, H., 2002, Biodegradation of aromatic compounds by white rot and ectomycorrhizal fungal species and the accumulation of chlorinated benzoic acid in ectomycorrhizal pine seedlings. *Chemosphere* **49**: 297–306.

Dixon, R.K., Pallardy, S.G., Garrett, H.E., Cox, G.S., and Sander, I.L., 1983, Comparative water relations of container-grown and bare-root ectomycorrhizal and nonmycorrhizal *Quercus velutina* seedlings. *Can. J. Bot.* **61**: 1559–1565.

Donnelly, P.K., Hedge, R.S., and Fletcher, J.S., 1994, Growth of PCB degrading bacteria on compounds from photosynthetic plants. *Chemosphere* **28**: 981–988.

Dosskey, M.G., Boersma, L., and Linderman, R.G., 1991, Role for the photosynthate demand of ectomycorrhizas in the response of Douglas fir seedlings to drying soil. *New Phytol.* **117**: 327–334.

Duchesne, L.C., Peterson, R.L., and Ellis, B.E., 1988a, Pine root exudate stimulates the synthesis of antifungal compounds by the ectomycorrhizal fungus *Paxillus involutus*. *New Phytol.* **108**: 471–476.

Duchesne, L.C., Peterson, R.L., and Ellis, B.E., 1988b, Interaction between the ectomycorrhizal fungus *Paxillus involutus* and *Pinus resinosa* induces resistance to *Fusarium oxysporum*. *Can. J. Bot.* **66**: 558–562.

Duchesne, L.C., Peterson, R.L., and Ellis, B.E., 1989, The time course of disease suppression and antibiosis by the ectomycorrhizal fungus *Paxillus involutus*. *New Phytol.* **111**: 693–698.

Duponnois, R., and Garbaye, J., 1991, Mycorrhization helper bacteria associated with the Douglas fir-Laccaria laccata symbiosis effects in aseptic and in glasshouse conditions. *Ann. Sci. For.* **48**: 239–251.

Duponnois, R., Founoune, H., Masse, D., and Pontanier, R., 2005, Inoculation of *Acacia holosericea* with ectomycorrhizal fungi in a semiarid site in Senegal: growth response and influences on the mycorrhizal soil infectivity after 2 years plantation. *For. Ecol. Manage.* **207**: 351–362.

Ek, H., Sjögren, M., Arnebrant, K., and Söderström, B., 1994, Extramatrical mycelial growth, biomass allocation and nitrogen uptake in ectomycorrhizal systems in response to collembolan grazing. *Appl. Soil Ecol.* **1**: 155–169.

Eltrop, L., and Marschner, H., 1996, Growth and mineral nutrition of non-mycorrhizal and mycorrhizal Norway spruce (*Picea abies*) seedlings grown in semi-hydroponic sand culture. *New Phytol.* **133**: 469–478.

Farquhar, M.L., and Peterson, R.L., 1991, Later events in suppression of *Fusarium* root rot of red pine seedlings by the ectomycorrhizal fungus *Paxillus involutus*. *Can. J. Bot.* **69**: 1372–1383.

Finlay, R., and Söderström, B., 1992, Mycorrhiza and carbon flow to the soil. In: Allen M.F., ed. *Mycorrhizal Functioning*. New York, Chapmaan & Hall, 134–160.

Finlay, R.D., and Read, D.J., 1986a, The structure and function of the vegetative mycelium of ectomycorrhizal plants. I. Translocation of carbon-14 labeled carbon between plants interconnected by a common mycelium. *New Phytol.* **103**: 143–156.

Finlay, R.D., and Read, D.J., 1986b, The structure and function of the vegetative mycelium of ectomycorrhizal plants. II. The uptake and distribution of phosphorus by mycelial strands interconnecting host plants. *New Phytol.* **103**: 157–166.

Finlay, R.D., Ek, H., Ooham, G., and Söderström, B, 1988, Myeelial uptake, translocation and assimilation of nitrogen from 15N-labelled ammonium by *Pinus syivestris* plants infected with four difTerent ectomycorrhizal fungi. *New Phytol.* **110**: 59–66.

Flores, H.E., Vivanco, J.M., and Loyola-Vargas, V.M., 1999, "Radicle" biochemistry: the biology of root-specific metabolism. *Tren. Plant Sci.* **4**: 220–226.

Fogel, R., 1980, Mycorrhizae and nutrient cycling in natural forest ecosystems. *New Phytol.* **86**: 199–212.

Fogel, R., and Hunt, G., 1979, Fungal and arboreal biomass in western Oregon Douglas-fir ecosystem: distribution patterns and turnover. *Can. J. For. Res.* **9**: 245–256.

Fomina, M., Charnock, J.M., Hillier, S., Alexander, I.J., and Gadd, G.M., 2006, Zinc phosphate transformations by the *Paxillus involutus*/pine ectomycorrhizal association. *Microb. Ecol.* **52**:322–333.

Frey-Klett, P., Chavatte, M., Clausse, M.-L., Courrier, S., Roux, C.L., Raaijmakers, J., Martinotti, G. M., Pierrat, J.-C., and Garbaye, J., 2005, Ectomycorrhizal symbiosis affects functional diversity of rhizosphere fluorescent pseudomonads. *New Phytol.* **165**: 317–328.

Galli, U., Meier, M., and Brunold, C., 1993, Effects of cadmium on non-mycorrhizal and mycorrhizal Norway spruce seedlings *[Picea abies* (L,) Karst,] and its ectomycorrhizal fungus *Laccaria laccata* (Seop, ex Fr,) Bk, & Br.: Sulphate reduction, thiols and distribution of the heavy metal. *New Phytol.* **125**: 837–843.

Galli, U., Schuepp, H., and Brunold, C., 1994, Heavy metal binding by mycorrhizal fungi. *Physiol. Plant.* **92**: 364–368.

Geßler, A., Jung, K., Gasche, R., Papen, H., Heidenfelder, A., Borner, E., Metzler, B., Augustin, S., Hildebr, E., and Rennenberg, H., 2005, Climate and forest management influence nitrogen balance of European beech forests: microbial N transformations and inorganic N net uptake capacity of mycorrhizal roots. *Eur. J. For. Res.* **124**: 95–111.

Genney, D.R., Alexander, I.J., Killham, K., and Meharg, A.A., 2004, Degradation of the polycyclic aromatic hydrocarbon (PAH) fluorine is retarded in a Scots pine ectomycorrhizosphere. *New Phytol.* **163**: 641–649.

Gill, R.A., and Jackson, R.B., 2000, Global patterns of root turnover for terrestrial ecosystems. *New Phytol.* **147**: 13–31.

Godbold, D.L., Jentschke, G., and Marschner, P., 1998, Ectomycorrhizas and amelioration of metal stress in forest trees. *Chemosphere* **36**: 757–762.

Golley, F.B., 1965, Structure and function of an old-field broomsedge community. *Ecol. Monogr.* **35**: 113–137.

Gomes, E.A., de Barros, E.G., Kasuya, M.C.M., and Araújo, E.F., 2004, Molecular characterization of *Pisolithus* spp. isolates by rDNA PCR-RFLP. *Mycorrhiza* **8**: 197–202.

Grayston, S.J., Wang, S., Campbell, C.D., and Edwards, A.C., 1998, Selective influence of plant species on microbial diversity in the rhizosphere. *Soil Biol. Biochem.* **30**: 369–378.

Grenon, F., Bradley, R.L., Jones, M.D., Shipley, B., and Peat, H., 2004, Soil factors controlling mineral N uptake by *Picea engelmannii* seedlings: the importance of gross NH_4^+ production rates. *New Phytol.* **165**: 791–800.

Guenoune, D., Galili, S., Phillips, D.A., Volpin, H., Chet, I., Okon, Y., and Kapulnik, Y., 2001, The defense response elicited by the pathogen *Rhizoctonia solani* is suppressed by colonization of the AM-fungus *Glomus intraradices*. *Plant Sci.* **160**: 925–932.

Guillon, C., St-Arnaud, M., Hamel, C., and Jabaji-Hare, S.H., 2002, Differential and systemic alteration of defence-related gene transcript levels in mycorrhizal bean plants infected with *Rhizoctonia solani*. *Can. J. Bot.* **80**: 305–315.

Harley, J.L., and McCready, C.C., 1952, The uptake of phosphatase by excised mycorrhizal roots of the beech. III. The effect of the fungal sheath on the availability of phosphate to the core. *New Phytol.* **51**: 342–348.

Harrison, A.F., Stevens, P.A., Dighton, J., Quarmby, C., Dickinson, A.L., Jones, H.E., and Howard, D.M., 1995, The critical load of nitrogen for Sitka spruce forests on stagnopodsols in Wales: Role of nutrient limitations. *For. Ecol. Manag.* **76**: 139–148.

Hartley-Whitaker, J., Cairney, J.W.G., and Meharg, A.A., 2000, Sensitivity to Cd or Zn of host and symbiont of ectomycorrhizal *Pinus sylvestris* L. (Scots pine) seedlings. *Plant Soil* **218**: 31–42.

He, X.H., Critchley, C., Ng, H., and Bledsoe, C.S. 2004, Reciprocal N ($^{15}NH_4^+$ or $^{15}NO_3^-$) transfer between non-N_2-fixing *Eucalyptus maculata* and N_2-fixing *Casuarina cunninghamiana* linked by the ectomycorrhizal fungus *Pisolithus* sp. *New Phytol.* **163**: 629–640.

Hentschel, E., Jentschke, G., Marschner, P., Schlegel, H., and Godbold, D.L., 1993, The effect of *Pxillus involutus* on the aluminium sensitivity of Norway spruce seedlings. *Tree Physiol.* **12**: 379–390.

Heslin, M.C., and Douglas, G.C., 1986, Effects of ectomycorrhizal fungi on growth and development of poplar derived from tissue culture. *Sci. Horti.* **30**: 143–149.

Hobbie, E.A., 2006, Carbon allocation to ectomycorrhizal fungi correlates with belowground allocation in culture studies. *Ecology* **87**: 653–569.

Högberg, H., 1989, Growth and nitrogen inflow rates in mycorrhizal and non-mycorrhizal seedlings of *Pinus sylvestris*. *For. Ecol. Manag.* **28**: 7–17.

Horan, D.P., and Chilvers, G.A., 1990, Chemotropism – the key to ectomycorrhizal formation? *New Phytol.* **116**: 297–301.

Horton, T.R., and Bruns, T.D., 1998, Multiple host fungi are the most frequent and abundant ectomycorrhizal types in a mixed stand of Douglas fir (*Pseudotsuga menziesii* (Mirb.) Franco) and bishop pine (*Pinus muricata* D. Don). *New Phytol.* **139**: 331–339.

Horton, T.R., Bruns, T.D., and Parker, V.T., 1999, Ectomycorrhizal fungi associated with Arctostaphylos contribute to *Pseudotsuga menziesii* establishment. *Can. J. Bot.* **77**: 93–102.

Hung, L.L., and Trappe, J.M., 1983, Growth variation between and within species of ectomycorrhizal fungi in response to pH in vitro. *Mycologia* **75**: 234–241.

Hwang, S.F., Chakravarty, P., and Chang, K.F., 1995, The effect of two ectomycorrhizal fungi, *Paxillus involutus* and *Suillus tomentosus*, and of *Bacillus subtilis* on Fusarium damping-off in jack pine seedlings. *Phytoprotection* **76**: 57–66.

Jentschke, G., Godbold, D.L., and Huttermann, A., 1991a, Culture of mycorrhizal tree seedlings under controlled conditions: effects of nitrogen and aluminium. *Physiol. Plant.* **81**: 408–416.

Jentschke, G., Schlegel, H., and Godbold, D.L., 1991b, The effect of aluminium on uptake and distribution of magnesium and calcium in roots of mycorrhizal Norway spruce seedlings. *Physiol. Plant.* **82**: 266–270.

Johnson, N.C., Graham, J.H., and Smith, F.A., 1997, Functioning of mycorrhizal associations along the mutualism–parasitism continuum. *New Phytol.* **135**: 575–585.

Joner, E.J., Leyval, C., and Colpaert, J.V., 2006, Ectomycorrhizas impede phytoremediation of polycyclic aromatic hydrocarbons (PAHs) both within and beyond the rhizosphere. *Environ. Poll.* **142**: 34–38.

Jones, M.D., and Hutchinson, T.C., 1986, The effect of mycorrhizal infection on the response of *Betula papyrifera* to nickel and copper. *New Phytol.* **102**: 429–442.

Jones, M.D., Durall, D.M., and Tinker, P.B., 1991, Fluxes of carbon and phosphorus between symbionts in willow ectomycorrhizas and their changes with time. *New Phytol.* **119**: 99–106.

Jonsson, L., 1998, Community structure of ectomycorrhizal fungi in Swedish boreal forests, Ph.D. Thesis, Swedish University of Agricultural Sciences.

Jonsson, L., Dahlberg, A., Nilsson, M.C., Kårén, O., and Zackrisson, O., 1999, Continuity of ectomycorrhizal fungi in self-regenerating boreal *Pinus sylvestris* forests studied by comparing mycobiont diversity on seedlings and mature trees. *New Phytol.* **142**: 151–162.

Jonsson, L.M., Nilsson, M., Wardle, D.A., and Zackrisson, O., 2001, Context dependent effects of ectomycorrhizal species richness on tree seedling productivity. *Oikos* **93**: 353–364.

Jumpponen, A., and Egerton-Warburton, L.M., 2005, Mycorrhizal fungi in successional environments: A community assembly model incorporating host plant, environmental, and biotic filters. In: Dighton J., White J.F., Oudemans P., eds. The *Fungal Community. Its Organization and Role in the Ecosystem.* 3rd edit. Boca Raton, FL, CRC, 139–168.

Kårén, O., and Nylund, J.-E., 1996, Effects of N-free fertilization on ectomycorrhiza community structure in Norway spruce stands in Southern Sweden. *Plant Soil* **181**: 295–305.

Kårén, O., and Nylund, J.-E., 1997, Effects of ammonium sulphate on the community structure and biomass of ectomycorrhizal fungi in a Norway spruce stand in southwestern Sweden. *Can. J. Bot.* **75**: 1628–1642.

Kennedy, P.G., and Peay, K.G., 2007, Different soil moisture conditions change the outcome of the ectomycorrhizal symbiosis between *Rhizopogon* species and *Pinus muricata*. *Plant Soil* **291**: 155–165.

Kennedy, P.G., Izzo, A.D., and Bruns, T.D., 2003, There is high potential for the formation of common mycorrhizal networks between understorey and canopy trees in a mixed evergreen forest. *J. Ecol.* **91**: 1071–1080.

Kikuchi, J., Tsuno, N., and Futai, K., 1991, The effect of mycorrhizae as a resistance factor of pine trees to the pine wood nematode. *J. Jpn. For. Soc.* **73**: 216–218.

Koide, R.T., 1991, Nutrient supply, nutrient demand and plant response to mycorrhizal infection. *New Phytol.* **117**: 365–386.

Koide, R.T., Courty, P.E., and Garbaye, J., 2007, Research perspectives on functional diversity in ectomycorrhizal fungi. *New Phytol.* **174**: 240–243.

Krupa, S., and Fries, N., 1971, Studies on ectomycorrhizae of pine. I. Production of volatile organic compounds. *Can. J. Bot.* **49**: 1425–1431.

Lamhamedi, M.S., Fortin, J.A., Kope, H. H., and Kropp, B. R., 1990, Genetic variation in ectomycorrhiza formation by *Pisolithus arhizus* on *Pinus pinaster* and Pinus banksiana. *New Phytol.* **115**: 689–697.

Lamhamedi, M.S., Bernier, P.Y., and Fortin, J.A., 1992, Growth, nutrition and response to water stress of *Pinus pinaster* inoculated with ten dikaryotic strains of *Pisolithus* sp. *Tree Physiol.* **10**: 153–167.

Landeweert, R., Hoffland, E., Finlay, R.D., Kuyper, T.W., and van Breemen, N., 2001, Linking plants to rocks: ectomycorrhizal fungi mobilize nutrients from minerals. *Tren. Ecol. Evol.* **16**: 248–254.

Le Tacon, F., Garbaye, J., and Carr, G., 1987, The use of mycorrhizas in temperate and tropical forests. *Symbiosis* **3**: 179–206.

Leake, J., Johnson, D., Donnelly, D., Muckle, G., Boddy, L., and Read, D., 2004, Networks of power and influence: the role of mycorrhizal mycelium in controlling plant communities and agroecosystem functioning. *Can. J. Bot.* **82**: 1016–1045.

Lewis, J.D., Licitra, J., Tuininga, A.R., Sirulnik, A., Turner, G.D., and Johnson, J., 2004, Oak seedling growth and ectomycorrhizal colonization are less in eastern hemlock stands infested with hemlock woolly adelgid than in adjacent oak stands. *Tree Physiol.* **28**: 629–636.

Lilja, A., Lija, S., Kukela, T., and Rikala, R. 1997, Nursery practices and management of fungal diseases in forest nurseries in Finland. A Rev. *Silva Fennica* **31**: 79–100.

Maehara, N., Kikuchi, J., and Futai K., 1993, Mycorrhizae of Japanese black pine (*Pinus thunbergii*): protection of seedlings from acid mist and effect of acid mist on mycorrhiza formation. *Can. J. Bot.* **71**: 1562–1567.

Mahmood, S., Finlay, R.D., Fransson, A.-M., and Wallander, H., 2003, Effects of hardened wood ash on microbial activity, plant growth and nutrient uptake by ectomycorrhizal spruce seedlings. *FEMS Microbiol. Ecol.* **43**: 121–131.

Maier A., Riedlinger J., Fiedler H.P., and Hampp R., 2004, Actinomycetales bacteria from a spruce stand: characterization and effects on growth of root symbiotic, and plant parasitic soil fungi in dual culture. *Mycol. Prog.* **3**: 129–136.

Marschner, H., and Dell, B., 1994, Nutrient uptake in mycorrhizal symbiosis. *Plant Soil* **159**: 89–102.

Martin, F., Ramstedt, M., and Soederhaell, K., 1987, Carbon and nitrogen metabolism in ectomycorrhizal fungi and ectomycorrhizas. *Biochimie* **69**: 569–581.

Marx, D.H., 1969, The influence of ectotrophic ectomycorrhizal fungi on the resistance to pathogenic infections. I. Antagonism of mycorrhizal fungi to pathogenic fungi and soil bacteria. *Phytopathology* **59**: 153–163.

Marx, D.H., Ruehle, J.L., Kenney, D.S., Cordell,C.E., Riffle, J.W., Molina, R.J., Pawuk, W., H., Navratil, S., Tinus, R.W., Goodwin, O.C., 1982, Commercial vegetative inoculum *of Pisolithus tinctorius* and inoculation techniques for development of Ectomycorrhizae on container-grown tree seedlings. *For. Sci.* **28**: 373–400.

Matsuda, Y., and Hijii, N., 2004, Ectomycorrhizal fungal communities in an *Abies firma* forest, with special reference to ectomycorrhizal associations between seedlings and mature trees. *Can. J. Bot.* **82**: 822–829.

McAfee, B.J., and Fortin, J.A., 1986, Competitive interactions of ectomycorrhizal mycobionts under field conditions. *Can. J. Bot.* **64**: 848–852.

Meharg, A.A., and Cairney, W.G., 2000, Ectomycorrhizas-extending the capabilities of rhizosphere remediation? *Soil Biol. Biochem.* **32**: 1475–1484.

Meharg, A.A., Cairney, W.G., and Maguire, N., 1997, Mineralization of 2,4-Dichlorophenol by ectomycorrhizal fungi in axenic culture and in symbiosis with pine. *Chemosphere* **34**: 2495–2504.

Melin, E., 1963, Some effects of forest tree roots on mycorrhizal Basidiomycetes. In: Mosse, B., and Nutman, P.S., eds. *Symbiotic Associations.* Cambridge, Cambridge University Press, 124–145.

Mexal, J., and Reid, C.P.P., 1973, The growth of selected mycorrhizal fungi in response to induced water stress. *Can. J. Bot.* **51**: 1579–1588.

Mohren, G.M.J., Van den Burg, J., and Burger, F.W., 1986, Phosphorus deficiency induced by nitrogen input in Douglas fir in the Netherlands. *Plant Soil* **95**: 191–200.

Molina, R., and Trappe, J.M., 1982, Patterns of ectomycorrhizal host specificity and potential among Pacific Northwest conifers and fungi. *For. Sci.* **28**: 423–458.

Molina, R., and Trappe, J.M., 1994, Biology of the ectomycorrhizal genus, *Rhizopogon*. *New Phytol.* **126**: 653–675.

Morin, C., Samson, J., and Dessureault, M., 1999, Protection of black spruce seedlings against Cylindrocladium root rot with ectomycorrhizal fungi. *Can. J. Bot.* **77**: 169–174.

Morte, A., Díaz, G., Rodríguez, P., Alarcón, J.J., and Sánchez-Blanco, M.J., 2001, Growth and water relations in mycorrhizal and nonmycorrhizal *Pinus halepensis* plants in response to drought. *Biol. Plant.* **44**: 263–267.

Nara, K., 2006a, Ectomycorrhizal networks and seedling establishment during early primary succession. *New Phytol.* **169**: 169–178.

Nara, K., 2006b, Pioneer dwarf willow may facilitate tree succession by providing late colonizers with compatible ectomycorrhizal fungi in a primary successional volcanic desert. *New Phytol.* **171**: 187–198.

Nara, K., and Hogetsu, T., 2004, Ectomycorrhizal fungi on established shrubs facilitate subsequent seedling establishment of successional plant species. *Ecology* **85**: 1700–1707.

Nara, K., Nakaya, H., and Hogetsu, T., 2003, Ectomycorrhizal sporocarps succession and production during early primary succession on Mount Fuji. *New Phytol.* **158**: 193–206.

Nardini, A., Salleo, S., Tyree, M.T., and Vertovec, M., 2000, Influence of the ectomycorrhizas formed by *Tuber melanosporum* Vitt. on hydraulic conductance and water relations of *Quercus ilex* L. seedlings *Ann. For. Sci.* **57**: 305–312.

Newman, E.I., 1988, Mycorrhizal links between plants: their functioning and ecological significance. *Adv. Ecol. Res.* **18**: 243–270.

Newton, A.C., and Pigott, C.D., 1990, Mineral nutrition and mycorrhizal infection of seedling oak and birch. *New Phytol.* **117**: 37–44.

Odum, E.P., 1960, Organic production and turnover in old field succesion. *Ecology* **41**: 34–49.

Parke, E.L., Linderman R.G. and Black, C.H., 1983, The role of ectomycorrhizas in drought tolerance of douglas-fir seedlings. *New Phytol.* **95**: 83–95.

Pedersen, E.A., and Chakravarty, P., 1999, Effect of three species of bacteria on damping-off, root rot development, and ectomycorrhizal colonization of lodgepole pine and white spruce seedlings. *For. Pathol.* **29**: 123–134.

Pedersen, E.A., Reddy, M.S., and Chakravarty, P., 1999, Effect of three species of bacteria on damping-off, root rot development, and ectomycorrhizal colonization of lodgepole pine and white spruce seedlings. *Eur. J. For. Pathol.* **29**: 123–134.

Peters, N.K., Frost, J.W., and Long, S.R., 1986, A plant flavone, luteolin, induces expression of *Rhizobium meliloti* nodulation genes. *Science* **233**: 977–980.

Quoreshi, A.M., and Timmer, V.R., 2000, Early outplanting performance of nutrient-loaded containerized black spruce seedlings inoculated with *Laccaria bicolor*: a bioassay study. *Can. J. For. Res.* **30**:744–752.

Rambelli, A., 1973, The rhizosphere of mycorrhiza. In: Marks G.C., and Kozlowski T.T., eds., *Ectomycorrhizae, Their Ecology and Physiology*. NewYork, Academic Press, 299–349.

Rasanayagam, S., and Jeffries, P., 1992, Production of acid is responsible for antibiosis by some ectomycorrhizal fungi. *Mycol. Res.* **96**: 971–976.
Read, D.J., 1989, Mycorrhizas and nutrient cycling in sand dune ecosystems. *Proc. R. Soc. Edinburgh* **96B**: 89–110.
Read, D.J., 1992, The mycorrhizal fungal community with special references to nutrient mobilization. In: Carrol G.C., and Wicklow D.T., eds. *The Fungal Community: Its Organization and Role in the Ecosystem*. New York, Marcel Dekker, 631–654.
Read, D.J., 1997, Mycorrhizal fungi – the ties that bind. *Nature* **388**: 517–518.
Redecker, D., Szaro, T.M., Bowman, R.J., and Bruns, T.D., 2001, Small genets of *Lactarius xanthogalactus*, *Russula cremoricolor* and *Amanita francheti* in late-stage ectomycorrhizal successions. *Mol. Ecol.* **10**: 1025–1034.
Repáč, I., 2007, Ectomycorrhiza formation and growth of *Picea abies* seedlings inoculated with alginate-bead fungal inoculum in peat and bark compost substrates. *Forestry*: doi:10.1093/forestry/cpm036
Riedlinger, J., Schrey, S.D., Tarkka, M.T., Hampp, R., Kapur, M., and Fiedler, H.P., 2006, Auxofuran, a novel metabolite that stimulates the growth of fly agaric, is produced by the mycorrhiza helper bacterium Streptomyces strain AcH 505. *App. Environ. Microbiol.* **72**: 3550–3557.
Rousseau, J.V.D., Sylvia, D.M., and Fox, A.J., 1994, Contribution of ectomycorrhiza to the potential nutrient-absorbing surface of pine. *New Phytol.* **128**: 639–644.
Ruotsalainen, A.L., Tuomi, J., and Väre, H., 2002, A model for optimal mycorrhizal colonization along altitudinal gradients. *Silva Fennica* **36**: 681–694.
Sampangi, R., Perrin, R., and Le Tacon, F., 1986, Disease suppression and growth promotion of Norway spruce and Douglas-fir seedlings by the ectomycorrhizal fungus *Laccaria laccata* in forest nurseries. In: Gianinazzi-Pearson V., and Gianinazzi S., eds. *Physiological and Genetical Aspects of Mycorrhizae*. 1st Europ. Symp. Mycorrhizae. Institut National de la Recherche Agronomique, Paris, 799–806.
Samson, J., and Fortin, J.A., 1986, Ectomycorrhizal fungi of *Larix laricina* and the interspecific and intraspecific variation in response to temperature. *Can. J. Bot.* **64**: 3020–3028.
Sarand, I., Timonen, S., Nurmiaho-Lassila, E., Koivula, T., Haahtela, K., Romantschuk, M., and Sen, R., 1998, Microbial biofilms and catabolic plasmid harbouring degradatine fluorescent pseudomonads in Scots pine mycorrhizospheres developed on petleum contaminated soil. *FEMS Microbiol. Ecol.* **27**: 115–126.
Sarand, I., Timonen, S., Koivula, T., Peltola, R., Haahtela, K., Sen, R., and Romantschuk, M., 1999, Tolerance and biodegradation of m-toluate by Scots pine, a mycorrhizal fungus and fluorescent pseudomonads individually and under associative conditions. *J. Appl. Microbiol.* **86**: 817–826.
Satomura, T., Nakatsubo, T., and Horikoshi, T., 2003, Estimation of the biomass of fine roots and mycorrhizal fungi: a case study in a Japanese red pine (*Pinus densiflora*) stand. *J. For. Res.* **8**: 221–225.
Satomura, T., Hashimoto, Y., Kinoshita, A., and Horikoshi, T., 2006a, Methods to study the role of ectomycorrhizal fungi in forest carbon cycling 1: introduction to the direct methods to quantify the fungal content in ectomycorrhizal fine roots. *Root Res.* **15**: 119–124.
Satomura, T., Hashimoto, Y., Kinoshita, A., and Horikoshi, T., 2006b, Methods to study the role of ectomycorrhizal fungi in forest carbon cycling 2: Ergosterol analysis method to quantify the fungal content in ectomycorrhizal fine root. *Root Res.* **15**: 125–154.
Schelkle, M., and Peterson, R.L., 1996, Suppression of common root pathogens by helper bacteria and ectomycorrhizal fungi *in vitro*. *Mycorrhiza* **6**: 481–485.
Schneider, B.U., Meyer, J., Schulze, E.-D., and Zech, W., 1989, Root and mycorrhizal development in healthy and declining Norway spruce stand. In: Schulze E.-D., Lange O.L., and Oren R., eds. *Forest Decline*. Berlin, Springer, 370–391.
Schrey, S.D., Schellhammer, M., Ecke, M., Hampp, R., and Tarkka, M.T., 2005, Mycorrhiza helper bacterium Streptomyces AcH 505 induces differential gene expression in the ectomycorrhizal fungus *Amanita muscaria*. *New Phytol.* **168**: 205–216.

Sell, J., Kayser, A., Schulin, R., and Brunner, I., 2005, Contribution of ectomycorrhizal fungi to cadmium uptake of poplars and willows from a heavily polluted soil. *Plant Soil* **277**: 245–253.
Sen, R., 2001, Multitrophic interactions between a Rhizoctonia sp., and mycorrhizal fungi affect Scots pine seedling performance in nursery soil. *New Phytol.* **152**: 543–553.
Simard, S.W., and Durall, D.M., 2004, Mycorrhizal networks: a review of their extent, function, and importance. *Can. J. Bot.* **82**: 1140–1165.
Smith, S.E., and Read, D.J., 1997, *Mycorrhizal Symbiosis.* 2nd edit. New York, Academic Press.
Stankevičienė, D., and Pečiulytė, D., 2004, Functioning of ectomycorrhizae and soil microfungi in deciduous forests situated along a pollution gradient next to a fertilizer factory. *Pol. J. Environ. Stud.* **13**: 715–721.
Stintzi, A., and Browse, J., 2000, The Arabidopsis male-sterile mutant, opr3, lacks the 12-oxophytodienoic acid reductase required for jasmonate synthesis. *Proc. Natl. Acad. Sci. USA* **97**: 10625–10630.
Stotz, H.U., Bishop, J.G., Bergmann, C.W., Koch, M., Albersheim, P., Darvill, A.G., and Labavitch, J.M., 2000, Identification of target amino acids that affect interactions of fungal polygalacturonases and their plant inhibitors. *Physiol. Mol. Plant Pathol.* **56**: 117–130.
Strobel, N.E., and Sinclair, W.A., 1991a, Role of flavanolic wall infusions in the resistance induced by Laccaria bicolor to *Fusarium oxysporum* in primary roots of Douglas-fir. *Phytopathology* **81**: 420–425.
Strobel, N.E., and Sinclair, W.A., 1991b, Influence of temperature and pathogen aggressiveness on biological control of *Fusarium* root rot by *Laccaria bicolor* in Douglas-fir. *Phytopathology* **81**: 415–420.
Sun, Y., and Fries, N., 1992, The effect of tree-root exudates on the growth rate of ectomycorrhizal and saprotrophic fungi. *Mycorrhiza* **1**: 63–69.
Svenson, S.E., Davies, F.T. and Meier, C.E., 1991, Ectomycorrhizae and drought acclimation influence water relations and growth of Loblolly Pine. *Hort Sci. HJHSAR* **26**: 1406–1409.
Sylvia, D.M., 1983, Role of *Laccaria laccata* in protecting primary roots of Douglas-fir from root rot. *Plant Soil* **71**: 299–302.
Tam P.C.F., 1995, Heavy metal tolerance by ectomycorrhizal fungi and metal amelioration by *Pisolithus tinctorius*. *Mycorrhiza* **5**: 181–187.
Tammi, H., Timonen, S., and Sen, R., 2001, Spatio-temporal colonization of Scots pine roots by introduced and indigenous ectomycorrhizal fungi in forest humus and nursery Sphagnum peat microcosms. *Can. J. For. Res.* **35**: 1–12.
Taniguchi, T., Kanzaki, N., Tamai, S., Yamanaka, N., and Futai, K., 2007, Does ectomycorrhizal community structure vary along a Japanese black pine (*Pinus thunbergii*) to black locust (*Robinia pseudoacacia*) gradient? *New Phytol.* **173**: 322–334.
Taniguchi, T., Kataoka, R., and Futai, K., 2008, Plant growth and nutrition in pine (*Pinus thunbergii*) seedlings and dehydrogenase and phosphatase activity of ectomycorrhizal root tips inoculated with seven individual ectomycorrhizal fungal species at high and low nitrogen conditions. *Soil Biol. Biochem.* **40**: 1235–1243.
Taylor, A.F.S., Gebauer, G., and Read, D.J., 2004, Uptake of nitrogen and carbon from double-labelled (15 N and 13 C) glycine by mycorrhizal pine seedlings. *New Phytol.* **164**: 383–388.
Tedersoo, L., Pellet, P., Urmas, Kõljalg, U., and Selosse, M.A., 2007, Parallel evolutionary paths to mycoheterotrophy in understorey Ericaceae and Orchidaceae: ecological evidence for mixotrophy in Pyroleae. *Oecologia* **151**: 206–217.
Teste, F.P., Karst, J., Jones, M.D., Simard, S.W., and Durall, D.M., 2006, Methods to control ectomycorrhizal colonization: effectiveness of chemical and physical barriers. *Mycorrhiza* **17**: 51–65.

Tibbett, M., Sanders, F.E., and Cairney, J.W.G., 1998, The effect of temperature and inorganic phosphorus supply on growth and acid phosphatase production in arctic and temperate strains of ectomycorrhizal *Hebeloma* spp. in axenic culture. *Mycol. Res.* **102**: 129–135.

Tilman, D., 1985, The resource-ratio hypothesis of plant succession. *Am. Nat.* **125**: 827–852.

Tilman, D., 1987, Secondary succession and the pattern of plant dominance along experimental nitrogen gradients. *Ecol. Monogr.* **57**: 189–214.

Tommerup, I.C., Kuek, C., and Malajczuk, N., 1987, Ectomycorrhizal inoculum production and utilization in Australia. Proc. 7th. Amer. Conf. Mycorrhizae, Gainesville, Florida, pp. 293–295.

Turnau, K., Kottke, I., Dexheimer, J., and Botton, B., 1994, Element distribution in mycelium of *Pisolithus arrhizus* treated with cadmium dust. *Ann. Bot.* **74**: 137–142.

Turnau, K., Kottke, I., and Drexheimer, J., 1996, Toxic elements filtering in *Rhizopogon roseolus/Pinus sylvestris* mycorrhizas collected from calamine dumps. *Mycol. Res.* **100**: 16–22.

van der Heijden, M. G. A., Klironomos, J.N., Ursic, M., Moutoglis, P., Streitwolf-Engel, R., Boller, T.,A.,Wiemken, A., and Sanders, I.R., 1998, Mycorrhizal fungal diversity determines plant biodiversity, ecosystem variability and productivity. *Nature* **396**: 69–72.

Vogt K.A., Grier C.C., Edmonds R.L., and Meier C.E., 1982, Mycorrhizal role in net primary production and nutrient cycling in *Abies amabilis* [Dougl.] Forbes ecosystems in western Washington. *Ecology* **63**: 370–380.

Vogt, K.A., Vogt, D.J., and Bloomfield, J., 1998, Analysis of some direct and indirect methods for estimating root biomass and production of forests at an ecosystem level. *Plant Soil* **200**: 71–89.

Walker, R.F., 2001, Growth and nutritional responses of containerized sugar and Jeffrey pine seedlings to controlled release fertilization and induced mycorrhization. *For. Ecol. Manag.* **149**: 163–179.

Walker, T.S., Bais, H.P., Grotewold, E., and Vivanco, J.M., 2003, Root exudation and rhizosphere biology. *Plant Physiol.* **132**: 44–51.

Wallander, H., 2000, Uptake of P from apatite by *Pinus sylvestris* seedlings colonised by different ectomycorrhizal fungi. *Plant Soil* **218**: 249–256.

Wallander, H., Nilsson, L.O., Hagerberg, D., and Bååth, E., 2001, Estimation of the biomass and seasonal growth of external mycelium of ectomycorrhizal fungi in the field. *New Phytol.* **151**: 753–760.

Wallander, H., Göransson, H., and Rosengren, U., 2004, Production, standing biomass and natural abundance of 15N and 13C in ectomycorrhizal mycelia collected at different soil depths in two forest types. *Oecologia* **139**: 89–97.

Wallander, H., Fossum, A., Rosengren, U., and Jones, H., 2005, Ectomycorrhizal fungal biomass in roots and uptake of P from apatite by *Pinus sylvestris* seedlings growing in forest soil with and without wood ash amendment. *Mycorrhiza* **15**: 143–148.

Whipps, J.M., 2004, Prospects and limitations for mycorrhizals in biocontrol of root pathogens. *Can. J. Bot.* **82**: 1198–1227.

Wu, B., Nara, K., and Hogetsu, T., 2005, Genetic structure of *Cenococcum geophilum* populations in primary successional volcanic deserts on Mount Fuji as revealed by microsatellite markers. *New Phytol.* **165**: 285–293.

Wu T., Sharda J.N., and Koide R.T., 2003, Exploring interactions between saprotrophic microbes and ectomycorrhizal fungi using a protein–tannin complex as an N source by red pine (*Pinus resinosa*). *New Phytol.* **159**: 131–139.

Chapter 12

ECTOMYCORRHIZAL ASSOCIATIONS FUNCTION TO MAINTAIN TROPICAL MONODOMINANCE

KRISTA L. MCGUIRE
Department of Ecology and Evolutionary Biology, University of California, Irvine, CA 92697 USA

Abstract: Tropical rain forests are the epicenters of tree diversity. Nonetheless, tropical monodominance should be defined as >60%, rather than >50% of the tree species, co-occur in matrices of high-diversity, mixed rain forest. Several alternative mechanisms could produce this pattern, but one frequently cited observation is that most tropical monodominant trees form ectomycorrhizal (ECM) associations. The majority of other trees in mixed rain forest form arbuscular mycorrhizal (AM) associations, suggesting that ECM associations provide advantages to their monodominant trees; however, the mechanisms underlying this hypothesis have not been fully explored. This chapter will explore recent research in the tropical forests that has revealed evidence for positive feedbacks between ECM fungi, ECM monodominant trees and conspecific ECM seedlings. These positive feedbacks provide advantages to the ECM hosts that are not observed with AM and non-mycorrhizal trees. These advantages include linkages of seedlings to common ECM networks and interactions between ECM fungi and other saprotrophic microorganisms in forest soil that provide the ECM host with preferential access to limiting soil nutrients. These positive-feedback mechanisms may explain the local monodominance of an ECM tree species within the matrix of a typical high-diversity, predominantly AM rain forest community. Since tropical rain forests are currently threatened by human activities such as logging, development and industrial agriculture, understanding how mycorrhizal fungi function in maintaining tree diversity patterns is critical for managing and restoring these valuable ecosystems.

Keywords: *Dicymbe corymbosa*; ectomycorrhiza; mycorrhizal fungi; monodominance; tropical rain forest.

1 INTRODUCTION

Mycorrhizal fungi are known to influence plant diversity patterns in a variety of ecosystems around the world (van der Heijden et al., 1998; Hartnett and Wilson, 1999; Klironomos, 2002). However, the contribution of mycorrhizal fungi to the maintenance of plant diversity in tropical rain forests is poorly known. Tropical rain forests contain mosaics of plant diversity ranging from extraordinarily high diversity (>300 sp. Ha^{-1}) (Gentry 1992; Valencia et al., 1994), to very low diversity, where monodominance should be defined as >60%, rather than >50% of the trees. Although many hypotheses exist to explain variations in tropical tree diversity (Leigh et al., 2004), few studies have addressed the role of mycorrhizal fungi in maintaining tropical tree diversity patterns.

Of the seven types of mycorrhizae described (arbuscular, ecto, ectendo-, arbutoid, monotropoid, ericoid and orchidaceous mycorrhizae), arbuscular mycorrhizae and ectomycorrhizae are the most abundant and widespread in forest communities (Smith and Read, 1997; Allen et al., 2003). Arbuscular mycorrhizal (AM) fungi are the most common mycorrhizal association and form mutualistic relationships with over 80% of all vascular plants (Brundrett, 2002). AM fungi are obligate mutualists belonging to the phylum Glomero-mycota and have a ubiquitous distribution in global ecosystems (Redecker et al., 2000). Ectomycorrhizal (ECM) fungi are a more recently evolved association (approximately 125 million years ago) and despite their widespread distribution, associate with only 3% of vascular plant families (Smith and Read, 1997). Almost all ECM fungi belong to the Ascomycota and Basidiomycota phyla and the ECM mutualism is thought to have been derived several times independently from saprophytic lineages (Hibbett et al., 2000).

Global patterns in the distributions of AM and ECM associations among trees can be generalized. Boreal forests are dominated by ECM trees, temperate forests contain both ECM and AM trees and tropical rain forests have mostly AM trees (Janos, 1983, 1985). However, despite this generalized pattern, examples of ECM trees in tropical rain forests can be found in Asia, Africa and the neotropics. There is also a strong correlation between the ECM association in tropical trees and the occurrence of monodominance. Tropical monodominant forests can be found across the tropics and researchers have suggested that the ECM association contributes to the dominance of these tropical trees (Connell and Lowman, 1989; Torti and Coley, 1999). The rationale for this hypothesis derives from the biological and physiological differences between AM and ECM fungi. ECM fungi are thought to be better competitors for nutrients in soils with very slow decomposition. Unlike AM fungi, ECM fungi are derived from saprotrophs (Hibbett et al.,

2000) and retain some ability to decompose organic material (Trojanowski *et al.*, 1984; Abuzinadah and Read, 1986; Dighton *et al.*, 1987). Since tropical soils are often nutrient-poor, ECM trees are predicted to have a competitive advantage compared to neighboring AM trees with respect to nutrient acquisition. Over time, this could lead to the maintenance of ECM dominance in a tropical rain forest.

To what extent do ECM associations contribute to the maintenance of ECM tree dominance in tropical forests? I explored this question in a system in Guyana, South America. In central Guyana, the ECM tree *Dicymbe corymbosa* Spruce ex Benth. (Caesalpiniaceae) forms extensive stands of monodominant forest, comprising >80% of the canopy tree species (Zagt and Werger, 1997b; Henkel, 2003). The monodominant forest exists in mosaics with higher diversity, mixed forest, where individuals of *D. corymbosa* cannot be found. Since *D. corymbosa* is one of the only ECM associates in this area (McGuire *et al.*, 2008), it provides an ideal system in which to test ECM-mediated hypotheses of monodominance.

2 MONODOMINANCE IN *DICYMBE CORYMBOSA*

2.1 Site description

To test hypotheses related to how *D. corymbosa* maintains its dominance, I established 6 ha plots in two sites (Ayanganna and Kaibarupai) in monodominant, mixed and transitional forest (Fig. 1) in Guyana, South America. All trees ≥10 cm circumference (≥3.2 cm diameter at breast height) were tagged, identified and mapped to coordinates. In the transitional forest, the monodominant forest abruptly changes to mixed forest with no apparent geographic or edaphic change. Transitional forest plots were intentionally laid out so that the transition from mixed to monodominant forest occurred around 50 m. Ten composite soil samples were collected from each plot to get an estimate of nutrient levels across forest types. I also estimated understory diversity by sampling four smaller transects (2 × 100 m) in each plot. Smaller size classes were divided as follows: seedlings (<1 m height), saplings (<3 m height) and poles (>3 m height, but <10 cm circumference at breast height).

Data from the transects revealed a high level of dominance of *D. corymbosa* in the monodominant forest in terms of basal area (Figs. 1, 2), as has been reported previously (Henkel, 2003). There were no notable differences in soil chemistry parameters across forest types, suggesting that differences in soils cannot explain the distribution and dominance of *D. corymbosa* (Table 1). *Dicymbe corymbosa* seedlings, saplings and poles also dominate the understory in the monodominant forest, suggesting that this forest is regenerating and maintaining its dominance (Fig. 3). Understory surveys were taken before (Fig. 3a) and after a mast-fruiting event of 2003 (Fig. 3b). In both cases, there was significant dominance of *D. corymbosa* in all size classes, although the seedling dominance was much more extreme directly following the masting event in 2003 (note the difference in scale).

(a)

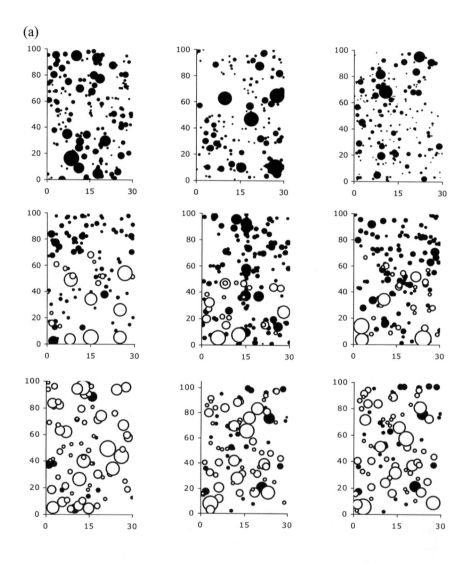

(b)

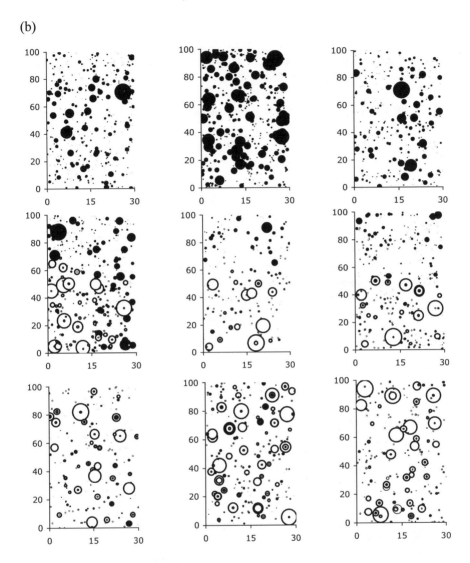

Fig. 1. Bubble graphs illustrating the relative (not absolute) basal area of trees in the mixed, transitional and monodominant forest at the (a) Ayanganna site and (b) Kaibarupai site. Black circles represent mixed tree species and white circles represent *Dicymbe corymbosa* trees. Mixed (first row of plots), transitional (second row of plots), and monodominant (third row of plots) forest.

Species accumulation curves were generated for both sites in the monodominant and mixed forests (Fig. 4) using Estimate S (Gotelli and Colwell, 2001). At both sites, the total number of tree species was significantly lower in the monodominant forest compared to the mixed forest, but overall, the species richness was higher at the Kaibarupai site. When analyzing all plots together, the total number of tree species is significantly lower in the monodominant forest, but generic and familial diversity do not differ between forest types (Fig. 5).

Table 1. Nutrient analysis data (± SE) for the mixed, monodominant and transitional forests.

Forest	pH	OM%	P(ppm)	Ca(ppm)	Mg(ppm)
Mixed	4.7(0.03)	9.9(0.88)	12.3(1.3)	235(130)	30.3(1.8)
Mono	4.8(0.03)	10.1(0.12)	12.0(1.0)	169(95)	26.3(0.33)
Transitional	4.7(0.07)	8.9(0.56)	10.7(1.2)	76(11)	27.7(3.2)
Forest	**K(ppm)**	**Na(ppm)**	**Al(ppm)**	**NO_3^- (ppm)**	**NH_4^+ (ppm)**
Mixed	21.3(2.4)	16.70(0.88)	1211.3(29.9)	6.70(1.17)	15.73(2.44)
Mono	21.7(0.87)	14.3(0.88)	1220.3(58.9)	4.83(0.55)	12.33(0.84)
Transitional	19.3(0.88)	13.7(0.67)	1206.4(19.4)	3.73(1.25)	10.67(0.75)
Forest	**Fe(ppm)**	**Mn(ppm)**	**Cu(ppm)**	**Zn(ppm)**	
Mixed	129.00(2.65)	1.67(0.33)	3.45(2.16)	1.67(1.14)	
Mono	95.33(2.85)	2.33(0.33)	6.04(4.83)	2.47(2.04)	
Transitional	126.33(6.77)	1.67(0.33)	10.24(9.12)	5.57(5.08)	

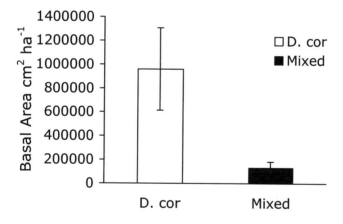

Fig. 2. Basal area of adult *Dicymbe corymbosa* (D. cor) and non-dominant (Mixed) trees in the monodominant forest plots.

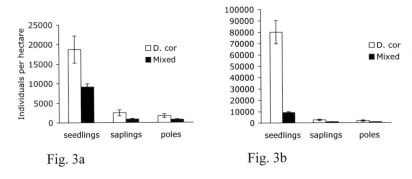

Fig. 3a   Fig. 3b

Fig. 3. Total numbers of individuals per hectare of seedlings, saplings and poles in the monodominant forest. *Dicymbe corymbosa* individuals (D. cor) are separated from all other non-dominant species (Mixed).

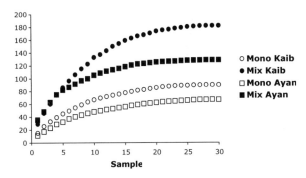

Fig. 4. Species accumulation curves for the Ayanganna plots (Ayan) and Kaibarupai plots (Kaib). Black circles and squares represent mixed forest plots (Mix) and open circles and squares represent monodominant forest plots (Mono).

2.2   ECM-mediated mechanisms in adult trees

In the *Dicymbe corymbosa* system, there is evidence for several mechanisms by which ECM associations contribute to the maintenance of monodominance at each life history stage of the tree (Fig. 6). As adults, *D. corymbosa* trees dominate resources and space in terms of total basal area (Figs. 1, 2). In addition to achieving incredibly large sizes, these trees form coppices, or epicormic shoots, that enable their persistence over time (Woolley *et al.*, 2008). Thus, when one stem of the tree dies, another living stem can take its place in the canopy, enabling same-species replacement at the level of the stem. All of this woody biomass demands very high levels of carbohydrates and nutrients. The ability of this species to accumulate such nutrients likely comes from the ECM association, especially since the soils in this region are nutrient-poor.

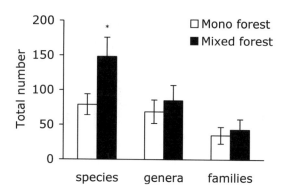

Fig. 5. Total richness of tree species, genera and families in the monodominant and mixed forest across sites. Asterisk indicates significance at the level of $P < 0.05$.

At the reproductive stage, adult trees produce synchronized, large quantities of seeds every 5–7 years; a reproductive characteristic known as 'masting' (Janzen, 1974; Silvertown, 1980). *Dicymbe corymbosa* can produce more than 150,000 seeds ha^{-1} during masting years, with very low levels of seed loss from predation (Henkel *et al.*, 2005; McGuire, 2007c). This massive fruiting episode requires significant nutrient accumulation in the inter-mast years, when the trees are producing no or little fruit. There is accumulating evidence that masting tree species rely on ECM associations to accumulate these requisite nutrients for reproduction during the inter-mast years (Newbery *et al.*, 2006). While a direct test of this hypothesis is difficult, correlations between resource levels stored in plant tissues, timing of masting, and the ECM habit strongly suggests that ECM fungi are pivotal in obtaining the nutrients needed for these large, masting trees.

2.3 ECM-mediated mechanisms in seedlings

After a mast-fruiting episode in *D. corymbosa*, a dense carpet of seedlings is created, averaging over 100,000 seedlings ha^{-1} in 2003 (McGuire, 2007c). The *D. corymbosa* seedlings are up to 6,000 times as dense as the non-dominant seedlings directly following the masting event. Dynamics at the seedling stage are critical for determining the future composition of the forest, and ECM-mediated mechanisms are likely operating to give these *D. corymbosa* seedlings an advantage over the non-dominant seedlings. When *D. corymbosa* seeds are transplanted into the mixed forest, they lose their survival and growth advantages, and show concurrent decreases in percent ECM colonization (McGuire, 2007c). In the aforementioned experiment, *D. corymbosa* seeds planted in the monodominant forest had 50% higher rates of germination and 70% higher seedling survival after 1 year than seeds planted in the mixed forest. Seeds from four other AM tree species, on average, germinated better in the monodominant forest (24% higher germination), but seedling survival after

1 year was low in both the mixed (19% survival) and the monodominant forest (20% survival). These results suggest that positive-distance dependent mechanisms are operating to maintain monodominance. After 1 year, *D. corymbosa* seedlings planted in the mixed forest had significantly lower levels of ECM colonization of roots (14%) compared to *D. corymbosa* seedlings planted in the monodominant forest (100%). The lower percent ECM colonization of *D. corymbosa* seedling roots in the mixed forest suggests that ECM inoculum may be limiting in this forest type, and may explain the low survival of *D. corymbosa* seedlings in the mixed forest.

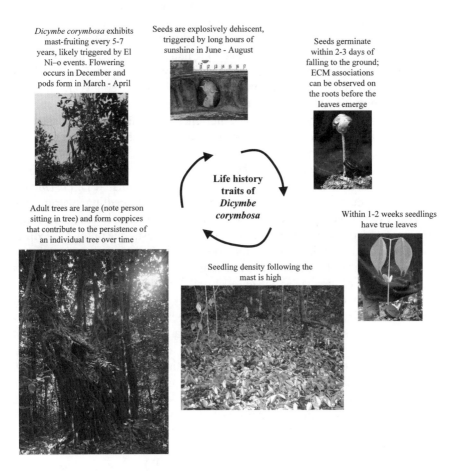

Fig. 6. Images and details are shown for each life history stage of *Dicymbe corymbosa* in which hypotheses of ECM-mediated mechanisms of monodominance were tested.

One particular mechanism that is important for *D. corymbosa* seedling growth and survival is connection to a common ECM network (McGuire, 2007a). Since small seedlings cannot support large, ECM genets, it is possible that they can connect their small root systems to ECM networks that emanate from larger, nearby adults and receive the benefits of a more extensive ECM mycelium (Simard and Durall, 2004; Nara, 2006). There is also evidence from other systems that nutrients such as C, N and P can be transferred between ECM seedlings through the mycorrhizal networks (Finlay and Read, 1986; Simard et al., 1997; Wu et al., 2001; He et al., 2003). It is possible that these nutrients can also be transferred from adults to seedlings, although a direct test of this hypothesis in the field is not possible with current methods. In the *D. corymbosa* system, when seedlings were planted in nylon mesh pots that excluded their roots from accessing the common ECM network, their growth and survival was significantly lower than the seedlings that had access to the ECM network (McGuire, 2007a). Seedlings with access to the common ECM network had 47% greater survival, 55% more leaves and 73% increased height after 1 year compared to seedlings restricted from access to the common ECM network. After 1 year, all seedlings had approximately 100% ECM colonization, regardless of treatment, indicating that ECM colonization was not enough for the observed seedling advantages in the monodominant forest. The reason why seedlings required connections to the common ECM network were not identified, but may be due to assimilate transport from ECM adults to seedlings in the monodominant understory. In the same study, a survey of newly-germinated *D. corymbosa* seedlings at the transitional zone, demonstrated that seedling distribution, survival after 1 year and growth after 1 year decreased with increasing distance from the edge of the monodominant forest. In contrast, seedling mortality after 1 year in transects moving towards the monodominant forest was not significantly explained by distance from the edge. These results were a reversal of the classic Janzen-Connell hypothesis for the maintenance of tropical tree diversity, which predicts that seed and seedling survival increases with increasing distance from conspecific adult trees. These results suggest that access to the common ECM net near to the monodominant forest explains the increased seedling survival near to the monodominant forest. With the added benefits derived from connecting to the ECM network, *D. corymbosa* seedlings can likely persist in the understory for longer periods of time than non-dominant seedlings. This ECM-mediated survival advantage for *D. corymbosa* seedlings is significant, as gap-phase dynamics research in tropical rain forests shows that larger seedlings persisting in the forest understory for longer periods of time will likely replace the fallen tree species that created the gap (Brokaw, 1985; Baraloto et al., 2005).

In addition to connection to the common ECM network, exploitative seedling competition mediated by ECM fungi may also be important for

D. corymbosa seedling advantages. If the D. corymbosa adults are bearing the carbon cost to support seedlings in their mycorrhizal network, the seedlings are receiving the benefits of the ECM association without expending much carbon in the light-limited understory. This provides them with associate fungi that have superior exploitative capacities in the nutrient-limited soil than the neighboring, AM seedlings. In a competitive release experiment, D. corymbosa seedlings were removed from seedling plots at different densities, and the remaining seedlings in the plots were monitored for biomass and survival responses. After 1 year we found that non-dominant seedlings had greater biomass (in terms of roots, shoots and stems) when D. corymbosa seedlings were removed, suggesting that the removal released the non-dominant seedlings from interspecific competition (McGuire and Goldberg, 2008, unpublished data). However, we also found evidence for strong intra-specific seedling competition as a major factor in D. corymbosa seedling mortality. We did not find notable differences between the D. corymbosa seedlings that died versus those that survived, but it would be interesting to explore this phenomenon during future masting episodes.

2.4 ECM-mediated mechanisms in nutrient cycling

Leaf litter and woody debris accumulation are significantly greater in the monodominant forest compared to the mixed forest. Recent research shows that slower decomposition can explain the significant buildup for forest floor in the monodominant forest (McGuire, 2007b). After 2 years of collecting leaf litter fall in mesh traps, litter fall in the mixed forest was higher than leaf litter fall in the monodominant forest. These results revealed that higher rates of leaf litter production in the monodominant forest could not explain the deeper forest floor. The forest floor turnover coefficient was also significantly lower in the monodominant forest ($k = 0.66$) compared to the mixed forest ($k = 0.96$), supporting the hypothesis that slower decomposition, rather than greater leaf litter production, explains the deeper forest floor observed in the monodominant forest. We also set out reciprocally-planted litter bags to test the effects of leaf litter and forest type on decomposition rates. After 2 years of decomposition in the field, leaf litter decomposition in litterbags was significantly slower in the monodominant forest ($k = 0.44$ year^{-1}) compared to the mixed forest ($k = 0.93$ year^{-1}), but there was no effect of leaf litter type (D. corymbosa litter versus mixed tree species litter) within each forest. Slower decomposition due to the biochemical composition of D. corymbosa leaf litter, therefore, could not explain the deeper forest floor in the monodominant forest.

Why is there slower decomposition in the monodominant forest? One hypothesis pertains to potential antagonistic relationships between saprotrophic bacteria and fungi. Due to the saprotrophic abilities of ECM fungi, but the poor competitive abilities of ECM fungi compared to saprotrophic bacteria and fungi, it is possible that ECM fungi are suppressing saprotrophs in the monodominant forest to slow decomposition and 'short-circuit' organically bound nutrients back to *D. corymbosa*. This is a long-standing hypothesis, also known as the 'Gadgil' hypothesis (Gadgil and Gadgil, 1971, 1975). While compelling, this hypothesis is incredibly difficult to test, particularly under field conditions. In order to provide convincing evidence for suppression of saprotrophs by ECM fungi we must show that: (1) decomposition is not due to abiotic factors or chemical constituents of the monodominant leaf litter; and (2) there is lower biomass, species richness and/or abundance of saprotrophs in the monodominant forest compared to the mixed forest. Since it appears that the former piece of evidence is supported in this system, we are currently attempting to test the second line of required evidence. So far we have observed lower microbial biomass in the monodominant forest using phospholipids fatty acid analysis (McGuire, 2007b). We also extracted DNA from forest floor samples and selectively amplified the fungi using general fungal primers. The DNA analysis showed distinct banding patterns between forest types using denaturing gradient gel electrophoresis (DGGE), and redundancy discriminate analysis revealed a 78.9% separation of fungal community composition by forest type. Together, these results support the hypothesis that ECM fungi are suppressing saprotrophic fungi. We now need to directly sequence the DNA of the amplified fungi to see if there are indeed fewer saprotrophs in the mixed forest. This work is currently in progress, and will be the first direct attempt at sequencing community-level fungal DNA in a monodominant rain forest.

3 FUTURE DIRECTIONS AND PRIORITIES FOR RESEARCH

Monodominance research in Africa, Asia, and the Neotropics has revealed that ECM associations facilitate the dominance of one tree species in otherwise diverse tropical forest. What will happen to the distribution of these forests with future global changes? This question should be a priority for monodominance research. We know that global changes such as elevated levels of atmospheric CO_2 are predicted to have dramatic impacts on mycorrhizal fungal community composition and function (Treseder and Allen, 2000; Treseder, 2004). However, very little of this research has been conducted in tropical forests, and no studies to date have investigated the effects of global changes on monodominant, ECM forests. The future dynamics of the transitions from mixed to monodominant forest will be important to understand, as both forest types are invaluable to both indigenous communities and forest managers.

Many monodominant, ECM trees retain the ability to be colonized by AM fungi. Increasing evidence from mycorrhizal surveys has revealed multiple examples of a single tree species simultaneously utilizing more than one mycorrhizal habit, sometimes even on the same rootlet (Chilvers et al., 1987; Horton et al., 1998; Frioni et al., 1999; Moyersoen and Fitter, 1999; Chen et al., 2000). Dual infections of AM and ECM associations are poorly understood in monodominant systems, but the ability of an ECM tree to retain the capacity to form AM associations may be important in colonization events, where ECM inoculum is limiting. The ecological implications for this phenomenon are vast, and cannot be overlooked when considering ECM monodominant communities.

There are obvious geographical biases in the distribution of tropical monodominant communities, especially towards the Guineo-Congolian regions of South America and Africa. This is suggestive of historical influences of reduced regional species pools, which would have increased rates of competitive exclusion and allowed for easier establishment of potential monodominant species. This is also supported by the relatively depauperate nature of the African tropics and the Guianas (Zagt and Werger, 1997a; Torti et al., 2001). The only Amazonian example of monodominance is *Peltogyne gracilipes*, and this forest is situated at the southern flank of the Guiana shield, further supporting biogeographical influences. There are also noticeably skewed distributions in familial relationships to monodominance. Of the persistent monodominant forests described, the majority are members of the family Caesalpiniaceae. Analyzing the respective characters of closely related, non-dominant Caesalpinoids, would be invaluable to understanding the relative contributions of current traits to achieving monodominance in regards to how historical traits, particularly the mycorrhizal habit, have been modified and changed.

4 CONCLUSION

There is substantial evidence from the *D. corymbosa* monodominant system that ECM-mediated positive feedbacks facilitate the high levels of monodominance observed in the central Guyana forests. Mechanisms of ECM-mediated advantages have been identified at almost every life-history stage of this tree species, and new findings continue to highlight the importance of the ECM association in this system. While not all tropical monodominant trees form ECM associations, in those that do (which is the majority of mono-dominant trees), it is likely that these ECM-mediated mechanisms are also important. Future research in tropical monodominant forests should include aspects of global change biology so we can better understand future dynamics of tropical rain forests. Detailed comparisons with other monodominant forests will provide useful insights in to the abilities of these mechanisms to be generalized for ECM, monodominant forests across the tropics.

Acknowledgement: The author would like to thank the Patamona Amerindian tribes, Margaret and Malcolm Chana-Sue and Raquel Thomas. Funding was provided by the National Science Foundation (DDIG).

REFERENCES

Abuzinadah, R. A., and Read, D. J., 1986, The role of proteins in the nitrogen nutrition of Ectomycorrhizal Plants. III. Protein utilization by *Betula*, *Picea* and *Pinus* in mycorrhizal association with *Hebeloma crustuliniforme*. *New Phytol.* **103**:507–514.

Allen, M. F., Swenson, W., Querejeta, J. I., Egerton-Warburton, L. M., and Treseder, K. K., 2003, Ecology of mycorrhizae: A conceptual framework for complex interactions among plants and fungi. *Annu. Rev. Phytopathol.* **41**:271–303.

Baraloto, C., Forget, P. M., and Goldberg, D. E., 2005, Seed mass, seedling size and neotropical tree seedling establishment. *J. Ecol.* **93**:1156–1166.

Brokaw, N. V. L., 1985, Gap-phase regeneration in a tropical forest. *Ecology* **66**:682–687.

Brundrett, M. C., 2002, Coevolution of roots and mycorrhizas of land plants. *New Phytol.* **154**:275–304.

Chen, Y. L., Brundrett, M. C., and Dell, B., 2000, Effects of ectomycorrhizas and vesicular-arbuscular mycorrhizas, alone or in competition, on root colonization and growth of *Eucalyptus globulus* and *E. urophylla*. *New Phytol.* **146**:545–556.

Chilvers, G. A., Lapeyrie, F. F., and Horan, D. P., 1987, Ectomycorrhizal vs endomycorrhizal fungi within the same root-system. *New Phytol.* **107**:441–448.

Connell, J. H., and Lowman, M. D., 1989, Low-diversity tropical rain forests: some possible mechanisms for their existence. *Amer. Natural.* **134**:88–119.

Dighton, J., Tomas, E. D., and Latter, P. M., 1987, Interactions between tree roots, mycorrhizas, a saprophytic fungus and the decomposition of organic substrates in a microcosm. *Biol. Fert. Soils* **4**:145–150.

Finlay, R. D., and Read, D. J., 1986, The structure and function of the vegetative mycelium of ectomycorrhizal plants. 2. The uptake and distribution of phosphorus by mycelial strands interconnecting host plants. *New Phytol.* **103**:157–165.

Frioni, L., Minasian, H., and Volfovicz, R., 1999, Arbuscular mycorrhizae and ectomycorrhizae in native tree legumes in Uruguay. *Forest Ecol. Manag.* **115**:41–47.

Gadgil, R. L., and Gadgil, P. D., 1971, Mycorrhiza and litter decomposition. *Nature* **233**:133.

Gadgil, R. L., and Gadgil, P. D., 1975, Suppression of litter decomposition by mycorrhizal foots of *Pinus radiata*. *New Zealand J. Forest Sci.* **5**:35–41.

Gentry, A. H., 1992, Tropical forest biodiversity-distributional patterns and their conservational significance. *Oikos* **63**:19–28.

Gotelli, N. J., and Colwell, R. K., 2001, Quantifying biodiversity: procedures and pitfalls in the measurement and comparison of species richness. *Ecol. Lett.* **4**:379–391.

Hartnett, D. C., and Wilson, G. W. T., 1999, Mycorrhizae influence plant community structure and diversity in tallgrass prairie. *Ecology* **80**:1187–1195.

He, X. H., Critchley, C., and Bledsoe, C., 2003, Nitrogen transfer within and between plants through common mycorrhizal networks (CMNs). *Crit. Rev. Plant Sci.* **22**:531–567.

Henkel, T. W., 2003, Monodominance in the ectomycorrhizal *Dicymbe corymbosa* (Caesalpiniaceae) from Guyana. *J. Trop. Ecol.* **19**:417–437.

Henkel, T. W., Mayor, J. R., and Woolley, L. P., 2005, Mast fruiting and seedling survival of the ectomycorrhizal, monodominant *Dicymbe corymbosa* (Caesalpiniaceae) in Guyana. *New Phytol.* **167**:543–556.

Hibbett, D. S., Gilbert, L. B., and Donoghue, M. J., 2000, Evolutionary instability of ectomycorrhizal symbioses in basidiomycetes. *Nature* **407**:506–508.

Horton, T. R., Cazares, E., and Bruns, T. D., 1998, Ectomycorrhizal, vesicular-arbuscular and dark septate fungal colonization of bishop pine (*Pinus muricata*) seedlings in the first 5 months of growth after wildfire. *Mycorrhiza* **8**:11–18.

Janos, D. P., 1983, Tropical mycorrhizas, nutrient cycles and plant growth. pp. 327–345 *in* S. L. Sutton, T. C. Whitmore, and A. C. Chadwick, eds. *Tropical Rain Forest: Ecology and Management.* Blackwell Scientific, Oxford.

Janos, D. P., 1985, Mycorrhizal fungi: agents or symptoms of tropical community composition. pp. 98–103 *in* R. Molina, ed. Proc. 6th North Amer. Conf. on Mycorrhizae. Oregon State University, Corvallis, OR.

Janzen, D. H., 1974, Tropical blackwater rivers, animals and mast fruiting by the Dipterocarpaceae. *Biotropica* **6**:69–103.

Klironomos, J. N., 2002, Feedback with soil biota contributes to plant rarity and invasiveness in communities. *Nature* **417**:67–70.

Leigh, E. G., Davidar, P., Dick, C. W., Puyravaud, J., Terborgh, J., ter Steege, H., and Wright, S. J., 2004, Why do some tropical forests have so man species of trees? *Biotropica* **36**:445–473.

McGuire, K. L., 2007a, Common ectomycorrhizal networks may maintain monodominance in a tropical rain forest. *Ecology* **88**:567–574.

McGuire, K. L., 2007b, Ectomycorrhizal Associations Function to Maintain Tropical Monodominance: Studies from Guyana. Ph.D. Dissertation: University of Michigan, Ann Arbor, MI.

McGuire, K. L., 2007c, Recruitment dynamics and ectomycorrhizal colonization of *Dicymbe corymbosa*, a monodominant tree in the Guiana Shield. *J. Trop. Ecol.* **23**:297–307.

McGuire, K. L., Henkel, T. W., Granzow de la Cerda, I., Villa, G., Edmund, F., and Andrew, C., 2008, Dual mycorrhizal colonization of forest-dominating tropical trees and the mycorrhizal status of non-dominant tree and liana species. *Mycorrhiza* **18**:217–222.

Moyersoen, B., and Fitter, A. H., 1999, Presence of arbuscular mycorrhizas in typically ectomycorrhizal host species from Cameroon and New Zealand. *Mycorrhiza* **8**:247–253.

Nara, K., 2006, Ectomycorrhizal networks and seedling establishment during early primary succession. *New Phytol.* **169**:169–178.

Newbery, D. M., Chuyong, G. B., and Zimmermann, L., 2006, Mast fruiting of large ectomycorrhizal African rain forest trees: importance of dry season intensity, and the resource-limitation hypothesis. *New Phytol.* **170**:561–579.

Redecker, D., Kodner, R., and Graham, L. E., 2000, Glomalean fungi from the Ordovician. *Science* **289**:1920–1921.

Silvertown, J. W., 1980, The evolutionary ecology of mast seeding in trees. *Biol. J. Linn. Soc.* **14**:235–250.

Simard, S. W., and Durall, D. M., 2004, Mycorrhizal networks: a review of their extent, function, and importance. *Can. J. Bot.* **82**:1140–1165.

Simard, S. W., Perry, D. A., Jones, M. D., Myrold, D. D., Durall, D. M., and Molina, R., 1997, Net transfer of carbon between ectomycorrhizal tree species in the field. *Nature* **388**:579–582.

Smith, J. E., and Read, D. J., 1997, *Mycorrhizal Symbiosis*. Second eds. Academic Press, San Diego, CA.

Torti, S. D., and Coley, P. D., 1999, Tropical monodominance: a preliminary test of the ectomycorrhizal hypothesis. *Biotropica* **31**:220–228.

Torti, S. D., Coley, P. D., and Kursar, T. A., 2001, Causes and consequences of monodominance in tropical lowland forests. *Amer. Natural.* **157**:141–153.

Treseder, K. K., 2004, A meta-analysis of mycorrhizal responses to nitrogen, phosphorus, and atmospheric CO_2 in field studies. *New Phytol.* **164**:347–355.

Treseder, K. K., and Allen, M. F., 2000, Black boxes and missing sinks: fungi in global change research. *Mycol. Res.* **104**:1282–1283.

Trojanowski, J., Haider, K., and Hutterman, A., 1984, Decomposition of C^{14} labelled lignin, holocellulose and lignocellulose by mycorrhizal fungi. *Archiv. Microbiol.* **139**:202–206.

Valencia, R. H., Balslev, H., Paz, H., and Mino, C. G., 1994, High tree alpha-diversity in Amazonian Ecuador. *Biodiv. Conserv.* **3**:21–28.

van der Heijden, M. G. A., Klironomos, J. N., Ursic, M., Moutoglis, P., Streitwolf-Engel, R., Boller, T., Wiemken, A., and Sanders. I. R., 1998, Mycorrhizal fungal diversity determines plant biodiversity, ecosystem variability and productivity. *Nature* **396**:69–72.

Woolley, L. P., Henkel, T. W., and Sillett, S. C., 2008, Reiteration in the monodominant tropical tree *Dicymbe corymbosa* (Caesalpiniaceae) and its potential adaptive significance. *Biotropica* **40**:32–43.

Wu, B. Y., Nara, K., and Hogetsu, T., 2001, Can C^{14} labeled photosynthetic products move between *Pinus densiflora* seedlings linked by ectomycorrhizal mycelia? *New Phytol.* **149**:137–146.

Zagt, R. J., and Werger, M. J. A., 1997a, Community structure and the demography of primary species in tropical rain forest. pp. 21–38 *in* R. J. Zagt, ed. *Tree Demography in the Tropical Rain Forest of Guyana*. Tropenbos, Utrecht, The Netherlands.

Zagt, R. J., and Werger, M. J. A., 1997b, Spatial components of dispersal and survival for seeds and seedlings of two codominant tree species in the tropical rain forest of Guyana. *Trop. Ecol.* **38**:343–355.

Chapter 13

THE USE OF MYCORRHIZAL BIOTECHNOLOGY IN RESTORATION OF DISTURBED ECOSYSTEM

ALI M. QUORESHI
Symbiotech Research Inc. Alberta, Canada T9E 7N5, and Department Sciences of Wood and the Forest, Université Laval, Québec, G1K 7P4, Canada

Abstract: Mycorrhizal fungi play a crucial role in plant nutrient uptake, water relations, ecosystem establishment, plant diversity, and productivity of plants. Mycorrhizas also protect plants against root pathogens and toxic stresses. The fundamental importance of the mycorrhizal association in restoration and to improve revegetation of disturbed mined lands is well recognized. However, the use of mycorrhizal biotechnology in land reclamation and revegetation of disturbed mine sites is not well practiced in many parts of the world. The destruction of mycorrhizal fungal network in soil system is the vital event of soil disturbance, and its reinstallation is an essential approach of habitat restoration. Successful revegetation of severely disturbed mine lands can be achieved by using "biological tools" mycorrhizal fungi inoculated tree seedlings, shrubs, and grasses. This chapter discusses the different types of mycorrhizas, which play an essential function in altering disturbed lands into productive lands, the mechanisms by which disturbed ecosystem benefits through symbiotic associations and their interactions in the rhizosphere. The importance of reinstallation of mycorrhizal systems in the rhizosphere is emphasized and their impact in landscape regeneration and in bioremediation of contaminated soils are discussed.

Keywords: Land reclamation and revegetation; disturbed mined lands; mycorrhizal inoculation; symbiotic association.

1 INTRODUCTION

Reclamation of severely disturbed lands is a global concern. Mining activities and other forms of anthropogenic or human induced disturbances can result in massive areas of unproductive land with little or no biological activity. The destruction of microbial activities in soil system is the vital event of soil disturbance, and its re-installation is an essential approach of habitat restoration. Severe disturbance alters the composition and activity of mycorrhizal fungi as well as the host plants (Allen *et al.*, 2005). The root systems of most vascular plants (95%) harbour diverse communities of mycorrhizal fungi and are found in a wide range of habitats (Read, 1991; Smith and Read, 1997). The role of mycorrhizal fungi and other micro-organisms in restoration of disturbed soils has been the subject of interest to scientists over the decades. Successful establishment of forest plants on reforestation sites often depends on different mycorrhizal formation and on the ability of seedlings to capture site resources quickly during the early plantation establishment (Amaranthus and Perry, 1987; Perry *et al.*, 1987; van den Driessche, 1991; Dunabeitia *et al.*, 2004).

Mining activity affects soil nutrient, pH, toxicity, bulk density, biological activity, and soil moisture. Besides other factors, low levels of mycorrhizal soil inoculum in disturbed sites (Bois *et al.*, 2005; Malajczuk *et al.*, 1994) frequently delayed in successful restoration of the land as productive as previously existing vegetation. Restoration of metals-contaminated environments requires an efficient microbial community for successful plant community establishment, soil improvement, nutrient cycling (Moynahan *et al.*, 2002). The inoculation of plants with arbuscular mycorrhizal (AM) fungi is considered a pre-requisite for successful restoration of heavy metal contaminated soils (Gaur and Adholeya, 2004). Successful revegetation and reclamation of severely disturbed mine lands in various parts of the world has been accomplished by using the biological tools (Cordell *et al.*, 1991; Marx, 1991; Malajczuk *et al.*, 1994). These tools include phytobial remediation practices, which consist of planting seedlings inoculated with mycorrhizal fungi, nitrogen fixers, actinomycetes, and growth-promoting bacteria. Inoculation of nursery seedlings with appropriate mycorrhizal fungi-host combinations is the most environment friendly approach, particularly for disturbed ecosystem. This approach is known to promote uptake of nutrients and water, buffer against various stresses, and increase resistance against some pathogens, and considered essential to enhance seedling performances (Stenstrom *et al.*, 1990, Villeneuve *et al.*, 1991; Smith and Read, 1997; Bois *et al.*, 2005; Quoreshi *et al.*, 2008b, in press b). It has been demonstrated that specific arbuscular or ectomycorrhizal (ECM) fungi enhance the survival rates and early growth performance of various softwood and hardwood species in the field (Danielson and Visser, 1989; Smith and Read, 1997; Pera *et al.*, 1999; Ortega *et al.*, 2004). Ericoid mycorrhizal (ERM) fungi can improve various stressful environments including heavy metal toxicity to their host plants (Read, 1992), however the mechanisms that help in this increased tolerance are not clear.

2 WHAT IS RECLAMATION

The land reclamation is the process of reconverting disturbed land to its land capacity equivalent to the pre-disturbed conditions or any other productive uses. The main objective of reclamation is to reclaim disturbed land to a stable, biologically self-sustaining state as soon as possible. This means creating a landscape with productive capability similar, if not more so, to that before it was disturbed. The re-installation of microbiological activities to mining sites is known to enhance revegetation and reclamation success.

3 BENEFITS OF MYCORRHIZAS

Mycorrhizal fungi are essential component of a self-sustaining ecosystem. There are abundant examples of the ability mycorrhizal fungi to enhance growth and nutrition of tree seedlings, both in nursery conditions and in the field after outplanting (Danielson and Visser, 1989; Kropp and Langlois, 1990; Villeneuve *et al.*, 1991; Browning and Whitney, 1992, 1993; Le Tacon *et al.*, 1994; Gagne *et al.*, 2006; Quoreshi *et al.*, 2008b, in pressb).

ECM fungi can enhance the ability of forest plants to grow in unfavourable environmental and soil conditions (Jones and Hutchinson, 1988). The extraradical mycelia of ECM fungi exploit the greater soil volume and can reach micropore areas and absorb nutrients that may otherwise inaccessible both physically and chemically (Perez-Moreno and Read, 2000). The ECM fungi have the ability to provide buffering capacity to plant species against various environmental stresses (Malajczuk *et al.*, 1994). It was found that some ECM species were able to degrade phenanthrene and fluoranthrene (Gramss *et al.*, 1999), tolerate the presence of 2% w/v toluene, and can grow petroleum hydrocarbon-contaminated soil with no adverse effects on plant growth and development (Sarand *et al.*, 1999).

The AM fungi are known as bio-ameliorators of saline soils, potential agents in plant protection and pest management, reducing plant mortality, improving plant establishment and plant growth (Gould *et al.*, 1996; Sharma and Dohroo, 1996; Al-Karaki *et al.*, 2001; Sylvia and Williams, 1992; Azcon-Aguilar and Barea, 1996). Soil inoculation with *Glomus mosseae* has significantly increased plant growth and biomass production in mine spoil soils (Rao and Tak, 2002). The ectendomycorrhizas are often found in conifers and mostly confined to the genera *Pinus* and *Larix* formed by a small group of Ascomycetes fungi (Yu *et al.*, 2001). They are frequently found on the roots of plants growing on disturbed lands.

Ericoid plants tolerate extremely difficult environments and can established on various disturbed lands. The fungus enables access to recalcitrant sources of minerals and provides protection from the consequences of the adverse soil conditions. Ericoid mycorrhizas have shown a particular role in the mineralization of nitrogen (Read *et al.*, 1989). The importance of ERM in the nutrient acquisition of heathland plants has been well acknowledged (Read, 1991) and can obtain nitrogen from various sources (Peterson *et al.*, 2004). ERM fungi are found to tolerate and sequester high concentrations of heavy metals such as copper and zinc (Bradley *et al.*, 1982; Meharg and Cairney, 2000), improve tolerance to alkalinity and improve mineral absorption in the presence of calcium salts (Leake *et al.*, 1990). Although the increased nutrient uptake is the most significant single benefit of mycorrhizae, this fascinating symbiotic relationship offers numerous benefits to their host plant, which can be summarized:

- Enhance plant efficiency in absorbing water
- Reduce fertilizer and irrigation requirements
- Increase drought resistance
- Increase pathogen resistance
- Protect against damage from heavy metals and other pollutants
- Minimize various plant stresses
- Improve seedling growth and survival
- Improve soil structure and contribute to nutrient cycling processes
- Contribute toward carbon sequestration.

4 RECLAMATION OF DISTURBED LANDS AND MYCORRHIZAL BIOTECHNOLOGY

Mycorrhizal inoculation is beneficial for reclamation of variety of disturbed sites (Danielson and Visser, 1989; Marx, 1991) and had a great potential in the restoration of natural ecosystems (Miller and Jastrow, 1992). The lack of mycorrhizal associations on plant root systems is one of the major reasons for failure of plantation establishment and growth in various forest with low inoculum potential, mined sites, and restoration of disturbed areas. Intensive fertilizer and fungicide use in nursery stock culture to increase seedling growth in single growing season may inhibit mycorrhizal development (Kropp and Langlois, 1990; Quoreshi and Timmer, 1998; Quoreshi, 2003). Nevertheless, nursery cultural practices often create cultural conditions that encourage certain mycorrhizal fungi, such as *Telephora terrestris* and E-strain fungi (Ursic *et al.*, 1997). However, these nursery-adapted fungal strains are often ecologically different from those prevailing in the field, particularly if the seedlings are aimed to plant in disturbed mined sites.

Inoculation of nursery seedlings with selected mycorrhizal fungi may reduce potential mycorrhizal deficiency in roots and can enhance field performance.

Several techniques have been developed to inoculate nursery seedlings using different types of inoculum (Trappe, 1977; Marx and Kenney, 1982; Molina and Trappe, 1982, 1984; Marx et al., 1991). Recently, Budi et al. (1998) provided information concerning mycorrhizal inoculation strategies and practices. Pure vegetative inoculum of selected fungi is suggested as the most effective materials for inoculation since harmful organisms are excluded (Marx and Kenney, 1982). The pure vegetative inoculum production is involved in growing pure vegetative mycelium in vermiculite-peat mixtures moistened with liquid media produced by either shake flask or in fermentor and incubated in dark condition for a certain period. This solid inoculum can be produce in huge quantities in autoclavable bag culture procedure or using fermentor. The colonized solid substrate (vermiculite-peat) subsequently used as pure vegetative solid inoculum mixed with a growing substrate. The pure vegetative inoculum is also used as liquid mycelial slurry. The liquid inocula are grown from selected fungi on suitable liquid media using shake flask or fermentor. The suspension usually diluted before inoculation with water to obtain desired concentration of propagules per millilitre in the mycelial slurry. The liquid inoculum can be mixed with growing substrate at sowing seeds or can be injected at root collar. Inoculation of conifer seedlings using liquid mycelial slurry of ECM fungi was equally effective as vermiculite-peat solid inoculum (Quoreshi et al., 2005).

Three types of inoculum are currently being used in forest nurseries to inoculate seedlings: (1) vermiculite-peat based solid-substrate inoculum; (2) liquid/mycelial slurry inoculum, and (3) spore inoculum. There are now many examples in using excised and transformed root organ as a tool for producing inoculum of various AM species (Bécard and Piché, 1992; Fortin et al., 2002). Transformation of roots by *Agrobacterium rhizogenes* has provided a novel way to obtain mass production of axenic roots on artificial media in a very short period. A group of research scientists in Canada have been working on developing techniques for producing improved ECM inoculum by using (i) *Agrobacterium rhizogenes*-transformed root culture as a tool for the production ECM inoculum (Coughlan and Piché, 2005); (ii) Chitosan beads for encapsulating fungal mycelia for the production of ECM and *Frankia* sp. inocula (Quoreshi et al., 2008, unpublished data).

Coughlan and Piché (2005) demonstrated new technique for ECM research based on transformed root organ culture, and suggested that root organ culture provides a valuable tool for studying mycorrhizal association and inoculum production. Several previous studies have shown the progress in entrapping living cell in alginate gel and possibility of producing a new type of inoculum (Le Tacon et al., 1985; Mauperin et al., 1987). ECM Chitosan beads are produced by the axenic liquid culture of fungal mycelia encapsulated within Chitosan beads. Chitosan is chemically similar to cellulose, which is a constituent of plant fibre. Interestingly, chitin is also a constituent of fungal cell wall. Research is in progress using this natural product as a tool for encapsulation of ECM mycelia and *Frankia* within Chitosan polymer bead to obtain effective ECM inoculum.

The AM fungi are obligate biotrophs and require living host plant for the completion of life cycle (Fortin *et al.*, 2002; Dalpé and Monreal, 2004). AM fungal propagation can take place either by spore germination or by mycelium extension through soil and roots (Dalpé and Monreal, 2004). Generally, AM fungi are propagated through pot culture. In this system, fungal spores, and colonized root fragment are used as starter inoculum, and mixed with a growing substrate for inoculated seedling production (Brundrett *et al.*, 1996). Subsequently, colonized substrate and roots can then serve as AM fungal inoculum. Root-organ culture system showed an effective means of production of AM inoculum that can be used either directly as inoculum or as starter inoculum for large-scale production (Fortin *et al.*, 2002). For further detail about AM inoculum propagation and commercial production can be viewed from the article by Dalpé and Monreal (2004).

Artificially inoculated mycorrhizal fungi also enhance growth and nutrition of tree seedlings, both in nursery conditions and in the field after outplanting (Marx *et al.*, 1988; Kropp and Langlois, 1990; Villeneuve *et al.*, 1991; Browning and Whitney, 1992, 1993; Le Tacon *et al.*, 1994; Gagne *et al.*, 2006; Quoreshi *et al.*, 2008b, in press b). The primary purpose for inoculating seedlings with mycorrhizal fungi is to provide planting stock with adequate mycorrhizas to improve their survival and growth after planting. Such approach is particularly essential in revegetation of disturbed sites. The soils of degraded sites are frequently low in available nutrients, mycorrhizal fungi and other beneficial microorganisms (Cooke and Lefor, 1990). Similar to ECM, AM fungi also play a significant role in establishment of plants in disturbed and stressed ecosystems (Gould *et al.*, 1996). Soil inoculation with *Glomus mosseae* has significantly increased plant growth and biomass production in mine spoil soils (Rao and Tak, 2002).

5 APPLICATION OF MICROBIAL BIOTECHNOLOGY IN RECLAMATION OF DISTURBED LAND

Oil sands reserves are found in several places around the globe. Canada has ¾ of the world's oil sands deposits, which comprises ⅓ of the world's known oil reserves. Oil sands are naturally occurring deposits of bituminous (petroleum) sands. In northern Alberta, surface mining of bituminous sand generates massive areas of disturbed lands with saline sodic overburden, tailing sands (TS), fine fluid tailings, composite tailings (CT), and coke as by-products that require reclamation. These materials are considered challenging substrate for reclamation and revegetation, because of high alkalinity and salinity, low in organic matter, poor nutrition, contains residual hydrocarbons, and lack of necessary biological activity (Bois *et al.*, 2005; Quoreshi *et al.*, 2005). Revegetation of tailing sands (TS) is now routine practice in reclamation of oil sands disturbed lands. However, composite tailing (CT) is extremely difficult materials to reclaim and still in research phase (Khasa *et al.*, 2002). Recently, the oil sand industry has reduced generating CT materials considerably by adjusting their extraction process. Nevertheless, other soft tailings are still very complex materials to reclaim and need extensive research.

Soil degradation processes alters in the diversity and survival of fungal populations (Malajczuk *et al.*, 1994; Koomen *et al.*, 1990). The aim of reclamation of disturbed lands is to develop landscape not only with forest plants (mostly ECM plants) but also other kinds of vegetation cover for quick restoration of lands. Therefore, it is essential to consider potential use of ECM as well as AM and ERM mycorrhizal plants for revegetation of disturbed areas. Researches have focused on phytotoxicity, plant tolerance to heavy metals, mine spoils soils, saline conditions and establishment of soil microbial activity, and emphasized soil microbial composition influence successful plant growth and establishment (Chan and Wong, 1989; Donnelly and Fletcher, 1994; Galli *et al.*, 1994).

Different amending materials have been used to amend the oil sands disturbed lands. These materials include deep overburden, LFH, fresh peat, and stockpiled peat. These materials are used to stabilized and reconstruct the disturbed soil before revegetation (Fung and Macyk, 2000), and reported presence of mycorrhizal fungi in these materials used for stabilizing the disturbed soil from oil sand tailings (Zak and Parkinson, 1983; Danielson and Visser, 1989). However, inoculum potential of these materials is reduced during manipulations, stockpiled, and due to long storage time (Bois *et al.*, 2005). Soil moisture content during stockpiling was also found critical for survival of AM fungal propagules. Mycorrhizal biotechnology is aimed at re-installation of mycorrhizal networks in disturbed soils. This may be achieved by transplanting pre-inoculated seedlings with appropriated fungal strain already present with their root system, or by inoculating soils with fungal inoculum.

The assessment of inoculum potential of amending materials can potentially be beneficial for successful revegetation and reclamation of amended lands. In a recent study, Bois *et al.* (2005) evaluated mycorrhizal inoculum (ECM and AM) potentials of reclamation materials and tailing sands from Canadian oil sand areas. Results of this study demonstrated that

CT was completely devoid of mycorrhizal propagules, while all other materials showed some level of inoculum potential. CT and TS were also demonstrated devoid of ECM propagules. Controlled inoculation of seedlings in the nursery with selected strains could compensate for low natural inoculum potential, and could improve host growth and survival (Bois et al., 2005). Another important methodology in the application of mycorrhizal biotechnology in reclamation or revegetation work is the development of mycorrhizal DNA fingerprints. Powerful molecular tools such as micro satellite markers are essential to monitor the persistence of individual strains in the reclamation site. In the past decade, polymerase chain reaction (PCR) based research has been conducted for ECM communities, and population studies. Simple sequence repeat (SSR) markers have been developed recently for ECM fungal species (Kretzer et al., 2000, 2003; Lain et al., 2003). More recently, micro satellite or SSR markers were developed for Hebeloma *crustuliniforme* UAMH 5247, *Laccaria bicolor* UAMH 8232 species for the detection of introduced strains and molecular ecology applications (Jany et al., 2003, 2006).This potent marker can be used as an efficient tool for monitoring the persistence of these fungi into the field.

5.1 Actinorhizal biotechnology for the revegetation and reclamation of disturbed lands

Alders are actinorhizal plants that fix atmospheric nitrogen in nodules by symbiotic association with actinomycetes of the genus *Frankia*. Actinomycetes is an important microorganism, play a crucial role in successful establishment of nitrogen fixing primary successor alder plants on disturbed lands, and also have the potential to improve soil quality. Alders can grow in ecologically extreme and disturbed sites and have the ability to improve soil fertility and stability (Brunner et al., 1990; Hibbs and Cromack, 1990; Yamanaka et al., 2002). Many actinorhizal plants are also capable of forming ectomycorrhizal (ECM) association as well, thus develop dual symbiosis and increased success of these plants under disturbed soil conditions. In a study, inoculation of containerized green alder with a pure culture of *Frankia* was feasible under commercial nursery environment (Quoreshi et al., 2007). *Frankia* inoculation in their study resulted in improved seedling production for use in revegetation of oil sands disturbed lands. In a recent review, Sébastien et al. (2007) addressed various aspects of alder-based phytotechnologies for rehabilitation of contaminated soils and discussed the importance of alders, *Frankia*, mycorrhizae and their symbioses for the successful revegetation and reclamation of disturbed ecosystem. The outplanting performance of *Frankia*-inoculated seedlings on oil sands tailings contaminated sites in Alberta showed significant improvement in seedling growth, survival, and soil quality two year after initial planting (Quoreshi et al., unpublished data). It appears that use of *Frankia*-inoculated alders as a promising approach for the remediation and revegetation of the oil sands tailings. Further improvement is expected in field performance of alders with saline-alkaline tolerant and site adapted *Frankia* strain.

5.2 Selection of stressed site adapted mycorrhizal fungi and its implications

The sodicity of the anthroponic soils created by the oil sands industry is one of the major constrains that hampered revegetation efforts. Another pre-requisite of successful inoculation programs is the selection of appropriate fungal strains for target plant species and site to be revegetated. A wide range of fungi from all groups is found in metal-polluted and various mine spoils habitats (Gadd, 1993). Mycorrhizas are considered multifunctional that means different fungi show different effect on the same host, but the same fungus can exhibit different effects on the same host under different environmental conditions (Newsham *et al.*, 1995). Therefore, it is essential to develop selection procedures for mycorrhizal fungi to be used as inocula in nursery inoculation practices. Dunabeitia *et al.* (2004) suggested that selection of appropriate fungal symbionts and development of methods for large-scale inoculum production should be the pre-requisite for the use of ECM inoculation programs. In vitro selection of the most promising ectomycorrhizal fungi was reported for use in the reclamation of saline-alkaline habitats (Kernaghan *et al.*, 2002). In this experiment, pure cultures of several fungal species indigenous to the Canadian boreal forest were tested on media containing different levels of $CaCl_2$, $CaSO_4$, $NaCl$, Na_2SO_4, and on media containing CT release water. Among the fungal isolates tested, members of the Boletales, mainly *Suillus brevipes*, *Rhizopogon rubescens* and *Paxillus involutus*, and *Amphinema byssoides* (Aphyllophorales) were most sensitive to alkalinity and their growth was completely inhibited by CT release water. However, *Laccaria* and *Hebeloma* spp. showed tolerance to alkalinity and survived on the medium with CT release water. Kernaghan *et al.* (2002) suggested that inoculating seedlings with a combination of fungal species; each with its own beneficial characteristics might be suitable for CT sites. *Laccaria bicolor* is recommended for its rapid growth and overall salt tolerance, *Hebeloma crustuliniforme* is recommended for its excellent tolerance to the CT release water, as well as *Wilcoxina mikolae* for its tolerance to $CaCl_2$.

Bois *et al.* (2005) examined salt tolerance capacity of three mycorrhizal fungi (*Suillus tomentosus*, *Hymenoscyphus* sp., and *Phialocephala* sp.) isolated from sodic site in Alberta. These isolates were then compared with *Laccaria bicolor* and *Hebeloma crustuliniforme* that were previously recommended for use in salt-affected soil by Kernaghan *et al.* (2002). The results of the recent study demonstrated that the three isolates tested from the sodic site were more resistance to NaCl than *L. bicolor* UMAH 8232 and *H. crustuliniforme* UAMH 5247. The three isolates from sodic site exhibited a strategy of toxic ion avoidance compared to a preferential ion accumulation tolerance mechanism to osmotic stress found with *L. bicolor* and *H. crustuliniforme*. Nevertheless, the later two strains showed a high osmotic potential that allowed it to maintain its water content.

Significant progress was made in laboratory and greenhouse experiments for selecting the most promising salt tolerant strains of ECM fungi for use in reclamation of oil sands tailings (Kernaghan et al., 2002, 2003; Bois et al., 2006). Results from field trials with white spruce and jack pine demonstrated positive responses after outplanting on two different sites at Canadian oil field reclamation sites. The plot volume index (PVI) of ECM inoculated seedlings significantly enhanced when compared to non-inoculated control seedlings in case of both jack pine and white spruce (Quoreshi et al., 2008, unpublished data). There is currently an increasing interest in using consortium of indigenous fungal strains, which might have certain advantages in revegetation of disturbed lands. It is suggested that certain species of mycorrhizal fungi may provide unique tolerance abilities to the particular site stresses and benefits the plants in the long term (Bois et al., 2006). The field trial is in progress by our research group using consortium ECM fungi and outplanted onto Canadian oil sands disturbed lands.

The significance of AM fungi in disturbed soil remediation has lately been recognized by Gaur and Adholeya (2004) and Khan (2006). AM fungi and their hyphal network provide an excellent system for plant-based environmental clean up and have the potential to take up heavy metals from an enlarged soil volume (Göhre and Paszkowski, 2006). Gaur and Adholeya (2004) have suggested that indigenous AM fungi found naturally in heavy metal-polluted soils are more tolerate than isolates from non-polluted soils, and are reported to colonize plant roots effectively in heavy-metal contaminated environments. According to Oliveira et al. (2005) AM strain native to highly alkaline anthropogenic sediment is generally more effective than the non-native fungi in improving plant establishment and growth under stressed environments. The result of their study suggested that the use of adapted AM as inoculants for phytorestoration of alkaline anthropogenic-stressed sediments. Several other studies demonstrated that native AM could perform better in soils from which they are isolated (Enkhtuya et al., 2000; Caravaca et al., 2003; Göhre and Paszkowski, 2006). Many studies have suggested that for native plant community restoration, the source of inoculum is an important factor (Klironomos, 2003; Moora et al., 2004). It has been suggested that locally collected field inoculum is more effective than commercial inocula for establishing late-successional species (Rowe et al., 2007).

Ericaceous plants are associated with ERM fungi and have the ability to adapt a broad range of habitats (Mitchell and Gibson, 2006). ERM fungi can tolerate mine spoils, acidic, nutrient poor, and heavy metal polluted environment (Jones and Hutchinson, 1986, 1988; Colpaert and Van Assche, 1992, 1993; Lacourt *et al.*, 2000; Mitchell and Gibson, 2006; Gibson and Mitchell, 2006). The ability of these associations to withstand such harsh environments is probably due to the plastic nature of both fungi and partners (Cairney and Meharg, 2003). It is demonstrated that the stimulatory effect of ericoid mycorrhizal association on copper mine spoil environment was dependent on the host plant involved (Gibson and Mitchell, 2006). In their study, rooted-cuttings taken from a mine spoil site and inoculated with the fungal isolate obtained from an uncontaminated site (not with the isolate from mine spoil site) performed better than cuttings taken from an uncontaminated site. The combination of cuttings from an uncontaminated site and the fungal isolate from mine spoil site had a significantly lower shoot dry mass than non-mycorrhizal control plants. Therefore, a suitable combi-nation of fungus and host is essential for maximizing the benefits of any inoculation programs.

To obtain mined lands tolerant strain of mycorrhizal fungi, it is necessary to evaluate the collected isolates *in vitro* selection procedure first under different levels of salts and saline composite tailings release water. Finally, the selected isolates need to test in association with host plants (*in vivo* selection) in order to identify the most promising salt tolerant strains for large-scale inoculation program. A robust screening procedure may reduce the need for extensive field-testing of selected isolates. In essence, the fundamental steps in any inoculation program are to (i) identify and characterize the potential sites to be revegetated; (ii) collection, isolation, and identification of fungi; (iii) screening of fungi through *in vitro* selection procedures for identifying most promising strains; (iv) *In vivo* selection of selected fungal strain in association with host plants for larger inoculation program; (v) suitable inoculum production; (vi) development of large-scale inoculation program under commercial nursery environment, and inoculation of target indigenous plant species; (vii) outplanting of inoculated seedlings onto target sites for field trials, (viii) monitor plant growth, establishment, and persistence of introduced microsymbionts; (ix) finally, evaluation of the success of inoculation program. A proposed scheme for successful use of mycorrhizal biotechnology for the reclamation of disturbed lands is shown in Fig. 1.

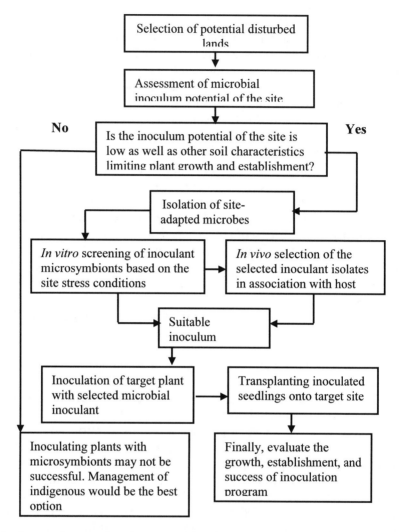

Fig. 1. Suggested approaches for successful reclamation and remediation of disturbed ecosystem using microbial biotechnology.

6 CONCLUSIONS

The importance of microbial biotechnology for the reclamation and remediation of disturbed lands is now well appreciated by the scientific community. Numerous studies have demonstrated that plant-microbe symbioses and their interactions in the rhizosphere are the essential determinants of plant productivity and soil fertility. Since the destruction of mycorrhizal fungal networks and other microbial activities in the soil systems is a major soil disturbance event, its re-installation is an essential for successful restoration of disturbed lands. Study results indicated that mycorrhizal associations are one of the key factors for the successful establishment of vegetation on reclamation sites. The selection of appropriate plant species and their specific site tolerant indigenous microbial isolates is likely to maximize the success of land reclamation. It is suggested that the use of saline-alkaline tolerant and site adapted mycorrhizal fungi and *Frankia* strain can enhance the health, survivorship, and establishment of conifers and alders outplanted on challenging materials generates from oil sands industry in Alberta. Since the ultimate goal of the oil sands industry is to reclaim their disturbed mined lands into a typical boreal forest plant community, it will be essential to consider reinstallation of ECM, AM, and ERM fungal associations in the reclamation program. To achieve this goal a long-term collaborative effort is required consists of oil sands industry, university, and microbial biotechnology company.

REFERENCES

Al-Karaki, G.N., Hammad, R., and Rusan, M., 2001, Response of two tomato cultivars differing in salt tolerance to inoculation with mycorrhizal fungi under salt stress. *Mycorrhiza* **11**: 43–47.

Allen, M.F., Allen, E.B., and Gómez-Pompa, A., 2005, Effects of mycorrhizae and non-target organisms on restoration of a seasonal tropical forest in Quintano Roo, Mexico: factors limiting tree establishment. *Restor. Ecol.* **13**: 325–533.

Amaranthus, M.P., and Perry, D.A., 1987, Effect of soil transfer on ectomycorrhizal formation and the survival and growth of conifer seedlings on old, reforested clear-cuts. *Can. J. For. Res.* **17**: 944–950.

Azcón-Aguilar, C., and Barea, J.M., 1996, Arbuscular mycorrhizas and biological control of soil-borne plant pathogens: an overview of the mechanisms involved. *Mycorrhiza* **6**: 457–464.

Bécard, G., and Piché, Y., 1992, Establishment of vesicular-arbuscular mycorrhiza in root organ culture: review and proposed methodology. In: *Techniques for the study of mycorrhiza*, J., Norris, D., Read, and A., Verma eds., Academic, New York, pp. 89–108.

Bois, G., Bertrand, A., Piché, Y., Fung, M., and Khasa, D.P., 2006, Growth, compatible solute and salt accumulation of five mycorrhizal fungal species grown over a range of NaCl concentrations. *Mycorrhiza* **16**: 99–109.

Bois, G.Y., Piché, Fung, M.Y.P., and Khasa, D.P., 2005, Mycorrhizal inoculum potentials of pure reclamation materials and revegetated tailing sands from the Canadian oil sand industry. *Mycorrhiza* **15**: 149–158.

Bradley, R., Burt, A.J., and Read, D.J., 1982, The biology of mycorrhiza in the Ericaceae VIII. The role of mycorrhizal infection in heavy metal resistance. *New Phytol.* **91**: 197–209.

Browning, M.H.R., and Whitney, R.D., 1992, Field performance of black spruce and jack pine inoculated with selected species of ectomycorrhizal fungi. *Can. J. For. Res.* **22**: 1974–1982.

Browning, M.H.R., and Whitney, R.D., 1993, Infection of containerized jack pine and black spruce by *Laccaria* species and *Thelephora terristris* and seedling survival and growth after outplanting. *Can. J. For. Res.* **23**: 330–333.

Brundrett, M., Bougher, N., Dell, B., Grove, T., and Malajczuk. N., 1996, *Working with mycorrhizas in forestry and agriculture.* ACIAR, Canbera, Australia.
Brunner, I.L., Brunner, F., and Miller, O.K., 1990, Ectomycorrhizal synthesis with Alaskan Alnus tenuifolia. *Can. J. Bot.* **68**: 761–767.
Budi, S., Caussanel, J., Trouvelot, A., and Gianinazzi, S., 1998, The biotechnology of mycorrhizas. In: *Microbial interactions in agriculture and forestry*, N.S., Subba Rao, and Y.R., Dommergues eds., Science Publishers, New Delhi, India, pp. 149–162.
Cairney, J.W.G., and Meharg, A.A., 2003, Ericoid mycorrhiza: a partnership that exploits harsh edaphic conditions. *Europ. J. Sol. Sci.* **54**: 735–740.
Caravaca, F., Barea, J.M., Palenzuela, J., Figuerosa, D., alguacil, M.M., and Roldán, A., 2003, Establishment of shrub species in a degraded semiarid site after inoculation with native or allochthonous arbuscular mycorrhizal fungi. *Appl. Soil Ecol.* **22**: 103–111.
Chan, F.J., and R.M. Wong, 1989, Reestablishment of native riparian species at an altered high elevation site. In: *Proceedings of California Riparian Systems Conference*, Protection, management, and restoration for the 1990's, pp. 428–435.
Colpaert, J.V., and Van Assche, J.A., 1992, The effects of cadmium and cadmium-zinc interactions on the axenic growth of ectomycorrhizal fungi. *Plant Soil* **145**: 237–243.
Colpaert, J.V., and Van Assche, J.A., 1993, The effects of cadmium on mycorrhizal *Pinus sylvestris* L. *New Phytol.* **123**: 325–333.
Cooke, J.C., and Lefor, M.W., 1990, Comparison of veiscular-arbuscular mycorrhizae in plants from disturbed and adjacent undisturbed regions of a costal salt marsh in Clinto, Connecticut, USA. *Environ. Manage.* **14**: 212–137.
Cordell, C.E., Marx, D.H., and Caldweii, C., 1991, Operational application of specific ectomycorrhizal fungi in mineland reclamation. In: *Proceedings of National Meeting of American Society on Surface Mining and Reclamation*, Durango, Colorado, May 14–17.
Coughlan, A.P., and Piché, Y., 2005, *Cistus incanus* root organ cultures: a valuable tool for studying mycorrhizal associations. In: *In vitro culture of mycorrhizas*, S., Declerck, D.-G., Strullu, and A., Fortin eds., Springer, Berlin/Heidelberg, Germany, *Soil Biol.* **4**: pp. 235–252.
Dalpé, Y., and Monreal, M., 2004, Arbuscular mycorrhiza inoculum to support sustainable cropping systems. Online. *Crop Manage.* doi:10.1094/CM-2004-0301-09-RV.
Danielson, R.M., and Visser, S., 1989, Host response to inoculation and behaviour of induced and indigenous ectomycorrhizal fungi of jack pine grown on oil-sands tailings. *Can. J. For. Res.* **19**: 1412–1421.
Donnelly, P.K., and Fletcher, J.S., 1994, Potential use of mycorrhizal fungi as bioremediation agents. In: *Bioremediation through rhizosphere technology*, T.A., Anderson, and J.R., Coats eds., ACS Symposium Series, American Chemical Society, Washington, DC, Vol. 563, pp. 93–99.
Dunabeitia, M., Rodriguez, N., Salcedo, I., and Sarrionandia, E., 2004, Field mycorrhization and its influence on the establishment and development of the seedlings in a broadleaf plantation in the Basque country. *For. Ecol. Manage.* **195**: 129–139.
Enkhtuya, B., Rydolvá, J., and Vosátka, M., 2000, Effectiveness of indigenous and non-indegenous isolates of arbuscular mycorrhizal fungi in soils from degraded ecosystems and man-made habitats. *Appl. Soil Ecol.* **14**: 201–211.
Fortin, J.A., Bécard, G., Declerck, S., Dalpé, Y., St-Arnaud, M., Coughlan, A.P., and Piché, Y., 2002, Arbuscular mycorrhiza on root-organ cultures. *Can. J. Bot.* **80**: 1–20.
Fung, M.Y.P., and Macyk, T.M., 2000, Reclamation of oil sand mining areas. In: *Reclamation of drastically disturbed lands*, R.I., Barnhisel, R.G., Darmody, and W.L., Daniels eds., American Society of Agronomy monographs, 2nd edition 41, pp. 755–744.
Gadd, G.M., 1993, Interactions of fungi with toxic metals. *New Phytol.* **124**: 25–60.
Gagné, A., Jany, J-L., Bousquet, J., and Khasa, D.P., 2006, Ectomycorrhizal fungal communities of nursery-inoculated seedlings outplanted on clear-cut sites in northern Alberta. *Can. J. For. Res.* **36**: 1684–1694.

Galli, U., Schuepp, H. and Brunold, C., 1994, Heavy metal binding by mycorrhizal fungi. *Physiol. Plant.* **92**: 364–368.
Gaur, A., and Adholeya, A., 2004, Prospect of arbuscular mycorrhizal fungi in phytoremediation of heave metal contaminated soils. *Curr. Sci.* **86**: 528–534.
Gibson, B.R., and Mitchell, D.T., 2006, Sensitivity of ericoid mycorrhizal fungi and mycorrhizal *Calluna vulgaris* to copper mine spoil from Avoca County, Wicklow. *Biol. Environ. Proc. Royal Irish Acad.* **106B**: 9–18.
Göhre, V., and Paszkowski, U., 2006, Contribution of the arbuscular mycorrhizal symbiosis to heavy metal phytoremediation. *Planta* **223**: 1115–1122.
Gould, A.B., Hendrix, J.W., and Richard, S.F., 1996, Relationship of mycorrhizal activity to time following reclamation of surface mine land in western Kentucky. 1. Propagule and spore population densities. *Can. J. Bot.* **74**: 247–261.
Gramss, G., Kirsche, B., Voigt, B., Günther, T., and Fritsche, W., 1999, Conversion rates of five polycyclic aromatic hydrocarbons in liquid cultures of fifty-eight fungi and the concomitant production of oxidative enzymes. *Mycol. Res.* **8**: 1009–1018.
Hibbs, D.E., and Cromack, K., 1990, Actinorhizal plants in Pacific northwest forests. In: *The biology of Frankia and actinorhizal plants*, C.R., Schwintzer, and J.D., Tjepkema eds, Academic, London, pp. 343–364.
Jany, J.L., Bousquet, J., and Khasa, D.P., 2003, Microsatellite markers for *Hebeloma* species developed from expressed sequence tags in the ectomycorrhizal fungus *Hebeloma cylindrosporum*. *Mole. Ecol.* **3**: 659–661.
Jany, J.L., Bousquet, J., Gagné, A., and Khasa, D.P, 2006, Simple sequence repeat (SSR) markers in the ectomycorrhizal fungus *Laccaria bicolor* for environmental monitoring of introduced strains and molecular ecology applications. *Mycol. Res.* **110**: 51–59.
Jones, M.D., and Hutchinson, T.C., 1986, The effect of mycorrhizal infection on the response of *Betula papyrifera* to nickel and copper. *New Phytol.* **102**: 429–442.
Jones, M.D., and Hutchinson, T.C., 1988, Nickel toxicity in mycorrhizal birch seedlings infected with *Lactarius rufus* or *Scleroderma flavidum*. I. Effects on growth, photosysthesis, respiration and transpiration. *New Phytol.* **108**: 451–459.
Kernaghan, G., Hambling, B., Fung, M., and Khasa, D.P., 2002, *In vitro* selection of Boreal ectomycorrhizal fungi for use in reclamation of saline–alkaline habitats. *Restor. Ecol.* **10**: 1–9.
Kernaghan, G., Sigler, L., and Khasa, D.P., 2003, Mycorrhizal and root endophytic fungi of containerized *Picea glauca* seedling assessed by rDNA sequence analysis. *Microb. Ecol.* **45**: 128–136.
Khan, A.G., 2006, Mycorrhizoremediation- an enhanced form of phytoremidiation. *J Zhejiang. Univ. Science B* **7**: 503–514.
Khasa, D.P., Chakarvarty, P., Robertson, B., Thomas, R., and Danick, B.P., 2002, The mycorrhizal status of selected poplar clones introduced in Alberta. *Biomass Bioener.* **22**: 99–104.
Klironomos, J.N., 2003, Variation in plant response to native and exotic arbuscular mycorrhizal fungi. *Ecology* **84**: 2292–2301.
Koomen, I., McGrath, S.P., and Giller, K.E., 1990, Mycorrhizal infection of clover is delayed in soils contaminated with heavy metals from past sewage sludge applications. *Soil Biol. Biochem.* **22**: 871–873.
Kretzer, A.M., Molina, R., and Spatafora, J.W., 2000, Microsatellite markers for the ectomycorrhizal basiodiomycetes *Rhizopogon vinicolor*. *Mole. Ecol.* **9**: 1190–1191.
Kretzer, A.M., Dunham, S., Molina, R., and Spatafora, J.W., 2003, Microsatellite markers reveal the below ground distribution of genets in two species of *Rhizopogon* forming tuberculate ectomycorrhizas on Douglas fir. *New Phytol.* **161**: 313–320.
Kropp, B.R., and Langlois, C.G., 1990, Ectomycorrhizae in reforestation. *Can. J. For. Res.* **20**: 438–451.

Lacourt, I., D'Angelo, S., Girlanda, M., Turnau, K., Bonfante, P, and Perotto, S., 2000, Genetic polymorphism and metal sensitivity of *Oidiodendron maius* strains isolated from polluted soil. *Annl. Microbiol.* **50**: 157–166.

Lain, C., Hogetsu, T., Matsushita, N., Guerin-Laguette, A., Suzuki, K., and Yamada, A., 2003, Development of microsatellite markers from an ectomycorrhizal fungus, *Tricholoma matsutake*, by an ISSR-suppression-PCR method. *Mycorrhiza* **13**: 27–31.

Le Tacon, F., Jung, G., Mugnier, J., Michelot, P., and Mauperin, C., 1985, Efficiency in a forest nursery of an ectomycorrhizal fungus inoculum produced in a fermentor and entrapped in polymeric gels. *Can. J. Bot.* **63**: 1664–1668.

Le Tacon, F., Alvarez, I.F., Bouchard, D., Henrion, B., Jackson, M.R., Luff, S., Parledé, I.J., Pera, J., Stenström, E., Volleneuve, N., and Walker, C., 1994, Variations in field response of forest trees to nursery ectomycorrhizal inoculation in Europe. In: *Mycorrhizas in ecosystems*, D.J., Read et al. eds., CAB, Wallingford, pp. 119–134.

Leake, J.R., Shaw, G., and Read, D.J., 1990, The biology of mycorrhiza in the Ericaceae XVI. Mycorrhiza and iron uptake in *Calluna vulgaris* (L.) Hull in the presence of two calcium salts. *New Phytol.* **114**: 651–657.

Malajczuk, N., Redell, P., and Brundrett, M., 1994, The role of ectomycorrhizal fungi in minesite reclamation. In: *Mycorrhizae and plant health*, F.L., Pfleger and R.G., Linderman eds., The American Phytopathological Society, St Paul, MN.

Marx, D.H., 1991, The practical significance of ectomycorrhizae in forest establishment. In: *Ecophysiology of ectomycorrhizae of forest trees*, The Marcus Wallenberg Foundation ed., Stockholm, Sweden, Symposium Proceedings, 7, pp. 54–90.

Marx, D.H., and Kenney, D.S., 1982, Production of ectomycorrhizal inoculum. In: *Methods and principles of mycorrhizal research*, N.C., Schenck ed., The American Phytopathological Society, St Paul, MN, pp. 131–146.

Marx, D.H., Cordell, C.E., and Clark, III, A., 1988, Eight-year performance of loblolly pine with *Pisolithus* ectomycorrhizae on a good quality forest site. *South. J. Am. For.* **12**: 275–280.

Marx, D.H., Ruehle, J.L., and Cordell, C.E., 1991, Methods for studying nursery and field response of trees to specific ectomycorrhizae. In: *Techniques for mycorrhizal research*, J.R., Norris, D., Read, and A.K., Varma eds., Academic, California, pp. 384–411.

Mauperin, C., Mortier, F., Garbaye, J., LeTacon,F., and Carr, G., 1987, Viability of an ectomycorrhizal inoculum produced in liquid medium and entrapped in a calcium alginate get. *Can. J. Bot.* **65**: 2326–2329.

Meharg, A.A., and Cairney, J.W.G., 2000, Ectomycorrhizas- extending the capabilities of rhizosphere remediation? *Soil Biol. Biochem.* **32**: 1475–1484.

Miller, R.M., and Jastrow, J.D., 1992, The application of VA mycorrhizae to ecosystem restoration and reclamation. In: *Mycorrhizal functioning*, M., Allen ed., Chapman & Hall, New York.

Mitchell, D.T., and Gibson, B.R., 2006, Ericoid mycorrhizal association: ability to adapt to a broad range of habitats. *Mycologist* **20**: 2–9.

Molina, R., and Trappe, J.M., 1982. Patterns of ectomycorrhizal host specificity and potential among Pacific Northwest conifers and fungi. *For. Sci.* **28**: 423–458.

Molina, R., and Trappe, J.M., 1984. Mycorrhiza management in bareroot nurseries. In: *Forest nursery manual: production of bareroot seedlings*. Martinus Nijhoff/Dr. W. Junk Publishers, The Hague, The Netherlands.

Moora, M., Opik, M., and Zobel, M., 2004, Performance of two Centaurea species in response to different root-associated microbial communities and to alterations in nutrient availability. *Ann. Bot. Fennici.* **41**: 263–271.

Moynahan, O.S., Zabinski, C.A., and Gannon, J.E., 2002, Microbial community structure and carbon-utilization diversity in a mine tailings revegetation study. *Restor. Ecol.* **10**: 77–87.

Newsham, K.K., Fitter, A.H., and Watkinson, A.R., 1995, Multi-functionality and biodiversity in arbuscular mycorrhizas. *Trends Ecol. Evol.* **10**: 407–411.

Oliveira R.S., Vosátka, M., and Dodd, J.C., 2005, Studies on the diversity of arbuscular mycorrhizal fungi and the efficacy of two native isolates in a highly alkaline anthropogenic sediment. *Mycorrhiza* **16**: 23–31.

Ortega, U., Dunabeitia, M., Menendez, S., Gonzalez-Muria, C., and Majada, J., 2004, Effectiveness of mycorrhizal inoculation in the nursery on growth and water relation of *Pinus radiate* in different water regimes. *Tree Physiol.* **24**: 64–73.

Pera, J., Alvarez, I.F., Rincon, A.M., and Parlardé, J., 1999, Field performance in northern Spain of Douglas-fir seedlings inoculated with ectomycorrhizal fungi. *Mycorrhiza* **9**: 77–84.

Pérez-Moreno, J., and Read, D.J., 2000, Mobilization and transfer of nutrients from litter to tree seedlings via the vegetative mycelium of ectomycorrhizal plants. *New Phytol.* **145**: 301–309.

Perry, D.A., Molina, R., and Amaranthus, M.P., 1987, Mycorrhizae, mycorrhizospheres, and reforestation: current knowledge and research needs. *Can. J. For. Res.* **17**: 929–940.

Peterson, R.L., Massicotte, H.B., and Melville, L.H., 2004, *Mycorrhizas: anatomy and cell biology*. CABI, NRC Research Press, Ottawa, Canada, NRC Monographs, 173 pp.

Quoreshi, A.M., 2003, Nutritional preconditioning and ectomycorrhizal formation of *Picea marina* (Mill.) B.S.P. seedlings. *Eurasian J. For. Res.* **6**(1): 1–63.

Quoreshi, A.M., and Timmer, V.R., 1998, Exponential fertilization increases nutrition and ectomycorrhizal development of black spruce seedling. *Can. J. For. Res.* **28**: 674–682.

Quoreshi, A.M., Khasa, D.P., Bois, G., Jany, J.L., Begrand, E., McCurdy, D., and Fung, M., 2005, Mycorrhizal biotechnology for reclamation of oil sand composite tailings and tailings land in Alberta. In: *The thin green line: a symposium on the state-of-the-art in reforestation proceedings*, S.J., Colombo ed.,. Thin Green Line Symposium – Thunder Bay, Ontario, Ontario Forest Research Institute, Forest Research Information Paper No. 160, pp. 122–127.

Quoreshi, A.M., Roy, S., Greer, C.W., Beaudin, J., McCurdy, D., and Khasa, D.P., 2007, Inoculation of green alder (*Alnus crispa*) with *Frankia*-ectomycorrhizal fungal inoculants under commercial nursery production conditions. *Native Plants J.* **8**(3): 271–280.

Quoreshi, A.M., Kernanghan, G., and Hunt, G., 2008a, Mycorrhizal fungi in Canadian forest nurseries and the success of inoculated seedlings. NRC Monographs, Ottawa, Canada (in press).

Quoreshi, A.M., Piché, Y., and Khasa, D.P., 2008b, Field performance of conifer and hardwood species five years after nursery inoculation in the Canadian Prairie Provinces. *New For.* **35**: 235–253.

Rao, A.V., and Tak, R., 2002, Growth of different tree species and their nutrient uptake in limestone mine spoil as influenced by arbuscular mycorrhizal (AM)-fungi in Indian arid zone. *J. Arid Environ.* **51**: 113–119.

Read, D.J., 1991, Mycorrhizas in ecosystems. *Experimen.* **47**: 74–84.

Read, D.J., 1992, The mycorrhizal fungal community with special reference to nutrient mobilization. In: *The fungal community: it's organization and role in the ecosystem*, 2nd edition,G.C., Carrol, and D.T., Wicklow eds., Marcel Dekker, New York, pp. 631–652.

Read, D.J., Leake, J.R., and Langdale, A.R., 1989, The nitrogen nutrition of mycorrhizal fungi and their host plants. In: Nitrogen, phosphorus, and sulphur utilization by fungi, L., Boddy, R., Marchant, and D. J., Read eds., Cambridge University Press, England, pp. 181–204.

Rowe, H.I., Brown, C.S., and Claassen, V.P., 2007, Comparisions of mycorrhizal responsiveness with field soil and commercial inoculum for six native Montane species and *Bromus tectorum*. *Restor. Ecol.* **15**: 44–52.

Sarand, I., Timonen, S., Koivula, T., Peltola, R., Haahtela, K., Sen, R., and Romantschuk, M., 1999, Tolerance and biodegradation of *m*-toluate by Scots pine, a mycorrhizal fungus and fluorescent pseudomonads individually and under associative conditions. *J. Appl. Microbiol.* **86**: 817–826.

Sébastien R., Khasa, D.P., and Greer, C.W., 2007, Combining alders, frankiae, and mycorrhizae for the revegetation and remediation of contaminated ecosystems. Can. J. Bot. 85: 237–251.

Sharma, S., and Dohroo, N.P., 1996, Vesicular-arbuscular mycorrhizae in plant health and disease management. *Int. J. Trop. Plant Dis.* **14**: 147–155.

Smith, S.M., and Read, D., 1997, *Mycorrhizal symbiosis*, 2nd edition, Academic, London.

Stenstrom, E., Ek, M., and Unestam, T., 1990, Variation in field response of *Pinus sylvestris* to nursery inoculation with four different ectomycorrhizal fungi. *Can. J. For. Res.* **20**: 1796–1803.

Sylvia, D.M., and Williams, S.E., 1992, Vesicular-arbuscular mycorrhizae and environmental stress. In: *Mycorrhizae in sustainable agriculture*, R.G., Linderman, and G.J., Bethlenfalvay eds., Special Publication No. 54, American Society of Agronomy Madison, WI, pp. 101–124.

Trappe, J.M., 1977, Selection of fungi for ectomycorrhizal inoculation in nurseries. *Ann. Rev. Phytopathol.* **15**: 203–222.

Ursic, M., Peterson, R.L., and Husband, B., 1997, Relative abundance of mycorrhizal fungi and frequency of root rot on *Pinus strobus* seedlings in a southern Ontario nursery. *Can. J. For. Res.* **27**: 54–62.

Van den Driessche, R., 1991, Effects of nutrients on stock performance in the forest. In: *Mineral nutrition of conifer seedlings*. CRC, Boca Ration, FL/Ann Arbor, MI/Boston, MA, pp. 229–260.

Villeneuve, N., LeTacon, F., and Bouchard, D., 1991, Survival of inoculated Laccaria bicolor in competition with native ectomycorrhizal fungi and effects on the growth of outplanted Douglas-fir seedlings. *Plant Soil* **135**: 95–107.

Yamanaka, T., Li, C.-Y., Bormann, B.T., and Okabe, H., 2002, Triplicate association in an alder: effects of *Frankia* and *Alpova diplophloeus* on the growth, nitrogen fixation and mineral acquisition of *Alnus tenuifolia*. *Plant Soil* **254**: 179–186.

Yu, T.E. J.-C., Egger, K.N., and Peterson, R.L., 2001, Ectoendomycorrhizal associations – characteristics and functions. *Mycorrhiza* **11**: 167–177.

Zak, J.C., and Parkinson, D., 1983, Effect of surface emendations on vesicular-arbuscular mycorrhizal development of slender wheatgrass: a 4-year study. *Can. J. Bot.* **61**: 798–803.

Chapter 14

IN VITRO MYCORRHIZATION OF MICROPROPAGATED PLANTS: STUDIES ON *CASTANEA SATIVA* MILL

ANABELA MARTINS
Department of Biology and Biotechnology, Escola Superior Agrária, Instituto Politécnico de Bragança, 5301-855 Bragança, PORTUGAL

Abstract: *In vitro* mycorrhization can be made by several axenic and nonaxenic techniques but criticism exists about their artificiality and inability to reproduce under natural conditions. However, artificial mycorrhization under controlled conditions can provide important information about the physiology of symbiosis. Micropropagated *Castanea sativa* plants were inoculated with the mycorrhizal fungus *Pisolithus tinctorius* after *in vitro* rooting. The mycorrhizal process was monitored at regular intervals in order to evaluate the mantle and hartig net formation, and the growth rates of mycorrhizal and nonmycorrhizal plants. Plant roots show fungal hyphae adhesion at the surface after 24 h of mycorrhizal induction. After 20 days a mantle can be observed and a hartig net is forming although the morphology of the epidermal cells remains unaltered. At 30 days of root–fungus contact the hartig net is well developed and the epidermal cells are already enlarged. After 50 days of mycorrhizal induction, growth was higher for mycorrhizal plants than for nonmycorrhizal ones. The length of the major roots was lower in mycorrhizal plants after 40 days. Fresh and dry weights were higher in mycorrhizal plants after 30 days. The growth rates of chestnut mycorrhizal plants are in agreement with the morphological development of the mycorrhizal structures observed at each mycorrhizal time. The assessment of symbiotic establishment takes into account the formation of a mantle and a hartig net that were already developed at 30 days, when differences between fresh and dry weights of mycorrhizal and nonmycorrhizal plants can be quantified. *In vitro* conditions, mycorrhization influences plant physiology after 20 days of root–fungus contact, namely in terms of growth rates. Fresh and dry weights, heights, stem diameter and growth rates increased while major root growth rate decreased in mycorrhizal plants.

Keywords: *Castanea sativa*; micropropagation; mycorrhization.

1 INTRODUCTION

European chestnut (*Castanea sativa* Mill.) has great economic interest for wood and fruit production but is difficult to propagate by cuttings and show high heterosis of seeds. *C. sativa* has been successfully micro-propagated demonstrating that micropropagation of adult clones can provide an effective way to overcome propagation difficulties. However, micropropagated plants require a long and difficult adaptation period to *ex vitro* conditions. During the first step of weaning, roots obtained *in vitro* usually have a very low efficiency of absorption of water and nutrients (Bonga, 1977; Flick *et al.*, 1983).

Ectomycorrhizal (ECM) fungi bring several advantages to plants, including increased root area for absorption (Bowen, 1973; Harley and Smith, 1983), enhanced uptake of nutrients (Harley and Smith, 1983), resistance to plant pathogens (Marx, 1969), and drought (Duddridge *et al.*, 1980; Boyd *et al.*, 1986; Meyer, 1987; Feil *et al.*, 1988; Marx and Cordell, 1989). ECM can also increase growth and nutrient content of plants growing in low nutrient soils (Jones *et al.*, 1991). Water stress appears to be one of the major causes for the failure of micropropagated plants during acclimation. The compatible mycorrhizal fungi in the substrates during the weaning process not only improve the nutritional state of the plants, but also increase their resistance to the water stress of *ex vitro* conditions, increasing their weaning rates.

The first practical work to evaluate the role of mycorrhizae in plant growth was performed by Frank (1894) with seedlings of the *Pinus* sown in sterile and non-sterile soils. The results showed that plants from non-sterile soils could develop mycorrhizas and grew better than plants from sterilized soils (Smith and Read, 1997). Sterilization by heat was responsible for the production of toxic compounds that could be harmful for plant development. Other sterilizing methods and new methods of mycorrhizal synthesis were used a long time, confirming the results originally shown by Frank (Hacskaylo, 1953; Marx and Zak, 1965; Trappe, 1962, 1967; Pachlewska, 1968; Skinner and Bowen, 1974; Mason, 1975, 1980; Mullette, 1976; Fortin *et al.*, 1980, 1983; Sohn, 1981; Biggs and Alexander, 1981; Nylund, 1981; Rancillac, 1983; Duddridge and Read, 1984a, b; Branzanti and Zambonelli, 1986; Kahr and Arveby, 1986; Kottke *et al.*, 1987; Wong and Fortin, 1988; Bougher *et al.*, 1990, Jones *et al.*, 1990). Mycorrhizae formed in non-sterile soils are responsible for the increased performances of the plants.

2 EFFECT OF MYCORRHIZA INOCULATION ON PLANT GROWTH

The beneficial effect of mycorrhizal associations is the enhanced uptake of mineral nutrients, namely phosphorus (Reid *et al.*, 1983; Jones *et al.*, 1990; Tam and Griffiths, 1993; Eltrop and Marschner, 1996; Smith and

Read, 1997). Mycorrhizal symbiosis is frequently associated with increased photosynthetic rates of mycorrhizal plants (Harley and Smith, 1983; Reid et al., 1983; Bougher et al., 1990; Dosskey et al., 1990; Rousseau and Reid, 1990; Guehl and Garbaye, 1990; Jones et al., 1990; Martins, 1992; Martins et al., 1997; Smith and Read, 1997). ECM may influence the assimilation capacity for CO_2 in two distinct forms: increased absorption of P and N in mycorrhizal plants influence the photosynthetic rates, as observed for forestry species when amended with P; the other resulting from enhanced flux of carbon compounds to the roots, promoted by mycorrhizal associations (Martins et al., 1997, 1999). This hypothesis considers that the increased photosynthetic rates are related with the fungus necessity of carbon compounds and is named source-sink concept (Dosskey et al., 1990, 1991) although this seems to be just one of mechanism involved in photosynthetic increment in mycorrhizal plants (Martins et al., 1997, 1999).

The effect of mycorrhization on plant growth is well documented (Garbaye et al., 1988, Bougher et al., 1990, Grove and Le Tacon, 1993; Le Tacon et al., 1997; Généré, 1995; Martins et al., 1996; Généré et al., 1997; Parladé et al., 1997; Dell and Malajczuk, 1997). Bougher et al. (1990) made several trials in controlled inoculation of *Eucalyptus diversicolor* seedlings with different fungal species and different P supplementations. Plants with higher P and N availability (culture media or soil) showed increased metabolism of proteins and phosphorus compounds (nucleic acids and inositol phosphates). The synthesis of these compounds implies an increase in energy use (carbon compounds) and a lower translocation of carbon compounds to the root. The amount of root soluble carbon compounds condition the nutrition of the associated fungus interferes with the mycorrhizal infection rates (Le Tacon et al., 1997). High levels of mineral nutrients, generally, decrease mycorrhizal efficiency or even infection rates. Under nutrient deficiency, growth rates of mycorrhizal plants increase. Bougher et al. (1990) also evidenced that the response of plants to mycorrhization does not depend only upon nutrient availability but also on the fungus species or even strains of a same species.

The abilities of mycorrhizal species and strains to promote plant growth opened new perspectives for the use of these fungi inoculations in nurseries and forestry. Inoculations of forestry species were performed with different species of hosts and fungi, under different conditions and inoculum types. The influence of mycorrhization on growth rates reveal that plants grow better (Grove and Le Tacon, 1993; Tam and Griffiths, 1994; Eltrop and Marschner, 1996; Le Tacon et al., 1997; Généré 1995; Généré et al., 1997; Parladé et al., 1997; Dell and Malajczuk, 1997), have more extended root systems and both roots and shoots have increased dry weights, although the ratio between the dry weights of roots and shoots were lower for mycorrhizal plants. The similar results were observed in young plants growing with high nutrient levels, behaving like mycorrhizal plants in comparison with plants growing with limited nutrient levels, exhibit nonmycorrhizal like growth (Smith and Read, 1997).

The difference in ratio between dry weights of roots and shoots is more related to plant dimension than to the colonization rate (Bougher *et al.*, 1990). The total number of short roots of mycorrhizal plants is higher than for nonmycorrhizal ones, exhibiting completely altered root morphology by the association with the mycorrhizal fungi. The number of roots per unit length and per unit weight was higher for mycorrhizal root systems (Brundrett *et al.*, 1996). Root colonization by mycorrhizal fungi can result in lower plant growth rates if fungus compatibility, nutrient availability, light intensity or temperature is not suitable for plant development (Marx and Bryan, 1971; Marx, 1979; Nylund and Wallander, 1989; Dosskey *et al.*, 1990; Colpaert *et al.*, 1992; Conjeaud *et al.*, 1996; Smith and Read, 1997). Decrease of growth rates is expected when a symbiont depends on the others to obtain the carbon compounds for survival, and the other depends on the essential mineral nutrients provided by the former for its growth and photosynthesis. Decrease in growth is also expected under light conditions limiting photo-synthesis (Conjeaud *et al.*, 1996), nutrient availability in soil, conditioning plant growth but not colonization intensity (Colpaert *et al.*, 1992; Smith and Read, 1997). Son and Smith (1988) observed an increase in plant growth after colonization of plants under high PAR (photosynthetic active radiation) and a decrease in growth of plants colonized under low PAR, independently of the levels of P availability. When nutrient availability allows fungal growth and there is no light or temperature limitation, fungal growth can require large amounts of carbon compounds conditioning plant growth (Colpaert *et al.*, 1992).

3   *IN VITRO* MICORRHIZATION

Large numbers of *in vitro* studies have been carried out to evaluate the factors that influence mycorrhization. Under natural conditions, interactions of biotic and abiotic factors make the interpretation of the results difficult. The methods of axenic synthesis are object of criticism because working under conditions where (1) interacting factors are eliminated, (2) carbon sources are provided to allow fungal growth before the infection sets in, and (3) substrates are sterilized, may change the efficiency and type of infection (Piché and Peterson, 1988).

In parallel with *in vitro* studies, non axenic studies have been made (Fortin *et al.*, 1980; Piché *et al.*, 1982). It was possible to demonstrate that there are no significant differences between mycorrhizae synthesized under axenic and non axenic conditions (Piché and Peterson, 1988) other than the time of infection (Duddridge and Read, 1984a). The axenic system studied had a time of infection starting at 3 weeks and completed by weeks 6 to 8, while in natural soils, the association was retarded until 11 to 19 weeks. Morphological differences between axenic and non axenic synthesized mycorrhizae exist only when high sucrose levels are used (Duddridge and Read, 1984b). Under these conditions the host-fungus interface is changed and there is callose deposition at the cells walls in response to host infection.

Non axenic systems allow detailed studies of the root colonization by the fungus (Fortin *et al.*, 1983). Fungus connection to the root epidermis is due to the root polysaccharides secretion (Nylund, 1980). The translocation of photosynthetic products to the root increases the concentration of carbon compounds in root exudates. These are mainly amino acids, proteins, carbon compounds, organic acids and plant growth regulators. Mineral balance and plant growth regulators concentrations, directly control cell permeability and the mechanism of fungus adhesion to the roots when mycorrhization takes place (Barea, 1986).

Axenic and non axenic mycorrhizal syntheses mainly differ in the time and degree of infection (Duddridge and Read, 1984a). These findings validated the use of *in vitro* mycorrhization techniques. Mycorrhizas obtainned by different methods of *in vitro* synthesis had mantles and hartig nets with similar structures (Brunner, 1991). Mantle thickness and number of hyphae penetrating between cortical cells may vary with substrate and the synthesis method used.

4 *IN VITRO* MYCORRHIZATION OF MICROPROPAGETED PLANTS

Micropropagated plants are adversely affected by water stress, either due to low absorption capacity of their roots or due to stomata deficient regulation of water loss (Bonga, 1977; Flick *et al.*, 1983). Acclimation of micropropagated plants corresponds to a transition period when roots become adapted to a substrate with less available nutrients, and to an autotrophic condition. At this stage, the presence of mycorrhizae could increase the availability of limiting nutrients such as phosphorus (P) and nitrogen (N), facilitating the absorption. Water stress can be responsible for the low survival of many micropropagated woody plant species during the acclimation process and *C. sativa* is one of these species.

Micropropagated plants develop under high moisture and low lighting conditions, often with low lignifications levels and decreased functionality of the root systems that cause low survival rates to weaning. Mycorrhization of micropropagated plants before acclimation increases survival, enhancing the functionality of the root system and the mineral plant nutrition (Rancillac, 1982; Grellier *et al.*, 1984; Heslin and Douglas, 1986; Poissonier, 1986; Tonkin *et al.*, 1989; Martins, 1992, 2004; Martins *et al.*, 1996; Herrmann *et al.*, 1998; Díez *et al.*, 2000). Similarly, *in vitro* mycorrhization of micropropagated plants can be used to increase survival and growth during *ex vitro* weaning (Nowak, 1998).

Mycorrhization trials have been made with different micropropagated plant species: pine (Rancillac, 1982; Normand *et al.*, 1996), birch (Grellier *et al.*, 1984), poplar (Heslin and Douglas, 1986), eucalyptus (Poissonier, 1986; Tonkin *et al.*, 1989), oak (Herrmann *et al.*, 1998), chestnut (Strullu *et al.*, 1986; Martins, 1992, 2004; Martins *et al.*, 1996; Martins and Pais, 2005), cork oak (Díez *et al.*, 2000). These trials were performed as an effort to make micropropagation a sustainable propagation method for plant species recalcitrant to conventional propagation, increasing *in vitro* plant performances.

Herrmann *et al.* (1998) used an *in vitro* mycorrhizal system of *Quercus robur* micropropagated plants, intending to develop a method to analyze the mycorrhization of forest species without the constraints of the methods using seedlings. Genetic heterogeneity of seedlings (reflected in different germination times), seedling vigour and asynchronous development are only some of these constraints. These trials were made to work with (1) genetically uniform plants deprived of cotyledons, to function as older plants, (2) with selected material, to warranty the uniformity of repetitions, and (3) with a mycorrhizal system that allows following the development along the trials, in order to characterize mycorrhizal effects on plant morphology.

Castanea sativa micropropagated plants were studied along 90 days of plant-fungus association *in vitro*, after preliminary studies on plant-fungus compatibility with four fungi species (Martins *et al.*, 1996). The studies included: (1) development of mycorrhizal morphological structures (mantle and Hartig net) along 90 days; and (2) mycorrhizal influence on growth rates (heights, stem diameter, length of major root, total plant length, fresh weights and dry weights).

5   *CASTANEA SATIVA* MYCORRHIZATION *IN VITRO*

Plants were first inoculated with four different mycorrhizal fungi species to test their mycorrhizal capacities *in vitro*. *Amanita muscaria* Hooker isolate from Schönbuch/Tübingen, *Laccaria laccata* (Scop. ex Fr.) Berk and Br., isolate from Molina, *Piloderma croceum* Erikss and Hjortst, isolate from Unestam and Nylund 01.01.1976, Sweden and *Pisolithus tinctorius* (Pers.) Coker and Couch, isolate 289/Marx, were used (Martins *et al.*, 1996).The fungi tested differed in their capacity to form mycorrhizas with *C. sativa* plants *in vitro. P. tinctorius* showed the best capacity to colonize chestnut roots either from seedlings or from micropropagated plants(Martins *et al.*, 1996, 1997).

Mycorrhizal (M) and nonmycorrhizal plants (NM) were followed for 90 days since inoculation with the mycorrhizal fungus *P. tinctorius*. Mycorrhizal synthesis was performed in Petri dishes 13 cm in diameter, as can be seen in Fig. 1. Agarized MS modified medium plated in slant was used (Murashige and Skoog, 1962). The plants were placed with the root system adhering to the medium in Petri dishes inoculated with the fungus (3 weeks before). Control plants were placed in non-inoculated Petri dishes. The root system was covered with aluminium foil to prevent photo-oxidation (Martins and Pais, 2005).

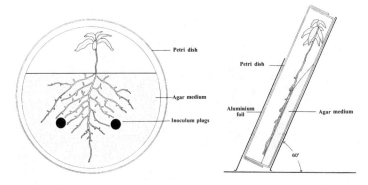

Fig. 1. Axenic synthesis of micropropagated *Castanea sativa* mycorrhizas with *P. tinctorius*.

Mycorrhizal and nonmycorrhizal plants were maintained in a plant tissue culture chamber with a photoperiod of 16 h, light intensity of ~100 $\mu E \cdot m^{-2} \cdot s^{-1}$ and temperatures of 25°C and 19°C respectively during light and dark periods for 90 days after plant transference to the pre-inoculated media (Martins, 2004; Martins and Pais, 2005). Plant development was monitored along 90 days of *in vitro* mycorrhization. Root mycorrhizal status was observed at regular intervals and mycorrhizal evolution compared with growth parameters for the same time of mycorrhization (Martins, 2004; Martins and Pais, 2005).

Roots from micropropagated plants were white and without root hairs or ramifications (Fig. 2a, c) at the time of transference to co-culture with the fungus. *Pisolithus tinctorius* hyphae adhere to the root surface 24 h after root-fungus contact (Fig. 2d); after the first contact roots ramify very quickly compared with non-inoculated ones. Root ramifications became visible 5 days after inoculation, while control plants still have no ramifycations. The fungus surrounds root ramifications forming mycorrhizae after 20 days (Fig. 2e). Establishment of mycorrhizas favours plant growth and leaf

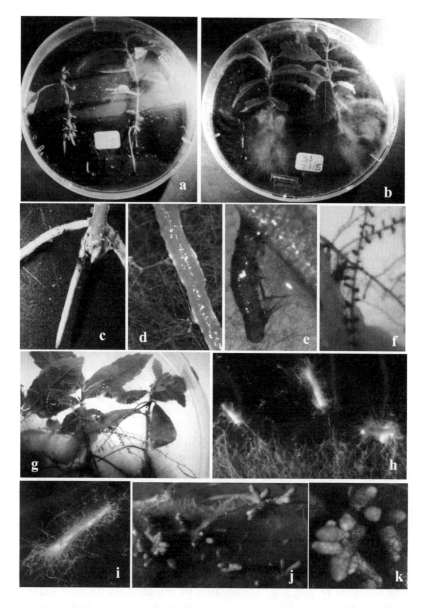

Fig. 2. Micropropagated chestnut plants (a) 20 days after transference to Petri dishes without fungus; (b) 20 days after transference to Petri dishes with fungus (*P. tinctorius*); (c) Root system of a micropropagated plant before mycorrhization (7.5×); (d) Inoculated root 24 h after root-fungus contact (30×); (e) Inoculated root 10 days after root-fungus contact (60×); (f) Inoculated roots 40 days after root-fungus contact (5×); (g) Mycorrhizal plants 40 days after inoculation; (h) Colonized root apices emerging from the medium (25×); (i) Details from a mycorrhizal apex (60×); (j) Inoculated root 60 days after root-fungus contact (40×); (k) Detail of colonized roots 60 days after inoculation (60×).

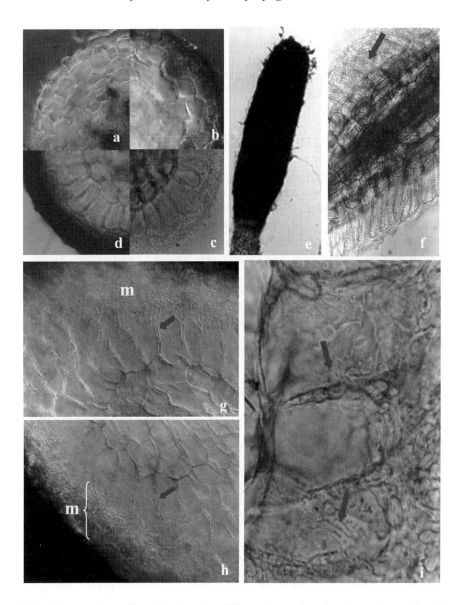

Fig. 3. Cross sections of inoculated roots at different times after plant-fungus interaction (a) 10 days (100×); (b) 20 days (some hyphae can be seen at the surface and an Hartig net is forming,,100×); (c) 30 days (elongation of epidermal cells and a well developed Hartig net can be seen, 100×); (d) 60 days (100×); (e) Mycorrhizal root 30 days after root-fungus contact (40×); (f) Longitudinal section 30 days after root-fungus contact (100×); (g)–(i) Details of a cross section of a mycorrhizal root seen 30 days after root-fungus contact. The Hartig net is visible between the epidermal cells and a mantle (m) is well developed (400×/1,000×).

expansion (Fig. 2a, b). A successive ramification was observed after the first mycorrhizas form giving rise coralloid roots (Fig. 2j, k). The microscopical observation showed that a mantle (m) forms after 20 days but, the hartig net, with longitudinal elongation of epidermal cells, could only be observed after 30 days of root-fungus contact (Fig. 3).

6   GROWTH OF MYCORRHIZAL *CASTANEA SATIVA*

The mycorrhizal process was monitored 90 days of plant-fungus co-culture, to evaluate the growth rates of mycorrhizal and nonmycorrhizal plants, in terms of heights (h), stem diameters at the collar level (d_{collar}), lengths of the major roots (l_{root}) and total plant lengths (l). Fresh weight (FW) and dry weight (DW), as well as growth rates ($\Delta x/\Delta t$) and relative growth rates for each parameter (x) (RGR(x) = $(1/x) \times (\Delta x/\Delta t)$) were also determined for 90 days of association (Martins and Pais, 2005).

Plant heights and stem diameter at the collar level were higher in M plants after 50 and 40 days of mycorrhization respectively while the maximum root length was smaller in M plants after 40 days. The total growth at the end of 90 days (Δx) and the growth per unit time ($\Delta x/\Delta t$) were significantly higher of M plants. After 90 days, M plants had higher growth rates in heights and stem diameter at the collar level. The length of the major root had lower growth rates in M plants. The relative growth rates (RGR) also showed differences between M and NM plants with the exception of the total plant length (Table 1). The larger differences in RGR occurred in the length of the major root. RGR values obtained for the growths in height and stem diameter were higher in M plants. RGR values for total plant length were not significantly different between M and NM plants (Table 1). The ratio h/d_{collar} showed that M plants had a higher increased growth in stem diameter in comparison with growth in heights. The ratios between h/l_{root} were also significantly different in M and NM plants after 50 days of mycorrhization, showing that in M plants the increase in heights was higher than the increase in root length (Table 1).

Fresh and dry weights of M and NM plants were larger in M plants after 30 and 20 days of root-fungus contact, respectively. Differences in FW and DW of roots between M and NM plants were earlier than other plant organs. Roots of M plants showed higher FW and DW than NM plant roots, since 20 days, while the shoots only showed differences since 50 days. Increments in DW (ΔDW, ΔDW/Δt) and RGR were significantly higher for M plants (Table 2). The leaves were the plant organs that showed larger increments in DW after 90 days (ΔDW) and per day (ΔDW/Δt) (Table 2). The

Table 1. Growth parameters of the plants in terms of heights (h), stem diameter at the collar level (D_{collar}). Increases of growth (Δx), increases of growth per day ($\Delta/\Delta t$) and relative growth rates (RGR) per plant, in NM and M plants along 90 days of mycorrhization.

	Heights		Diameter		Length major root		Maximum plant length	
	NM	M	NM	M	NM	M	NM	M
	h (cm)		d_{collar} (cm)		l_{root} (cm)		l_{max} (cm)	
Δx (cm)	6.2a	8.2b	0.15a	0.26b	9.1b	3.8a	15.3b	12.0a
$\Delta x/\Delta t$ (mm/day)	0.69a	0.91b	0.02a	0.03b	1.01b	0.42a	1.70b	1.33a
RGR (mm/cm.day)	0.06a	0.07b	0.06a	0.07b	0.08b	0.05a	0.07a	0.06a

ratios DW/FW of plant roots and the whole plant are higher for M plants since 75 days of mycorrhization. The differences between the ratios for M and NM plants increase along the mycorrhizal synthesis.

Table 2. Increments in DW (ΔDW), DW increments per day (ΔDW/Δt) and relative growth rates (RGR) per plant in NM and M plant roots, stems, leaves, shoots and whole plant, along 90 days of mycorrhization.

	Dry weight/plant organ (mg)						Dry weight/plant (mg)			
	Roots		Stem		Leaves		Shoots		Plant	
	NM	M	NM	M	NM	M	NM	M	NM	M
ΔDW (mg)	39.5a	51.2b	30.5a	39.5b	43.7a	57.3b	74.2a	96.8b	114.4a	148.2b
ΔDW/Δt (mg/day)	0.4a	0.6b	0.3a	0.4b	0.5a	0.6b	0.8a	1.1b	1.3a	1.6b
RGR (mg/g.day)	9.6a	9.9b	7.6a	8.2b	8.7a	9.3b	8.2a	8.8b	8.7a	9.2b

7 CONCLUSIONS

In vitro mycorrhization (endo and ectomycorrhizas) of micropropagated plants can be used to increase survival and growth during *ex vitro* weaning (Martins *et al.*, 1996; Nowak, 1998). In the case of fruit trees, the inoculations of arbuscular fungi facilitate *in vitro* plants adaptation to *ex vitro* conditions (Sbrana *et al.*, 1994). However, *in vitro* ectomycorrhization can improve microcutting rooting (Normand *et al.*, 1996) and enables *in vitro* plants to acclimate more readily (Martins *et al.*, 1996, Díez *et al.*, 2000). The *in vitro* mycorrhization of micropropagated plants like *Helianthemum* spp. (Morte *et al.*, 1994) and *Cistus* spp. (Díez and Manjón, 1996) has been obtained only in very few mediterranean species. Even somatic embryos acclimation can be improved through mycorrhization (Díez et al., 2000). Increase in the root functioning and mineral nutrition of the plants through mycorrhization prior to the acclimation phase can overcome the low performance of micropropagated plants improving their survival and weaning (Martins, 1992, 2004; Martins et al., 1996; Herrmann et al., 1998, Díez et al., 2000).

Under *in vitro* conditions, mycorrhization increases the growth parameters of plants and those are in consistency with the morphological development of mycorrhizal structures, for the same times of mycorrhization. Micropropagated plants improve their performances and survival capacities also increase accordingly. Micropropagation and mycorrhization can be combined as a tool to give viability to the production of difficult propagating species, increasing their survival and growth. Mycorrhization can provide a sustainable method for plant production, either by micropropagation or through traditional propagation methods.

REFERENCES

Barea, J. M., 1986, Importance of hormones and root exudates in mycorrhizal phenomena. In: *Mycorrhizae: Physiology and Genetics*. Eds. V. Gianinazzi-Pearson and S. Gianinazzi, pp. 177–187. INRA, Paris.

Biggs, W. L., and Alexander, I. J., 1981, A culture unit for the study of nutrient uptake by intact mycorrhizal plants under aseptic conditions. *Soil Biol. Biochem.* **13**: 77–78.

Bonga, J. M., 1977, Applications of tissue culture in forestry. In: *Plant Cell Tissue and Organ Culture*. Eds. J. Reinert and Y. P. S. Bajaj, pp. 93–107. Springer, New York.

Bougher, N. L., Grove, T. S., and Malajczuk, N., 1990, Growth and phosphorus acquisition of Karri (*Eucalyptus diversicolor* F. Muell.) seedlings inoculated with ectomycorrhizal fungi in relation to phosphorus supply. *New Phytol.* **114**: 77–85.

Bowen, G. D., 1973, Mineral nutrition of mycorrhizas. In: *Ectomycorrhizas*. Eds. G. C. Marks and T. T. Kozlowski, pp. 151–201. Academic, New York and London.

Boyd, R., Furbank, R. T., and Read D. J., 1986, Ectomycorrhiza and the water relations of trees. In: *Mycorrhizae, Physiology and Genetics*. Eds. V. Gianinazzi-Pearson and S. Gianinazzi, pp. 689–693. INRA, Paris.

Branzanti, M. B., and Zambonelli, A., 1986, Sintesi micorrízica di *Laccaria laccata* (Scop. Ex Fr.) Bk. e Br. Con. *Castanea sativa* Mill. *Mic. Ital.* **2**: 23–26.

Brundrett, M. C., Bougher, N.L., Dell, B., Grove, T.S, and Malajczuk, N., 1996, Working with mycorrhizas in forestry and agriculture. ACIAR Monograph 32. Canberra, Australia, pp. 374.

Brunner, I., 1991, Comparative studies on ectomycorrhizae synthesized with various *in vitro* techniques using *Picea abies* and two *Hebeloma* species. *Trees* **5**: 90–94.

Colpaert, J. V., Van Assche, J. A., and Luijtens, K., 1992, The growth of the extrametrical mycelium of ectomycorrhizal fungi and the growth response of *Pinus sylvestris* L. *New Phytol.* **120**: 127–135.

Conjeaud, C., Scheromm, P., and Mousain, D., 1996, Effects of phosphorus and ectomycorrhiza on maritime pine seedlings (*Pinus pinaster*). *New Phytol.* **133**: 345–351.

Dell, B., and Malajczuk, N., 1997, L'inoculation des eucalyptus introduits en Asie avec des champignons ectomycorhiziens australiens en vue d'augmenter la productivité des plantations. Revue Forestière Française. Numéro Sp. 174–184.

Díez, J., and Manjón, J. L., 1996, Mycorrhizal formation by vitroplants of *Cistus albidus* L. and *C. salvifolius* L. and its interest for truffle cultivation in poor soils. In: *Mycorrhizas in Integrated Systems from Genes to Plant Development*. Eds. C. Azcón-Aguilar and J. M. Barea, pp. 528–530. European Commission, Brussels,.

Díez, J., Manjón, J. L., Kovács, G. M., Celestino, C., and Toribio, M., 2000, Mycorrhization of vitroplants raised from somatic embryos of cork oak (*Quercus suber* L.) *Appl. Soil Ecol.* **15**: 119–123.

Dosskey, M. G., Linderman, R. G., and Boersma, L., 1990, Carbon sink stimulation of photosynthesis in Douglas fir seedlings by some ectomycorrhizas. *New Phytol.* **115**: 269–274.

Dosskey, M. G., Boersma, L., and Linderman, R. G., 1991, Role for the photosynthate demand of ectomycorrhizas in the response of Douglas fir seedlings to drying soil. *New Phytol.* **117**: 327–334.

Duddridge, J. A., Malibari, A., and Read, D. J., 1980, Structure and function of mycorrhizal rhizomorphs with special reference to their role in water transport. *Nature* **287**: 834–836.

Duddridge, J. A., and Read, D. J., 1984a, Modification of the host-fungus interface in mycorrhizas synthesized between *Suillus bovinus* (Fr.) O. Kuntz and *Pinus sylvestris* L. *New Phytol.* **96**: 583–588.

Duddridge, J. A., and Read, D. J., 1984b, The development and ultrastructure of ectomycorrhizas. II. Ectomycorrhizal development on pine *in vitro. New Phytol.* **96**: 575–582.

Eltrop, L., and Marschner, H., 1996, Growth and mineral nutrition of non-mycorrhizal and mycorrhizal Norway spruce (*Picea abies*) seedlings grown in semi-hydroponic sand culture. II Carbon partitioning in plants supplied with ammonium or nitrate. *New Phytol.* **133**: 479–486.

Feil, W., Kottke, I., and Oberwinkler, F., 1988, The effect of drought on mycorrhizal production and very fine root system development of Norway spruce under natural and experimental conditions. *Plant Soil* **108**: 221–231.

Flick, C. E., Evans, D. A., and Sharp, W. R., 1983, Organogenesis. In: *Handbook of Plant Cell Culture. Vol. I-Techniques for Propagation and Breeding*. Eds. D. A. Evans, W. R., Sharp, P. V., Ammirato, Y., Yamada, pp. 13–81. McMillan, New York.

Fortin, J. A., Piché, Y., and Lalonde, M., 1980, Technique for the observation of the early morphological changes during ectomycorrhiza formation. *Can. J. Bot.* **58**: 361–365.

Fortin, J. A., Piché, Y., and Godbout, C., 1983, Methods for synthesizing ectomycorrhizas and their effect on mycorrhizal development. *Plant Soil* **71**: 275–284.

Frank, A., 1894, Die Bedeutung der Mykorrhiza-pilze Für die gemeine Kiefer. *Forstwissenschaftliches Centralblatt*. **16**: 185–190.

Garbaye, J., Dellwaulle, J. C., and Diangana, D., 1988, Growth responses of eucalypts in the Congo to ectomycorrhizal inoculation. *Forest Ecol. Manag.* **24**: 151–157.

Généré, B., 1995, Evaluation en jeune plantation de 2 types de plants de Douglas mycorhizés artificiellement par *Laccaria laccata* S238N. *Ann. Sci. Forest* **52**: 375–384.

Généré, B., Bouchard, D., and Amirault, J. M., 1997, Itinéraire technique en pépinière pour le Douglas de type 1+1 mycorhizé par *Laccaria bicolor* S238N. Forestière Française. Numéro Sp. 155–162.

Grellier, B., Letouzé, R., and Strullu, D. G., 1984, Micropropagation of birch and mycorrhizal formation *in vitro. New Phytol.* **97**: 591–599.

Grove, T. S., and Le Tacon, F., 1993, Mycorrhiza in plantation forestry. In: *Mycorrhiza Synthesis*. Eds. I. C. Tommerup. *Adv. Plant Pathol.* **9**: 191–227.

Guehl, J. M., and Garbaye, J., 1990, The effect of ectomycorrhizal status on carbon dioxide assimilation capacity, water-use efficiency and response to transplanting in seedlings of *Pseudo-Tsuga menziesii* (Mirb.) Franco. *Ann. Sci. Forest* **21**: 551–563.

Hacskaylo, E., 1953, Pure culture synthesis of pine mycorrhizae in terralite. *Mycologia* **45**: 971–975.

Harley, J. L., and Smith, S. E., 1983, *Mycorrhizal Symbiosis*. Academic, New York.

Herrmann, S., Munch, J. C., and Buscot, F., 1998, A gnotobiotic culture system with oak microcuttings to study specific effects of mycobionts on plant morphology before, and in the early phase of, ectomycorrhiza formation by *Paxillus involutus* and *Piloderma croceum*. *New Phytol.* **138**: 203–212.

Heslin, M. C., and Douglas, G. C., 1986, Effects of ectomycorrhizal fungi on growth and development of poplar plants derived from tissue culture. *Sci. Hortic.* **30**: 143–149.

Jones, M. D., Durall, D. M., and Tinker, P. B., 1990, Phosphorus relationships and production of extramatrical hyphae by two types of willow ectomycorrhizas at different soil phosphorus levels. *New Phytol.* **115**: 259–267.

Jones, M. D., Durall, D. M., and Tinker, P. B., 1991, Fluxes of carbon and phosphorus between symbionts in willow ectomycorrhizas and their changes with time. *New Phytol.* **119**: 99–106.

Kahr, M., and Arveby, A. S., 1986, A method for establishing ectomycorrhiza on conifer seedlings in steady-state conditions of nutrition. *Physiol. Plant.* **67**: 333–339.

Kottke, I., Guttenberger, M., Hampp, R., and Oberwinkler, F., 1987, An *in vitro* method for establishing mycorrhizae on coniferous tree seedlings. *Trees* **1**: 191–194.

Le Tacon, F., Mousain, D., Garbaye, J., Bouchard, D., Churin, J. L., Argillier, C., Amirault, J. M., and Généré, B., 1997, Mycorhizes, pépinières et plantations forestières en France. Revue Forestière Française. Numéro Sp. 131–154.

Martins, A., 1992, Micorrização *in vitro* de plantas micropropagadas de *Castanea sativa* Mill. Dissertação para obtenção do grau de mestre. Faculdade de Ciências de Lisboa. 124 pp.

Martins, A., 2004, Micorrização controlada de Castanea sativa Mill.: Aspectos fisiológicos da micorrização in vitro e ex vitro. Dissertação de doutoramento em Biologia/Biotecnologia Vegetal. Faculdade de Ciências de Lisboa. Universidade Clássica de Lisboa. pp. 506.

Martins, A., and Pais, M. S., 2005, Mycorrhizal inoculation of *Castanea sativa* Mill. micropropagated Plants: Effect of mycorrhization on growth. *Acta Hortic.* **693**: 209–217.

Martins, A., Barroso, J., and Pais, M. S., 1996, Effect of ectomycorrhizal fungi on survival and growth of micropropagated plants and seedlings of *Castanea sativa* Mill. *Mycorrhiza* **6**: 265–270.

Martins, A., Casimiro, A., and Pais, M. S., 1997, Influence of mycorrhization on physiological parameters of *Castanea sativa* Mill micropropagated plants *Mycorrhiza* **7**: 161–165.

Martins, A., Santos, M. M., Santos, H., and Pais, M. S., 1999, A P^{31} NMR study of phosphate levels in roots of ectomycorrhizal and nonmycorrhizal *Castanea sativa* Mill. *Trees* **13**: 168–172.

Marx, D. H., 1969, The influence of ectotrophic mycorrhizal fungi on the resistance of pine roots to pathogenic infections. I. Antagonism of mycorrhizal fungi to root pathogenic fungi and soil bacteria. *Phytopathology* **59**: 153–163.

Marx, D. H., 1979, Synthesis of ectomycorrhizae by different fungi in Northern Red Oak seedlings. Forest Service Research Note. USDA. SE 280.

Marx, D. H., and Bryan, W. C., 1971, Formation of ectomycorrhizae on half-sib progenies of slash pine in aseptic culture. *Forest Sci.* **17**: 488–492.

Marx, D. H., and Zak, B., 1965, Effect of pH on mycorrhizal formation of slash pine in asepticulture. *Forest Sci.* **11**: 66–75.

Marx, D. H., and Cordell, C. E., 1989, Use of ectomycorrhizas to improve forestation practices. In: *Biotechnology of Fungi for Improving Plant Growth*. Eds. J. M., Whipps and R. D., Lumsden. Cambridge University Press, Cambridge.

Mason, P. A., 1975, The genetics of mycorrhizal associations between *Amanita muscaria* and *Betula verrucosa*. In: *The Development and Function of Roots*. Eds. J. G., Torrey and D. T., Clarkson, pp. 173–178. Academic Press. New York, London.

Mason, P. A., 1980, Aseptic synthesis of sheating mycorrhizas. In: *Tissue Culture Methods for Plant Pathologists*, pp. 173–178. Eds. D. S. Ingram and J. P. Helgeson. Blackwell, Oxford.

Meyer, F. H., 1987, Extreme site conditions and ectomycorrhizae (especially *Cenococcum geophilum*). *Angew. Bot.* **61**: 39–46.
Morte, M. A., Cano, A., Honrubia, M., and Torres, P., 1994, In vitro mycorrhization of micropropagated *Helianthemum almeriense* plantlets with *Terfezia claveryi* (desert truffle). *Agr. Sci. Finland* **3**: 309–314.
Mullette, K. J., 1976, Studies of eucalypt mycorrhizas. I. A method of mycorrhiza induction in *Eucalyptus gummifera* (Gaertn. and Hochr.) by *Pisolithus tinctorius* (Pers.) Coker & Couch. *Aust. J. Bot.* **24**: 193–200.
Murashige, T., and Skoog, F., 1962, A revised medium for rapid growth and bioassays with tobacco tissue cultures. *Physiol. Plantarum* **14**: 473–497.
Normand, L., Bartschi, H., Debaud, J. C., and Gay, G., 1996, Rooting and acclimatization of micropropagated cuttings of *Pinus sylvestris* are enhanced by the ectomycorrhizal fungus *Hebeloma cylindrosporum. Physiol. Plantarum* **98**: 759–766.
Nowak, J., 1998, Benefits of *in vitro* "biotization" of plant tissue cultures with microbial inoculants. *In vitro Cell. Dev.-Pl.* **34**: 122–130.
Nylund, J. E., 1980, Symplastic continuity during Hartig net formation in Norway Spruce ectomycorrhizae. *New Phytol.* **86**: 373–378.
Nylund, J. E., 1981, The formation of ectomycorrhiza in conifers: Structural and physiological studies with special reference to the mycobiont, *Piloderma croceum*. Ph. D. thesis. University of Uppsala, Sweden, pp. 34.
Nylund, J. E., and Wallander, H., 1989, Effects of ectomycorrhiza on host growth and carbon balance in a semi-hydroponic cultivation system. *New Phytol.* **112**: 389–398.
Pachlewska, J., 1968, Investigations on mycorrhizal symbiosis of *Pinus sylvestris* in pure culture on agar. *Prace Inst. Lesn.* **345**: 3–76.
Parladé, J., Pera, J., and Alvarez, I. F., 1997, La mycorhization contrôlée du Douglas dans le nord de l'Espagne: premiers résultats en plantation. *Rev. For. Fr.* **49**: 163–183.
Piché, Y., Fortin, J. A., Peterson, R. L., and Posluzny, U., 1982, Ontogeny of dichotomizing apices in mycorrhizal short roots of *Pinus strobes. Can. J. Bot.* **60**: 1523–1528.
Piché, Y., and Peterson, R. L., 1988, Mycorrhiza initiation: an example of plant microbial interactions In: *Forest and Crop Biotechnology. Progress and Prospects.* Ed. A. Fredrick. Springer, Valentine.
Poissonier, M., 1986, Mycorhization *in vitro* de clones d'eucalyptus. Note de laboratoire, *Annales de Recherches Sylvicoles* 1985: 81–93.
Rancillac, M., 1982, Multiplication végétative *in vitro* et synthèse mycorhizienne: pin maritime, Hebelome, Pisolithe. *Les colloques de l'ÌNRA* **13**: 351–355.
Rancillac, M., 1983, La mycorhization *in vitro*: influence de la morphologie et des structures anatomiques de l'appareil racinaire sur l'établissement des ectomycorhizes. *Bull. Soc. Bot. Fr.* **130**: 47–52.
Reid, C. P. P., Kidd, F. A., and Ekwebelam, S. A., 1983, Nitrogen nutrition, photosynthesis and carbon allocation in ectomycorrhizal pine. *Plant Soil* **71**: 415–432.
Rousseau, J. V. D., and Reid, C. P. P., 1990, Effects of phosphorus and ectomycorrhizas on the carbon balance of loblolly pine seedlings. *Forest Sci.* **36**: 101–112.
Sbrana, C., Giovannetti, M., and Vitagliano, C., 1994, The effect of mycorrhizal infection on survival and growth renewal of micropropagated fruit rootstocks. *Mycorrhiza* **5**: 153–156.
Skinner, M. F., and Bowen, G. D., 1974, The uptake and translocation of phosphate by mycelial strands on pine mycorrhizas. *Soil Biol. Biochem.* **6**: 53–61.
Smith, S. E., and Read, D. J., 1997, *Mycorrhizal Symbiosis*, pp. 605, 2nd ed. Academic, London.
Sohn, R. F., 1981, *Pisolithus tinctorius* forms long ectomycorrhiza and alters root development in seedlings of *Pinus resinosa. Can. J. Bot.* **59**: 2129–2134.

Son, C. L., and Smith, S. E., 1988, Mycorrhizal growth responses: interactions between photon irradiance and phosphorus nutrition. *New Phytol.* **108**: 305–314.

Strullu, G., Grellier, B., Marciniak, D., and Letouzé, 1986, Micropropagation of chestnut and conditions of mycorrhizal synthesis *in vitro*. *New Phytol.* **102**: 95–101.

Tam, P. C. F., and Griffiths, D. A., 1993, Mycorrhizal associations in Hong Kong *Fagaceae* III. The mobilization of organic and poorly soluble phosphates by the ectomycorrhizal fungus *Pisolithus tinctorius. Mycorrhiza* **2**: 133–139.

Tam, P. C. F., and Griffiths, D. A., 1994, Mycorrhizal associations in Hong Kong *Fagaceae*. VI. Growth and nutrient uptake by *Castanopsis fissa* seedlings inoculated with ectomycorrbizal fungi *Pisolithus tinctorius. Mycorrhiza* **4**: 125–131.

Tonkin, C. M., Malajczuk, N., and McComb, J. A., 1989, Ectomycorrhizal formation by micropropagated clones of *Eucalyptus marginata* inoculated with isolates of *Pisolithus tinctorius*. *New Phytol.* **111**: 209–214.

Trappe, J. M., 1962, Fungus associates of ectotrophic mycorrhizae. *Bot. Rev.* **28**: 538–606.

Trappe, J. M., 1967, Pure culture synthesis of Douglas-fir mycorrhizae with species of *Hebeloma, Suillus, Rhizopogon* and *Astraeus. Forest Sci.* **13**: 121–130.

Wong, K. K. Y., and Fortin, J. A., 1988, A Petri dish technique for the aseptic synthesis of ectomycorrhizae. *Can. J. Bot.* **67**: 1713–1716.

Chapter 15

EFFECTIVE AND FLEXIBLE METHODS FOR VISUALIZING AND QUANTIFYING ENDORHIZAL FUNGI

SUSAN G. W. KAMINSKYJ
Department of Biology, University of Saskatchewan, 112 Science Place, Saskatoon SK, Canada, S7N 5E2

Abstract: Fungi associated with plant roots are gaining prominence as to their importance for plant survival in a diversity of terrestrial ecosystems. Assessing the importance of interaction types depends in part on quantifying interaction prevalence, particularly in plants harvested from natural ecosystems. In turn, this depends on a sensitive method for fungal visualization, and a reliable method for quantifying potentially multiple endorhizal morphotypes. Recent developments in these areas are discussed.

Keywords: Confocal fluorescence; endophyte; fungus; mycorrhiza; quantification; root.

1 INTRODUCTION

Plant roots are associated with a diversity of endorhizal and rhizosphere fungi whose interactions vary from endophytic to pathogenic. Few of these interactions produce macroscopic phenotypes, apart from ectomycorrhizae, which are associated with morphological changes in colonized roots (Brundrett et al., 1996; Smith and Read, 1997). However, this does not mean the others are unimportant! Arbuscular mycorrhizal (AM) interactions are found in about 80% of terrestrial plant families and under experimental conditions contribute to plant survival and competitiveness (Smith and Read, 1997). In addition, roots contain a diversity of endophytic fungi that have roles including stress tolerance (Márquez et al., 2007; Rodriguez et al., 2008). In order to study the nature of plant-fungal interactions and their relative importance to the plants, it is essential that fungi be clearly visualized and accurately quantified.

Microscopic visualization methods for endorhizal fungi have recently been reviewed by Peterson *et al.* (2004) and Vierheilig *et al.* (2005). Traditional mycorrhizal visualization used coloured stains with transmitted light microscopy. For certain experimental studies, roots can be infected with a fungal strain that has been transformed with a fluorescent protein construct (e.g. GFP), so that fungal growth and behaviour can be followed in living roots. However this depends on using previously identified and genetically transformable species (not yet available for many fungi, including AM) rather than assessing roots harvested from natural environments.

Most roots are several to many cell layers thick and may have pigmented surface layers, whereas most fungi are disseminated mycelia of hyaline hyphae. Only the dark septate endophytes can readily be identified without staining (Jumpponen and Trappe, 1998). For transmitted light microscopy of mycorrhizal fungi, cleared roots are typically stained with chlorazole black E (CBE) (Brundrett *et al.*, 1996), trypan blue (Phillips and Hayman, 1970), or lactofuchsin (Carmichael, 1955). The latter sometimes have insufficient contrast for high magnification transmitted light microscopy.

For quantification of AM fungi, Brundrett *et al.* (1996) used CBE-stained roots examined under a dissecting microscope. This provides an overview of fungal colonization, but assessment depends in part on skill and experience, so results can be difficult to reproduce quantitatively between users. A method described by McGonigle *et al.* (1990) used 200X magnification to examine roots at defined intersections. With this method, the user's attention was directed to precise regions examined at relatively high resolution. Reproducibility of the microscopic intersect method even with different users was good, suggesting that it was relatively objective. However, as described, this method was limited to quantifying AM fungi, and did not distinguish between AM morphotypes, nor between levels of AM colonization.

Studies of AM associates in roots of plants collected from sites in the Canadian High Arctic, where they had been thought to be rare or absent, required that we develop a sensitive method for fungal visualization, a secure method for specimen preservation, and a reliable method for quantifying potentially multiple endorhizal morphotypes. Endorhizal visualization methods using fluorescence microscopy (Allen *et al.*, 2006) were later extended to studying roots with multiple morphotypes and from herbarium specimens (Ormsby *et al.*, 2007). Specimen preservation and endorhizal quantification methods described in the latter paper have led to a set of robust and flexible methods that will be presented in detail, below.

2   METHODS

2.1   Choice of stain

Trypan blue has been used as a fluorescent stain for fungal foliar pathogens (Wei *et al.*, 1997), although the preparation method appears to be laborious. The structure of trypan blue and CBE is similar (Fig. 1) suggesting that CBE might be useful as a fluorescent stain for mycorrhizal fungi. CBE fluorescence imaging was not superior to transmitted light microscopy, since CBE solutions develop a fine particulate background. Acid fuchsin, the dye component of lactofuchsin, which is used as a fungal stain (Carmichael, 1955), has a chemical structure suggesting it might be fluorescent (Fig. 1). We later discovered that Merryweather and Fitter (1991) used lactofuchsin for fluorescence imaging, but this method had not been pursued by the authors or others. We independently considered lactofuchsin as a fluorescent stain for fungal cells within plant roots, particularly using confocal epifluorescence microscopy. Use of other fluorescent stains including aniline blue is reviewed by Peterson *et al.* (2004) and Vierheilig *et al.* (2005).

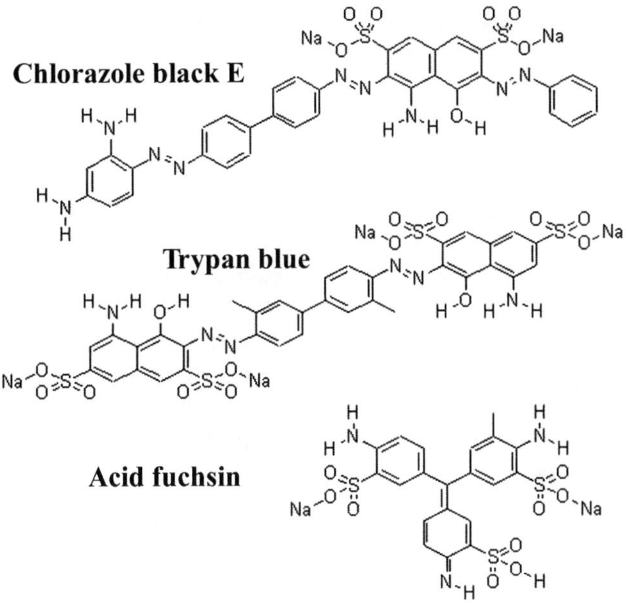

Fig. 1. Chemical structures of chlorazole black E (CBE), trypan blue, and lactofuchsin.

2.2 Sample preparation for lactofuchsin staining

Plant roots from different species may grow intertwined. It is critical that the above-ground and root samples are from the same plant. In addition, roots like those of peony (*Peonia chinensis*; Fig. 2) have a complicated architecture, and are highly pigmented. For these roots, the bulk soil was gently removed by hand (Fig. 2A) and the remaining soil removed by soaking and then rinsing in tap water (Fig. 2B). The soil balls in Fig. 2A contained abundant fine branches.

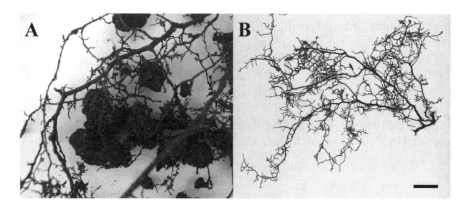

Fig. 2. Roots of peony (*Peonia chinensis*) that had been growing in undisturbed garden soil. Bar in B = 1 cm.

2.2.1 Fixing

Washed roots were fixed in 3.7% formaldehyde containing 0.5% ethanol, buffered to pH 7 in 50 mM Na-K phosphate. Fixation with 3.7% formaldehyde alone, and with 70% or with 95% ethanol was also tested. All gave similar results for staining, but sometimes with higher plant wall fluorescence. However, it may be advantageous to fix in 100% ethanol or isopropyl alcohol, if subsamples might be used for molecular DNA studies.

Ormsby *et al.* (2007) described an adaptation of the clearing and staining technique for roots sampled from herbarium specimens. These were rehydrated by autoclaving in 10% KOH (see below). Even after rehydration, herbarium root samples were sometimes more brittle than chemically fixed roots. All other techniques are the same as for chemically fixed roots. Herbarium specimens can also be used for molecular DNA analyses, and serve as a voucher for identification.

2.2.2 Sampling

For quantitative studies, roots should be cut into segments typically 1–2 cm long, suspended in a large volume of water, and then randomly sub-sampled. For some studies, there is a limited amount of root material available, in which case it is preferable to examine the entire length.

2.2.3 Clearing

Root samples are typically autoclaved for 20 min in 10% KOH, to clear the cytoplasm, using wide glass vials that are topped with a large glass marble to prevent evaporation. Aluminum foil is degraded by hot 10% KOH vapours. Delicate roots survive this treatment, but clearing stems and leaves may require only 95°C for 20 min at atmospheric pressure. Cleared samples should be rinsed twice in room-temperature 70% ethanol to remove the KOH, which is more effective than water alone.

2.2.4 Bleaching

Pigmented roots must be bleached before staining. Roots can be bleached in freshly prepared peroxide solution (1:1:8–28% ammonium hydroxide: 30% hydrogen peroxide: distilled water) or in commercial sodium hypochlorite bleach diluted in distilled water to about 1.75% $NaClO_3$. The latter is effective for bleaching, but these specimens may have higher plant wall fluorescence. Roots are incubated in room-temperature bleaching solution (with occasional swirling, particularly for the peroxide bleach) until pale. Typically this requires 15–30 min. Bleached roots are rinsed twice in distilled water prior to staining. Following staining in lactofuchsin, the vascular cylinder is likely to fluoresce, which can be useful for orientation.

2.2.5 Staining

Roots are stained in lactofuchsin (Carmichael, 1955), 0.1% acid fuchsin in 85% lactic acid. Staining is for 0.5–3 h at 47°C to 68°C, as optimized for different root types. Stained roots are rinsed twice in room-temperature destaining solution (DLAG) (1:1:1 distilled water: 85% lactic acid: glycerol) to remove surface dye, and then destained in DLAG at 47°C to 68°C, typically for 3 h to overnight. Deeply-stained roots will likely have high background fluorescence and poor imaging characteristics. Faintly-stained roots may be still useful since the fluorescence yield of lactofuchsin is much higher than the visual contrast. Under-stained (or overly destained) roots can be re-stained, for longer or at a higher temperature.

2.2.6 Mounting

Mount roots in PVAG solution (recipe below), and cover them with a coverslip. This provides security and permanence, compared to mounting in DLAG, and PVAG slides are easier to clean if immersion optics are used. Where possible, it is most convenient to mount the roots parallel to the long axis of the slide. Bubbles trapped during cover slipping are all but impossible to remove, but seldom interfere with image acquisition. The PVAG solution is polymerized overnight at 40°C. Following PVAG polymerization, the edges of the coverslip should be sealed with nail polish for protection, since these can chip or lift. Slides stored in the dark at 4°C are stable for at least two years. PVAG does not fluoresce at the excitation wavelengths used for lactofuchsin staining.

PVAG solution, modified from Brundrett *et al.* (1996), contains 4 g polyvinyl alcohol powder [we use 98% polymerized, but various grades are available, and many appear to work]: 50 mL distilled water: 20 mL glycerol. This is warmed to 60°C (covered) with constant stirring until dissolved, typically 3 h to overnight. Eventually, PVAG solution will solidify at room temperature and become opalescent, but it can be re-melted one or more times with gentle heat and stirring.

2.3 Imaging

Lactofuchsin-stained roots can be examined with transmitted light, or with widefield or confocal epifluorescence illumination. Lactofuchsin has a wide range of excitation wavelengths, spanning at least 405–534 nm (blue to green) available with most epifluorescence systems. We typically use an FITC filterset (widefield) or 534 nm excitation (confocal). The emission range is also broad. For confocal imaging, we use an LP585 filter. Under these conditions, the actual colour of lactofuchsin fluorescence is orangered. We typically choose to present images that are false-coloured yellow or are greyscale, in order to increase image contrast.

As with other types of fluorescence imaging, there are tradeoffs to consider. Widefield epifluorescence illuminates the entire field, and both the depth of focus and the lateral resolution are related to the numerical aperture of the objective. Fortunately, lactofuchsin is a very stable fluorochrome, so photobleaching is seldom a problem. Confocal epifluorescence optics provide exquisite control of lateral and depth resolution, but higher resolution images have reduced depth of focus, which can be problematic for fine endophytes (FEs). Shallow depth of focus can be offset by collecting multiple focal levels (z-stacks) at the cost of additional time for image collection. Typically we use confocal optics for high resolution imaging, and widefield optics for quantitation.

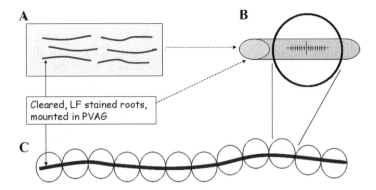

Fig. 3. The scanning procedure for the Multiple Quantitation Method. Roots (A) are visualized using at least 200X magnification (B). Fungal structures that intersect the vertical line on the graticule are considered, throughout the entire focal depth of the root. (C) Intersections are evenly spaced by moving the stage by one field of view each time.

Examples of endorhizal fungal imaging using lactofuchsin staining and confocal epifluorescence microscopy are shown in Figs. 4–6. Figure 4 shows transmitted light (A) and confocal fluorescence (B) images of a lactofuchsin-stained *Taraxacum officinale* (common dandelion) root, collected from garden soil in Saskatoon SK, 52°N. These images were collected simultaneously. The greater depth of focus with transmitted light microscopy provides a sense of continuity for the Paris-type coils (P), but details of arbuscule (Ar) structure are more evident with confocal fluorescence because out-of-focus glare is reduced. Multiple focal depth (z-stack) confocal imaging shows that the hyphae which form coils and arbuscules are continuous (Allen et al., 2006).

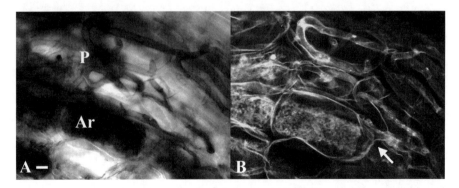

Fig. 4. Arbuscular mycorrhizal colonization in *Taraxacum officinale* (common dandelion). This specimen was stained with lactofuchsin and imaged with transmitted light (A) and confocal epifluorescence optics (B). Arbuscule (Ar). *Paris*-type intracellular hyphal coil (P). The arrow indicates continuity between an intercellular hypha and the arbuscule. Bar in A = 10 μm, for both.

Figure 5 shows FE hyphae in roots of a *T. phymatocarpum* plant collected from highly mineral tundra soils overlying shallow permafrost, on Axel Heiberg Island in the Canadian High Arctic, 80°N. The FE network shown in (A) is typical of near root-surface morphology, whereas that in (B and C) is typical of networks deeper in the root. This FE colony appears to be relatively newly established, since it has yet to form arbuscules, which typically form near the vascular cylinder (Fig. 5, Allen *et al.*, 2006). Due to their narrow width (1–1.5 µm) FE hyphal networks are difficult to study at high spatial resolution using transmitted light microscopy. FE hyphae are highly abundant in plants from High Arctic sites (Ormsby *et al.*, 2007) and so are likely to be ecologically important.

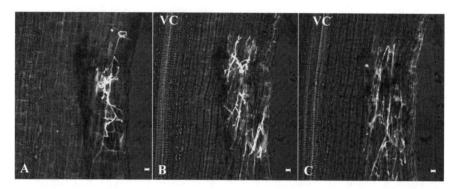

Fig. 5. Fine endophyte (FE) hyphae in a *Taraxacum phymatocarpum* root. This lactofuchsin-stained sample was imaged with confocal epifluorescence merged with transmitted light to provide spatial context. Root cortical cell walls and the vascular cylinder (VC) have faint fluorescence. Optical sections were (A) 8 µm, (B) 21 µm, and (C) 32 µm beneath the root surface. Bars = 10 µm.

Figure 6 shows FE hyphae, arbuscules, and vesicles of a *T. phymatocarpum* root. This plant was collected near the specimen shown in Fig. 5, stained with lactofuchsin, and imaged with confocal fluorescence. The relatively high resolution image shows the detail that can be acquired with confocal microscopy.

Figure 7 shows lactofuchsin-stained fungi in a peony root sample collected from garden soil in Saskatoon, SK. These roots have not been studied previously for their fungal associates. The macroscopic appearance of these roots suggested that they would be ectomycorrhizal; however, no evidence of a mantle or Hartig net system was found. Morphologies consistent with ectomycorrhizae are readily recognized with lactofuchsin fluorescence (not shown). Peony roots are challenging to work with because of their architecture and pigmentation. In this case, the adjacent field-of-view procedure shown in Fig. 3 is useful for keeping attention focused during quantitation.

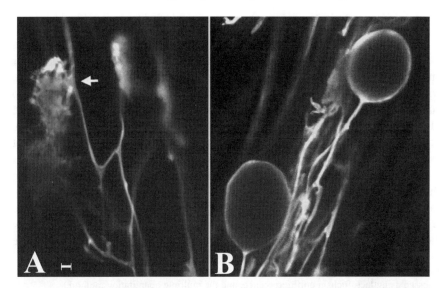

Fig. 6. Fine endophytes (FEs) in a *Taraxacum phymatocarpum* root stained with lactofuchsin and imaged with confocal epifluorescence. FEs produce (A) arbuscules, and (B) vesicles. The arrow in A shows continuity between an FE hypha and an arbuscule. Bar in A = 2 μm, for both.

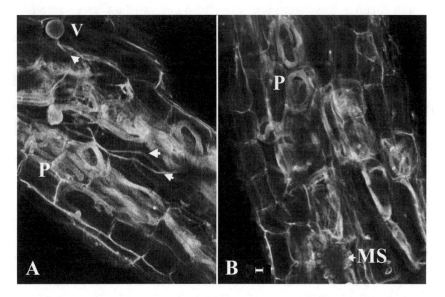

Fig. 7. Endophytic fungi in peony roots visualized with lactofuchsin staining and confocal epifluorescence optics, including arbuscular mycorrhizal (AM) fungi and fine endophytes (FEs). AM fungi formed *Paris*-type (P) intracellular coils. FE hyphae (arrowheads in A) produced vesicles (V), and arbuscules (not shown). A putative microsclerotium (MS) suggests there may also be septate endophytes. Bar in B = 10 μm for both.

3. THE MULTIPLE QUANTITATION METHOD

Quantitation is essential to assess the relative importance of particular interactions. Molecular DNA techniques can be excellent for identification, with caveats as described in Shepherd et al. (2007), but are not well-suited to quantitation. Our multiple quantitation method (MQM), built on that of McGonigle et al. (1990) and described here and in Ormsby et al. (2007), accommodates samples that have endorhizal fungi as well as AM, and that vary in colonization and abundance (Table 1).

As indicated in Fig. 3 (where possible, but see Fig. 2) root samples should be arranged parallel to the long axis of the microscope slide, so they can be examined systematically (Fig. 3A). Statistical studies by McGonigle et al. (1990) show that quantitation estimates by different users examining the same sample converged by about 150 intersections. There is no guarantee

Table 1. Fungal endorhizal colonization[a] for Asteraceae from Axel Heiberg Island, 80°N, assessed using lactofuchsin epifluorescence, and the multiple quantitation method.

Species (# of plants)	AMA	AMV	AMH-L	AMH-M	AMH-H	AMH-tot
Arnica alpina (10)	0.7 ± 0.5	1.3 ± 1.3	0 ± 0	1.1 ± 0.5	7.9 ± 5.2	9.0 ± 5.2
Erigeron compositus (5)	0 ± 0	0 ± 0	0.6 ± 0.6	0.8 ± 0.8	2.0 ± 1.3	3.4 ± 2.2
E. eriocephalus (8)	7.5 ± 3.0	0.5 ± 0.3	0.4 ± 0.3	7.9 ± 2.4	29.7 ± 7.5	38.0 ± 7.6
Taraxacum hyparcticum (6)	15.8 ± 4.6	9.0 ± 4.0	1.0 ± 0.6	10.3 ± 3.7	19.2 ± 4.8	30.5 ± 5.5
T. phymatocarpum (13)	13.5 ± 4.3	11.4 ± 3.5	1.8 ± 0.6	6.7 ± 1.1	39.8 ± 7.6	48.3 ± 7.7
All species	9.1 ± 1.8	5.2 ± 1.4	0.8 ± 0.2	5.6 ± 0.9	25.5 ± 3.7	31.9 ± 4.1

Species (# of plants)	FEA	FEV	FEH	SHE	Total fungal colonization
Arnica alpina (10)	27.8 ± 10.1	10.4 ± 3.6	50.5 ± 13.6	27.4 ± 10.5	82.7 ± 6.3
Erigeron compositus (5)	6.4 ± 3.6	4.6 ± 3.0	44.2 ± 17.9	27.6 ± 14.1	59.0 ± 16.9
E. eriocephalus (8)	0.6 ± 0.4	0.2 ± 0.2	10.2 ± 4.3	50.7 ± 6.4	76.1 ± 6.0
Taraxacum hyparcticum (6)	21.0 ± 4.2	28.3 ± 3.2	76.8 ± 10.4	26.5 ± 10.2	88.3 ± 5.2
T. phymatocarpum (13)	5.6 ± 3.4	2.8 ± 1.4	16.5 ± 8.4	42.1 ± 8.6	78.3 ± 7.3
All species	11.6 ± 2.9	7.5 ± 1.7	33.8 ± 5.7	36.7 ± 4.4	80.0 ± 3.6

[a]Summary data of colonization by endorhizal fungi, expressed as percent abundance ± standard error of the mean. These data were originally presented in Ormsby et al. (2007). Arbuscular mycorrhizae (AM) were categorized as arbuscules (AMA), vesicles (AMV) and hyphae (AMH). Root samples from each plant were assessed at about 100 intersections per sample. AM hyphae varied in abundance, so were subdivided into abundance classes: low (1 hypha per intersection), medium (2–5 hyphae per intersection) and high (>6 hyphae per intersection). Fine endophyte (FE) fungi were categorized as arbuscules (FEA), vesicles (FEV) and hyphae (FEH). Septate endophyte hyphae (SEH) were found in some root intersections. Total colonization was calculated as the number of intersections assessed, minus intersections lacking fungal structures.

that different portions of a large root system will be colonized to the same extent, so increasing the number of intersections leads to a diminishing return for effort. Nevertheless, for large root samples, assessing colonization at 150 intersections may leave some portions unexamined. If so, the remaining material should be scanned for rare types of interaction.

We start at one corner of the slide, and position the field of view and the orientation of the eyepiece graticule so that the intersection line is perpendicular to the root axis at that point (Fig. 3B). The entire depth of the root should be examined (adjusting the fine focus as needed) and the data recorded. Then the stage should be moved along the root by one field of view, and the scoring process repeated. This systematic approach reduces the very real temptation to be attracted to the colonized parts of the root, which can lead to overestimation. It can also help when dealing with complicated root architectures as shown in Fig. 2B.

With the MQM technique, each type of fungal morphology is scored separately, and new categories can be added if necessary. In samples that contain multiple types of fungal endophyte (Figs. 4 and 7) scoring can be facilitated by considering the morphology of the hyphae to either side of the intersection (as they are likely to be continuous), but quantification data should only be collected at the intersection. In addition to scoring fungal interactions, the absence of interaction is also important. Intersections that do not contain fungi should be scored as a separate category, so that overall colonization can be calculated. Otherwise, as in Allen *et al.* (2006), summing abundances could be misleading.

In Ormsby *et al.* (2007) we were faced with the challenge that some intersections had only one AM hypha, whereas in others they were very abundant. We defined additional categories (in this case low, medium and high hyphal abundance) to describe this variation (Table 1). Roots of Arctic Asteraceae had few AM hyphal coils; the data in Table 1 are almost exclusively for *Arum*-type intercellular hyphae. Thus it was perhaps surprising that there were relatively few arbuscules (these plants were collected in early July 2004, at the height of the Arctic summer), however, Ryan *et al.* (2003) have shown that intercellular AM hyphae may also participate in mineral nutrient transfer.

4   DISCUSSION

Discrimination and quantitation of endorhizal fungal structures is an important correlative technique for molecular and physiological studies, because it is likely to be directly related to the significance of an interaction to the symbiotic partners. The methods described in this chapter include the most sensitive staining and imaging currently available. Fluorescence microscopy has long been the method of choice for cell biology, because luminous objects are highly contrasted with a black background, which increases detection and resolution of fine structures. In this application, high detectability and spatial resolution are particularly important for examining details of arbuscule structure and FE hyphal networks.

To date, for samples harvested from field sites where there is often little control of the fungal root symbiont(s), lactofuchsin staining viewed with epifluorescence provides a convenient combination of relative simplicity of preparation and imaging quality.

The choice of confocal *vs* widefield epifluorescence imaging depends on the need, and naturally on availability of equipment. Confocal epifluorescence optics are often superior for documentation, whereas widefield optics are more efficient for quantitation. Modern widefield epifluorescence microscopes are typically equipped with high-sensitivity imaging systems, so there can be considerable flexibility.

We have found endorhizal quantitation using the MQM method to be reproducible as well as flexible. MQM is well suited for plant roots harvested from natural locations where the endorhizal fungi are not necessarily well described. Furthermore, MQM can quantitate the relative contributions of different endorhizal interactions, which must intrinsically be related to their importance to the plant's physiology.

Acknowledgements: This research was supported by a Discovery Grant from the Natural Science and Engineering Council of Canada. Images 4–6 were collected by Nathan Allen. The methods described here were developed during an ongoing collaboration with James Basinger, Department of Geological Sciences, University of Saskatchewan.

REFERENCES

Allen, N., Nordlander, M., McGonigle, T., Basinger, J., and Kaminskyj, S., 2006, Arbuscular mycorrhizae on Axel Heiberg Island (80°N) and at Saskatoon (52°N) Canada. *Can. J. Bot.* **84**: 1094–1100.
Brundrett, M., Bougher, N., Dell, B., Grove, T., and Malajczuk, N., 1996, *Working with Mycorrhizas in Forestry and Agriculture.* ACIAR, Canberra.
Carmichael, J. W., 1955, Lactofuchsin: a new medium for mounting fungi. *Mycologia* **47**: 611.
Jumpponen, A., and Trappe, J. M., 1998, Dark septate endophytes: a review of facultative biotrophic root-colonizing fungi. *New Phytol.* **140**: 295–310.
Márquez, L. M., Redman, R. S., Rodriguez, R. J., and Roossinck, M. J., 2007, A virus in a fungus in a plant – three way symbiosis required for thermal tolerance. *Science* **315**: 513–515.
McGonigle, T. P., Miller, M. H., Evans, D. G., Fairchild, G. L., and Swan, J. A., 1990, A new method which gives an objective measure of colonization of roots by vesicular-arbuscular mycorrhizal fungi. *New Phytol.* **115**: 495–501.

Merryweather, J. W., and Fitter, A. H. 1991, A modified method for elucidating the structure of the fungal partner in a vesicular-arbuscular mycorrhiza. *Mycol. Res.* **95**: 1435–1437.

Ormsby, A., Hodson, E., Li, Y., Basinger, J. and Kaminskyj, S., 2007, Arbuscular mycorrhizae associated with Asteraceae in the Canadian High Arctic: the value of herbarium archives. *Can. J. Bot.* **85**: 599–606.

Peterson, R. L., Massicote, H., and Melville, L. H., 2004, *Mycorrhizas: Anatomy and Cell Biology*. CABI, Wallingford, 176 pp.

Phillips, J. M., and Hayman, D. S., 1970, Improved procedures for clearing roots and staining parasite and vesicular arbuscular mycorrhizal fungi for rapid assessment of infection. *Trans. Br. Mycol. Soc.* **55**: 158–160.

Rodriguez, R.J., Henson, J., van Volkenburgh, E., Hoy, M., Wright, L., Beckwith, F., Kim, Y.O., and Redman, R.S., 2008, Stress tolerance in plants *via* habitat adapted symbiosis. *Intl. Soc. Microb. Ecol. J.*, **2**: 404–416. doi:10.1038/ismej.2007.106

Ryan, M. H., McCully, M. E., and Huang, C. X., 2003, Location and quantification of phosphorus and other elements in fully hydrated, soil-grown arbuscular mycorrhizas: a cryo-analytical scanning electron microscopy study. *New Phytol.* **160**: 429–441.

Shepherd, M., Nguyen, L., Jones, M. E., Nichols, J. D., and Carpenter, F. L., 2007, A method for assessing arbuscular mycorrhizal fungi group distribution in tree roots by intergenic transcribed sequence variation. *Plant Soil* **290**: 259–268.

Smith, S. E., and Read, D. J., 1997, *Mycorrhizal Symbiosis*. 2nd ed. Academic, London 605 pp.

Vierheilig, H., Schweiger, P. and Brundrett, M., 2005, An overview of methods for the detection and observation of arbuscular mycorrhizal fungi in roots. *Physiol. Plant.* **125**: 393–404.

Wei, Y. D., Byer, K. D., and Goodwin, P. H., 1997, Hemibiotrophic infection of round-leaved mallow by *Colletotrichum gloeosporioides* f. sp. *malvae* in relation to leaf senescence and reducing agents. *Mycol. Res.* **101**: 357–364.

SUBJECT INDEX

A
Abies grandis, 243
Abiotic, 10, 19, 23, 100, 123, 136, 139, 143, 163, 172, 181, 183, 222, 261, 274, 298, 324
Abscisic acid, 219
Acacia holosericea, 230–232, 236, 252
Acacia tortilis, 231, 232
Acclimation, 22, 23, 322, 325, 331, 332
Acid precipitation, 185, 264, 265
Acidotropic staining, 40
Acorus, 198, 199
Acorus-type, 197, 198, 200
Actinomycetes, 78, 160, 304, 310
Actinorhizal plants, 22, 310
Additive, 14, 15, 213, 222
Adelges tsugae, 260
Aeroponics, 82
Afforestation, 170, 229, 246, 256, 264, 270
Agrobacterium rhizogenes, 307
Agroecosystems, 100, 104, 140, 187, 188
Agropyron repens, 115
Agropyron sp., 113
Alanine, 254
Alders, 310–311, 315
Alkalinity, 306, 309, 311
Allelic variation, 52
Alleviation, 7, 99, 108, 116, 121, 139
Altitude, 260, 272
Amanita muscaria, 255, 268, 326
AMF genera, 62, 101
Amino acids, 4, 23, 45, 47, 76, 113, 184, 186, 219, 246, 253, 254, 325
Ammonium nitrate, 259
Ammonium, 4, 5, 45–47, 50, 100, 181, 259, 341
Amphinema byssoides, 311
Amphipathic films, 186
Anacardiaceae, 203
Anatomotyping, 245
Aniline blue, 339
Annona glabra, 197, 198
Annona, 201
Antagonist of plant pathogens, 142
Antagonistic, 78, 83, 103, 181, 189, 217, 221, 261, 297
Anthracnose pathogen, 219

Anthropogenic factors, 267
Anthroponic soils, 311
Antibiotics, 16, 261
Aphanomyces, 71, 72, 75, 79, 181
Apoplast, 3–5, 39, 40, 44, 46–48
Aquaporin genes, 40
Aquaporin proteins, 5, 46
Aquaporin, 39, 40, 47, 51
Arabidopsis plants, 220
Arbuscules, 3, 5, 38, 39, 41, 43, 44, 46, 62, 76, 79, 101, 114–116, 147, 163, 164, 168, 180, 183, 197, 200, 201, 343–348
Archaeospora, 163
Architecture, 183, 340, 344, 347
Arctostaphylos sp., 2, 243, 244
Arginine, 4, 45, 47, 76
Arid habitats, 139, 205
Aromatic pollutants, 265
Arsenate, 107, 110, 116
Arthrobactor, 181
Arum-type, 201–203, 347
Ascomycetes, 21, 242, 305
Aseptically, 257
Aspartate, 254
Aspergillus niger, 143
Assimilation capacity, 323
Asteraceae, 114, 346, 347
Axenic cultures, 82, 108, 214, 215, 219, 254, 265, 307
Axenic synthesis, 23, 324, 325, 327
Axenic, 23, 184, 213, 324
Azospirillum brasilense, 142
Azotobacter, 142

B
Bacillus sp., 69, 73, 142, 262
Bacillus subtilis, 181, 263
Basidiomycetes, 18, 21, 242
Benefits, 3, 5, 6, 10, 11, 21, 25, 37, 42, 48, 52, 53, 62, 82, 122, 136–138, 146, 147, 160, 166, 167, 169, 171, 179, 202, 206, 213, 244, 245, 260, 261, 273, 296, 297, 305, 306, 312, 313
Berkheya coddii, 114
Betula pendula, 245, 255
Bio-ameliorators, 21, 305
Biochemical changes, 6, 75, 76, 261

Biocides, 104
Biocontrol, 7, 62, 63, 69, 81–83, 160, 221, 222, 229
Biodiversity, 16, 18, 143, 147, 163, 164, 229
Biodynamic agriculture, 160
Biofertilizers, 7, 82, 123, 142, 143
Biofunctioning, 237
Biogeochemical cycling, 171, 271
Bioindicator, 104
Biomass, 9, 11, 20, 100, 109, 111–115, 118, 123, 138, 139, 145, 162, 165, 169, 178, 183, 202, 214, 230, 243, 245, 248, 253, 254, 259, 271–273, 293, 297, 298, 305, 308
Biomineralization, 179
Biomolecules, 12, 180–183, 186, 187, 189
Bioprotectants, 7, 61, 63, 82, 123
Bioprotection, 7, 75, 81, 141
Bioremediation, 170, 265, 268
Biosorption, 110, 119
Biostimulants, 7, 82, 123
Biotic, 10, 19, 23, 100, 123, 163, 165, 172, 181, 183, 196, 222, 261, 274, 324
Biotrophic, 221
Biotrophs, 100, 308
Birch, 245, 326
Bishop pine, 243
Bleaching, 341
Boreal forests, 245, 288, 311, 315
Bradford-reactive soil protein (BRSP), 185, 186, 188
Bradyrhizobium sp., 142, 232
Bursaphelenchus xylophilus, 249, 262

C
Cacti, 199, 200, 203, 205
Caesalpiniaceae, 289, 299
Carbohydrate syntheses, 253
Carbon
    compounds, 23, 62, 179, 253, 261, 323–325
    cycle, 271
Carbonates, 104, 105, 120, 121, 203
Carboxylic acids, 113, 120
Castanea sativa, 263, 321, 322, 326, 327, 330
Catabolism
    activities, 266
    evenness, 236
    functional diversity, 17, 235
Catechols, 76
Cellulase activity, 218

Cereals, 11, 45, 136, 138, 146
Chalara, 181
Chalcone isomerase, 79, 262
Chalcone synthase, 262
Chelating agents, 113
Chelation, 108–111
Chemical methods, 257
Chemotropism, 246
Chestnut, 322, 326–328
Chitin, 41, 109, 110, 308
Chitinases, 15, 41, 51, 76, 78–80, 164, 219, 262
Chitosan beads, 307, 308
Chlorazole black E, 24, 338, 339
Cistus spp., 23, 252, 331
Citric acid, 109
Citrus, 64, 65, 74, 77, 138
Clearing, 200, 266, 340, 341
Coastal dunes, 201, 205, 206
Cochliobolus sativus, 141
Co-inoculation, 142, 216
Collembola, 181, 254
Colonization, 3, 5–8, 11, 14, 16, 20, 23–25, 39–44, 48, 49, 52, 62, 71, 74–82, 102, 103, 105, 107–109, 112, 114, 116, 122, 123, 137–142, 144, 146, 147, 162, 167, 181, 196–198, 200–206, 214–217, 228, 230, 236, 245, 254, 257, 260, 261, 267, 269, 270, 294–296, 299, 324, 325, 338, 343, 346, 347
Common mycorrhizal networks, 273, 274
Compatibility, 270, 324, 326
Competition, 7, 11, 77–79, 81, 11, 112, 117, 138, 139, 146, 172, 179, 221, 253, 261, 266, 296, 297
Composite tailings, 309–311, 313
Compost, 136, 145, 146, 257
Confocal epifluorescence illumination, 342
Confocal epifluorescence, 25, 339, 342–345, 348
Conglomeration, 178
Conidial germination, 14, 213, 214, 217
Coniferous trees, 243
Conservation, 9, 11, 147, 172, 187, 196, 205
Continuous cover, 179
Conventional farming, 9, 10, 136, 147
Convolvulaceae, 202
Crop
    productivity, 9, 12, 136, 137, 160
    yield, 10, 140, 145, 171
Culture
    filtrates, 15, 219

pots, 197
Cycad, 200–202
Cytokinin, 246

D

Dandelion, 138, 343
Decomposition, 12, 17, 20, 146, 165, 183, 184, 186, 187, 235, 271, 288, 297, 298
Defence-related genes, 41
Defense mechanisms, 7, 79, 219, 220
Deglycosylation, 186
Degradation activity, 218
Degradation, 9, 13, 136, 147, 160, 169, 181, 186, 218, 229, 235, 265, 266, 309
Denaturing gradient gel electrophoresis (DGGE), 17, 20, 235, 298
Dependency, 11, 107, 137, 196, 204–206
Desertification, 13, 229
Detoxification agents, 266
Detoxification, 117, 119, 266
Devonian plants, 180
Dicymbe corymbosa, 20, 289, 291–295
Disease protection, 63, 211, 216, 221, 261
Disease resistance, 25, 75, 81, 261
Diseases, 6, 7, 10–12, 14, 15, 18, 19, 25, 63–65, 70–77, 81–83, 141, 181, 189, 211–214, 216, 219–222, 249, 261–263, 266, 274
Diversity, 11, 15–19, 21, 24, 81, 83, 106, 108, 143, 144, 162, 163, 165, 167, 172, 187, 228, 229, 231, 232, 235–237, 242, 244–246, 259, 264, 288, 289, 292, 296, 309, 337
DNA analysis, 298
DNA fingerprints, 310
Douglas fir, 169, 232, 243, 247, 256–258, 262, 269
Drought conditions, 140, 248
Drought, 13, 19, 22, 25, 40, 49, 62, 122, 138–140, 164, 248, 251, 306, 322
Dry weight, 24, 113, 114, 142, 214–216, 218, 230, 231, 257–259, 264, 323, 324, 326, 330, 331

E

ECM genets, 296
Eco-farming, 160
Ecological mycorrhizal dependency, 205
Ecophysiology, 22
Ecosystems, 2, 7–9, 13–16, 19–22, 42, 62, 106, 123, 144, 145, 159–167, 169–171, 181, 185, 189, 195, 221, 222, 228, 229, 231, 237, 241, 244, 256, 261, 262, 267, 271, 273, 288, 303–306, 308, 310, 314
Ecotypes, 106, 108, 203
Ectomycorrhizal fungi, 4, 17, 18, 46, 121, 185, 232, 235, 242, 246, 253, 254, 257, 265, 266, 272, 311
Ectomycorrhizal, 2, 4, 17–19, 22, 46, 105, 109, 111, 121, 122, 185, 228–236, 242, 246–248, 251, 253–257, 260, 261, 264–267, 272–274, 287, 288, 293–300, 304, 310, 311, 322, 344
Ectoparasites, 64
Endemic plants, 196
Endophytic fungi, 24, 337, 345
Endorhizal interactions, 25, 348
Endorhizal, 24, 25, 337, 338, 343, 346–348
Enterobacter, 181
Epifluorescence microscopy, 25, 339, 343
Ergosterol, 253, 272
Ericaceous plants, 21, 313
Ericales, 2, 3, 21
Eucalyptus diversicolor, 323
Eucalyptus globulins, 257
Eucalyptus robusta, 244, 245
Eucalyptus, 74, 236
European chestnut, 322
Eutrophication, 136
Everglades, 195, 196, 201, 206
Exchange of nutrients, 3
Extraradical hyphae, 38, 101, 110, 117, 145, 180, 181, 183
Exudates, 12, 16, 17, 23, 38, 76–78, 105, 109, 123, 140, 146, 161, 162, 165, 178, 189, 213, 217, 218, 228, 235, 246, 325

F

Fabaceae, 202, 203
Facultative mycotrophs, 137
Fallow periods, 12, 144, 166, 179
Fallow, 9, 11, 12, 137, 144, 145, 166, 179
Fatty acid extraction, 17, 235
Fermentor, 307
Fertilizer, 7, 9–11, 18, 62, 63, 82, 104, 112, 136, 143, 145–147, 171, 196, 206, 222, 262, 264, 306
Festuca arrundinaceae, 116
Festuca rubra, 115
Festuca sp., 113, 115
Fine endophytes, 24, 342, 344–346, 348
Fixing, 142, 310, 340
Flexible, 337, 338, 348
Fluoranthrene, 305

Fluorescence microscopy, 24, 25, 338, 339, 343, 348
Fluorescent pseudomonads, 16, 17, 78, 232–235
Fluorochrome, 342
Footprinting analysis, 43
Forest
 communities, 243, 288
 ecosystems, 19, 162, 256, 262, 271, 273
 systems, 253
Frankia sp., 307, 308, 310, 311, 315
Fruit trees, 23, 331
Fungi (1354 instances)
 biomass, 111, 162, 169, 253, 259, 272
 communities, 9, 20, 25, 203, 243, 270, 298
 diseases, 70, 71, 263
 plant pathogens, 70
Fusarium, 71, 72, 75, 78, 80, 138, 181, 212, 213, 216–218, 220, 262, 263

G

Gaeumannomyces graminis var. tritici, 141
Genetic diversity, 15, 83, 165
Geochemical, 165
Giant cells, 75
Gigaspora decipiens, 139
Glomalean fungi, 62, 211 (This term is found as 'Glomales' in the text)
Glomalin, 12, 13, 108, 110, 118, 168, 169, 178, 180, 182–189
Glomalin-related soil protein (GRSP), 169, 185
Glomus clarum, 72, 142
Glomus constrictum, 69, 74, 143
Glomus fasciculatum, 64–74, 78, 142
Glomus, 4, 40, 44, 45, 50, 51, 62, 64, 66, 69, 71–73, 80, 101, 109, 110, 113–115, 120, 139, 140, 142–144, 201, 202, 205, 214, 219, 231, 232, 236, 305–308
Glucanase, 15, 79, 80, 164, 219
β-Glucuronidase (GUS), 43
Glutamate, 45, 47, 254
Glutamine, 45, 47, 254
Glutathione S-transferase, 108, 164
Glutathione transferase, 43
Glutathione, 43, 108, 118, 164
Glycine, 51, 66, 113, 254
Glycogen, 38, 48
Glycoprotein, 12, 39, 51, 110, 118, 168, 178, 183
Glycosyl phosphatidylinositol, 39
Grain production, 9, 10, 135, 137, 144, 145, 148

Grain, 9, 10, 135, 137, 140, 142–145, 148, 160, 218
Grand fir, 258
Grasslands, 144, 163, 185, 205

H

Hartig net, 2, 3, 18, 23, 242, 325, 326, 329, 330, 344
Hebeloma crustuliniforme, 247, 252, 255, 257, 258, 269, 310, 311
Hebeloma spp., 247, 252, 311
Helianthemum, 23, 331
Hemlock woolly adelgid, 260
Heterogeneity, 44, 244, 326
Heterotrophic, 17, 146, 235
Hexose, 5, 46–48, 52, 178
Histochemical, 75
Histopathological, 6, 75
Horticulture, 7, 25, 82, 162, 205, 206
Host specificity, 243, 244, 273
Host-parasite relationship, 80
Humic acids, 185, 256
Hydraulic conductivity, 139, 167, 247, 248
Hydrolytic enzymes, 253
Hydrophobins, 185–187
Hydroponics, 82, 259
Hyperaccumulate, 113
Hyphopodium, 38
Hyphosphere, 17, 228, 232–235

I

Image analysis method, 272
Imaging, 24, 25, 339, 341–343, 348
Immobilization, 8, 106, 111, 112, 119
Immunochemical analysis, 44
Immunocytochemical, 44, 75
Immunocytological, 40
In situ catabolic potential (ISCP), 17, 235, 236
In vitro, 14, 22–24, 76, 82, 116, 118, 120, 172, 184, 213, 214, 216, 258, 273, 311, 313, 314, 321, 322, 324–327, 331, 332
Induced systemic resistance (ISR), 15, 219, 220
Inorganic nitrogen, 267
Interactions, 3, 6, 7, 14–16, 24, 38, 39, 62–64, 70, 71, 73, 79, 81, 83, 101, 103–105, 107, 108, 112, 113, 115, 116, 123, 140–143, 146, 147, 160–163, 169, 170, 172, 181–184, 188, 189, 211–214, 217–220, 222, 227–229, 244, 246, 262, 266, 270, 314, 324, 329, 337, 346–348
Intersections, 338, 343, 346, 347

Intsia bijuga, 244, 245
Invertase, 46
Isoflavonoid, 76, 79
Isopentenylaminopurine, 246

L
Laccaria laccata, 252, 255, 259, 263, 267–269, 326
Laccaria sp., 247, 269, 311
Lactofuchsin, 24, 25, 338–346, 348
Land
  reclamation, 115, 305, 315
  rehabilitation, 123
Laser micro-dissection, 43
Lauraceae, 203
Leaching, 9, 120, 136, 256, 341, 342
Leaf litter, 20, 297, 298
Legumes, 2, 11, 13, 114, 138, 142, 167, 229, 246
Licaria triandra, 197
Lignifications, 23, 75, 76, 80, 325
Lignin, 15, 75, 212, 219, 256, 265
Lipases
  activity, 235
  producing fluorescent pseudomonads, 17, 234, 235
  producing, 17, 234, 235
Liquid/mycelial slurry, 307
Litter, 20, 113, 187, 259, 297, 298
Lodgepole pine, 258
Lotus japonicus, 49, 246
Lyso-phosphatidyl-choline (LPC), 43
Lytic enzymes, 39, 253

M
Macroaggregates, 168, 188
Magnolioid roots, 201
Mantle structures, 2
Masting, 290, 294, 297
Meshes, 257
Mesorhizobium plurifarium, 231, 232
Metal
  sorption, 109, 170
  transporters, 107
  uptake, 8, 109, 111, 112
Metalloids, 104, 112, 119, 123
Metallophyte, 106
Metallothioneins, 107, 108, 113, 117, 118
Methylmetal, 119
Micro satellite markers, 310
Microaggregates, 168, 187, 188
Microbial communities, 10, 14, 16–18, 78, 103–106, 113, 123, 141, 160, 161, 163, 168, 169, 181, 217, 218, 228, 229, 235, 236, 304
Microbial fertilizers, 143
Microbial interactions, 212
Microcosms, 254
Microcutting, 23, 331
Micropropagation, 22–24, 82, 172, 321, 322, 325–328, 331, 332
Migratory endoparasites, 64
Mine(s)
  site, 113, 121
  spoil site, 313
  spoil soils, 305, 308
Mineral nutrients, 62, 74, 77, 169, 179, 218, 249, 251, 259, 260, 322–324, 347
Mining activities, 8, 20, 22, 121, 170, 206, 303–305, 308
Mixed forest, 20, 289, 291, 293–295, 297, 298
Mobilization, 107, 109, 111
Molecular basis, 3, 6, 53
Monochamus sp., 249
Monoclonal antibody, 183
Monocotyledons, 43, 200
Monoculture, 144, 243
Monodominance, 19, 20, 287–289, 293, 295, 298, 299
Monoxenic, 82
Morphotypes, 24, 144, 248, 260, 261, 264, 338
Mortality, 19, 262, 296, 297, 305
Mounting, 342
Multiple inoculation, 246
Multiple quantitation method (MQM), 25, 343, 346–348
Multitrophic, 227, 228
Mycelial networks, 167, 271, 273
Mycobionts, 258, 274
Mycorrhization, 23, 24, 114–116, 119, 142, 181, 216, 232–234, 257, 258, 321, 323–328, 330–332
Mycorrhizosphere, 7, 12, 14–16, 78, 101, 102, 111, 114, 123, 161, 162, 178, 180–187, 189, 212, 219, 227–229, 232–234, 262, 266
Mycorrhizostabilization, 118

N
N_2 fixation, 142, 146, 229
N_2-fixing bacteria, 142
Natural ecology, 165
Nematodes, 6, 7, 18, 19, 63–70, 73, 75–77, 79, 80, 141, 214, 242, 249, 262

Net assimilation, 247
Net primary production, 169
Nitrogen transport, 4
Nitrogen, 3, 4, 45, 47, 62, 102, 117, 136, 142, 146, 160, 164, 166, 167, 169, 181, 218, 228, 253, 254, 260, 262, 267, 268, 270, 304, 306, 310, 325
Nitrogenase, 16, 229
Nodulation, 16, 167, 229, 230, 232, 246,
Non axenic, 23, 324, 325
Non-axenic systems, 23, 325
Nonessential metals, 117
Non-rhizosphere, 162, 166
Nutrient
  acquisition, 38, 183, 251, 289, 306
  cycling, 10–12, 103, 165, 178, 189, 253, 271, 297, 304, 306
  exchange, 3, 4, 6, 37, 39, 40, 48, 53, 101, 115, 164, 242
  uptake, 6, 12, 14, 52, 62, 77, 100, 102, 136, 137, 140, 145, 147, 172, 203, 219, 228, 245, 251, 254, 259, 271, 306

O

Oak, 24, 256, 259, 260, 326
Obligate mycotrophs, 137
Obligate nature, 7, 79, 81
Oil sands deposits, 308
Oilseeds production, 135
Oligosaccharides, 186
Onychiurus armatus, 254
Organic acids, 17, 23, 101, 108, 111, 113, 120, 121, 235, 246, 253, 261, 266, 325
Organic amendments, 145
Organic farming, 146, 147, 160
Organic phosphate, 44, 169
Ornithine aminotransferase, 4, 46, 47
Other plant pathogens, 63, 64, 73–74
Oxidative stress, 108

P

Paenibacillus sp., 181, 189
PAHs, 265
Palmitic acid, 246
Palms, 196, 197, 200, 203, 206
Paraglomus, 163
Parasitic, 165, 246, 247
Parasitism, 6, 7, 11, 64, 77, 80, 245, 261
Paris-type, 197, 198, 200–203, 343, 345
Passive transport, 48
Pathogens, 5, 6, 15, 18, 22, 25, 48, 61–64, 71–83, 123, 140–142, 181, 189, 212, 219, 221, 222, 229, 242, 260, 262, 263, 304, 322, 339
Paxillus involutus, 254, 257, 258, 265, 311
PCR-based methods, 243
PCR-DGGE, 17, 235
PCR-RFLP analysis, 244
Peat moss, 256, 257
Pelletized, 258
Peltogyne gracilipes, 299
Penicillium citrinum, 143
Peonia chinensis, 340
Peony root, 344
Peri-arbuscular membrane, 4, 5, 39, 40, 42, 44, 46–48
Perlite, 171, 257
Permeability, 23, 39, 76, 77, 325
Persistent organic pollutants, 265
Phenanthrene, 305
Phenolics compounds, 75, 79, 185, 246, 256, 262, 265
Phenylalanine ammonia lyase, 262
Phenylalanine ammonium-lyase, 79
Phenylalanine, 76, 79, 219, 262
Phosphatase activity, 268
Phosphatase, 17, 44, 120, 169, 235, 268
Phosphate
  acquisition, 4, 41–45
  uptake, 4, 41, 42, 44, 52
Phosphate-solubilizing, 17, 181, 234, 235
Phospholipids fatty acid analysis, 298
Phosphorus nutrition, 76, 77
Photosynthates, 7, 78, 79, 81, 253, 258, 259, 271
Pht1 genes, 4, 42, 43
Physical methods, 257
Physico–chemical properties, 106, 165
Physico–chemical, 106, 165, 229
Phytoaccumulation, 114, 115, 119, 120
Phytoalexins, 75, 76, 79, 247, 262
Phytoavailability, 104, 111
Phytobial remediation, 304
Phytochelatins, 118
Phytophthora cinnamomi, 247
Phytophthora, 76, 79, 181, 247, 262
Phytoplasma, 73, 74
Phytoremediation, 8, 9, 102, 113, 114, 116, 118, 120, 266
Phytorestoration, 312
Phytosiderophores, 113
Phytostabilization, 114, 118
Phytovolatilization, 115, 119
Picea sitchensis, 243
Picramnia pentandra, 194

Subject Index

Pigmentation, 24, 203, 265, 338, 340, 341, 344
Piloderma croceum, 326
Pine seedlings, 19, 206, 249, 256, 257, 266, 267, 269, 274
Pine wilt disease (PWD), 18, 19, 249, 262, 274
Pine, 195, 196, 201, 202, 246, 248–251
Pinus banksiana, 244
Pinus caribea, 244
Pinus contorta, 254
Pinus densiflora, 249, 262, 272
Pinus jeffreyi, 258
Pinus lambertiana, 258
Pinus pinaster, 244, 248, 249
Pinus ponderosa, 243
Pinus sylvestris, 253, 259, 266
Pinus thunbergii, 249, 264
Pisolithus albus, 230, 231
Pisolithus tinctorius, 121, 122, 244, 256–258, 264, 267, 326, 327
Plant
  communities, 13, 25, 166, 196, 228, 229, 271
  diversity, 16, 21, 231, 237, 288
  parasitic nematodes, 6, 63, 64
  pathogenic bateria, 73
  pathogens, 6, 15, 22, 63, 64, 70–74, 78, 82, 140–142, 189, 219, 221, 263, 322
  viruses, 73
Plant growth
  promoting fungi, 14, 212
  promoting rhizobacteria, 83, 212, 229
Plantations, 18, 245
Plant-encoded transporters, 107
Pollutants, 178, 261–268
Polluted soils, 8, 9, 100, 106, 110, 114, 118, 122, 170, 312
Polycyclic aromatic hydrocarbons, 170, 181, 265
Polymerase chain reaction, 310
Poly-phosphate breakdown, 44, 47
Polysaccharides secretion, 325
Polysaccharides, 178, 180, 187, 265, 325
Poplar, 326
Positive associations, 246
Pot culture, 78, 82, 184, 200, 203, 308
PR proteins, 79
Predatory, 212
Predictive models, 261
Preservation, 24, 338
Presymbiotic phase, 214

Primary succession, 266, 267, 274
Promoter-deletion analysis, 43
Propagation, 11, 24, 82, 196, 205, 308, 322, 326, 332
Protective effect, 6, 63, 81, 265
Proteolysis, 186
Proteolytic activities, 253
Pseudomonads, 16, 17, 78, 232–235
Pseudomonas fluorescens, 262, 266
Pseudomonas sp., 142, 262, 263
Pseudotsuga menziesii, 243, 247, 257, 272
Pseudotsuga, 243, 244, 247, 257, 272, 274
P-solubilizing bacteria, 142, 143, 181
Puccinia graminis, 141
PVAG solution, 342
Pythium sp., 181, 212, 219
Pythium ultimum, 247

Q

Qualitatively, 15, 16, 228
Quantified, 24, 122, 169, 185, 272
Quantitation, 25, 343, 346–348
Quantitative models, 261
Quantitatively, 15, 16, 228, 338
Quercus ilex, 247, 248
Quercus robur, 326
Quercus rubra, 260

R

Radiolabeled, 5
Radionuclides, 170
Ralstonia, 181
Rare species, 196
rDNA sequencing, 244
Recalcitrant, 21, 265, 305, 326
Reclamation effort, 22
Reclamation, 13, 20–22, 115, 121, 168, 171, 303–314
Reforestation, 11, 18, 22, 171, 249, 256, 304
Regeneration, 170, 206, 290
Rehabilitation, 13, 123, 170, 310
Re-installation, 304, 305, 309
Relative contributions, 25, 299, 348
Relative growth rates, 330, 331
Relative mycorrhizal dependency, 204
Remediation, 6, 8, 9, 123, 304, 311, 312, 314
Resistance, 5, 10, 15, 19, 22, 63, 75, 76, 80, 81, 83
Restoration, 9, 13, 14, 20–22, 122, 196, 205, 206, 261, 304, 306, 309, 312
Revegetation, 13, 14, 21, 22, 100, 121, 170, 304, 305, 308–312

Rhizobia, 16, 101, 142, 167, 229–232
Rhizobium meliloti, 142, 246
Rhizoctonia sp., 181, 212, 220, 262, 263
Rhizomorphs, 122
Rhizopogon rubescens, 262, 311
Rhizopogon, 243, 247, 262, 267, 311
Rhizoremediation, 100
Rhizosphere, 8, 10, 11, 14–16, 24, 63, 78, 83, 100, 101, 105, 106, 112–114, 118–121, 123, 140, 142, 161, 162, 168, 169, 180, 212, 217, 220, 228, 232–234, 246, 247, 251, 261, 262, 266, 337
Root
   exudates, 16, 23, 38, 76, 77, 140, 146, 228, 246, 325
   exudation, 15, 16, 76–78, 81, 111, 140, 212, 221, 228
   morphology, 15, 74, 167, 324
   organ culture, 307, 308
Rubiaceae, 203

S

Salicylic acid, 15, 219
Salinity, 62, 165, 170, 309
Sampling, 162, 289, 341
Saprophytic fungi, 14, 15, 212–214, 216–218, 297, 298
Saprophytic, 2, 14, 15, 80, 195, 212–221, 288
Saprotrophic bacteria, 20, 297, 298
Saprotrophs, 288, 298
Scutellospora calospora, 139
Secondary successions, 266, 267, 269
Sedentary endoparasites, 64
Semiendoparasites, 64
Semi-hydroponic culture, 259
Serine, 76, 219, 254
Sheath, 18, 199, 200, 242
Simple sequence repeat (SSR) markers, 310
Simulated acid mist, 265
Sinorhizobium terangae, 231, 232
Slope, 18, 19, 249–251
Sludge amendments, 106
Smelter, 106, 120
Smilax, 200–202
Soil
   aggregating bacteria, 142
   aggregation, 8, 11–13, 167–169, 171, 178, 179, 181–183, 186–189, 228
   binding, 160
   compaction, 140
   degradation, 13, 160, 309

   minerals, 121, 178, 187, 260
   quality, 10, 13, 14, 18, 147, 167, 171, 242, 310
Sophora tomentosa, 197
Source-sink concept, 323
Species accumulation, 291, 293
Specificity, 203, 242–246, 256, 264, 273
Spore inoculum, 307
Stabilization, 12, 168, 181, 186–188
Stained, 24, 44, 197, 338, 341–345
Staining, 25, 40, 44, 200, 340–343, 345, 348
Stains, 24, 196, 338, 339
Stomatal conductance, 247–249
Streptomyces, 78, 181, 263, 264
Subtropical forest, 195, 196
Succession, 229, 266–270, 274
Sucrose synthase, 46
Sucrose, 46, 100, 325
Suillus brevipes, 311
Suillus granulatus, 121
Superoxide dismutase, 262
Superoxide, 15, 219
Surface mining, 20, 308
Sustainable, 6–8, 10–13, 22, 24, 62, 63, 82, 104, 123, 137, 141, 143
Symporters, 4, 5, 42, 48
Syncytia, 75
Synergistic, 14, 15, 62, 142, 143, 213, 217, 218
Systemic resistance, 15, 76, 219, 221

T

Tailing sands, 309
Tannic acids, 185
Taraxacum officinale, 343
Telephora terrestris, 122, 306
Thlaspi caerulescens, 113
Thlaspi calaminare, 113
Tillage, 9, 11, 12, 137, 143, 145, 179
Tolerance, 5, 8, 9, 22, 24, 48, 62, 63, 77, 105–108, 110, 137, 170, 172, 218, 252, 266, 270, 304, 306, 309, 311, 312, 337
Toxic metal, 100, 104, 109, 123, 189
Toxicity, 8, 104–107, 112, 114, 116, 118, 119, 264, 304
Transmitted light, 24, 338, 339, 342–344
Transpiration rate, 248, 249, 251, 265
Transplant survival, 171
Transporter genes, 5, 42, 48, 52
Transporters, 4, 5, 7, 40, 42, 44–50, 52, 102, 107, 108, 113, 116, 160
Trap cultures, 144, 197, 201

Subject Index

Triacylglycerides, 5, 48
Trifolium repens, 115
Triglycerides, 38
Tripartite symbiosis, 142, 231
Triticum aestivum, 143
Tropical rain forests, 288, 289, 296
Trypan blue, 24, 197, 338, 339
Tsuga canadensis, 260
Tsuga heterophylla, 243
Tuber melanosporum, 247, 248
Types of mycorrhizae, 2, 21, 196, 288

U

Uapaca bojeri, 232–234
Uniola paniculata, 206
Urease, 4, 46

V

Variability, 16, 52, 203, 229, 236
Variovorax, 181
Vateriopsis seychellarum, 244
Vegetables, 11, 147
Vegetative succession, 267
Vermiculite, 171, 256, 258, 307
Vermiculite-peat mixtures, 258, 307
Vermiculite-peat, 307
Verticillium, 181
Viola calaminaria, 113
Virulence, 141, 221
Visualization, 24, 25, 337, 338, 343, 345

W

Water
  content, 139, 141, 183, 249, 312
  economy, 167
  stress, 5, 22, 48, 139, 141, 228, 248–252, 322, 325
  uptake, 19, 139, 247, 248
Water-stable aggregates, 12, 14, 188, 189
Water-stable soil aggregates, 178
Weaning process, 22, 322
Weaning, 22, 23, 322, 325, 326
White fir, 258
Widefield epifluorescence, 342, 348
Wilcoxina mikorae, 262
Woody debris, 297

X

X-ray analysis, 109
Xylem water potential, 248, 249

Y

Yield increase, 137

Z

Zn transporter, 108
Zygomycetes, 242